BIOCHEMICAL, PHYSIOLOGICAL AND MOLECULAR AVENUES FOR COMBATING ABIOTIC STRESS IN PLANTS

BIOCHEMICAL, PHYSIOLOGICAL AND MOLECULAR AVENUES FOR COMBATING ABIOTIC STRESS IN PLANTS

Edited by

SHABIR HUSSAIN WANI

Academic Press is an imprint of Elsevier
125 London Wall, London EC2Y 5AS, United Kingdom
525 B Street, Suite 1650, San Diego, CA 92101, United States
50 Hampshire Street, 5th Floor, Cambridge, MA 02139, United States
The Boulevard, Langford Lane, Kidlington, Oxford OX5 1GB, United Kingdom

Library of Congress Cataloging-in-Publication Data
A catalog record for this book is available from the Library of Congress

British Library Cataloguing-in-Publication Data
A catalogue record for this book is available from the British Library

ISBN 978-0-12-813066-7

For information on all Academic Press publications
visit our website at https://www.elsevier.com/books-and-journals

Publisher: Andre Gerhard Wolff
Acquisition Editor: Nancy Maragioglio
Editorial Project Manager: Mary Preap
Production Project Manager: Bharatwaj Varatharajan
Cover Designer: Victoria Pearson

Typeset by SPi Global, India

CONTENTS

CONTRIBUTORS

Shruti Ahlawat
Department of Microbiology, Maharshi Dayanand University, Rohtak, India

Nudrat A. Akram
Department of Botany, Government College University, Faisalabad, Pakistan

Naser A. Anjum
Department of Botany, Aligarh Muslim University, Aligarh, India

Abid A. Ansari
Department of Biology, Faculty of Science, University of Tabuk, Tabuk, Saudi Arabia

Muhammad Tehseen Azhar
Department of Plant Breeding and Genetics, University of Agriculture, Faisalabad, Pakistan; School of Biological Sciences M084, The University of Western Australia, Perth, Australia

Sahana Basu
Department of Biotechnology, Assam University, Silchar, India

Abhishek Bohra
Crop Improvement Division, ICAR-Indian Institute of Pulses Research, Kanpur, India

Surekha Challa
Department of Biochemistry and Bioinformatics, Gandhi Institute of Technology and Management (GITAM), Deemed-to-be-University, Visakhapatnam, India

Narsingh Chauhan
Department of Biochemistry, Maharshi Dayanand University, Rohtak, India

Titash Dutta
Department of Biochemistry and Bioinformatics, Gandhi Institute of Technology and Management (GITAM), Deemed-to-be University, Visakhapatnam, India

Manu P. Gangola
Department of Plant Sciences, University of Saskatchewan, Saskatoon, SK, Canada

Mansour Ghorbanpour
Department of Medicinal Plants, Faculty of Agriculture and Natural Resources, Arak University, Arak, Iran

Ritu Gill
Centre for Biotechnology, Maharshi Dayanand University, Rohtak, India

Sarvajeet S. Gill
Centre for Biotechnology, Maharshi Dayanand University, Rohtak, India

Parul Goel
National Agri-Food Biotechnology Institute, Sahibzada Ajit Singh Nagar, Punjab, India

Aarti Gupta
National Institute of Plant Genome Research, New Delhi, India

Mirza Hasanuzzaman
Department of Agronomy, Faculty of Agriculture, Sher-e-Bangla Agricultural University, Dhaka, Bangladesh

Venura Herath
Department of Agricultural Biology, Faculty of Agriculture, University of Peradeniya, Peradeniya, Sri Lanka

Rohit Joshi
Stress Physiology and Molecular Biology Laboratory, School of Life Sciences, Jawaharlal Nehru University, New Delhi, India

Dharanipathi Kamalachandran
National Institute of Plant Genome Research, New Delhi, India

Tushar Khare
Department of Biotechnology, Modern College of Arts, Science and Commerce (Savitribai Phule Pune University), Pune, India

Gautam Kumar
Department of Life Science, Central University of South Bihar, Patna, India

Vinay Kumar
Department of Biotechnology, Modern College of Arts, Science and Commerce (Savitribai Phule Pune University); Department of Environmental Science, Savitribai Phule Pune University, Pune, India

Punam Kundu
Centre for Biotechnology, Maharshi Dayanand University, Rohtak, India

Bendangchuchang Longchar
National Institute of Plant Genome Research, New Delhi, India

Mahmood Maleki
Department of Biotechnology, Institute of Science and High Technology and Environmental Science, Graduate University of Advanced Technology, Kerman, Iran

Shefali Mishra
ICAR-National Research Centre on Plant Biotechnology,
New Delhi, India

Nageswara R.R. Neelapu
Department of Biochemistry and Bioinformatics, Gandhi Institute of Technology and Management (GITAM), Deemed-to-be-University, Visakhapatnam, India

Bharathi R. Ramadoss
Department of Plant Sciences, University of Saskatchewan, Saskatoon, SK, Canada

Akula Ramakrishna
Monsanto Crop Breeding Station, Bangalore, India

Abdul Rehman
School of Biological Sciences M084, The University of Western Australia, Perth, Australia

Muhammad Sadiq
Department of Botany, Government College University, Faisalabad, Pakistan

Maryam Sarwat
Amity Institute of Pharmacy, Amity University, Noida, India

Muthappa Senthil-Kumar
National Institute of Plant Genome Research, New Delhi, India

Krishna K. Sharma
Department of Microbiology, Maharshi Dayanand University, Rohtak, India

Varsha Shriram
Department of Botany, Prof. Ramkrishna More College (Savitribai Phule Pune University), Pune, India

Kadambot H.M. Siddique
The UWA Institute of Agriculture, The University of Western, Australia, Perth, WA, Australia

Pratika Singh
Department of Life Science, Central University of South Bihar, Patna, India

Anil K. Singh
ICAR-Indian Institute of Agricultural Biotechnology, Ranchi, India

Balwant Singh
ICAR-National Research Centre on Plant Biotechnology, New Delhi, India

Narendra Tuteja
Plant Molecular Biology Group, International Centre for Genetic Engineering and Biotechnology (ICGEB), New Delhi, India

Shabir H. Wani
Mountain Research Centre for Field Crops, Sher-e-Kashmir University of Agricultural Sciences and Technology, Kashmir, India; Department of Plant, Soil, and Microbial Sciences, Michigan State University, East Lansing, MI, United States

FOREWORD

I am happy to learn that Dr. Shabir Hussain Wani has edited this volume entitled, "Biochemical, Physiological, and Molecular Avenues for Combating Abiotic Stress in Plants," for the reputed publisher, Elsevier. Dr. Wani has been in contact with me through email for the last several years. I met him in January of 2017 at the Plant and Animal Genome Conference in San Diego, California, United States. He had good experience working in the area of drought and salinity in rice. I was impressed with his zeal and commitment to science including research, teaching, and dissemination of scientific knowledge. Therefore, a book coming from him in the area of abiotic stress tolerance in plants is really a welcome initiative.

The challenges of abiotic and biotic stress on plant growth and development are evident from changing climate, as plants evolve different mechanisms to cope with stress effects. These mechanisms include biochemical, physiological, molecular, and genetic changes. Abiotic stress is one of the great challenges for global food security and will become even more serious in the coming years. The severe effects of several abiotic stresses reduce the yields of several crop plants. Furthermore, plant responses to abiotic stresses (such as drought, heat, salinity, etc.) are highly complex, so there is an urgent need to employ and integrate various novel approaches to understand the molecular basis of stress responses and to create avenues for developing stress-tolerant cultivars. With the advent of genomic resources and next-generation sequencing technologies, research can be directed toward precise understanding of the target genes responsible for controlling important traits for tolerance to abiotic stresses. Systematic research and deployment of modern technologies (including molecular breeding, genetic engineering, and genome editing) will lead to development of high-yielding crop varieties with abiotic stress tolerance.

Dr. Wani has done a marvelous job by having high-quality chapters from the experts in different research areas. Among 14 chapters in the volume, four of them (Chapters 1–4) extensively discuss the role of biochemical factors, especially the critical role of sugars, polyamine metabolism, the essence of targeting the redox regulatory mechanisms, and engineering crop plants with compatible solutes to overcome abiotic stresses. The next two chapters (Chapters 5 and 6) delve into physiological approaches, crop phenomics, and the effects of water stress on crop productivity. I must congratulate Dr. Wani for articulating the volume with chapters on genomics (Chapters 7 and 8), and transgenic approaches (Chapters 9–12) being deployed for mitigating yield losses attributable to abiotic stresses. In addition, Chapter 12 highlights the field performance of transgenics with enhanced drought tolerance, providing an example of translating research into application in the field.

This book is a timely reference material for academicians, researchers, and graduate students working in the area of plant abiotic stress and biotechnology. I congratulate Dr. Wani for producing this outstanding volume. I am sure that it will be read, liked, and cited by a large number of scientists, students, and policymakers.

Rajeev K. Varshney
ICRISAT, Hyderabad, India

PREFACE

In the current context of global climate change, with erratic weather conditions (temperature, rainfall, humidity, particularly during critical plant growth periods) exposing plants to various abiotic stresses (such as drought, salinity, extreme temperatures, and flooding), it is crucial that we hasten efforts to clearly understand the biochemical, physiological, and molecular mechanisms underlying abiotic stress tolerance in plants. While several initiatives have begun in important research institutes at the global level, the speed at which abiotic stress-tolerant cultivars are being developed is not keeping pace with the ever-increasing pressure of abiotic stresses attributable to climate change. In addition, the complex genetic mechanisms involved in plant adaptation to abiotic stresses have been a major impediment for crop improvement using conventional plant breeding tools. Molecular biology advances have opened new vistas in understanding these complex mechanisms. Throughout this book, "Biochemical, Physiological, and Molecular Avenues for Combating Abiotic Stress in Plants," I have tried my best to include chapters describing the significance of abiotic stress in plants under the current climate change scenario and the prospective interventions being applied at the global level, utilizing current high-throughput technologies to crack the code of abiotic stress traits, and digging deep into the molecular mechanisms responsible for these complex traits in plants. This book is a tremendous, inclusive reference source for researchers, teachers, and graduate students involved in the study of abiotic stress tolerance in plants. This book uses plant molecular biology tools by revealing principles and applications of newly developed technologies and their application in the development of resilient, stress-resistant plants to combat abiotic stresses. The chapters are written by world-renowned researchers and academicians in the field of plant stress biology. I express sincere thanks and gratefulness to my revered authors, because without their untiring efforts, this project would not have been possible. I am also thankful to Elsevier for providing such an opportunity to complete this book. I am thankful to all my family members, especially my wife for her support during the language-editing process.

Last, I bow in reverence to Almighty Allah, who gave me the intellect and strength to complete this book project.

Shabir Hussain Wani
Srinagar, India

CHAPTER 1

Transcription Factors Based Genetic Engineering for Abiotic Tolerance in Crops

Venura Herath
Department of Agricultural Biology, Faculty of Agriculture, University of Peradeniya, Peradeniya, Sri Lanka

Contents

1. INTRODUCTION

Plants are exposed to different intensities and durations of changing environmental conditions such as drought, salinity, and temperature extremes throughout their life cycle. These conditions place immense stress on plant growth and survival. The impact of these harsh conditions can cause irreversible damage to plants. When conditions are severe enough, they can cause irreversible damage that ultimately results in plant death. These environmental stresses not only threaten plant life, but also the survival of humans by affecting the world's agriculture production. Environmental extremes in some cases have caused a reduction in food production by more than 50%. To make the situation worse, nearly 800 million people are undernourished and require better access to food, especially in many low-income countries. On top of that, to feed the rapidly increasing world population, which is expected to reach 10 billion by the year 2050, world food production needs to be doubled. This underscores the urgency of ensuring a stable

Biochemical, Physiological and Molecular Avenues for Combating Abiotic Stress in Plants
https://doi.org/10.1016/B978-0-12-813066-7.00001-2

supply of food using conventional and novel strategies while minimizing the environmental footprint.

Plants have developed various morphological, physiological, and genetic adaptations to cope with abiotic stress conditions throughout evolution. When exposed to stresses, plants alter their biological processes at all levels, including molecular, cellular, and the whole plant level. Stresses induce the expression of a large number of stress-responsive genes by altering the transcription, while many growth- and development-related genes are repressed (Fig. 1). This helps plants conserve energy until the stress condition is over. There are two broad categories of stress-responsive genes: regulatory and functional. Regulatory genes code for proteins, including transcription factors (TFs), receptors, protein kinases, and many proteins involved in regulating detoxification and degradation of proteins. The role of regulatory genes is to facilitate the stress-response signal transduction pathways by modulating the expression levels of downstream stress-responsive genes. Functional genes code for proteins that are involved in functions such as osmolyte synthesis, redox regulation, and ubiquitination. These genes are essential for stress alleviation and recovery in plants.

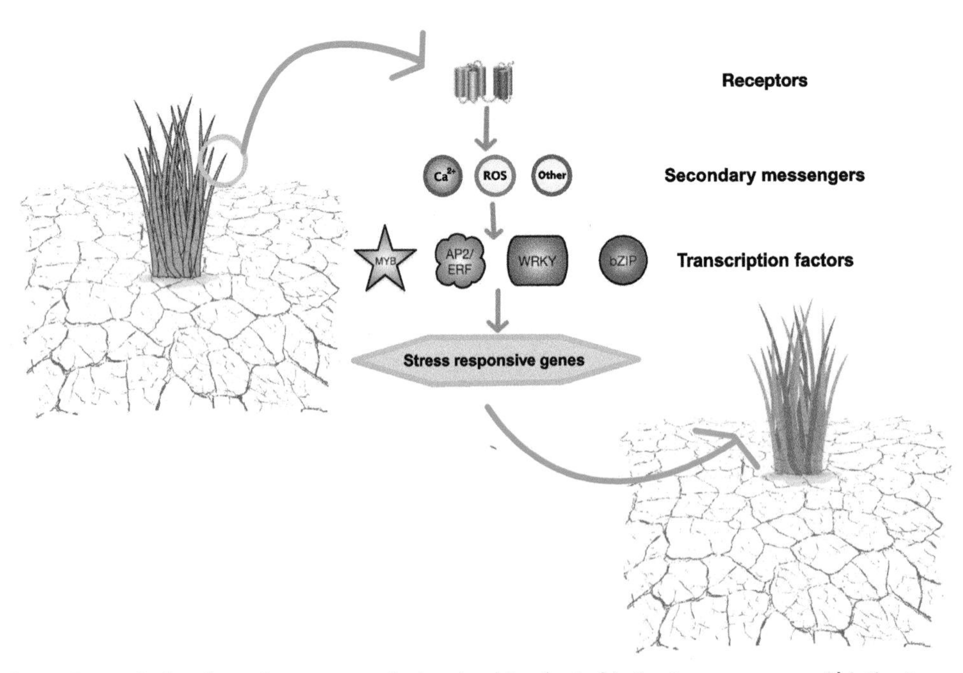

Fig. 1 Transcriptional regulatory networks involved in plant abiotic stress response. Abiotic stresses like drought, salinity, low temperature induce the expression of transcription factors with the help of secondary messengers. These TFs will bind to the specific cis-elements of their target genes causing an altered expression. These downstream genes are involved in different functions related to stress response and tolerance.

Advances in forward and reverse genetic studies, coupled with the increasing number of model systems, has shed light on major regulatory networks involved in plant stress responses (Fig. 1). Those studies helped in the identification of major transcription factors driving unique regulons, as well as the complex crosstalk between such regulons. The APETELA 2/ethylene-responsive element binding factor (AP2/ERF), basic leucine zipper (bZIP), myelob lastosis (MYB), no apical meristem (NAM), arabidopsis transcription activation factor (ATAF), and cup-shaped cotyledon (CUC), (NAC) and WRKY are the major TF families responsible for abiotic stress tolerance in plants. Some of the TFs that belong to the preceding families were found to be crucial for regulating downstream stress-responsive genes by binding to specific cis-elements located in their promoter.

Genetic engineering has shown to be one of the most promising futuristic approaches to breed "climate-ready crops" with multiple-stress tolerance. Early efforts to generate transgenic plants were mostly focused on the modification of single-functional genes that are involved in increased tolerance to abiotic stresses. However, this approach comes with inherent disadvantages, such as suboptimal availability of the protein product due to the plant's effort to regain homeostasis. Also, the expression level might not be sufficient to acquire stress tolerance due to the complex networked nature of the stress response pathways.

Compared with the modification of functional genes, TF engineering provides an opportunity to target unique stress response regulons (which consist of many functional and regulatory genes), with a more specific method of preventing nonspecific transcription of off-target genes. The first of such examples is the overexpression of *CBF1/DREB1B* and *CBF3/DREB1A* in the model species *Arabidopsis thaliana*. Their overexpression resulted in tolerance to abiotic stresses such as drought and low temperature. Thereafter, thousands of studies have been published validating the potential of TF engineering toward achieving stress tolerance in plants, including staple food crops such as rice, wheat, and maize using more precise and efficient tools.

2. DISCOVERY OF CANDIDATE TFs

The development of effective transgenic strategies depends on the evidence of how the abiotic stress signals are received and relayed, and how plants respond at the molecular, cellular, and whole plant level. There are different genomic, transcriptomic, proteomic, and phenomic techniques for characterizing such responses. Analysis of transcriptomic and proteomic profiles under different abiotic stress conditions is the most direct approach taken to identify key transcriptional regulators involved in the stress response. That provides a platform for manipulating key regulators in order to produce abiotic stress-tolerant crops.

Microarray and RNA-Seq are the two most commonly used large-scale expression profiling techniques that facilitate simultaneous identification of the whole stress response transcriptome at once. These approaches helped identify thousands of abiotic stress-response TFs in crop species (Nakashima et al., 2009, 2014; Zhu, 2016). Out of the two techniques, the novel RNA-Seq approach provides advantages over the microarray approach. It provides the ability to profile transcripts, even though the whole genome of the particular organism is not sequenced. Also, RNA-Seq provides information about introns and exons, abundance of transcripts, and alternative splicing forms. As an example, it helped to identify ~750 new candidate genes that were not previously identified as drought-response genes by other expression analysis platforms in rice (Yoo et al., 2017). Similarly, RNA-seq aided in identification of 18 new TFs in rice for drought stress (Shin et al., 2016). Small- to medium-scale expression profiling is frequently carried out using a real-time quantitative reverse transcriptase polymerase chain reaction (real-time qRT-PCR). The same technique is used to validate the results obtained from microarray and RNA-Seq techniques. These techniques helped to identify many unique and common regulatory and functional genes involved in the stress response.

Functional characterization of identified TFs is carried out using the ab initio promoter analysis of stress-response genes. It helps establish the role of the identified TFs in the regulation of stress response by virtue of the TF and its putative binding cis-element/s. There is a wide array of software available for the identification of already characterized and novel cis-elements such as new PLACE, PLANTCARE, and TRANSFAC. Validation of novel cis-elements is carried out by techniques such as electroporotic mobility shift assays (EMSA), chromatin immunoprecipitation with massively parallel DNA sequencing (ChIP-Seq) and DNAse I footprinting followed by identification of their putative TFs using southwestern blotting or yeast one-hybrid assays.

Uncovering how the identified TFs are regulated is of prime importance in TF engineering. It helps in identifying other interacting proteins, and how they regulate the function of TFs of interest. The list of regulators include co-activators, repressors, kinases, ubiquitinatinases, E3 ligases, SUMO, other TFs, and so forth (Gonzalez, 2016). Yeast two-hybrid assays and affinity purification mass spectrophotometry (AP-MS) techniques are used in identification of interacting partners of TFs. This understanding greatly helps in developing TF engineering strategies toward the crop improvement.

3. MAJOR TRANSCRIPTION FACTOR FAMILIES

TFs are involved in diverse functions related to plant growth, development, and stress responses. They play a key role in spatiotemporal regulation of abiotic stress responses in plants. To date, there are 58 TF families identified in angiosperms (Jin et al., 2014). With the help of expression profiling strategies described in the previous section, thousands of abiotic stress response TFs are identified in plants. These abiotic stress-response TFs are enriched with major families of AP2/ERF, bZIP, MYB, NAC, WRKY and some other TF families such as basic helix-loop-helix (bHLH), Homeobox, trihelix, and many more (Bhattacharjee et al., 2015; Carretero-Paulet et al., 2010; Liu et al., 2014b; Mizoi et al., 2012; Qin et al., 2014; Rushton et al., 2012; Smita et al., 2015; Sornaraj et al., 2016). The following section focuses on five major abiotic stress-response TFs. Some members of these TFs have shown promising outcomes that can be effectively used toward crop improvement, which are briefly described as follows.

3.1 AP2/ERF TF Family

The AP2/ERF TF family is characterized by the presence of the AP2 DNA binding domain (Sakuma et al., 2002). This family consists of four subfamilies, namely AP2, ERF, related to ABI3/VP1(RAV), and dehydration-responsive element binding (DREB) (Herath, 2016; Mizoi et al., 2012). Their role in plant development, hormonal response, and biotic and abiotic stress response is well documented (Dietz et al., 2010). Among the four subfamilies, members of the DREB and ERF families are recognized as major TFs that are involved in abiotic stress signaling. The DREB subfamily specifically binds to the DRE/C-repeat (DRE/CRT) cis-elements (A/GCCCGAC) upon receiving the abiotic stress signal, and thereby regulating the expression of dehydration/cold regulated (RD/COR) genes. Overexpression of *OsDREB2A* results in enhanced tolerance toward both drought and salinity in rice while heterologus expression of *AtDREB1A* results in drought tolerance in rice, indicating the highly conserved nature of DREB genes in plants (Mallikarjuna et al., 2011; Ravikumar et al., 2014). Similarly, overexpression of *StDREB1* provides the tolerance to salinity conditions (Bouaziz et al., 2013). In addition, overexpression of *DREB* genes in many crops have shown their ability to conquer tolerance to drought, salinity, and low temperature stresses (Todaka et al., 2012).

The ERF subfamily TFs are expressed in response to various abiotic stress conditions. These TFs specifically bind to the GCC-box cis-elements (AGCCGCC) and regulate the expression of downstream target genes. Overexpression of ERF genes results in enhanced tolerance to drought and salinity in rice and tomatoes; drought, salinity, and disease tolerance in soybeans; and cold tolerance in wheat (Joo et al., 2013; Pan et al., 2012; Wei et al., 2016; Zhang et al., 2009; Zhu et al., 2014). Although the ERF TF subfamily is the largest subfamily of the AP2/ERF TF family, the function of many ERF genes is yet to be discovered.

3.2 bZIP TF Family

The bZIP TF family is involved in developmental and stress responses in eukaryotic organisms. They are characterized by the presence of the bZIP domain, consisting of a DNA-binding domain and a dimerization domain (leucine zipper). Absicic acid (ABA) has been shown to enhance the expression of bZIP genes in many species.

These TFs bind to the ABA responsive cis-elements (ABRE) (Foster et al., 1994; Zong et al., 2016). Overexpression of bZIP genes such as *OsbZIP16*, *OsbZIP46CA1* results in drought tolerance, and overexpression of *OsbZIP23*, *OsbZIP71* led to drought and salinity tolerance in rice (Chen et al., 2012a; Liu et al., 2014a; Tang et al., 2012; Xiang et al., 2008). Heterologous expression of *AtAREB1* causes drought tolerance in soybeans (Barbosa et al., 2013). Similar to the ERF subfamily, functions of many bZIP genes are yet to be identified.

3.3 MYB TF Family

MYB TF family members consist of the MYB DNA binding domain and activation domain located in the C-terminus region. Out of the four subfamilies, R2R3-MYB subfamily genes are more common in plants. They are involved in cell cycle, cell development, and hormonal regulation in addition to their role in plant stress response. MYB TFs have shown to be promising targets for genetic engineering because of their involvement in multiple stress responses. Overexpression of *OsMYB2* improves tolerance to drought, salinity, and cold in rice by increased production of proline and soluble sugars (Yang et al., 2012). Similar responses were observed by overexpressing *TaMYB2A* in wheat.

3.4 NAC TF Family

The NAC TF family represents a plant-specific TF with a unique NAC domain containing DNA binding, nuclear localization, and dimerization motifs. They act as both activators and repressors. These TFs can recognize the CAGG core DNA binding sequence of the NAC recognition sequence (NACRS) cis-elements. NAC TFs are involved in both developmental and stress responses, including formation of shoot apical meristem, cell walls, branching of shoots, development of lateral roots, stress-induced flowering, and biotic and abiotic stress responses (Nuruzzaman et al., 2013; Olsen et al., 2005; Tran et al., 2010). Overexpression of many NAC TFs resulted in stress tolerance. As an example, overexpression of *OsSNAC2* resulted in dehydration, cold, and salinity tolerance in rice (Hu et al., 2008). Also, *OsNAC5* has shown drought and salinity tolerance when it overexpressed in rice (Takasaki et al., 2010). Interestingly, *OsNAC10* overexpression lines have resulted in not only drought tolerance, but also increased grain yield when driven by a root-specific promoter (Jeong et al., 2010).

3.5 WRKY TF Family

WRKY TFs represent one of the largest families of TFs found in plants. They are characterized by the presence of highly conserved WRKY domains of 60 residues long. This domain binds to the W-box cis-elements with the core TTGACC/T sequence (Agarwal et al., 2011; Eulgem et al., 2000; Rushton et al., 2012). WRKY TFs are also involved in embryogeneis, seed development, plant growth, flowering time, leaf senescence, hormonal signaling, biotic and abiotic stress responses (Rushton et al. 2012; Cai et al. 2014; Chen et al. 2012b). Overexpression of *OsWRKY11* resulted in drought and heat tolerance, while overexpression of *OsWRKY30* resulted in drought tolerance in rice (Shen et al., 2012; Wu et al., 2009). Heterologous overexpression of two wheat WRKY TFs, namely *TaWRKY2* and *TaWRKY19*, resulted in salt, drought, and freezing (only in *TaWRKY19*) tolerance in Arabidopsis (Niu et al., 2012).

4. DEVELOPMENT OF ENGINEERED TF CROPS

Transcription-factor engineering is considered the most powerful tool to produce crops with increased yield, increased nutrient use efficiency, and stress tolerance. There are a multitude of approaches used in engineering TF into plants (Fig. 2). Overexpression is the most popular strategy used to study and generate transgenic plants. Overexpression counts for more than 95% of the transgenic lines produced to date. In this approach, TFs are expressed in a single, independent manner with the help of strong constitutive promoters such as CAMV35S, Actin, and ubiquitin. The first attempt at TF engineering was carried out using this technique in Arabidopsis by overexpressing the *CBF/DREB* TF genes in Arabidopsis thaliana. Thereafter, overexpression became the popular choice

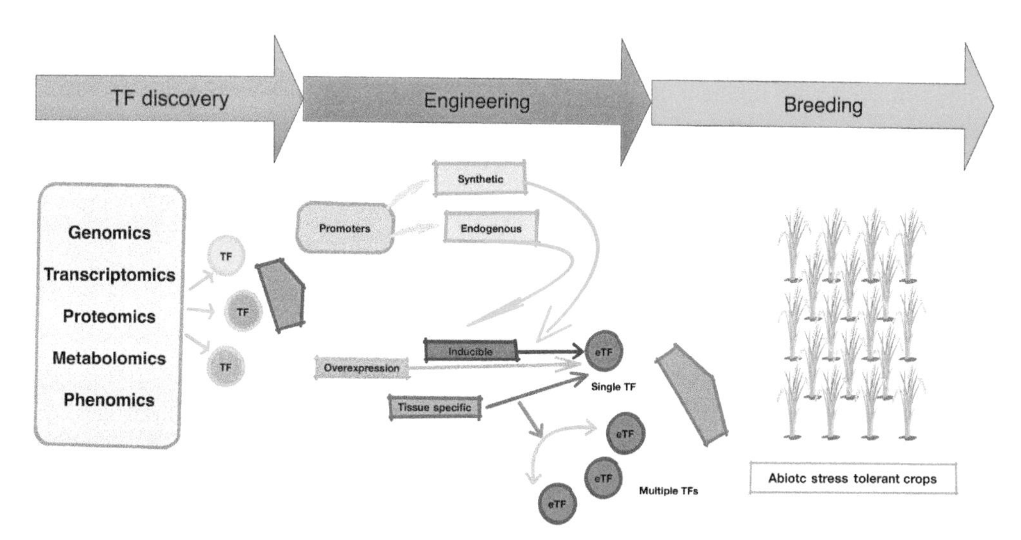

Fig. 2 TF engineering strategies for the production of abiotic stress tolerant crops. Discovery of key abiotic stress related transcription factors is done using different omic approaches. After the identification, different engineering approaches are followed in order to assure the spatiotemporal expression of the TF or TFs in the target crop. Engineered TF or TFs are then transformed in to the target crop. After a series of field evaluations and safety evaluations, TF engineered crops are released to the consumers. *eTF*, engineered transcription factors.

among researchers for regulon engineering. In addition to the use of constitutive promoters, inducible promoters, such as RD29A (abiotic stress response), and alcA (chemically induced) are used. Inducible promoters provide the opportunity to express the TFs only when either a stress or a chemical signal is present (Kasuga et al., 2004; Moore et al., 2006; Tang et al., 2004). Also, the expression of TFs can be confined to cells or tissues with the help of cell- or tissue-specific promoters. As an example, the rice LP2 promoter is shown to be expressed specifically in leaf tissue in a light-dependent manner (Thilmony et al., 2009). There is a growing list of promoters that provide a toolkit for scientists to engineer TFs to express in a spatiotemporal manner (Jeong and Jung, 2015).

Interestingly, constitutive overexpression of the intact form of some TFs will not result in stress tolerance. Overexpression of *AREB1* provides an example for such a situation. When the intact form *AREB1* is overexpressed, transgenic lines fail to show the drought tolerance. However, constitutive overexpression of the active form of AREB results in activation of its downstream targets, including late embryogenic abundant (LEA) genes and other regulatory genes conferring drought tolerance (Fujita, 2005). Also, the modification of sites of posttranslational modification results in the production of active forms of the TFs. A missense mutation in the phosphorylation domain of the TRAB1 gene causing the change from serine to aspartic acid results in an enhanced level of activation without the trigger molecule ABA (Kagaya et al., 2002). This indicates the potential of induction of site-specific mutations in TFs in the regulation engineering.

The knock-out of TF can also be used in TF engineering. Here, TF downregulation can be achieved with the help of RNA interference, co-suppression, or loss-of-function mutants (Cominelli and Tonelli, 2010). A loss-of-function mutant derived from an ethyl meth-anesulfonate (EMS)-mutagenized M2 rice seedling carrying a mutation in drought and salt tolerance (DST) TF has shown that lack of its expression can enhance the stomatal closure and reduce the stomatal density. As a result, the plants are showing enhanced drought and salt tolerance (Huang et al., 2009). Also, a knock-out mutation in *AtMYB60* resulted in a 30% reduction in stoma size, leading to enhanced drought tolerance (Cominelli et al., 2005; Gray, 2005).

Transformation of an engineered TF-containing cassette into a plant host is a major step in transgenic technology. There are three popular transformation methods used in genetic transformation. Among the three methods, *Agrobacterium*- mediated gene transfer is considered the easiest and most convenient method to transfer both monocots and dicots. The gene gun, alternatively known as the particle bombardment method, is also used in the transformation of plants. When comparing the two methods, *Agrobacterium*-mediated gene transfer usually results in a random integration of one copy of the transgene, while the gene gun method can cause random integration of multiple copies of the transgene, which is considered a disadvantage. Protoplast transformation is another method used in gene transformation. Both protoplast transformation and gene gun methods can be effectively used in order to overcome the host dependence of the *Agrobacterium*-mediated gene transfer method. The major disadvantage of the protoplast transformation is the dependence of transformation efficiency on the handling skills of the operator (Hansen and Wright, 1999). The rapidly advancing field of genome editing is reaching new heights in TF engineering.

Genome editing technology is based on sequence-specific DNA cleavage using programmable endonucleases. These endonucleases cause double strand breaks at specific locations of the target gene, activating cellular DNA repair mechanisms. This provides a platform for introducing necessary genetic modifications via homology-directed repair and error-prone nonhomologous end-joining breaks (Belhaj et al., 2013). The zinc finger nuclease (ZFNs) transcription activator-like effector nucleases (TALENs), and clustered regulatory interspaced short palindromic repeats (CRISPRs)/CRISPR-associated (Cas) are the three most popular genome editing approaches currently used in the production of improved crops. Most importantly, minimal biosafety regulations are applied to genome edited crops because these techniques are inserting minimal or no foreign DNA. These techniques are successfully used in many plants species such as *Arabidopsis*, rice, tobacco, soybean, maize, and poplar to generate various traits of interest (Baltes et al., 2017; Khandagale and Nadaf, 2016). Genome editing has great potential for use in TF engineering strategies involved in overexpression, removal of inhibitory domains, and knock-out/knock-down of TF expression, and most importantly, the modification of multiple TFs involved in abiotic stress-response networks.

5. LIMITATIONS

The occurrence of unwanted phenotypes is considered one of the major limitations associated with the overexpression of target TFs. As an example, overexpression of AtAREB1 in soybeans results in reduced plant height, caused by the reduction of internode lengths (Barbosa et al., 2013). Similarly, overexpression of SIERF5 in tomatoes results in stunted growth (Pan et al., 2012). The occurrence of these negative phenotypes may be due to the pleiotropic effects of the selected TFs on traits other than the stress tolerance. Also, the use of strong constitutive promoters such as CAMV35s and ubiquitin interfere with the cellular homeostasis and energy balance, which ultimately results in unwanted phenotypes. These promoter effects can be controlled to a greater extent by the use of inducible and tissue-specific promoters (Jeong and Jung, 2015).

When using *Agrobacterium*-mediated gene transfer or gene gun methods, there is a chance of activation/inhibition of off-target genes at the site of insertion due to the random nature of integration. Such alterations will lead to the expression of undesirable traits. Novel genome editing techniques can minimize the occurrence of such incidences.

However, usage of genome editing is greatly limited by the availability of reference genome sequences.

Unfortunately, transgenic and genome-edited crops are facing increased opposition from public and environmental organizations due to unpredictable risks to food safety and the environment. However, a majority of the claims made are not based on scientific evidence, but rather, on personal beliefs. This hurdle is hindering the production and approval process of these crops, sometimes up to 15–20 years (Azadi et al., 2015; Khandagale and Nadaf, 2016).

6. SYNTHETIC TFs

As described in the previous section, successful implementation of TF engineering in the development of abiotic stress-tolerant crops is limited for several reasons. The beneficial feature of TFs for TF engineering, that is, the ability to regulate multiple genes, can become a problem when the TFs interfere with plant growth and development and cellular energy balance, and cause secondary effects. Therefore, it is essential to come up with a system that functions independently from the native regulatory networks enabling the minimization of off-target effects of TFs that lead to altered phenotypes and an imbalance of cellular bioenergetics. The synthetic TFs (sTFs) provide an ideal system for this purpose (Liu and Stewart, 2016).

The sTFs are generated by incorporating engineered DNA-binding domains (DBDs). These DBDs consist of a nuclear localization signal and effector (activation or repression) domains. The herpes simplex virus virion protein 16 (VP16) is one of the most widely used activators in sTFs. Krupple associated box 1 (KRAB1) and nSin3A interacting domain (SID) are widely used as repressor domains (Mehrotra et al., 2017).

The sTFs are designed using C2H2 zinc finger (ZF) proteins, transcriptional activator-like effectors (TALEs) and dCas9s. These sTFs provide the ability to target any locus of interest (endogenous and transgene). ZF-sTF carrying VP16 domain is successfully used in identification of the genes involved in the homologous recombination in *Arabidopsis* (Jia et al., 2013). TALE-sTFs are used in the induction of Xa genes involved in resistance to *Xanthomonas oryzae* in rice (Li et al., 2013). Also, RNA-guided transcriptional regulation of genes is demonstrated using dCas9-sTF in tobacco (Piatek et al., 2015). These sTFs can be effectively combined with synthetic promoters maintaining a basal level of expression until the correct signal is received. Most importantly, complete synthetic transcriptional regulatory networks can be designed by combining multiple sTFs and synthetic promoters providing full control over abiotic stress responses (Liu and Stewart, 2016). Even though minimal research has been conducted, it is clear that sTFs have great potential for designing abiotic stress tolerance in crops.

7. CONCLUDING REMARKS

The world population is going to reach 10 billion by 2050. The world temperature is rapidly rising, causing erratic weather patterns as a result of global warming. To feed the increasing population under these conditions, we need to pursue innovative food production strategies, creating another green revolution. Biotech crops have already shown us that they can play a significant role in increasing crop production using limited land area with a minimal environmental footprint. Among the different genetic engineering strategies used in developing biotech crops, TF engineering is shown to be one of the most promising approaches. Using the inherent advantage of controlling multiple genes of a regulon, TF engineering can effectively regulate many processes related to plant growth and development. TF engineering is further supported by the increasing availability of big data coming from high-throughput technologies such as DNA-Seq and RNA-seq. Also, the ever-expanding toolkit facilitates more precise spatiotemporal expression of TFs. Most importantly, with the availability of sTFs coupled with synthetic promoters, the development of future abiotic stress-tolerant crops with optimal yield and minimal off-target effects is going to be a reality. Finally, it is essential to have interdisciplinary cooperation to bring this science-based asset to farmers so they can provide a food supply sufficient to end the world hunger.

ACKNOWLEDGMENT

The author acknowledges support provided by grants from the International Foundation of Science, Sweden (grant no.: C/5267-1).

REFERENCES

Agarwal, P., Reddy, M.P., Chikara, J., 2011. WRKY: its structure, evolutionary relationship, DNA-binding selectivity, role in stress tolerance and development of plants. Mol. Biol. Rep. 38, 3883–3896. https://doi.org/10.1007/s11033-010-0504-5.

Azadi, H., Samiee, A., Mahmoudi, H., Jouzi, Z., Rafiaani Khachak, P., De Maeyer, P., Witlox, F., 2015. Genetically modified crops and small-scale farmers: main opportunities and challenges. Crit. Rev. Biotechnol. 36, 434–446. https://doi.org/10.3109/07388551.2014.990413.

Baltes, N.J., Gil-Humanes, J., Voytas, D.F., 2017. Genome engineering and agriculture: opportunities and challenges. In: Progress in Molecular Biology and Translational Science. first ed. Elsevier, Amsterdam. https://doi.org/10.1016/bs.pmbts.2017.03.011.

Barbosa, E.G.G., Leite, J.P., Marin, S.R.R., Marinho, J.P., de Fátima Corrêa Carvalho, J., Fuganti-Pagliarini, R., Farias, J.R.B., Neumaier, N., Marcelino-Guimarães, F.C., de Oliveira, M.C.N., Yamaguchi-Shinozaki, K., Nakashima, K., Maruyama, K., Kanamori, N., Fujita, Y., Yoshida, T., Nepomuceno, A.L., 2013. Overexpression of the ABA-dependent AREB1 transcription factor from Arabidopsis thaliana improves soybean tolerance to water deficit. Plant Mol. Biol. Report. 31, 719–730. https://doi.org/10.1007/s11105-012-0541-4.

Belhaj, K., Chaparro-Garcia, A., Kamoun, S., Nekrasov, V., 2013. Plant genome editing made easy: targeted mutagenesis in model and crop plants using the CRISPR/Cas system. Plant Methods 9, 39. https://doi.org/10.1186/1746-4811-9-39.

Bhattacharjee, A., Ghangal, R., Garg, R., Jain, M., 2015. Genome-wide analysis of homeobox gene family in legumes: identification, gene duplication and expression profiling. PLoS One 10, 1–22. https://doi.org/10.1371/journal.pone.0119198.

Bouaziz, D., Pirrello, J., Charfeddine, M., Hammami, A., Jbir, R., Dhieb, A., Bouzayen, M., Gargouri-Bouzid, R., 2013. Overexpression of StDREB1 transcription factor increases tolerance to salt in transgenic potato plants. Mol. Biotechnol. 54, 803–817. https://doi.org/10.1007/s12033-012-9628-2.

Cai, Y., Chen, X., Xie, K., Xing, Q., Wu, Y., Li, J., Du, C., Sun, Z., Guo, Z., 2014. Dlf1, a WRKY transcription factor, is involved in the control of flowering time and plant height in rice. PLoS One 9, 1–13. https://doi.org/10.1371/journal.pone.0102529.

Carretero-Paulet, L., Galstyan, A., Roig-Villanova, I., Martínez-García, J.F., Bilbao-Castro, J.R., Robertson, D.L., 2010. Genome-wide classification and evolutionary analysis of the bHLH family of transcription factors in Arabidopsis, poplar, rice, moss, and algae. Plant Physiol. 153, 1398–1412. https://doi.org/10.1104/pp.110.153593.

Chen, H., Chen, W., Zhou, J., He, H., Chen, L., Chen, H., Deng, X.W., 2012a. Basic leucine zipper transcription factor OsbZIP16 positively regulates drought resistance in rice. Plant Sci. 193–194, 8–17.

Chen, L., Song, Y., Li, S., Zhang, L., Zou, C., Yu, D., 2012b. The role of WRKY transcription factors in plant abiotic stresses. Biochim. Biophys. Acta, Gene Regul. Mech. 1819, 120–128. https://doi.org/10.1016/j.bbagrm.2011.09.002.

Cominelli, E., Tonelli, C., 2010. Transgenic crops coping with water scarcity. New Biotechnol. 27, 473–477. https://doi.org/10.1016/j.nbt.2010.08.005.

Cominelli, E., Galbiati, M., Vavasseur, A., Conti, L., Sala, T., Vuylsteke, M., Leonhardt, N., Dellaporta, S.L., Tonelli, C., 2005. A guard-cell-specific MYB transcription factor regulates stomatal movements and plant drought tolerance. Curr. Biol. 15, 1196–1200. https://doi.org/10.1016/j.cub.2005.05.048.

Dietz, K.J., Vogel, M.O., Viehhauser, A., 2010. AP2/EREBP transcription factors are part of gene regulatory networks and integrate metabolic, hormonal and environmental signals in stress acclimation and retrograde signalling. Protoplasma 245, 3–14. https://doi.org/10.1007/s00709-010-0142-8.

Eulgem, T., Rushton, P.J., Robatzek, S., Somssich, I.E., 2000. The WRKY superfamily of plant transcription factors. Trends Plant Sci. 5, 199–206. https://doi.org/10.1016/S1360-1385(00)01600-9.

Foster, R., Izawa, T., Chua, N., 1994. Plant bZIP proteins gather at ACGT elements. FASEB J. 8, 192–200.

Fujita, Y., 2005. AREB1 is a transcription activator of novel ABRE-dependent ABA signaling that enhances drought stress tolerance in Arabidopsis. Plant Cell 17, 3470–3488. https://doi.org/10.1105/tpc.105.035659.

Gonzalez, D.H., 2016. Introduction to transcription factor structure and function. In: Plant Transcription Factors. Elsevier, Amsterdam, pp. 3–11. https://doi.org/10.1016/B978-0-12-800854-6.00001-4.

Gray, J., 2005. Guard cells: transcription factors regulate stomatal movements. Curr. Biol. 15, R593–R595. https://doi.org/10.1016/j.cub.2005.07.039.

Hansen, G., Wright, M.S., 1999. Recent advances in the transformation of plants. Trends Plant Sci. 4, 226–231. https://doi.org/10.1016/S1360-1385(99)01412-0.

Herath, V., 2016. Small family, big impact: in silico analysis of DREB2 transcription factor family in rice. Comput. Biol. Chem. 65, 128–139. https://doi.org/10.1016/j.compbiolchem.2016.10.012.

Hu, H., You, J., Fang, Y., Zhu, X., Qi, Z., Xiong, L., 2008. Characterization of transcription factor gene SNAC2 conferring cold and salt tolerance in rice. Plant Mol. Biol. 67, 169–181. https://doi.org/10.1007/s11103-008-9309-5.

Huang, X.-Y., Chao, D.-Y., Gao, J.-P., Zhu, M.-Z., Shi, M., Lin, H.-X., 2009. A previously unknown zinc finger protein, DST, regulates drought and salt tolerance in rice via stomatal aperture control. Genes Dev. 23, 1805–1817. https://doi.org/10.1101/gad.1812409.

Jeong, H.J., Jung, K.H., 2015. Rice tissue-specific promoters and condition-dependent promoters for effective translational application. J. Integr. Plant Biol. 57, 913–924. https://doi.org/10.1111/jipb.12362.

Jeong, J.S., Kim, Y.S., Baek, K.H., Jung, H., Ha, S.-H., Do Choi, Y., Kim, M., Reuzeau, C., Kim, J.-K., 2010. Root-specific expression of OsNAC10 improves drought tolerance and grain yield in rice under field drought conditions. Plant Physiol. 153, 185–197. https://doi.org/10.1104/pp.110.154773.

Jia, Q., van Verk, M.C., Pinas, J.E., Lindhout, B.I., Hooykaas, P.J.J., Van der Zaal, B.J., 2013. Zinc finger artificial transcription factor-based nearest inactive analogue/nearest active analogue strategy used for the identification of plant genes controlling homologous recombination. Plant Biotechnol. J. 11, 1069–1079. https://doi.org/10.1111/pbi.12101.

Jin, J., Zhang, H., Kong, L., Gao, G., Luo, J., 2014. PlantTFDB 3.0: a portal for the functional and evolutionary study of plant transcription factors. Nucleic Acids Res. 42, 1182–1187. https://doi.org/10.1093/nar/gkt1016.

Joo, J., Choi, H.J., Lee, Y.H., Kim, Y.K., Song, S.I., 2013. A transcriptional repressor of the ERF family confers drought tolerance to rice and regulates genes preferentially located on chromosome 11. Planta 238, 155–170. https://doi.org/10.1007/s00425-013-1880-6.

Kagaya, Y., Hobo, T., Murata, M., Ban, A., Hattori, T., 2002. Abscisic acid-induced transcription is mediated by phosphorylation of an abscisic acid response element binding factor, TRAB1. Plant Cell 14, 3177–3189. https://doi.org/10.1105/tpc.005272.constitute.

Kasuga, M., Miura, S., Shinozaki, K., Yamaguchi-Shinozaki, K., 2004. A combination of the Arabidopsis DREB1A gene and stress-inducible rd29A promoter improved drought- and low-temperature stress tolerance in tobacco by gene transfer. Plant Cell Physiol. 45, 346–350. https://doi.org/10.1093/pcp/pch037.

Khandagale, K., Nadaf, A., 2016. Genome editing for targeted improvement of plants. Plant Biotechnol. Rep. 10, 327–343. https://doi.org/10.1007/s11816-016-0417-4.

Li, T., Huang, S., Zhou, J., Yang, B., 2013. Designer TAL effectors induce disease susceptibility and resistance to Xanthomonas oryzae pv. oryzae in rice. Mol. Plant 6, 781–789. https://doi.org/10.1093/mp/sst034.

Liu, W., Stewart, C.N., 2016. Plant synthetic promoters and transcription factors. Curr. Opin. Biotechnol. 37, 36–44. https://doi.org/10.1016/j.copbio.2015.10.001.

Liu, C., Mao, B., Ou, S., Wang, W., Liu, L., Wu, Y., Chu, C., Wang, X., 2014a. OsbZIP71, a bZIP transcription factor, confers salinity and drought tolerance in rice. Plant Mol. Biol. 84, 19–36. https://doi.org/10.1007/s11103-013-0115-3.

Liu, G., Li, X., Jin, S., Liu, X., Zhu, L., Nie, Y., Zhang, X., 2014b. Overexpression of rice NAC gene SNAC1 improves drought and salt tolerance by enhancing root development and reducing transpiration rate in transgenic cotton. PLoS One. 9. https://doi.org/10.1371/journal.pone.0086895.

Mallikarjuna, G., Mallikarjuna, K., Reddy, M.K., Kaul, T., 2011. Expression of OsDREB2A transcription factor confers enhanced dehydration and salt stress tolerance in rice (Oryza sativa L.). Biotechnol. Lett. 33, 1689–1697. https://doi.org/10.1007/s10529-011-0620-x.

Mehrotra, R., Renganaath, K., Kanodia, H., Loake, G.J., Mehrotra, S., 2017. Towards combinatorial transcriptional engineering. Biotechnol. Adv. 35, 390–405. https://doi.org/10.1016/j.biotechadv.2017.03.006.

Mizoi, J., Shinozaki, K., Yamaguchi-Shinozaki, K., 2012. AP2/ERF family transcription factors in plant abiotic stress responses. Biochim. Biophys. Acta, Gene Regul. Mech. 1819, 86–96. https://doi.org/10.1016/j.bbagrm.2011.08.004.

Moore, I., Samalova, M., Kurup, S., 2006. Transactivated and chemically inducible gene expression in plants. Plant J. 45, 651–683. https://doi.org/10.1111/j.1365-313X.2006.02660.x.

Nakashima, K., Ito, Y., Yamaguchi-Shinozaki, K., 2009. Transcriptional regulatory networks in response to abiotic stresses in Arabidopsis and grasses. Plant Physiol. 149, 88–95. https://doi.org/10.1104/pp.108.129791.

Nakashima, K., Yamaguchi-Shinozaki, K., Shinozaki, K., 2014. The transcriptional regulatory network in the drought response and its crosstalk in abiotic stress responses including drought, cold, and heat. Front. Plant Sci. 5, 170. https://doi.org/10.3389/fpls.2014.00170.

Niu, C.F., Wei, W., Zhou, Q.Y., Tian, A.G., Hao, Y.J., Zhang, W.K., Ma, B., Lin, Q., Zhang, Z.B., Zhang, J.S., Chen, S.Y., 2012. Wheat WRKY genes TaWRKY2 and TaWRKY19 regulate abiotic stress tolerance in transgenic Arabidopsis plants. Plant Cell Environ. 35, 1156–1170. https://doi.org/10.1111/j.1365-3040.2012.02480.x.

Nuruzzaman, M., Sharoni, A.M., Kikuchi, S., 2013. Roles of NAC transcription factors in the regulation of biotic and abiotic stress responses in plants. Front. Microbiol. 4, 1–16. https://doi.org/10.3389/fmicb.2013.00248.

Olsen, A.N., Ernst, H.A., Leggio, L.L., Skriver, K., 2005. NAC transcription factors: structurally distinct, functionally diverse. Trends Plant Sci. 10, 79–87. https://doi.org/10.1016/j.tplants.2004.12.010.

Pan, Y., Seymour, G.B., Lu, C., Hu, Z., Chen, X., Chen, G., 2012. An ethylene response factor (ERF5) promoting adaptation to drought and salt tolerance in tomato. Plant Cell Rep. 31, 349–360. https://doi.org/10.1007/s00299-011-1170-3.

Piatek, A., Ali, Z., Baazim, H., Li, L., Abulfaraj, A., Al-Shareef, S., Aouida, M., Mahfouz, M.M., 2015. RNA-guided transcriptional regulation in planta via synthetic dCas9-based transcription factors. Plant Biotechnol. J. 13, 578–589. https://doi.org/10.1111/pbi.12284.

Qin, Y., Ma, X., Yu, G., Wang, Q., Wang, L., Kong, L., Kim, W., Wang, H.W., 2014. Evolutionary history of trihelix family and their functional diversification. DNA Res. 1–12. https://doi.org/10.1093/dnares/dsu016.

Ravikumar, G., Manimaran, P., Voleti, S.R., Subrahmanyam, D., Sundaram, R.M., Bansal, K.C., Viraktamath, B.C., Balachandran, S.M., 2014. Stress-inducible expression of AtDREB1A transcription factor greatly improves drought stress tolerance in transgenic indica rice. Transgenic Res. 23, 421–439. https://doi.org/10.1007/s11248-013-9776-6.

Rushton, D.L., Tripathi, P., Rabara, R.C., Lin, J., Ringler, P., Boken, A.K., Langum, T.J., Smidt, L., Boomsma, D.D., Emme, N.J., Chen, X., Finer, J.J., Shen, Q.J., Rushton, P.J., 2012. WRKY transcription factors: key components in abscisic acid signalling. Plant Biotechnol. J. 10, 2–11. https://doi.org/10.1111/j.1467-7652.2011.00634.x.

Sakuma, Y., Liu, Q., Dubouzet, J.G., Abe, H., Shinozaki, K., Yamaguchi-Shinozaki, K., 2002. DNA-binding specificity of the ERF/AP2 domain of Arabidopsis DREBs, transcription factors involved in dehydration- and cold-inducible gene expression. Biochem. Biophys. Res. Commun. 290, 998–1009. https://doi.org/10.1006/bbrc.2001.6299.

Shen, H., Liu, C., Zhang, Y., Meng, X., Zhou, X., Chu, C., Wang, X., 2012. OsWRKY30 is activated by MAP kinases to confer drought tolerance in rice. Plant Mol. Biol. 80, 241–253. https://doi.org/10.1007/s11103-012-9941-y.

Shin, S.-J., Ahn, H., Jung, I., Rhee, S., Kim, S., Kwon, H.-B., 2016. Novel drought-responsive regulatory coding and non-coding transcripts from Oryza sativa L. Genes Genomics 38, 949–960. https://doi.org/10.1007/s13258-016-0439-x.

Smita, S., Katiyar, A., Chinnusamy, V., Pandey, D.M., Bansal, K.C., 2015. Transcriptional regulatory network analysis of MYB transcription factor family genes in rice. Front. Plant Sci. 6, 1–19. https://doi.org/10.3389/fpls.2015.01157.

Sornaraj, P., Luang, S., Lopato, S., Hrmova, M., 2016. Basic leucine zipper (bZIP) transcription factors involved in abiotic stresses: a molecular model of a wheat bZIP factor and implications of its structure in function. Biochim. Biophys. Acta, Gen. Subj. 1860, 46–56. https://doi.org/10.1016/j.bbagen.2015.10.014.

Takasaki, H., Maruyama, K., Kidokoro, S., Ito, Y., Fujita, Y., Shinozaki, K., Yamaguchi-Shinozaki, K., Nakashima, K., 2010. The abiotic stress-responsive NAC-type transcription factor OsNAC5 regulates stress-inducible genes and stress tolerance in rice. Mol. Gen. Genomics. 284, 173–183.

Tang, W., Luo, X., Samuels, V., 2004. Regulated gene expression with promoters responding to inducers. Plant Sci. 166, 827–834. https://doi.org/10.1016/j.plantsci.2003.12.003.

Tang, N., Zhang, H., Li, X., Xiao, J., Xiong, L., 2012. Constitutive activation of transcription factor OsbZIP46 improves drought tolerance in rice. Plant Physiol. 158, 1755–1768. https://doi.org/10.1104/pp.111.190389.

Thilmony, R., Guttman, M., Thomson, J.G., Blechl, A.E., 2009. The LP2 leucine-rich repeat receptor kinase gene promoter directs organ-specific, light-responsive expression in transgenic rice. Plant Biotechnol. J. 7, 867–882. https://doi.org/10.1111/j.1467-7652.2009.00449.x.

Todaka, D., Nakashima, K., Shinozaki, K., Yamaguchi-Shinozaki, K., 2012. Toward understanding transcriptional regulatory networks in abiotic stress responses and tolerance in rice. Rice 5, 6. https://doi.org/10.1186/1939-8433-5-6.

Tran, L.-S.P., Nishiyama, R., Yamaguchi-Shinozaki, K., Shinozaki, K., 2010. Potential utilization of NAC transcription factors to enhance abiotic stress tolerance in plants by biotechnological approach. GM Crops 1, 32–39. https://doi.org/10.4161/gmcr.1.1.10569.

Wei, X., Jiao, G., Lin, H., Sheng, Z., Shao, G., Xie, L., Tang, S., Xu, Q., Hu, P., 2016. GRAIN INCOMPLETE FILLING 2 regulates grain filling and starch synthesis during rice caryopsis development. J. Integr. Plant Biol. 59 (2), 134–153. https://doi.org/10.1111/jipb.12510.

Wu, X., Shiroto, Y., Kishitani, S., Ito, Y., Toriyama, K., 2009. Enhanced heat and drought tolerance in transgenic rice seedlings overexpressing OsWRKY11 under the control of HSP101 promoter. Plant Cell Rep. 28, 21–30. https://doi.org/10.1007/s00299-008-0614-x.

Xiang, Y., Tang, N., Du, H., Ye, H., Xiong, L., 2008. Characterization of OsbZIP23 as a key player of the basic leucine zipper transcription factor family for conferring abscisic acid sensitivity and salinity and drought tolerance in rice. Plant Physiol. 148, 1938–1952.

Yang, A., Dai, X., Zhang, W.H., 2012. A R2R3-type MYB gene, OsMYB2, is involved in salt, cold, and dehydration tolerance in rice. J. Exp. Bot. 63, 2541–2556. https://doi.org/10.1093/jxb/err431.

Yoo, Y.-H., Nalini Chandran, A.K., Park, J.-C., Gho, Y.-S., Lee, S.-W., An, G., Jung, K.-H., 2017. OsPhyB-mediating novel regulatory pathway for drought tolerance in rice root identified by a global RNA-Seq transcriptome analysis of rice genes in response to water deficiencies. Front. Plant Sci. 8, 580. https://doi.org/10.3389/fpls.2017.00580.

Zhang, G., Chen, M., Li, L., Xu, Z., Chen, X., Guo, J., Ma, Y., 2009. Overexpression of the soybean GmERF3 gene, an AP2/ERF type transcription factor for increased tolerances to salt, drought, and diseases in transgenic tobacco. J. Exp. Bot. 60, 3781–3796. https://doi.org/10.1093/jxb/erp214.

Zhu, J.-K., 2016. Abiotic stress signaling and responses in plants. Cell 167, 313–324. https://doi.org/10.1016/j.cell.2016.08.029.

Zhu, X., Qi, L., Liu, X., Cai, S., Xu, H., Huang, R., Li, J., Wei, X., Zhang, Z., 2014. The wheat ethylene response factor transcription factor pathogen-induced ERF1 mediates host responses to both the necrotrophic pathogen Rhizoctonia cerealis and freezing stresses. Plant Physiol. 164, 1499–1514. https://doi.org/10.1104/pp.113.229575.

Zong, W., Tang, N., Yang, J., Peng, L., Ma, S., Xu, Y., Li, G., Xiong, L., 2016. Feedback regulation of ABA signaling and biosynthesis by a bZIP transcription factor targets drought resistance related genes. Plant Physiol. 171, 2810–2825. https://doi.org/10.1104/pp.16.00469.

CHAPTER 2

Sugars Play a Critical Role in Abiotic Stress Tolerance in Plants

Manu P. Gangola, Bharathi R. Ramadoss
Department of Plant Sciences, University of Saskatchewan, Saskatoon, SK, Canada

Contents

1. INTRODUCTION

Plants are complex, but sessile, organisms, therefore, they are often exposed to challenging environmental conditions or abiotic factors, including decreased water availability (drought), extreme temperatures (heat/cold/freezing), extreme light conditions, decreased and/or excessive availability of ions in soil, and soil structure/texture (Rosa et al., 2009a,b; Cramer et al., 2011; Keunen et al., 2013; Wang et al., 2015). The random and unheralded disturbance(s) in the environmental or abiotic factor(s) limiting plant performance and/or productivity below the optimum level are considered abiotic stress(es), which is one of the major causes of crop loss across the globe (Krasensky and Jonak, 2012; Duque et al., 2013). Abiotic stresses have been predicted to reduce the average global crop yield by >50%, and to affect >90% of the total global land area (Cramer et al., 2011). Abiotic stress in plants has three primary phases (Rosa et al., 2009a,b; Duque et al., 2013), including sensing, signaling, and exhaustion. *Sensing* is the foremost phase that a plant experiences through various mechanisms when there is

Biochemical, Physiological and Molecular Avenues for Combating Abiotic Stress in Plants
https://doi.org/10.1016/B978-0-12-813066-7.00002-4

disturbance in any of the abiotic factors. Abiotic stress primarily causes ion imbalance and hyperosmotic stress in a plant cell. The *signaling* cascade [reactive oxygen species (ROS), calcium, and others] of the plant cell senses the changes in the cell and induces the resistance machinery leading to the third phase, *exhaustion*. This phase involves changes in the physiological functions in the plant cells. A fourth phase, regeneration, involves the partial or full normalization of plant cell functions, but can only be observed when the stresses are removed. Consequently, abiotic stress results in decreased photosynthesis, water transport inhibition, osmotic/ion/nutrient imbalance, plasma-membrane instability, oxidative stress (outburst of ROS or imbalance between ROS and antioxidant system of the plant cell), and other unfavorable changes in the plant cell machinery that collectively reduce plant growth and development (Rosa et al., 2009a,b; Krasensky and Jonak, 2012; Van den Ende and El-Esawe, 2014). Therefore, abiotic stress is one of the major problems constraining sustainable agricultural crop production in various parts of the world, and it needs to be addressed to insure food security for the growing world population, which is estimated to be more than 9 billion by 2050 (Jaggard et al., 2010; Bevan et al., 2017).

To counteract the impact of abiotic stress, plants have evolved complex physio-biochemical and molecular strategies, including antioxidant systems and resistance gene(s). Many reports have been published explaining the mechanism underlying the antioxidant system and resistance gene(s), and efforts have been made to modulate these systems to improve plant performance in abiotic-stressed conditions. In recent years, sugars and carbohydrates have emerged as potential agents for improving plant tolerance against abiotic stress (Gupta and Kaur, 2005; Sami et al., 2016). Sugars and carbohydrates are two essential plant cell constituents (Hernandez-Marin and Martínez, 2012), and have been named after their basic chemical formula $[C_x(H_2O)_y]$, which defines them as hydrates of carbon with hydrogen and oxygen in the same ratio as in water. Sugars are polyhydroxy aldehydes, or ketones, and are primarily categorized based on molecular size, which is determined by degree of polymerization (DP), the type of linkage (α or non-α), and characteristics of individual monomers. Such classification divides sugars into four classes: mono- (DP 1), di- (DP 2), oligo- (DP 3–9), and poly-saccharides (DP $\geq$10) (Cummings and Stephen, 2007). In plants, sugars participate in various structural, biochemical, and physiological properties (Peshev and Van den Ende, 2013; Gangola et al., 2016). Being chemically diverse molecules, sugars, especially during the past decade, have been studied for their crucial role in abiotic stress tolerance. Therefore, this chapter summarizes the metabolism and diverse roles of sugars during abiotic stress tolerance in plants. The chapter also includes some of the important targets of sugars' metabolic pathways to modulate their concentration in plant cells to improve salt, drought, and cold tolerance in different crop species, along with their limitations and challenges.

2. METABOLISM OF SUGARS IMPORTANT FOR ABIOTIC STRESS TOLERANCE: A BRIEF INTRODUCTION

Plants are autotrophic organisms; therefore, they utilize light energy to fix carbon dioxide and water through a photosynthetic mechanism in the chloroplast (Baker et al., 2012; Tarkowski and Van den Ende, 2015). This process helps plant cells maintain two major pools of metabolites that can be converted into one another via reversible enzymatic reactions or transporters, as per the requirement of the plant cell. These two major pools consist of the following intermediates: the triose phosphate pool, which includes 3-phosphoglycerate (3-PGA) and dihydroxyacetone phosphate (DHAP), and the hexose phosphate pool (glucose 1-phosphate, glucose 6-phosphate, fructose 1-phosphate, and ADP-glucose) (Fig. 1; Granot et al., 2013; Tarkowski and Van den Ende, 2015; Griffiths et al., 2016). The hexose phosphates play the leading role in the sugars' metabolism; whereas the triose phosphates act as major transporters of carbon from chloroplast to cytosol, where they are converted into hexose phosphates to synthesize sugar molecules (Griffiths et al., 2016).

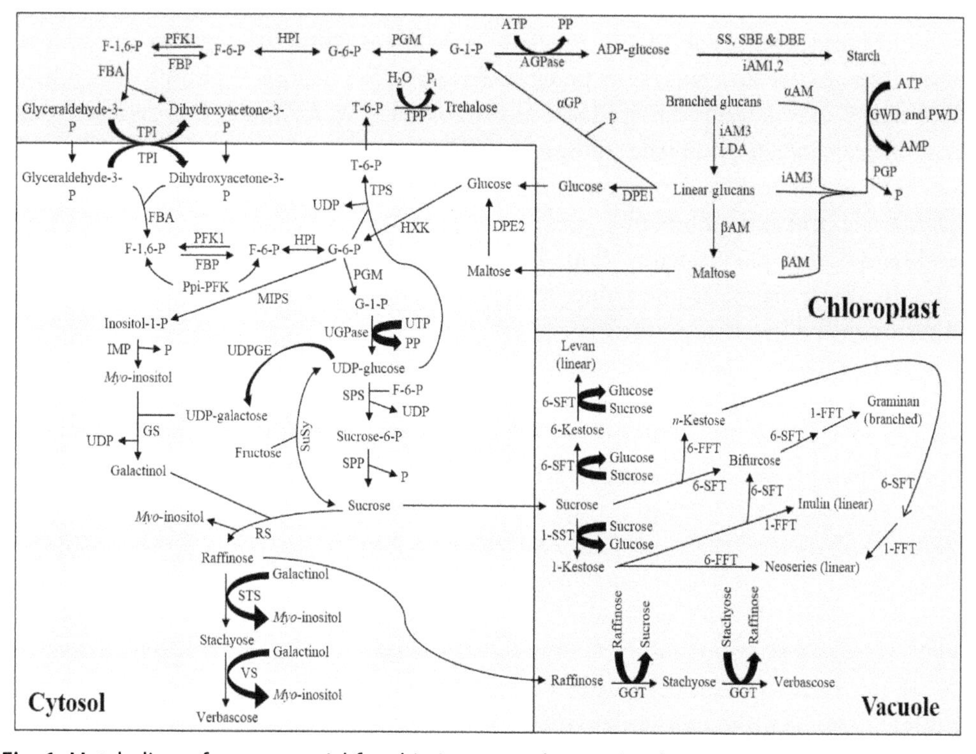

Fig. 1 Metabolism of sugars crucial for abiotic stress tolerance in plants.

Sucrose, being a nonreducing sugar with restricted chemical activity, is the major transport and storage molecule in most of the plants. A sucrose molecule is composed of one glucose and one fructose molecule joined by $\alpha(1 \rightarrow 2)$ glycosidic linkage (Chibbar et al., 2016). It is mainly synthesized in cytosol from triose phosphates (photosynthetic product) in two consecutive steps catalyzed by sucrose-phosphate synthase (SPS) and sucrose-phosphate phosphatase (SPP), respectively (Ruan, 2014). Alternatively, sucrose can also be synthesized by starch degradation, or through a reversible reaction between NDP-glucose (nucleotide diphosphate like uridine diphosphate) and fructose catalyzed by sucrose synthase (SuSy) enzyme (Fig. 1) (Nguyen et al., 2016). SPS is the limiting step of sucrose biosynthesis, which can be induced allosterically by glucose 6-phosphate and inhibited by inorganic phosphate (Pi). SPS also has various phosphorylation sites regulating its activity, leading to modulated sucrose biosynthesis (Ruan, 2014). SuSy is present in soluble and membrane-bound form in the plant cell, and can both synthesize and degrade sucrose. Sucrose can be translocated synplastically or apoplastically to the phloem cells and in sink tissues, where it can be either stored in the vacuole by the transporters at the tonoplast or hydrolyzed into glucose + fructose by invertase (Roitsch and González, 2004; Lemoine et al., 2013).

Trehalose is a disaccharide in which two glucose molecules are connected by $\alpha(1 \rightarrow 1)$ glycosidic bond (Lunn et al., 2014). In plants, trehalose is synthesized in two consecutive steps: UDP-glucose and glucose-6-phosphate react in a trehalose-6-phosphate synthase (TPS) catalyzed reaction yielding trehalose-6-phosphate, which dephosphorylates to trehalose by trehalose-6-phosphate phosphatase (TPP) (Fig. 1) (Delorge et al., 2014). The former reaction is proposed to occur in cytosol, whereas the later one occurs in the chloroplast. The trehalose does not break down in chloroplast, and shows a plasma membrane-bound phenomenon (Fig. 1). Although no transporter has been reported to the date for trehalose-6-phospahte or trehalose yet, one of the hexose/sucrose transporters might be involved in the process (Wingler and Paul, 2013).

Raffinose family oligosaccharides (RFO) are the second most abundant soluble sugars after sucrose (Frias et al., 1999). RFO also act as a major photosynthate transporter in the members of Cucurbitaceae, Verbenaceae, Lamiaceae, Oleaceae, and Scrophulariaceae families (Sprenger and Keller, 2000). RFO, having raffinose, stachyose, and verbascose as consecutive members, are galactosyl derivatives of sucrose. RFO biosynthesis begins with the formation of galactinol in a galactinol synthase (GS) catalyzed reaction between *myo*-inositol and UDP-galactose (Kannan et al., 2016). Galactinol interacts with sucrose, raffinose, and stachyose to synthesize raffinose, stachyose, and verbascose in reactions catalyzed by raffinose synthase (RS), stachyose synthase (STS), and verbascose synthase (VS), respectively. An alternate pathway catalyzed by galactan:galactan galactosyltransferase (GGT) has also been suggested for the biosynthesis of stachyose and verbascose (Fig. 1) (Gangola et al., 2016). RFO is mainly synthesized in the cytosol and can be transported to chloroplast by a hypothesized active transporter (Schneider and Keller, 2009) and vacuole by an unknown mechanism. There is controversy about the biosynthesis of RFO, as some studies also suggest that they are synthesized in vacuole as well; however, the biosynthetic steps are similar, irrespective of the site of biosynthesis. In seeds, RFO are mainly synthesized during later stages of development and can be broken down during seed germination by α-galactosidase, providing carbon and energy to the growing seedling (Peterbauer et al., 2001; Gangola et al., 2016).

Fructans, the fructose polymers, serve as soluble carbohydrate reserves in about 12%–15% of all flowering plants belonging to Liliaceae, Amaryllidaceae, Gramineae, and Compositae (Livingston et al., 2009; da Silva et al., 2013; Apolinário et al., 2014). Fructans in plants are classified into four distinct categories based on the position of the glucosyl unit and linkage types between fructosyl residues: levan, neoseries, inulin, and graminan. Fructan biosynthesis is determined by the photosynthetic demand and sucrose concentration of the sink. Fructans are synthesized in vacuole from sucrose molecules in reactions catalyzed by various fructosyltransferases (Fig. 1) (Cimini et al., 2015). The two predominant precursors for the higher members of the fructan family in most of the plants are 1-kestose and 6-kestose that are synthesized from sucrose in sucrose:sucrose-1-fructosyltransferase (1-SST) and sucrose:fructan-6-fructosyltransferase (6-SFT) catalyzed reactions, respectively. The elongation of the precursor molecules is catalyzed by fructan:fructan-1-fructosyltransferase (1-FFT), 6-SFT, or fructan:fructan-6G-fructosyltransferase (6-FFT) enzymes (Livingston et al., 2009).

Starch is the predominant storage polysaccharide in seeds, and is composed of two glucan polymers, amylose [linear ($\alpha 1 \rightarrow 4$)-glucan polymer with rare ($\alpha 1 \rightarrow 6$) branching] and amylopectin (highly branched glucan polymer). Starch biosynthesis is a complex mechanism, and mainly involves ADP-glucose pyrophosphorylase (AGPase), starch synthases (SSs), starch branching enzymes (SBEs), and starch debranching enzymes (DBEs; pullulanase and isoamylase) (Chibbar et al., 2016). Starch can be broken down into branched glucans, maltose and linear glucans by α-amylase, β-amylase and isoamylse-3, respectively, which can either be converted into monosaccharides, or transferred to cytosol, to synthesize other sugars, as shown in Fig. 1.

3. DIVERSE ROLES OF SUGARS DURING ABIOTIC STRESS TOLERANCE

Sugars are chemically active biomolecules and are involved in crucial physio-chemical mechanisms such as photosynthesis, respiration, seed germination, flowering, senescence, and so forth. Therefore, modulating sugar composition or concentration in plants may improve their responses or adaptation to abiotic stress (Fig. 2). Sugars have been characterized for their following diverse roles during abiotic stress tolerance.

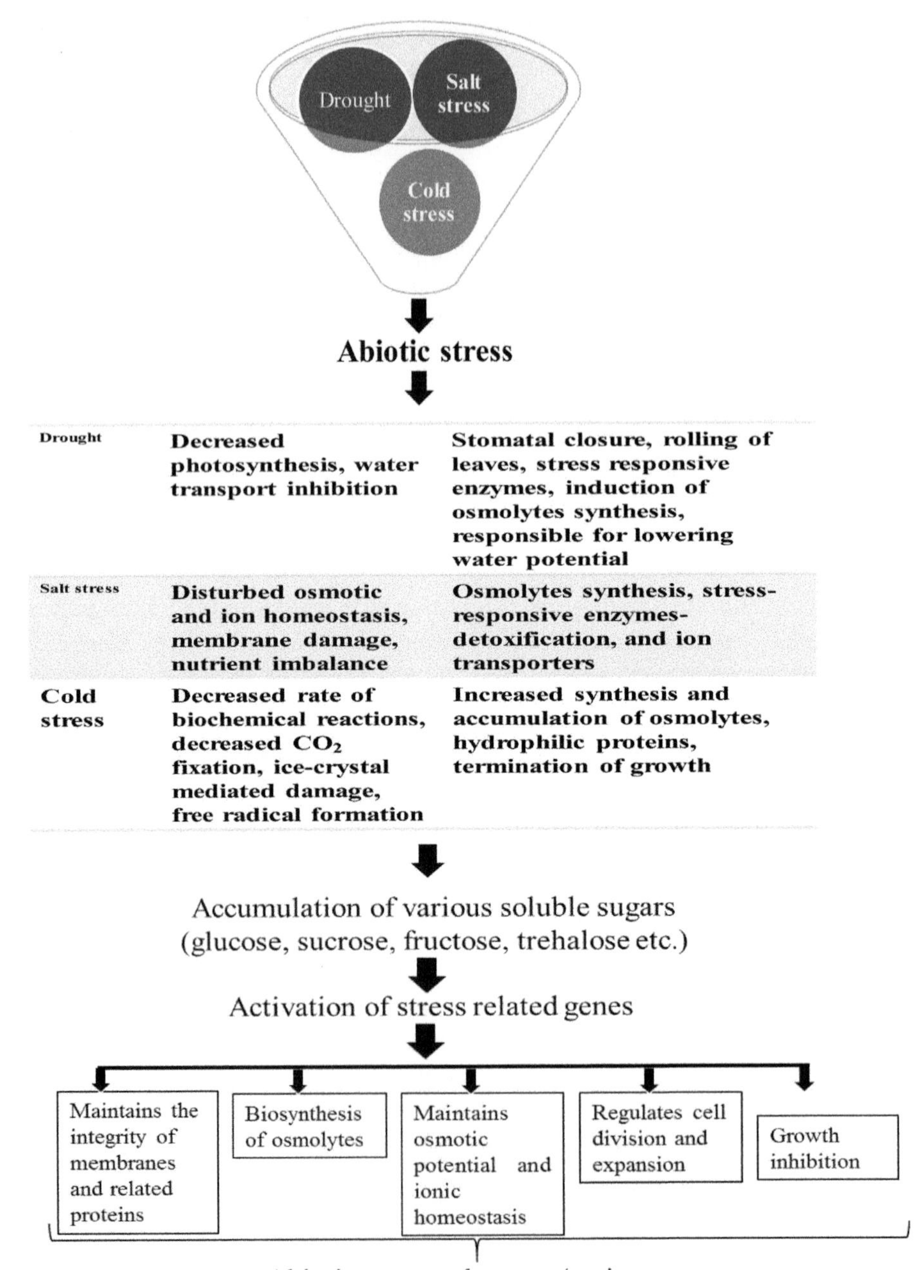

Fig. 2 Summarization of role of sugars during abiotic stress tolerance.

3.1 Scavenging Reactive Oxygen Species

Reactive oxygen species (ROS) are highly reactive forms of oxygen, and mainly include singlet oxygen ($^1O_2$), superoxide ion radical ($O_2^{\bullet-}$), hydroxyl radical ($HO^{\bullet}$), and hydrogen peroxide (H_2O_2) (Blokhina et al., 2003; Edreva, 2005; Karuppanapandian et al., 2011). ROS are by-products of the aerobic metabolism, and their concentrations are in balance with the antioxidant system of the plant cell during normal growth conditions (Kwak et al., 2006). However, exposure of the plant to abiotic stress increases the production of ROS in the cell, causing disruption of the cellular redox homeostasis and degradation of important biomolecules such as lipids, proteins, and nucleic acids (Torres et al., 2006). This condition in plant cells is termed as oxidative stress. The antioxidant system in plants traditionally includes vitamins C and E, various classes of phytochemicals (flavonoids, terpenoids, carotenoids, etc.) and enzyme- based systems such as catalase, superoxide dismutase, and peroxidases (Gill and Tuteja, 2010; Foyer and Shigeoka, 2011; Gangola et al., 2013). However, in recent years, the sugars in plants have emerged as a new class of antioxidant molecules.

The involvement of monosaccharides as direct antioxidants is rare. Monosaccharides are more likely to affect the antioxidative property of a plant cell in an indirect approach, either by contributing through their polymers, or as secondary messengers to induce the expression or activity of other antioxidants. Common disaccharides sucrose, trehalose, maltose, and lactose show a significant free-radical quenching effect in vitro (Wehmeier and Mooradian, 1994; Morelli et al., 2003). Disaccharides were shown to be more active toward, and less damaged by hydroxyl radicals than monosaccharides, which seemed to be dependent on the number of •OH residues (Morelli et al., 2003). Compounds such as insulin (fructan) (Peshev et al., 2013) have been found to have better ROS scavenging capability than assessed disaccharides. However, disaccharides such as sucrose seem to be placed at a level that is moderate in antioxidative capacity, but due to their small size and ease of transport, they may play a greater role in ROS control. RFO, along with galactinol, were reported as hydroxyl radical scavengers in Arabidopsis (Nishizawa et al., 2008). RFO might also have the capability to scavenge other forms of ROS. During this detoxification process, RFO are proposed to convert in their oxidized radical forms that are further regenerated by reacting with other antioxidants such as ascorbic acid or flavonoids (Van den Ende and Valluru, 2009; Bolouri-Moghaddam et al., 2010). Fructans stabilize tonoplast by scavenging hydroxyl radicals, thus preventing lipid peroxidation of the membrane (Van den Ende and Valluru, 2009). Consequently, fructans convert into fructan radicals that may be reprocessed by classical vacuolar antioxidants (Bolouri-Moghaddam et al., 2010; Peshev et al., 2013). Besides this, fructans have also been associated with increased accumulation of ascorbate and glutathione, and thus are hypothesized to be connected to a cytoplasmic antioxidant network (Bolouri-Moghaddam et al.,

2010). A tonoplast vesicle-derived exocytosis (TVE) system might be operative to transport fructans from vacuole to apoplast, where it acts as a direct (scavenging hydroxyl radical), as well as indirect, antioxidant (stabilizing membrane) during abiotic stress conditions (Van den Ende and Valluru, 2009).

Antioxidants scavenge ROS by following three different mechanisms: (a) electron transfer, (b) hydrogen atom transfer, and (c) radical addition reaction (Hernandez-Marin and Martínez, 2012). Theoretically, and based on previous reports, the second mechanism is hypothesized as the predominant ROS scavenging mechanism for sugars, during which hydrogen is preferably used from C—H instead of O—H due to lower bond energy (Matros et al., 2015). Consequently, the outcomes of the reactions are a carbon-centered radical and water. However, a recent report by Peshev et al. (2013) demonstrated carbon-centered radicals and free hexoses (lower DP sugars) as the outcomes of the reaction between sugars and ROS (hydroxyl radical specifically). The resulting free sugars were glucose/fructose, melibiose/sucrose/fructose/galactose/glucose, and inulobiose/sucrose/glucose/fructose when sucrose, raffinose, and 1-kestotriose (a type of fructan) reacted with hydroxyl radicals, respectively. It confirmed the partial breakdown of the sugars during the scavenging reaction (Peshev et al., 2013). In 2015, Matros et al. provided the *in-planta* evidence of radical reactions of sucralose (an artificial analog of sucrose) in Arabidopsis tissues, and predicted the mechanisms of sucrose's ROS scavenging mechanism. In brief, sucrose reacts with hydroxyl radicals and forms sucrosyl radicals that may undergo four different reactions. In two reactions, sucrosyl radicals may convert into monosaccharide radicals and nonradicals with and without keto groups; whereas, sucrosyl radicals may oxidize during the third reaction, yielding hydrate products. In the fourth reaction, sucrosyl radicals may recombine to form unique oligosaccharides of a higher degree of polymerization. The same mechanisms may also be true for other sugars present in plants, but lack the experimental evidence to date.

3.2 Sugars as Osmo-Protectants

Abiotic stresses, especially drought, heat, and salinity, induce dehydration of plant cells, causing osmotic stress that may sequentially lead to disrupted hydrophilic interaction, degradation of biomolecules structure (especially protein denaturation), collapse of organelles, and destabilization of cell membranes (Garcia et al., 1997; Hare et al., 1998). Salt stress induces the toxicity of specific ions such as Na^+ and Cl^-, which thus reduces the uptake of crucial minerals, including potassium, phosphorus, nitrogen, and calcium. Na^+ toxicity also disturbs the Na^+/K^+ ratio in the plant cell, which is crucial for normal cellular operations (Singh et al., 2015). Likewise, drought has been associated with the disruption of the K^+/H^+ ratio in plant cells. Therefore, to protect cells from the increased dehydration during abiotic stress, osmo-protectants, or osmolyte concentrations, need to be increased to maintain the turgor pressure of the cell and enhance stress tolerance in plants.

Many osmo-protectants have been identified in plants, among which sugars, including sucrose, trehalose, RFO, and fructans constitute an important group (Slama et al., 2015). The hydroxyl groups of the sugars can replace water molecules to maintain the hydrophilic interactions in plant cells, which is crucial to stabilize native macromolecules and membrane structure during dehydration (Koster, 1991; Pukacka et al., 2009). The accumulation of osmo-protective sugars has been attributed to maintain the ion partitioning and homeostasis in the plant cell, thus helps in maintaining proper cell functions and enhancing the abiotic stress- tolerance. Trehalose is the most promising osmo-protective sugar in terms of concentration required (Nahar et al., 2016) and can be replaced by sucrose and other sugars in plants. The important genes of the sugars' biosynthetic pathways have been targeted in or transferred to plant species, including tomatoes (*Lens esculentum* Mill.; Cortina and Culiáñez-Macià, 2005), arabidopsis (*Arabidopsis thaliana*; Han et al., 2005; Miranda et al., 2007; Nishizawa et al., 2008; Liu et al., 2007; Zhifang and Loescher, 2003), rice (*Oryza sativa* L.; Garg et al., 2002; Jang et al., 2003; Kawakami et al., 2008; Pujni et al., 2007), tobacco (*Nicotiana tabacum* L.; Han et al., 2005; Zhang et al., 2005; Pilon-Smits et al., 1998; Parvanova et al., 2004; Pilon-Smits et al., 1995; Li et al., 2007; Shen et al., 1997; Sheveleva et al., 1997; Oberschall et al., 2000; Fukushima et al., 2001), petunias (*Petunia* × *hybrid*; Pennycooke et al., 2003; Chiang et al., 2005), beets (*Beta vulgaris* L.; Pilon-Smits et al., 1999), wheat (*Triticum aestivum* L.; Abebe et al., 2003), loblolly pines (*Pinus taeda* L.; Tang, et al., 2005), and persimmons (*Diospyros kaki* L.f.; Gao et al., 2001; Deguchi et al., 2004) to enhance tolerance against oxidation, drought, salinity, and extreme temperature stresses (Table 1).

Sugars also assist in the development of desiccation tolerance and dehydration-tolerant structures such as seed and pollen in plants. The first mechanism by which sugars provide desiccation tolerance is water replacement, as discussed earlier. The other mechanism of desiccation tolerance by sugars is the "vitrification," or glass formation in the plant cell, during which the cell solution acts like a plastic solid or highly viscous solution. Vitrified cell solution ensures: (i) stability by precluding mechanisms required for diffusion, (ii) no cellular collapse, as it fills in the blanks within the organelles or molecules, and (iii) hydrogen bonding within the cell (Koster and Leopold, 1988; Koster, 1991; Martínez-Villaluenga et al., 2008; Angelovici et al., 2010). RFO, along with late embryogenesis-abundant (LEA) proteins and small heat shock proteins (sHSP) synthesize the glassy state of the cytosol that limits the monosaccharides' biosynthesis, leading to decreased respiration and inhibited Maillard's reaction (Martínez-Villaluenga et al., 2008; Pukacka et al., 2009).

3.3 Sugars as Signaling Molecules

Sugars, besides being storage/transport, structural, and energy molecules, also act as signaling mechanisms during abiotic stress tolerance in plants (Li and Sheen, 2016).

Table 1 List of sugar compounds and their role in various abiotic stress tolerance in several crop plants

S. No.	Sugar	Transgene	Species	Enhanced tolerance to	Reference
Disaccharides					
1.	Trehalose	*Trehalose-6-phosphate synthase*	*Lycopersicon esculentum*	Drought, oxidative stress, salinity	Cortina and Culiáñez-Macià (2005)
		Trehalose-6-phosphate synthase and phosphatase	*Arabidopsis thaliana*	Drought, salinity, temperature stress	Miranda et al. (2007)
		Trehalose phosphorylase	*Nicotiana tabacum*	Drought	Han et al. (2005)
		Trehalose synthase	*N. tabacum*	Drought, salinity	Zhang et al. (2005)
		otsA, otsB	*N. tabacum*	Drought	Pilon-Smits et al. (1998)
		TPS1	*N. tabacum*	Drought and water deficit	Pilon-Smits et al. (1998)
		otsA, otsB	*Oryza sativa*	Salt and drought	Garg et al. (2002)
		Trehalose- 6-phosphate and phosphatase	*Oryza sativa*	Salt, drought and cold	Jang et al. (2003)
Raffinose family oligosaccharides (RFO)					
2.	Galactinol	*Galactinol synthase*	*A. thaliana*	Oxidative stress, chilling, drought, salinity	Nishizawa et al. (2008)
3.	Raffinose	*α-Galactosidase*	*Petunia × hybrida cv Mitchell*	Freezing	Pennycooke et al. (2003)
		UDP-glucose 4-epimerase	*A. thaliana*	Drought, freezing, salinity	Liu et al. (2007)
Fructans					
4.	Fructans	*Levansucrase*	*N. tabacum*	Freezing	Parvanova et al. (2004)
		SacB	*N. tabacum*	Drought stress	Pilon-Smits et al. (1995)
		SacB	*Beta vulgaris*	Drought stress	Pilon-Smits et al. (1999)
		Sucrose:sucrose 1-fructosyltransferase	*N. tabacum*	Freezing	Li et al. (2007)

		Sucrose:sucrose 1-fructosyltransferase and sucrose:fructan 6-fructosyltransferase	*O. sativa*	Chilling	Kawakami et al. (2008)
Sugar alcohols					
5.	Mannitol	*Mannitol-1-phosphate dehydrogenase*	*N. tabacum*	Oxidative stress (paraquat)	Shen et al. (1997)
		Mannitol-1-phosphate dehydrogenase	*Triticum aestivum* L.	Water stress and salinity	Abebe et al. (2003)
		Mannitol-1-phosphate dehydrogenase	*O. sativa*	Drought, salinity	Pujni et al. (2007)
		Mannitol-1-phosphate dehydrogenase	*Petunia x hybrida (Hook) Vilm. cv. Mitchell*	Chilling	Chiang et al. (2005)
		Mannitol-1-phosphate dehydrogenase	*Pinus taeda*	Salinity	Tang, et al. (2005)
		Mannose 6-phosphate reductase	*A. thaliana*	Salinity	Zhifang and Loescher (2003)
6.	D-Ononitol	*IMT1 (myo-inositol O-methyl transferase) of common ice plant*	*N. tabacum*	Drought and salinity stress	Sheveleva et al. (1997)
Sorbitol					
7.	Sorbitol	*S6PDH*	*Diospyros kaki*	Salinity	Gao et al. (2001)
		Glucitol-6-phosphate dehydrogenase	*P. taeda*	Salinity	Tang et al. (2005)
		Sorbitol-6-phosphate dehydrogenase	*D. kaki*	Salinity	Deguchi et al. (2004)
8.	Aldose/Aldehyde reductase	*MsALR*	*N. tabacum*	Chemical and drought stress	Oberschall et al. (2000)
9.	Invertase	*Apoplastic Invertase*	*N. tabacum*	Salt tolerance	Fukushima et al. (2001)

Sugar signaling in plants includes sugar sensing, signal transduction, and target gene(s) expression. Glucose is sensed predominantly by either hexokinase (HXK) dependent or independent pathways in plant cells. Sensing by HXK involves the phosphorylation of sugars, whereas the HXK-independent pathways can sense the sugar as such. Hexokinase (HXK) is the most widely studied glucose-sensor in plants (Van den Ende and El-Esawe, 2014). Hexokinase represent a multigene family with two members in *Solanum tuberosum*, four in *Solanum lycopersicum*, five in *Vitis vinifera*, six in *A. thaliana*, nine in *Zea mays/N. tabacum*, 10 in *O. sativa* and 11 in *Physcomitrella patens* (Aguilera-Alvarado and Sánchez-Nieto, 2017). HXKs are categorized into four groups based on their subcellular localization: (a) type A HXKs are characterized by a hydrophobic sequence of 30 amino acids with chloroplast signal at the N-terminus end, (b) type B HXKs have a highly hydrophobic helix of 24 amino acids that adhere to the mitochondrion, (c) type C HXKs (present in monocots and moss) are cytosolic, and hence lack any signal peptide or membrane attachment, and (d) type D HXKs are also mitochondrial, but have a different peptide compared with type B HXKs (Aguilera-Alvarado and Sánchez-Nieto, 2017). Some of the type B HXKs with nuclear directing signals are translocated to the nucleus and are the most studied HXKs among all four groups, as they are important for sugar signaling in plants during normal and stressed environmental conditions. The nuclear-localized HXK, together with proteasome, synthesizes a glucose-signal complex that represses the photosynthesis when the glucose level is abundant. Conversely, a low glucose level interrupts the HXK-mediated signal from abiotic stress. The site of intracellular sugar sensing by HXK still needs to be explored (Valluru and Van den Ende, 2011). Although evidence for HXK- independent glucose sensing pathways has been reported, the process still requires additional investigation (Ramon et al., 2008).

A sucrose-specific signaling pathway has also been documented influencing photosynthesis, and biosynthesis of fructans and anthocyanins (a class of flavonoids in plants that participate in plant development and stress responses). Sucrose accumulation is determined by the balance between sucrose synthesizing (SPS and SPP) and degrading (SuSy/Invertase) enzymes that are regulated by hormones and circadian clocks in plants. Sucrose signaling has also shown interconnection with other signaling pathways activated by light, abscisic acid, and other phytohormones that might also be associated with calcium signaling pathways in plants. Although no sucrose sensor has been identified in plants to date, it is hypothesized that sucrose-signaling might be converted into a trehalose-6-phospahate-signal that modulates anthocyanin biosynthesis through *MYB75* (a transcription factor involved in regulating anthocyanin biosynthesis) (Van den Ende and El-Esawe, 2014). Invertases catabolize sucrose into glucose and fructose units, and are mainly of two types: glycosylated acid (present in apoplast or vacuole), and nonglycosylated alkaline/neutral (present on mitochondrion, plastid, or cytosol) invertases. The alkaline/neutral invertases have been associated with abiotic stress- tolerance in plants (Valluru and Van den Ende, 2011). Glucose synthesized during invertase-catalyzed reactions in mitochondrion and/or cytosol maintains the HXK activity supporting the reactive oxygen species' homeostasis (Valluru and Van den Ende, 2011).

Abiotic stress also causes energy deprivation in plant cells activating SnRK1 [sucrose nonfermenting1 (SNF1)-related protein kinase 1, which has a catalytic domain similar to Snf1 (Sucrose nonfermenting-1) of yeast and AMPK (AMP-activated protein kinase) in animals], which has been reported to link various signaling (inositol, sugar, and stress signals) pathways and modulate the activity of transcription factors regulating the biosynthesis of stress-related compounds in plants (Van den Ende and El-Esawe, 2014). SnRK1 is also an important regulator of carbon metabolism, and provides alternate carbon/energy/metabolite sources during abiotic stress tolerance (Lin et al., 2014; Emanuelle et al., 2016). Sugars, or their derivatives, especially glucose, glucose-6-phosphate, and trehalose-6-phosphate, have shown their impact of SnRK1. SnRK1-binding proteins have shown a glucose-dependent activity regulating SnRK1 functioning in plant cells, whereas glucose-6-phosphate and trehalose-6-phosphate regulate SnRK1 activity by changing the SnRK1 conformation via an uncharacterized intermediate compound, respectively. Sugar or energy deprivation in plant cells during stress induces the activity of SnRK1 that reprograms the plant cell metabolism or energy production influencing growth, organogenesis, abiotic stress responses, and interaction with pathogens (Valluru and Van den Ende, 2011; Yu et al., 2015).

Sugars and hormones are potential candidates for long-distance signaling in plants. The prominent members of sugar signaling cascades including hexokinase (HXK) and SnRK1, which also interconnects with phytohormones, thus helping protect the plant from abiotic stresses (Ljung et al., 2015). Sugar abundance in plant cells may trigger ABA signaling, or ABA may activate a target of a distinct sugar signaling pathway. There are two important components of sugar-ABA interconnection: a transcription factor *ABI4* and an *ANAC060* gene (Arabidopsis NAC family transcription factor 060). ABI4 regulates the expression of sugar-responsive genes by binding to their promoters. The sugar-ABA signaling cascade also utilizes *ABI4* in inducing the expression of *ANAC060*, whose nuclear localization represses the sugar-ABA signaling pathway (Li et al., 2014; Ljung et al., 2015).

Sucrose and glucose are important for auxin synthesis or signaling in plants, respectively. Sucrose also stabilizes DELLA proteins that are important for various developmental and stress responses in plants, and connects sucrose-GA signaling cascade to brassinosteroids (BR). However, DELLA proteins are negative regulators of gibberellin (GA) signaling, which explains the repression of sucrose-dependent induced anthocyanin biosynthesis by GA. Moreover, starch metabolism has also been connected to BR signaling through β-amylases, acting as maltose sensors in plant cells (Ljung et al., 2015).

4. TARGETING SUGARS TO DEVELOP ABIOTIC STRESS TOLERANT CROP VARIETIES

The increasing human population, coupled with the loss of agricultural land due to industrialization, urbanization, desertification, and climatic changes pose challenges to world agriculture. Breeding crop plants to increase yield and feed the growing population has been efficient in feeding the growing population so far. However, to feed the 9 billion people expected by 2050, 44 million metric tons of food will be needed per year (Godfray et al., 2010). These yield discrepancies are even more challenging with respect to the projected results of global warming.

Sugars are essential components of abiotic stress tolerance in plants as described herein (Fig. 2). The accumulation of sugars in plants has been widely reported as a response to abiotic stresses (Table 1). Initially, conventional breeding programs were being practiced, exploiting the genetic variation of crops at different gene pools to produce abiotic stress tolerant/resistant cultivars/varieties. Consequently, few abiotic stress-tolerant cultivars/breeding lines have been developed in various crop species, most of which could not perform well when tested in the field experiments. This limits the success of producing abiotic-tolerant cultivars of different agriculturally important crops using conventional breeding approaches. Therefore, it is worthwhile to utilize wild relatives as donors for resistance gene(s) for crop improvement to enhance abiotic stress tolerance. However, transferring abiotic-tolerant gene(s) from wild relatives to domesticated crops is time-consuming and labor-intensive. Moreover, reproductive barriers restrict the transfer of favorable alleles from wild relatives. Therefore, genetic engineering has emerged as an effective alternative approach, and it is being used worldwide to improve abiotic stress tolerance.

4.1 Salt Stress

Salt stress modifies the physiology of plants through reduction of photosynthesis, cell division, nitrogen assimilation, and eventually growth and development. Worldwide, 800 million hectares of soil is affected by salinity (FAO, 2008). In addition, salinity problems are increasing at a rate of ~10% annually worldwide, mostly in Asia (Rains and Goyal, 2003; Ashraf and Foolad, 2007). Furthermore, modern agriculture and inadequate agronomic practices are causing substantial salinization of agricultural land. Increased irrigation in cultivated areas has created more salination when irrigation water contains a high concentration of solutes, and, fewer drainage systems can also increase the level of salinity that is harmful to crop plants. Therefore, salinity is one of the major abiotic stresses influencing crop productivity. Usually, saline soil contains excessive amounts of Na^+ and Cl^- ions, leading to reduced ion imbalance, water potential, and reduced growth and development of the plants. Sugars in plants serve as osmolytes to alleviate the negative effects of salt stress (Almodares et al., 2008a,b). The enhanced concentration of glucose, sucrose, and fructose takes place under salinity conditions that play a vital role in carbon storage, osmoprotection, and osmotic homeostasis, as well as scavenging of free radicals (Rosa et al., 2009a,b).

Transgenic rice expressing the trehalose gene has shown enhanced tolerance against salinity, drought, and cold stresses (Ashraf et al., 2008; Garg et al., 2002). Likewise, rice plants expressing the chimeric gene Ubi1: TPSP showed increased trehalose accumulation, resulting in improved tolerance to salt and cold stresses (Jang et al., 2003). However, most of the transgenic plants expressing trehalose showed pleiotropic effects modulating other plant development processes (Ashraf et al., 2008). The *mt1D* gene expressed in tobacco and wheat plants showed increased mannitol accumulation and enhanced salt stress tolerance (Tarczynski et al., 1993; Abebe et al., 2003).

4.2 Drought Stress

Breeding for drought tolerance is probably the most complex and challenging task scientists encounter while attempting to improve the genetic potential of different crop species. The abiotic stresses are responsible for ~89% of all crop failure, and among them, drought is responsible for >40% of the losses (Ort and Long, 2003).

Glucose plays a key role in stomal closure and enhances a plant's adaptability under heat/drought conditions (Osakabe et al., 2014). Several studies have demonstrated that increased levels of raffinose, stachyose, and verbascose were observed during seed desiccation (Peterbauer and Richter, 1998; Seki et al., 2007; Mohammadkhani and Heidari, 2008). Accumulation of sugar during drought/heat stress prevents the oxidation of cell membranes under drought conditions (Arabzadeh, 2012). Furthermore, sugars also maintain the turbidity of leaves and maintain the water level from dehydration of membranes (Sawhney and Singh, 2002). Soluble sugars also maintain leaf water content and osmotic potential during drought conditions (Xu et al., 2007). The introduction of bi-function gene-encoding TPS and TPP enzymes of trehalose biosynthesis into rice increased the accumulation of trehalose, thus enhancing cold, drought, and salt tolerance (Jang et al., 2003).

4.3 Cold Stress

Temperature is one of the primary environmental factors that limit plant distribution and crop productivity. Low temperature affects the rates of biochemical processes differently and thus induces imbalances between partial processes in metabolic pathways. In plants, a variety of soluble sugars, such as glucose, sucrose, fructose, and RFOs are known to provide freeze tolerance (Yuanyuan et al., 2009). The soluble sugars not only act as osmoprotectants, but also play a vital role in providing acclimatization during chilling stress through contact with lipid bilayers (Garg et al., 2002). Normally, trehalose presence is very low, but during exposure to cold stress, its concentration increases considerably (Fernandez et al., 2010).

Changes in the levels of soluble sugars have been shown to affect cold tolerance in plants. During cold stress, sugars can act as osmotic potentials, as well as functioning as signaling molecules. On the other hand, sugars also affect other housekeeping functions during plant development. Further studies about the actual role of each sugar in cold response can be done using advanced technologies. These findings might provide new insights into mechanisms by which sugar response pathways react during cold stress response.

5. LIMITATIONS AND CHALLENGES

The transformation experiment has been carried out to understand the role of individual sugars on various abiotic stress resistance/tolerance in crop plants in which most of the experiments were carried out in model plants such as Arabidopsis and tobacco, where there has been remarkable success in terms of tolerance to many of the abiotic stresses. However, these model plants cannot be predictive for the agriculturally important crop plants. Therefore, the practical approach is to employ transformation technology directly to a crop of interest so that we can access the actual potential of the gene in the desirable background. Although rice and wheat have also been used in transformation experiments, the experiments were conducted in controlled conditions, and most of them were tested in the early stage of germination or vegetative stage of the plants. Further research is needed to understand the expression pattern under natural conditions through multilocation trials. Even though considerable efforts have been made so far to develop abiotic-tolerant cultivars of various crop plants using conventional plant breeding methods, there has been limited success in achieving the desired goal of developing promising varieties. With the advent of molecular biology and genetic transformation techniques, it was assumed that developing varieties resistant to various abiotic stresses would be suitable and create a relatively high throughput, but the results are not very encouraging to date. Most genetic engineering programs focus on single-gene transfer, although it is now widely known that abiotic tolerance traits are complex and controlled by multiple genes with a multitude of physiological, biochemical, and molecular processes, which are involved in the mechanism of abiotic stress resistance/tolerance.

6. CONCLUSIONS

Sugars play diverse roles during plant growth and development, therefore, their abundance, direct participation in stress tolerance as osmo-protectants/antioxidants, signaling function, and significant correlation with photosynthesis or source-sink association make them a potential target for modulating plant tolerance against abiotic stresses. Efforts have been made to understand the mechanisms underlying sugars' protective role against abiotic stress and to develop crop varieties with improved tolerance by modulating their biosynthetic pathway. Recent advances in molecular biology, especially next-generation sequences, has alleviated the problem of identifying crucial compounds or genes participating in abiotic stress tolerance, yet there are very few examples of developing a consistent or stable crop variety against abiotic stresses. Therefore, agricultural or plant scientists must focus on translating the available genomic/proteomic/metabolomic information into developing abiotic stress-tolerant crop varieties.

REFERENCES

Abebe, T., Guenzi, A.C., Martin, B., Cushman, J.C., 2003. Tolerance of mannitol-accumulating transgenic wheat to water stress and salinity. Plant Physiol. 131 (4), 1748–1755.

Aguilera-Alvarado, G.P., Sánchez-Nieto, S., 2017. Plant hexokinases are multifaceted proteins. Plant Cell Physiol. 58, 1151–1160.

Almodares, A., Hadi, M.R., Ahmadpour, H., 2008a. Sorghum stem yield and soluble carbohydrates under different salinity levels. Afr. J. Biotechnol. 7, 4051–4055.

Almodares, A., Hadi, M.R., Dosti, B., 2008b. The effects of salt stress on growth parameters and carbohydrates contents in sweet sorghum. Res. J. Environ. Sci. 2, 298–304.

Angelovici, R., Galili, G., Fernie, A.R., Fait, A., 2010. Seed desiccation: a bridge between maturation and germination. Trends Plant Sci. 15, 211–218.

Apolinário, A.C., de Lima Damascenoa, B.P.G., de Macêdo Beltrãob, N.E., Pessoac, A., Convertid, A., da Silva, J.A., 2014. Inulin-type fructans: a review on different aspects of biochemical and pharmaceutical technology. Carbohydr. Polym. 101, 368–378.

Arabzadeh, N., 2012. The effect of drought stress on soluble carbohydrates (sugars) in two species of Haloxylon persicum and Haloxylon aphyllum. Asian J. Plant Sci. 11, 44e51.

Ashraf, M., Foolad, M.R., 2007. Improving plant abiotic-stress resistance by exogenous application of osmoprotectants glycine betaine and proline. Environ. Exp. Bot. 59, 206–216.

Ashraf, M., Athar, H.R., Harris, P.J.C., Kwon, T.R., 2008. Some prospective strategies for improving crop salt tolerance. Adv. Agron. 97, 45–110.

Baker, R.F., Leach, K.A., Braun, D.M., 2012. SWEET as sugar: new sucrose effluxers in plants. Mol. Plant 5, 766–768.

Bevan, M.W., Uauy, C., Wulff, B.B., Zhou, J., Krasileva, K., Clark, M.D., 2017. Genomic innovation for crop improvement. Nature 543, 346–354.

Blokhina, O., Virolainen, E., Fagerstedt, K.V., 2003. Antioxidants, oxidative damage and oxygen deprivation stress: a review. Ann. Bot. 91, 179–194.

Bolouri-Moghaddam, M.R., Roy, K.L., Xiang, L., Rolland, F., Van den Ende, W., 2010. Sugar signaling and antioxidant network connections in plant cells. FEBS J. 277, 2022–2037.

Chiang, Y.J., Stushnoff, C., McSay, A.E., Jones, M.L., Bohnert, H.J., 2005. Overexpression of mannitol-1-phosphate dehydrogenase increases mannitol accumulation and adds protection against chilling injury in petunia. J. Am. Soc. Hort. Sci. 130, 605–610.

Chibbar, R.N., Jaiswal, S., Gangola, M., Båga, M., 2016. Carbohydrate metabolism. In: Wrigley, C., Corke, H., Seetharaman, K., Faubion, J. (Eds.), Encyclopedia of Food Grains. In: vol. 2. Elsevier Ltd., Oxford, UK, pp. 161–173.

Cimini, S., Locato, V., Vergauwen, R., Paradiso, A., Cecchini, C., Vandenpoel, L., Verspreet, J., Courtin, C.M., D'Egidio, M.G., Van den Ende, W., De Gara, L., 2015. Fructan biosynthesis and degradation as part of plant metabolism controlling sugar fluxes during durum wheat kernel maturation. Front. Plant Sci. 6, 89.

Cortina, C., Culiáñez-Macià, F.A., 2005. Tomato abiotic stress enhanced tolerance by trehalose biosynthesis. Plant Sci. 169, 75–82.

Cramer, G.R., Urano, K., Delrot, S., Pezzotti, M., Shinozaki, K., 2011. Effects of abiotic stress on plants: a systems biology perspective. BMC Plant Biol. 11, 163.
Cummings, J.H., Stephen, A.M., 2007. Carbohydrate terminology and classification. Eur. J. Clin. Nutr. 61, S5–S18.
da Silva, F.G., Cangussu, L.M.B., de Paula, S.L.A., Melo, G.A., Silva, E.A., 2013. Seasonal changes in fructan accumulation in the underground organs of *Gomphrena marginata* Seub (Amaranthaceae) under rock-field conditions. Theor. Exp. Plant. Physiol. 25, 46–55.
Deguchi, M., Koshita, Y., Gao, M., Tao, R., Tetsumura, T., Yamaki, S., Kanayama, Y., 2004. Engineered sorbitol accumulation induces dwarfism in Japanese persimmon. J. Plant Physiol. 161, 1177–1184.
Delorge, I., Janiak, M., Carpentier, S., Van Dijck, P., 2014. Fine tuning of trehalose biosynthesis and hydrolysis as novel tools for the generation of abiotic stress tolerant plants. Front. Plant Sci. 5, 147.
Duque, A.S., de Almeida, A.M., da Silva, A.B., da Silva, J.M., Farinha, A.P., Santos, D., Fevereiro, P., de Sousa Araújo, S., 2013. Abiotic stress responses in plants: unraveling the complexity of genes and networks to survive. In: Vahdati, K., Leslie, C. (Eds.), Abiotic Stress—Plant Responses and Applications in Agriculture. InTech, London, UK, pp. 49–101.
Edreva, A., 2005. Generation and scavenging of reactive oxygen species in chloroplasts: a submolecular approach. Agric. Ecosyst. Environ. 106, 119–133.
Emanuelle, S., Doblin, M.S., Stapleton, D.I., Bacic, A., Gooley, P.R., 2016. Molecular insights into the enigmatic metabolic regulator, SnRK1. Trends Plant Sci. 21, 341–353.
FAO, 2008. FAO Land and Plant Nutrition Management Service. Available from:http://www.fao.org/ag/agl/agll/spush (Verified 24 May 2010).
Fernandez, O., Béthencourt, L., Quero, A., Sangwan, R.S., Clément, C., 2010. Trehalose and plant stress responses: friend or foe? Trends Plant Sci. 15, 409–417.
Foyer, C.H., Shigeoka, S., 2011. Understanding oxidative stress and antioxidant functions to enhance photosynthesis. Plant Physiol. 155, 93–100.
Frias, J., Bakhsh, A., Jones, D.A., Arthur, A.E., Vidal-Valverde, C., Rhodes, M.J.C., Hedley, C.L., 1999. Genetic analysis of the raffinose oligosaccharide pathway in lentil seeds. J. Exp. Bot. 50, 469–476.
Fukushima, E., Arata, Y., Endo, T., Sonnewald, U., Sato, F., 2001. Improved salt tolerance of transgenic tobacco expressing apoplastic yeast-derived invertase. Plant Cell Physiol. 42, 245–249.
Gangola, M.P., Parkash, J., Ahuja, P.S., Dutt, S., 2013. Components of antioxidant system of *Picrorhiza kurrooa* exhibit different spatio-temporal behavior. Mol. Biol. Rep. 40, 6593–6603.
Gangola, M.P., Jaiswal, S., Kannan, U., Gaur, P.M., Båga, M., Chibbar, R.N., 2016. Galactinol synthase enzyme activity influences raffinose family oligosaccharides (RFO) accumulation in developing chickpea (*Cicer arietinum* L.) seeds. Phytochemistry 125, 88–98.
Gao, M., Tao, R., Miura, K., Dandekar, A.M., Sugiura, A., 2001. Transformation of Japanese persimmon (*Diospyros kaki* Thunb.) with apple cDNA encoding NADP-dependent sorbitol-6-phosphate dehydrogenase. Plant Sci. 160, 837–845.
Garcia, A.B., Engler, J.D.A., Iyer, S., Gerats, T., Van Montagu, M., Caplan, A.B., 1997. Effects of osmoprotectants upon NaCl stress in rice. Plant Physiol. 115, 159–169.
Garg, A.K., Kim, J.K., Owens, T.G., Ranwala, A.P., Do Choi, Y., Kochian, L.V., Wu, R.J., 2002. Trehalose accumulation in rice plants confers high tolerance levels to different abiotic stresses. Proc. Natl. Acad. Sci. U. S. A. 99, 15898–15903.
Gill, S.S., Tuteja, N., 2010. Reactive oxygen species and antioxidant machinery in abiotic stress tolerance in crop plants. Plant Physiol. Biochem. 48, 909–930.
Godfray, H.C.J., Beddington, J.R., Crute, I.R., Haddad, L., Lawrence, D., Muir, J.F., Pretty, J., Robinson, S., Thomas, S.M., Toulmin, C., 2010. Food security: the challenge of feeding 9 billion people. Science 327, 812–818.
Granot, D., David-Schwartz, R., Kelly, G., 2013. Hexose kinases and their role in sugar-sensing and plant development. Front. Plant Sci. 4.
Griffiths, C.A., Paul, M.J., Foyer, C.H., 2016. Metabolite transport and associated sugar signalling systems underpinning source/sink interactions. Biochim. Biophys. Acta 1857, 1715–1725.
Gupta, A.K., Kaur, N., 2005. Sugar signalling and gene expression in relation to carbohydrate metabolism under abiotic stresses in plants. J. Biosci. 30, 761–776.

Han, S.E., Park, S.R., Kwon, H.B., Yi, B.Y., Lee, G.B., Byun, M.O., 2005. Genetic engineering of drought-resistant tobacco plants by introducing the *trehalose phosphorylase* (TP) gene from *Pleurotus sajor-caju*. Plant Cell Tiss. Org. Cult. 82, 151–158.

Hare, P.D., Cress, W.A., Van Staden, J., 1998. Dissecting the roles of osmolyte accumulation during stress. Plant Cell Environ. 21, 535–553.

Hernandez-Marin, E., Martínez, A., 2012. Carbohydrates and their free radical scavenging capability: a theoretical study. J. Phys. Chem. B 116, 9668–9675.

Jaggard, K.W., Qi, A., Ober, E.S., 2010. Possible changes to arable crop yields by 2050. Philos. Trans. R. Soc. B 365, 2835–2851.

Jang, I.C., Oh, S.J., Seo, J.S., Choi, W.B., Song, S.I., Kim, C.H., Kim, Y.S., Seo, H.S., Do Choi, Y., Nahm, B.H., Kim, J.K., 2003. Expression of a bifunctional fusion of the *Escherichia coli* genes for trehalose-6-phosphate synthase and trehalose-6-phosphate phosphatase in transgenic rice plants increases trehalose accumulation and abiotic stress tolerance without stunting growth. Plant Physiol. 131, 516–524.

Kannan, U., Sharma, R., Khedikar, Y., Gangola, M.P., Ganeshan, S., Båga, M., Chibbar, R.N., 2016. Differential expression of two galactinol synthase isoforms *LcGolS1* and *LcGolS2* in developing lentil (*Lens culinaris* Medik. Cv CDC Redberry) seeds. Plant Physiol. Biochem. 108, 422–433.

Karuppanapandian, T., Moon, J.C., Kim, C., Manoharan, K., Kim, W., 2011. Reactive oxygen species in plants: their generation, signal transduction, and scavenging mechanisms. Aust. J. Crop. Sci. 5, 709–725.

Kawakami, A., Sato, Y., Yoshida, M., 2008. Genetic engineering of rice capable of synthesizing fructans and enhancing chilling tolerance. J. Exp. Bot. 59, 793–802.

Keunen, E.L.S., Peshev, D., Vangronsveld, J., Van Den Ende, W.I.M., Cuypers, A.N.N., 2013. Plant sugars are crucial players in the oxidative challenge during abiotic stress: extending the traditional concept. Plant Cell Environ. 36, 1242–1255.

Koster, K.L., 1991. Glass formation and desiccation tolerance in seeds. Plant Physiol. 96, 302–304.

Koster, K.L., Leopold, A.C., 1988. Sugars and desiccation tolerance in seeds. Plant Physiol. 88, 829–832.

Krasensky, J., Jonak, C., 2012. Drought, salt, and temperature stress-induced metabolic rearrangements and regulatory networks. J. Exp. Bot. 63, 1593–1608.

Kwak, J.M., Nguyen, V., Schroeder, J.I., 2006. The role of reactive oxygen species in hormonal responses. Plant Physiol. 141, 323–329.

Lemoine, R., La Camera, S., Atanassova, R., Dédaldéchamp, F., Allario, T., Pourtau, N., Bonnemain, J.L., Laloi, M., Coutos-Thévenot, P., Maurousset, L., Faucher, M., 2013. Source-to-sink transport of sugar and regulation by environmental factors. Front. Plant Sci. 4, 272.

Li, L., Sheen, J., 2016. Dynamic and diverse sugar signaling. Curr. Opin. Plant Biol. 33, 116–125.

Li, H.J., Yang, A.F., Zhang, X.C., Gao, F., Zhang, J.R., 2007. Improving freezing tolerance of transgenic tobacco expressing sucrose: sucrose 1-fructosyltransferase gene from *Lactuca sativa*. Plant Cell Tiss. Org. Cult. 89, 37–48.

Li, P., Zhou, H., Shi, X., Yu, B., Zhou, Y., Chen, S., Wang, Y., Peng, Y., Meyer, R.C., Smeekens, S.C., Teng, S., 2014. The ABI4-induced Arabidopsis ANAC060 transcription factor attenuates ABA signaling and renders seedlings sugar insensitive when present in the nucleus. PLoS Genet. 10. e1004213.

Lin, C.R., Lee, K.W., Chen, C.Y., Hong, Y.F., Chen, J.L., Lu, C.A., Chen, K.T., Ho, T.H.D., Yu, S.M., 2014. SnRK1A-interacting negative regulators modulate the nutrient starvation signaling sensor SnRK1 in source-sink communication in cereal seedlings under abiotic stress. Plant Cell 26, 808–827.

Liu, H.L., Dai, X.Y., Xu, Y.Y., Chong, K., 2007. Over-expression of OsUGE-1 altered raffinose level and tolerance to abiotic stress but not morphology in Arabidopsis. J. Plant Physiol. 164, 1384–1390.

Livingston, D.P., Hincha, D.K., Heyer, A.G., 2009. Fructan and its relationship to abiotic stress tolerance in plants. Cell. Mol. Life Sci. 66, 2007–2023.

Ljung, K., Nemhauser, J.L., Perata, P., 2015. New mechanistic links between sugar and hormone signalling networks. Curr. Opin. Plant Biol. 25, 130–137.

Lunn, J.E., Delorge, I., Figueroa, C.M., Van Dijck, P., Stitt, M., 2014. Trehalose metabolism in plants. Plant J. 79, 544–567.

Martínez-Villaluenga, C., Frias, J., Vidal-Valverde, C., 2008. Alpha-galactosides: antinutritional factors or functional ingredients? Crit. Rev. Food Sci. Nutr. 48, 301–316.

Matros, A., Peshev, D., Peukert, M., Mock, H.P., Van den Ende, W., 2015. Sugars as hydroxyl radical scavengers: proof-of-concept by studying the fate of sucralose in Arabidopsis. Plant J. 82, 822–839.
Miranda, J.A., Avonce, N., Suárez, R., Thevelein, J.M., Van Dijck, P., Iturriaga, G., 2007. A bifunctional TPS–TPP enzyme from yeast confers tolerance to multiple and extreme abiotic-stress conditions in transgenic Arabidopsis. Planta 226, 1411–1421.
Mohammadkhani, N., Heidari, R., 2008. Drought-induced accumulation of soluble sugars and proline in two maize varieties. World Appl. Sci. J. 3, 448–453.
Morelli, R., Russo-Volpe, S., Bruno, N., Scalzo, R.L., 2003. Fenton-dependent damage to carbohydrates: free radical scavenging activity of some simple sugars. J. Agric. Food Chem. 51, 7418–7425.
Nahar, K., Hasanuzzaman, M., Fujita, M., 2016. Roles of osmolytes in plant adaptation to drought and salinity. In: Iqbal, N., Nazar, R., Khan, N.A. (Eds.), Osmolytes and Plants Acclimation to Changing Environment: Emerging Omics Technologies. Springer, New Delhi, pp. 37–68.
Nguyen, Q.A., Luan, S., Wi, S.G., Bae, H., Lee, D.S., Bae, H.J., 2016. Pronounced phenotypic changes in transgenic tobacco plants overexpressing sucrose synthase may reveal a novel sugar signaling pathway. Front. Plant Sci. 6, 1216.
Nishizawa, A., Yabuta, Y., Shigeoka, S., 2008. Galactinol and raffinose constitute a novel function to protect plants from oxidative damage. Plant Physiol. 147, 1251–1263.
Oberschall, A., Deák, M., Török, K., Sass, L., Vass, I., Kovács, I., Fehér, A., Dudits, D., Horváth, G.V., 2000. A novel aldose/aldehyde reductase protects transgenic plants against lipid peroxidation under chemical and drought stresses. Plant J. 24 (4), 437–446.
Ort, D., Long, S.P., 2003. Converting solar energy into crop production. In: Chrispeels, M.J., Sadava, D.E. (Eds.), Plants, Genes, and Crop Biotechnology. Jones and Bartlett Publisher International, Masschusetts, USA, pp. 240–269.
Osakabe, Y., Yamaguchi-Shinozaki, K., Shinozaki, K., Tran, L.S.P., 2014. ABA control of plant macroelement membrane transport systems in response to water deficit and high salinity. New Phytol. 202, 35–49.
Parvanova, D., Popova, A., Zaharieva, I., Lambrev, P., Konstantinova, T., Taneva, S., Atanassov, A., Goltsev, V., Djilianov, D., 2004. Low temperature tolerance of tobacco plants transformed to accumulate proline, fructans, or glycine betaine. Variable chlorophyll fluorescence evidence. Photosynthetica 42, 179–185.
Pennycooke, J.C., Jones, M.L., Stushnoff, C., 2003. Down-regulating α-galactosidase enhances freezing tolerance in transgenic petunia. Plant Physiol. 133 (2), 901–909.
Peshev, D., Van den Ende, W., 2013. Sugars as antioxidants in plants. In: Tuteja, N., Gill, S.S. (Eds.), Crop Improvement Under Adverse Conditions. Springer Science + Business Media, New York, pp. 285–307.
Peshev, D., Vergauwen, R., Moglia, A., Hideg, É., Van den Ende, W., 2013. Towards understanding vacuolar antioxidant mechanisms: a role for fructans? J. Exp. Bot. 64, 1025–1038.
Peterbauer, T., Richter, A., 1998. Galactosylononitol and stachyose synthesis in seeds of adzuki bean. Purification and characterization of stachyose synthase. Plant Physiol. 117, 165–172.
Peterbauer, T., Lahuta, L.B., Blöchl, A., Mucha, J., Jones, D.A., Hedley, C.L., Gòrecki, R.J., Richter, A., 2001. Analysis of the raffinose family oligosaccharide pathway in pea seeds with contrasting carbohydrate composition. Plant Physiol. 127, 1764–1772.
Pilon-Smits, E.A., Ebskamp, M.J., Paul, M.J., Jeuken, M.J., Weisbeek, P.J., Smeekens, S.C., 1995. Improved performance of transgenic fructan-accumulating tobacco under drought stress. Plant Physiol. 107, 125–130.
Pilon-Smits, E.A., Terry, N., Sears, T., Kim, H., Zayed, A., Hwang, S., van Dun, K., Voogd, E., Verwoerd, T.C., Krutwagen, R.W., Goddijn, O.J., 1998. Trehalose-producing transgenic tobacco plants show improved growth performance under drought stress. J. Plant Physiol. 152, 525–532.
Pilon-Smits, E.A., Terry, N., Sears, T., van Dun, K., 1999. Enhanced drought resistance in fructan-producing sugar beet. Plant Physiol. Biochem. 37, 313–317.
Pujni, D., Chaudhary, A., Rajam, M.V., 2007. Increased tolerance to salinity and drought in transgenic indica rice by mannitol accumulation. J. Plant Biochem. Biotechnol. 16, 1–7.
Pukacka, S., Ratajczak, E., Kalemba, E., 2009. Non-reducing sugar levels in beech (*Fagus sylvatica*) seeds as related to withstanding desiccation and storage. J. Plant Physiol. 166, 1381–1390.
Rains, D.W., Goyal, S.S., 2003. Strategies for managing crop production in saline environments: an overview. J. Crop. Prod. 7, 1–10.
Ramon, R., Rolland, F., Sheen, J., 2008. Sugar sensing and signaling. Arabidopsis Book. 6, e0117.

Roitsch, T., González, M.C., 2004. Function and regulation of plant invertases: sweet sensations. Trends Plant Sci. 9, 606–613.

Rosa, M., Hilal, M., González, J.A., Prado, F.E., 2009a. Low-temperature effect on enzyme activities involved in sucrose–starch partitioning in salt-stressed and salt-acclimated cotyledons of quinoa (*Chenopodium quinoa* Willd.) seedlings. Plant Physiol. Biochem. 47, 300–307.

Rosa, M., Prado, C., Podazza, G., Interdonato, R., González, J.A., Hilal, M., Prado, F.E., 2009b. Soluble sugars: metabolism, sensing and abiotic stress: a complex network in the life of plants. Plant Signal. Behav. 4, 388–393.

Ruan, Y.L., 2014. Sucrose metabolism: gateway to diverse carbon use and sugar signaling. Annu. Rev. Plant Biol. 65, 33–67.

Sami, F., Yusuf, M., Faizan, M., Faraz, A., Hayat, S., 2016. Role of sugars under abiotic stress. Plant Physiol. Biochem. 109, 54–61.

Sawhney, V., Singh, D.P., 2002. Effect of chemical desiccation at the post-anthesis stage on some physiological and biochemical changes in the flag leaf of contrasting wheat genotypes. Field Crop Res. 77, 1–6.

Schneider, T., Keller, F., 2009. Raffinose in chloroplasts is synthesized in the cytosol and transported across the chloroplast envelope. Plant Cell Physiol. 50, 2174–2182.

Seki, M., Umezawa, T., Urano, K., Shinozaki, K., 2007. Regulatory metabolic networks in drought stress responses. Curr. Opin. Plant Biol. 10, 296–302.

Shen, B.O., Jensen, R.G., Bohnert, H.J., 1997. Increased resistance to oxidative stress in transgenic plants by targeting mannitol biosynthesis to chloroplasts. Plant Physiol. 113, 1177–1183.

Sheveleva, E., Chmara, W., Bohnert, H.J., Jensen, R.G., 1997. Increased salt and drought tolerance by D-ononitol production in transgenic *Nicotiana tabacum* L. Plant Physiol. 115, 1211–1219.

Singh, M., Kumar, J., Singh, S., Singh, V.P., Prasad, S.M., 2015. Roles of osmoprotectants in improving salinity and drought tolerance in plants: a review. Rev. Environ. Sci. Biotechnol. 14, 407–426.

Slama, I., Abdelly, C., Bouchereau, A., Flowers, T., Savouré, A., 2015. Diversity, distribution and roles of osmoprotective compounds accumulated in halophytes under abiotic stress. Ann. Bot. 115, 433–447.

Sprenger, N., Keller, F., 2000. Allocation of raffinose family oligosaccharides to transport and storage pools in *Ajuga reptans*: the roles of two distinct galactinol synthases. Plant J. 21, 249–258.

Tang, W., Peng, X., Newton, R.J., 2005. Enhanced tolerance to salt stress in transgenic loblolly pine simultaneously expressing two genes encoding *mannitol-1-phosphate dehydrogenase* and *glucitol-6-phosphate dehydrogenase*. Plant Physiol. Biochem. 43, 139–146.

Tarczynski, M.C., Jensen, R.G., Bohnert, H.J., 1993. Stress protection of transgenic tobacco by production of the osmolyte mannitol. Science 259, 508–510.

Tarkowski, Ł.P., Van den Ende, W., 2015. Cold tolerance triggered by soluble sugars: a multifaceted countermeasure. Front. Plant Sci. 6, 203.

Torres, M.A., Jones, J.D.J., Dangl, J.L., 2006. Reactive oxygen species signaling in response to pathogens. Plant Physiol. 141, 373–378.

Valluru, R., Van den Ende, W., 2011. *Myo*-inositol and beyond—emerging networks under stress. Plant Sci. 181, 387–400.

Van den Ende, W., El-Esawe, S.K., 2014. Sucrose signaling pathways leading to fructan and anthocyanin accumulation: a dual function in abiotic and biotic stress responses? Environ. Exp. Bot. 108, 4–13.

Van den Ende, W., Valluru, R., 2009. Sucrose, sucrosyl oligosaccharides, and oxidative stress: scavenging and salvaging? J. Exp. Bot. 60, 9–18.

Wang, T., McFarlane, H.E., Persson, S., 2015. The impact of abiotic factors on cellulose synthesis. J. Exp. Bot. 67, 543–552.

Wehmeier, K.R., Mooradian, A.D., 1994. Autooxidative and antioxidative potential of simple carbohydrates. Free Radic. Biol. Med. 17, 83–86.

Wingler, A., Paul, M., 2013. The role of trehalose metabolism in chloroplast development and leaf senescence. In: Biswal, B., Krupinska, K., Biswal, U.C. (Eds.), Plastid Development in Leaves During Growth and Senescence. Springer Science, New York, pp. 551–565.

Xu, S.M., Liu, L.X., Woo, K.C., Wang, D.L., 2007. Changes in photosynthesis, xanthophyll cycle, and sugar accumulation in two North Australia tropical species differing in leaf angles. Photosynthetica 45 (3), 348–354.

Yu, S.M., Lo, S.F., Ho, T.H.D., 2015. Source–sink communication: regulated by hormone, nutrient, and stress cross-signaling. Trends Plant Sci. 20, 844–857.

Yuanyuan, M., Yali, Z., Jiang, L., Hongbo, S., 2009. Roles of plant soluble sugars and their responses to plant cold stress. Afr. J. Biotechnol. 8, 2004–2010.
Zhang, S.Z., Yang, B.P., Feng, C.L., Tang, H.L., 2005. Genetic transformation of tobacco with the *trehalose synthase* gene from *Grifola frondosa* Fr. enhances the resistance to drought and salt in tobacco. J. Integr. Plant Biol. 47, 579–587.
Zhifang, G., Loescher, W.H., 2003. Expression of a celery mannose 6-phosphate reductase in *Arabidopsis thaliana* enhances salt tolerance and induces biosynthesis of both mannitol and a glucosyl-mannitol dimer. Plant Cell Environ. 26, 275–283.

FURTHER READING

Li, Y., Lee, K.K., Walsh, S., Smith, C., Hadingham, S., Sorefan, K., Cawley, G., Bevan, M.W., 2006. Establishing glucose-and ABA-regulated transcription networks in Arabidopsis by microarray analysis and promoter classification using a relevance vector machine. Genome Res. 16, 414–427.
Molinari, H.B.C., Marur, C.J., Daros, E., De Campos, M.K.F., De Carvalho, J.F.R.P., Pereira, L.F.P., Vieira, L.G.E., 2007. Evaluation of the stress-inducible production of proline in transgenic sugarcane (*Saccharum* spp.): osmotic adjustment, chlorophyll fluorescence and oxidative stress. Physiol. Plant. 130 (2), 218–229.
Salerno, G.L., Curatti, L., 2003. Origin of sucrose metabolism in higher plants: when, how and why? Trends Plant Sci. 8, 63–69.
Zhu, B., Su, J., Chang, M., Verma, D.P.S., Fan, Y.L., Wu, R., 1998. Overexpression of a *Δ1-pyrroline-5-carboxylate synthetase* gene and analysis of tolerance to water-and salt-stress in transgenic rice. Plant Sci. 139, 41–48.

CHAPTER 3

Polyamines Metabolism: A Way Ahead for Abiotic Stress Tolerance in Crop Plants

Pratika Singh*, Sahana Basu†, Gautam Kumar*
*Department of Life Science, Central University of South Bihar, Patna, India
†Department of Biotechnology, Assam University, Silchar, India

Contents

Abbreviations

ACC 1-aminocyclopropane-1-carboxylic acid
ADC arginine decarboxylase
AIH agmatine iminohydrolase
Cad cadaverine
DAO diamine oxidase
GPX glutathione peroxidise
GSH reduced glutathione
GSSG oxidized glutathione
GST glutathione *S*-transferase
HR hypersensitive response
NO nitric oxide
ODC ornithine decarboxylase
PAO polyamine oxidase
Pro proline
Put putrescine
ROS reactive oxygen species
SAM *S*-adenosyl methionine

Biochemical, Physiological and Molecular Avenues for Combating Abiotic Stress in Plants
https://doi.org/10.1016/B978-0-12-813066-7.00003-6

SAMDC *S*-adenosyl methionine decaboxylase
SDPS spermidine synthase
SMO spermine oxidase
Spd spermidine
Spm spermine

1. INTRODUCTION

1.1 Abiotic Stress and Polyamines–Assessment of the Relationship

Plants live in continuously changing environments that often hinder their development and physiological activities. Abiotic stresses, including salinity (Kumar et al., 2009, 2012b), drought (Basu et al., 2017), heat (Dwivedi et al., 2017), cold (Sanghera et al., 2011), and heavy metals (Kumar et al., 2012a) severely affect growth and productivity of several crop plants. Abiotic stresses have been found to reduce crop yields by up to 70% (Gosal et al., 2009). Therefore, understanding the stress tolerance mechanism in plants has become one of the most challenging jobs for plant biologists for development of stress tolerant plants with sustainable yield. Plants' exposure to abiotic stresses can be classified into three random stages: stress perception, stress response, and stress outcome. Polyamines (PA) play a significant role in improving abiotic stress tolerance in plants (Alcazar et al., 2010). Research has revealed the involvement of PAs in modulating the defense response of plants to diverse environmental stresses.

PAs are low-molecular-weight, ubiquitous, phytohormone-like aliphatic amine compounds containing unsaturated hydrocarbon with two or more primary amino groups and organic nonprotein polycations governing several fundamental processes of plant growth and development (Groppa and Benavides, 2008). They occur in free, conjugated (associated with small molecules such as phenolic acids), or bound forms (associated with various macromolecules). PAs play a major role in plant organogenesis, embryogenesis, flower initiation, floral development, and fruit development (Gill and Tuteja, 2010). They have acid neutralizing and antioxidant properties, thereby act as antisenescence and antistress agents. They also stabilize membranes and cell walls by binding with negatively-charged DNA, RNA, and different protein molecules (Zhao and Yang, 2008). PAs were first explained in 1678 by Antonie van Leeuwenhoek as compounds found in seminal fluid, leading to the naming of two of its members—spermine (Spm) and spermidine (Spd) (Bachrach, 2010). They are present in all living organisms, with the most familiar PAs being Spm, Spd, and putrescine (Put), followed by cadaverine (Cad) and 1,3-diaminopropane (1,3-DAP). Concentrations of PAs vary markedly depending on the plant species or organ and developmental stages, and are much higher in plants than endogenous phytohormones (Liu et al., 2007). Heavy metal stress (Zn, Cu, Cd, Mn, Pb, and Fe) has also been shown to induce PA biosynthesis in plants (Franchin et al., 2007; Groppa et al., 2007). However, the effect of heavy metal stress on the concentration of PAs or their role in particular stress alleviation is not well understood. Recent research has confirmed the contribution of PAs (Cad) in metal uptake (Soudek et al., 2016). Exogenous applications of PAs have also been shown to be effective in enhancing the tolerance of crops for various environmental stresses such as salinity (Ndayiragije and Lutts, 2006; Pathak et al., 2014), drought, cold (Nayyar and Chander, 2004), high temperatures (Murkowski, 2001), and flooding stress (Yiu et al., 2009). The ubiquity of PAs in cells indicates their importance in stress tolerance. This has been confirmed by the depletion of PAs, which affect an enormous number of biological processes within the cell.

2. BIOSYNTHESIS OF POLYAMINES UNDER ABIOTIC STRESSES

Chief classes of PAs are the triamine—spermidine [$NH_2(CH_2)_3NH(CH_2)_4NH_2$], tetramine—spermine [$NH_2(CH_2)_3NH(CH_2)_4NH(CH_2)_3NH_2$], and their diamine obligate precursor—putrescine [$NH_2(CH_2)_4NH_2$], which are present in plant cells. Their structures are illustrated in Fig. 1. The homeostasis of PAs within the cell is primarily accomplished through the regulation of its biosynthesis. The plant PA synthesis pathway has been found to differ from animals as it involves two precursors, L-ornithine and L-arginine, to generate putrescine. However, L-ornithine is exclusively employed in animals (Gupta et al., 2013). The biosynthesis pathways of main PAs such as Put, Spd, and Spm are shown in Fig. 2. The diamine Put synthesis ensues through either arginine decarboxylase (ADC) (EC 4.1.1.19) via agmatine (Agm) or ornithine decarboxylase (ODC) (EC 4.1.1.17). Moreover, the ODC pathway is also active in the early stages of plant growth, development, organ differentiation, and reproductive stages. Two

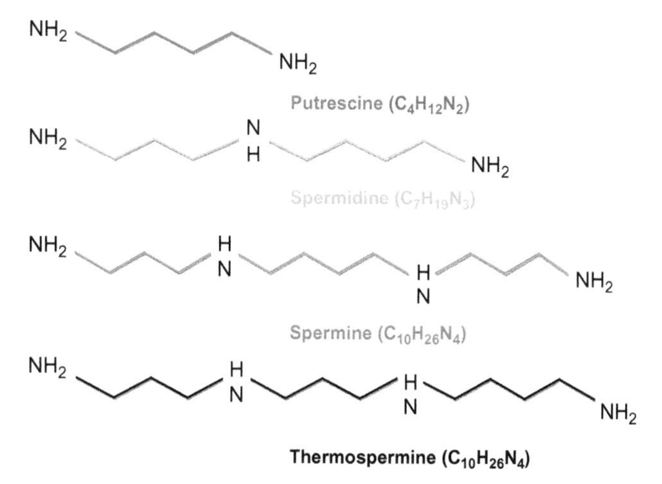

Fig. 1 Chemical structures of the major polyamines.

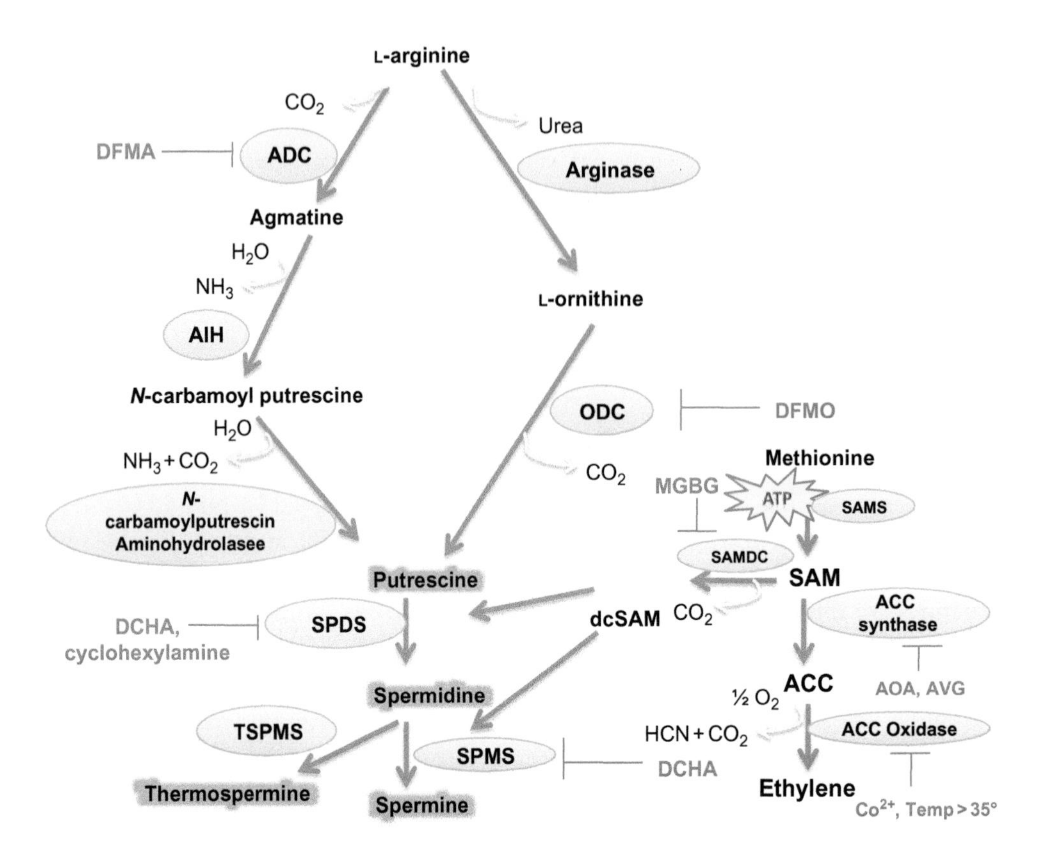

Fig. 2 Biosynthesis of polyamines in plants (inhibitors shown in *red* while enzymes encircled in a *green box*).

separate enzymes: *N*-carbamoyl putrescine amidohydrolase (CPA) (EC 3.5.1.53) and AIH (AIH) (EC 3.5.3.12) play a part in the conversion of Agm into Put. Studies suggest that ADC is the limiting step for Put biosynthesis in plants because the overexpression of homologous ADC2 in *Arabidopsis* is sufficient to promote Put accumulation (Alcazar et al., 2005). ODC, SAMDC, and Spd synthase have been reported to be localized in the cytoplasm. The oat ADC is localized in chloroplasts related to the thylakoid membranes (Borrell et al., 1995); whereas a nuclear or chloroplast localization is detected in different tobacco tissues (Bortolotti et al., 2004), resulting in clarification that subcellular compartmentalization of the ADC pathway happens in plants, which could result in gradient concentrations of Put within the cell. Transport or shuttle mechanisms for PAs in plants have not yet been reported. Recently, it has been proposed that ADC/ODC alternative pathways reflect their completely different biological process origins. ADC, AIH, and CPA in plants could originate from a cyanobacterial ancestor of chloroplast; whereas

ODC could derive from bacterial genes present in a common ancestor of plants and animals that acquired the cyanobacterial endosymbiont (Illingworth et al., 2003), thus showing different evolutionary origins. Spd functions as a substrate for the synthesis of the higher PA-Spm. Spd and Spm are synthesized by the successive attachment of aminopropyl, with Put first to synthesize Spd and then Spd to synthesize spermine (Spm). These reactions are catalyzed by amino propyl transferases such as Spd synthase (SPDS) (EC 2.5.1.16) and Spm synthase (SPMS) (EC 2.5.1.22). Aminopropyl is made due to the decarboxylation of *S*-adenosylmethionine (SAM) by *S*-adenosylmethionine decarboxylase (SAMDC) (EC 4.1.1.50). SAM is produced from the amino acid L-methionine and ATP by *S*-adenosylmethionine synthetase (SAMS) (EC 2.5.1.6). SAM is a common precursor for both PAs and ethylene, and SAMDC regulates both biosynthetic pathways, as illustrated in Fig. 2. So there is an associated antagonistic relationship within the synthesis pathway between ethylene and PAs. It should be noted that there is no known gene encoding ODC in the sequenced genome of the model plant *Arabidopsis thaliana* until now (Hanfrey et al., 2001), suggesting that this species could solely produce Put via the ADC pathway. PA synthesis may vary between tissues/organs, one example being that the shoot apical meristem of tobacco (*Nicotiana tabacum*) serves as the predominant site of Spd and Spm synthesis, while Put is mostly synthesized in roots (Moschou et al., 2008). A number of investigators have used PA inhibitors to modulate the cellular PA titer to determine their role in varied plant processes. Some commonly used inhibitors of PA synthesis are as follows.

1. Difluoromethyl ornithine (DFMO), an irreversible inhibitor of ODC (Bey et al., 1987). Difluoro methyl ornithine acts specifically on ornithine decarboxylase and has basically no action on any other enzyme. Consequently, it has been used to inhibit putrescine biosynthesis in a variety of cells, each in vitro and in vivo. Once difluoro methyl ornithine is administered, the amount of putrescine and spermidine fall quickly; and particularly in rapidly replicating cells, there is an intense inhibition of growth and replication. As a result of DFMO that inhibits cell replication strongly, there has been significant interest in the potential therapeutic use of this compound.
2. Difluoromethylarginine (DFMA), an irreversible inhibitor of ADC (Bitonti et al., 1987).
3. Cyclohexylamine (CHA), a competitive inhibitor of spermidine synthase (Hibasami et al., 1980).
4. Methylglyoxal bis (guanyl hydrazone), MGBG, continues to be used extensively for the inhibition of *S*-adenosylmethionine decarboxylase and consequently of spermidine biogenesis, each in vivo and in vitro. However, it is conditionally specific and conjointly inhibits some other enzymes such as diamine oxidase. A series of analogs have similarly been synthesized and studied, one of which is an irreversible inhibitor [(1,1′ methylethanediylidine) dinitrolo-bis(3-aminoguanidine)]. MGBG has been used for remedy of neoplasms; however, its clinical use is limited by its toxicity. While

in ethylene biosynthesis, the common inhibitors are Aminoethoxy-vinylglycine (AVG) and aminooxyacetic acid (AOA), which block the conversion of AdoMet to ACC. The Cobalt ion (Co^{2+}) is also an inhibitor of the ethylene biosynthetic pathway, blocking the conversion of ACC to ethylene by ACC oxidase.

3. POLYAMINES IN RESPONSE TO DIFFERENT ABIOTIC STRESSES

3.1 Polyamines and Cold Stress

One of the chief environmental factors restraining the geographical distribution of plants is low temperature, as it accounts for important reductions in the yield of agriculturally important crops (Sanghera et al., 2011). Low temperature harms many plant species, specifically those adapted to tropical climates. In contrast, some species from temperate regions are able to grow in response to low-nonfreezing temperature, an adaptive process referred to as cold acclimation. Several molecular, biochemical, and physiological changes occur during cold acclimation, most of them being associated with major changes in gene expression and metabolite profiles. During recent years, transcriptomic and metabolomic approaches have permitted the identification of cold-responsive genes and main metabolites that gather in plants exposed to cold. The obtained data support the previously-held idea that PAs are involved in plant responses to cold, though their specific role is still not well understood. Global methodologies aiming to identify correspondence between genes and/or metabolites with cold treatments very frequently serve a prominent role in the PA biosynthetic pathway in the cold response. Accumulation of Put in several species has been reported to be induced by their exposure to low temperatures (Martin-Tanguy, 1985). An increase in Put induction parallel to an increase in frost resistance was seen in the case of wheat (Racz et al., 1996). Certain data clearly validates that several plants responded to low temperature acclimation with uniform and sizable increase in Spd (Flores, 1991). The chilling injury of zucchini (*Cucurbito pepo*) can be abridged by preconditioning the squash for 2 days at 10°C. Preconditioning leads to a significant increase in Spd and Spm, but not in Put levels. The increases in Spd and Spm are interrelated with elevated SAM decarboxylase activity (Kramer and Wang, 1990). Spd and Spm may prevent chilling injury in squash by a mechanism involving fortification of membrane lipids. In variance, Lee et al. (1997) recounted that in rice seedlings of a chilling-tolerant cultivar, levels of Put and activity of ADC in both shoots and roots and levels of Spd and Spm and activity of SAM decarboxylase in shoots are augmented after exposure to chilling. In a chilling-sensitive cultivar level of Put of ADC in shoots are found to be amplified slightly after exposure to chilling, while those of roots declined drastically. Both the cultivars remains unchanged in terms of the activity of ODC, after exposure to chilling in rice. However, the increase in PA content in the wheat has been reported from the increased transcription of the both ADC and ODC genes (Kovacs and Simon-Sarkadi, 2010). DFMA, but not DFMO-inhibited free Put accumulation in chilled seedlings of the chilling-tolerant cultivar, led the reduction in chilling tolerance, together with a decrease in survival and an increase in electrolyte leakage. These effects were found to be reversed by the addition of Put. Fascinatingly, the authors showed that in the chilling-tolerant cultivar, chilling induced an increase of free ABA levels first, followed by ADC activity, and last, free Put levels. Transcript profiling has also confirmed that cold enhances the expression of *ADC1, ADC2*, and *SAMDC2* genes (Urano et al., 2003; Cuevas et al., 2009). Reduced expression of *NCED3* and numerous ABA-regulated genes was sensed in the *adc1* mutants at low temperature. All these results supported the theory that Put and ABA are assimilated in a positive feedback loop, in which ABA and Put communally promote each other's biosynthesis and in response to abiotic stress, free Put levels are increased in cold treatment, which correlates with the induction of *ADC* genes. According to a proposed model (Alcazar et al., 2011), an increase in ADC1 transcript levels is detectable, which increases the biosynthesis of PAs. The Put also initiates the ABA-dependent signaling pathway, activating ABRE COR (cold responsive genes), thus resulting in a cold acclimation response after cold treatment. It has been noted that the changes in PA level during stress conditions varies from species to species, even from strain to strain. Upon treatment with cold stress in winter wheat, an increase in Put and Spd was observed; however, the Put level decreased in spring wheat (Szalai et al., 1997). A link exists between PAs and NO for the duration of chilling stress in tomato seedlings. The exogenous application of Spd and Spm has been found to induce the production of NO in an H_2O_2 dependent manner by NOS and NR pathways. Put could progress chilling tolerance by means of the activation of ABA synthesis. Additionally, under chilling stress conditions, the application of NO has been reported to improve endogenous Put and Spd levels through upregulation of the relatable PA biosynthetic genes, expounding cross-talk among PAs, abscisic acid (ABA), nitric oxide (NO), and hydrogen peroxide (H_2O_2) (Diao et al., 2017).

3.2 Polyamines and Drought

Drought is one of the most disastrous abiotic stresses, severely reducing crop productivity (Basu et al., 2017). The relationship of PAs to drought stress has been described by several researchers. Osmotic stress induced by mannitol, increases the Put, Spd and Spm contents in wheat (Galiba et al., 1993). Correspondingly, the osmotic stress has been reported to introduce a drastic increase in Put and Spd contents in the tolerant species, *Lycopersicon pennellii* than in sensitive species, *Lycopersicon esculentum* (Santa-Cruz et al., 1997). The scarcity of water has been found to induce a larger increase of the Put synthesis in drought-tolerant sugarcane varieties, compared to the sensitive ones (Zhang et al., 1996).

3.3 Polyamines and Heat Stress

Heat stress is one of the most destructive abiotic stresses, imposing a negative impact on the production of major crops (Dwivedi et al., 2017). High temperature stress results in the accumulation of PAs for beans (Kuznetsov and Shevyakova, 1997). Heat stress has been reported to induce larger PA synthesis and accumulation within the tolerant rice genotype than in sensitive ones (Roy and Ghosh, 1996). Additionally, the inhibition of chickpea seed germination at supra-optimal temperature has been found to be relieved by exogenous Put (Gallardo et al., 1996). The participation of PAs in heat stress response has been moreover established in the rice plants, during which high temperature treatment improved the levels of cadaverine, Put, and Spd contents (Shevyakova et al., 2001). Similar to PAs, proline (Pro) also plays a significant role in reducing injuries caused by the deficit in water level and high temperature (De Ronde et al., 2004). Moreover, it behaves as an antioxidant. Transgenic soybean plants that contain the gene coding for the last enzyme of Pro biosynthesis, L-Δ1-pyrroline-5-carboxylatereductase (P5CR) (EC 1.5.1.2), within the sense direction, has advanced Pro content and minimum damage during instantaneous water deficit and heat stress compared with that of the wild type plants. Pro and PAs have two common precursors, arginine (Arg) and glutamate (Glu), therefore it may be likely that the influence of Pro concentration also results in the changes within the process of PA synthesis. During water shortages, certain plants tend to accumulate Put, which is strengthened by the fact that the transcript profiling under the preceding conditions induces the expression of certain genes that are concerned and involved in the biosynthesis. The expression of these genes is additionally induced by ABA treatment (Urano et al., 2003; Alcazar et al., 2010). This raises questions about the fact that the accumulation of Put and upregulation of PA-biosynthetic genes under water stress are chiefly ABA-dependent responses. Several drought-inducible genes are responsive to ABA, though the ABA-independent pathways are also activated in response to drought conditions. So to work out the role and therefore the involvement of ABA in the transcriptional regulation of the PA synthesis pathway in response to drought, expressions of PA biogenesis genes *ADC1*, *ADC2*, *AIH*, *CPA*, *SPDS1*, *SPDS2*, *SPMS*, *ACL5*, *SAMDC1*, and *SAMDC2* have been analyzed in *Arabidopsis thaliana* wild type and mutant plants for impaired ABA biosynthesis (*aba2-3*) or signaling (*abi1-1*). *ADC2*, *SPDS1*, and *SPMS* genes are found to be the foremost responsive ones to drought stress. The improved and enhanced expression of these three PA biosynthesis genes (Alcazar et al., 2006) have been recommended to play a possible role for *ADC2*, *SPDS1*, and *SPMS* in the drought response. Remarkably, whereas, *ADC2* and *SPDS1* expressions have been found to be amplified many times by drought treatment, the expression of their gene paralogs, *ADC1* and *SPDS2*, did not change strikingly. These observations are consistent with the acquisition of sure stress-specificity, perhaps due to divergent evolution of cis-regulatory elements in their promoters. Indeed, different cis elements are found in the promoters of PA biosynthesis genes. ABA-responsive elements (ABRE) or ABRE-related motifs also are found in the promoters of *ADC2*, *SPDS1*, and *SPMS*, which are highly upregulated in response to drought. The analysis in *aba2-3* and *abi1-1* mutants are reported to exhibit much more moderate rise in

ADC2, *SPDS1*, and *SPMS* expressions. These results demonstrate that the transcriptional upregulation of *ADC2*, *SDPS1*, and *SPMS* by drought stress is mediated by ABA. Hence, ABA is an upstream regulator of PA biosynthesis in response to drought stress. ROS is produced and accumulated under drought stress due to an imbalance between production and utilization of photo generated reductants, as a result of water desiccation-induced stomata closure that limits CO_2 accessibility and diminishes fixation through the Bensen cycle. Each Put and Spd levels displayed substantial correlation with the activities of SOD, CAT, APX, GR and the Put level is correlated with GSH, demonstrating that the antioxidant system is regulated by PAs in centipede grass (Liu et al., 2017). PA treatment increases the activities of antioxidant enzymes and reduces the oxidative damages in chickpeas *(Cicer arietinum* L.) (Nayyar and Chander, 2004), *Brassica juncea* (Verma and Mishra, 2005) and white clover. Involvement of Spd in osmotic stress-induced transient rice of H_2O_2, Ca_2C, and NO signal molecules activate antioxidant enzyme activities and gene expression (Peng et al., 2016). Exogenous PAs increase the extent of tolerance to drought and salt stresses in Bermuda grass (*Cynodon dactylon*), while considerably increasing the profusion of antioxidant enzymes and several other stress-related proteins (Shi et al., 2013), whereas, downregulation of PA synthesis causes reduced antioxidant enzyme activities and drought tolerance in transgenic rice (Chen et al., 2014).

3.4 Polyamines and Salinity Stress

An important abiotic stress factor that keeps a tight harness on growth and productivity of crop plants in areas of the world affected by soil salinization is salinity (Kumar et al., 2009, 2012b). Differences in PA (Put, Spd, Spm) response under salt-stress have been re-counted among and within several species. For example, endogenous levels of PAs (Put, Spd, and Spm) decrease in rice seedlings under NaCl stress (Prakash and Prathapsenan, 1988). In contrast, salinity has also been shown to increase the accumulation of PAs in plants (Basu et al., 1988). Santa-Cruz et al. (1997) reported that the (Spd + Spm): Put ratios increase with salinity in the salt-tolerant tomato species (*L. pennellii*), but not in the salt-sensitive species (*L. esculentum*). In both the species, stress treatments has been found to reduce the levels of Put and Spd. The Spm levels were not decreased with the levels of salinity in *L. pennellii*, whereas, they greatly decreased in *L. esculentum*. The effects are unlike NaCl concentrations on maize embryogenic cells, and have also been reported at the specific areas where increased salt concentration remarkably decreased the growth of the calluses and showed a significant increase in the total PA (Put, Spd) content, particularly caused by a rise in Put. Yamaguchi et al. (2006) even suggested the shielding role of Spm when its addition repressed the salt sensitivity in Spm-deficient mutants. Bouchereau et al. (1999) proposed that PA responses to salt stress are also ABA-dependent, because both *ADC2* and *SPMS* are induced by ABA.

Alcazar et al. (2006) claimed that stress-responsive, drought responsive (DRE), low temperature-responsive (LTR), and ABA-responsive elements (ABRE and/or ABRE-related motifs) are present in the promoters of PA biosynthetic genes. This also strengthens the view that in response to drought and salt treatments, the expression of some of the genes convoluted in PA biosynthesis are regulated by ABA (Alcazar et al., 2010). The results gathered from the loss-of-function mutations in PA biosynthetic genes further supported the protective role of PAs in plant response to salt stress. For example, EMS mutants of *Arabidopsis thaliana spe1-1* and *spe2-1* (which map to *ADC2*) demonstrating reduced ADC activity are scarce in PA accumulation after acclimation to high NaCl concentrations and display additional sensitivity to salt stress (Kasinathan and Wingler, 2004). It is worth mentioning that while some plants accumulate PAs, others have persistent or even depressed endogenous PA content when exposed to salt or any other stress conditions, and an individual plant species exhibits varied responses in terms of PA levels. For example, salt tolerance can be positively related with spermidine, whereas it can be negatively correlated with Spm levels in rice. Exogenous application of Spd is efficaciously associated with salt stress enhancement in terms of growth and productivity (Todorova et al., 2013). In addition to PA anabolism, research has shown that the increased activities of CuAO and PAO during salt stress help in maintaining the intracellular PA concentration and provide an important signal molecule, H_2O_2, as a catabolic by-product of PA oxidation, due to considerable salt tolerance. One of the possible elucidations to a reduction of the negative salinity effect on plants is the accumulation of calcium ions (Ca^{2+}). Ca^{2+} ions play an important role in plant physiology and metabolism as they put notable impact on the permeability of the cell wall and membranes, exercise a stabilizing effect on protein conformation and influence of the enzyme activity. High salinity results in PA exodus to the cell wall and by stimulating apoplastic DAO and PAO, a quick increase in the levels of ROS ($*O^{2-}$, H_2O_2, and *OH). This activates a Ca^{2+} influx across the plasma membrane and increases cytosolic free $[Ca^{2+}]$cyt. The former source of apoplastic ROS is PM NADPH oxidase (NOX) (Pottosin et al., 2014a), which is stimulated by the salt stress-induced cytosolic Ca^{2+} signals in seconds.

4. INTERCONNECTION BETWEEN POLYAMINES CATABOLISM, ROS GENERATION, AND METABOLIC ROUTES

As much as their cellular functions are varied and infrequently conflicting, so are their roles in plant stress. They have been deemed significant in preparing the plant for stress tolerance and directly aiding in ameliorating the causes of stress, and at the same time, their own catabolic products are accountable for causing stress damage. Several aspects of the link between PAs and abiotic stress in plants and their apparently clashing roles in the progression have been reviewed over the years (Alcazar et al., 2006). Put is catabolized by diamine oxidases diamine oxidases. DAOs are copper-containing amine oxidases (CuAO) during a reaction that converts Put into Δ1-pyrroline and generates ammonia and H_2O_2 as byproducts (Fig. 3). DAOs are preferentially localized in plant cell walls and hydrogen peroxide resulting from Put catabolism, which could also be very important in lignifications and cross-linking reactions under normal and stress conditions. The noteworthy feature of PAs is their ambiguous role, as being sources of ROS and potential ROS scavengers and playing the role of redox homeostasis regulators in plants. Following the oxidation of Put, Δ1-pyrroline is catabolized into γ-aminobutyric acid (GABA), which is ultimately changed into succinic acid, a component of the TCA cycle. Gamma-aminobutyric acid (GABA) is also involved in defense mechanisms, shielding plants from stress through the regulation of cellular pH, acting as an osmoregulator or signaling molecule. 4-Aminobutanal is produced by DAO and PAOs can be converted into GABA via 1-pyrroline. Thus, an increase in PA levels may be followed by an increase in GABA build up. Other classes of amine oxidases are flavin-containing polyamine oxidases (PAO) involved in the terminal catabolism of Spd and Spm producing 4-aminobutanal or 1-(3-aminopropyl)-pyrroline, 1,3-diaminopropane and H_2O_2. Large amounts of DAO and PAO are reported in dicotyledonous and monocotyledonous

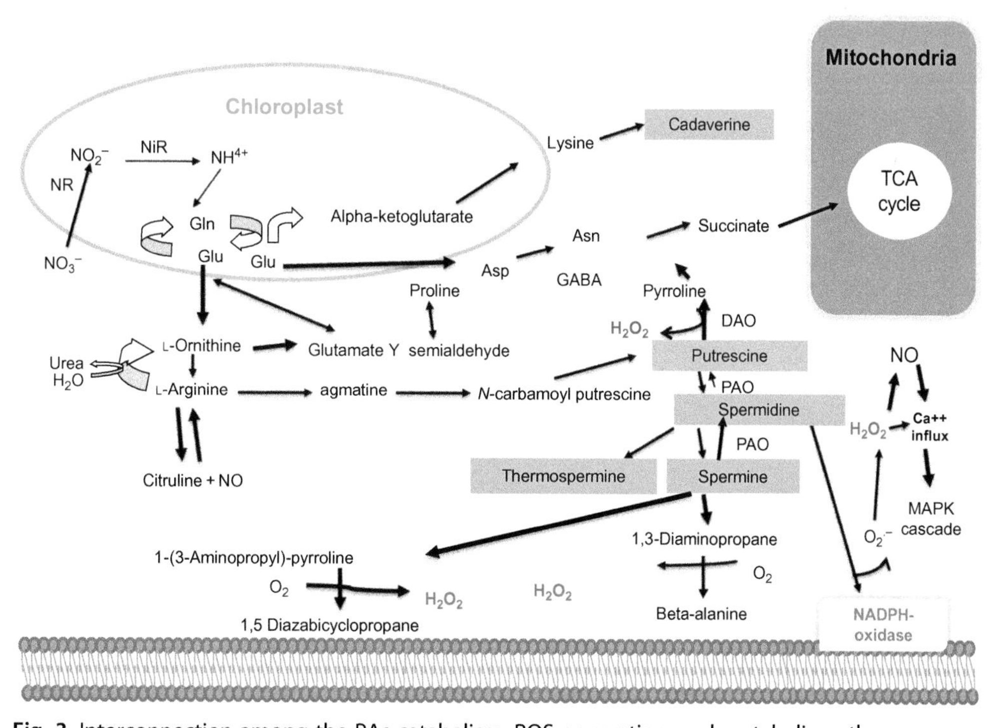

Fig. 3 Interconnection among the PAs catabolism, ROS generation, and metabolic pathways.

plants, respectively (Cona et al., 2006). However, most of the roles for GABA under stress is still unclear and thus needs to be explored. In soybean roots exposed to salinity, the degradation of PAs has been associated with increased levels of GABA (Xing et al., 2007). Conversely, throughout the recovery from stress, the level of GABA are reduced, along with an increase of Pas, consequently these studies suggest that PA catabolism might contribute to an increase GABA levels during salinity. PAs, particularly Spd, also induce superoxide anion (O^{2-}) production by the activation of NADPH-oxidase. However, O^{2-} dismutates spontaneously or enzymatically to H_2O_2. The ratio of O^{2-} to H_2O_2 is a significant signal in transcription (Andronis et al., 2014) and may be the intermediary of PAs in plant adaptation to unfavorable conditions. H_2O_2 has long been known as a signal molecule. It is able to mediate different processes, such as stomatal closure, directly due to its ability to influence ion channels; whereas, it may also activate specific stress response processes through the MAPK cascade (Moschou et al., 2008). Stress adaptive responses are closely associated to the capability of the plant to regulate ion transport and ion homeostasis. One of the best examples of the PA action mechanism in signaling is their influence on ion channels, which they exert by direct binding and through PA-induced signaling molecules (ROS and NO). PAs may regulate the activity of ion channels indirectly by membrane depolarization. The hyperpolarization-activated Ca^{2+} influx and the NO-induced release of intracellular Ca^{2+} result in a higher cytoplasmic Ca^{2+} concentration, which is a major component in general stress responses, such as stomatal movements; the cytosolic Ca^{2+} level additionally regulates several plasma membrane channels (Pottosin et al., 2014b). PA oxidation is a cause of H_2O_2 in the apoplast, which can augment to the defense response against pathogens. There is evidence that some PAOs participate within the hypersensitive response (HR) in *Nicotiana tabacum* plants immune to the tobacco mosaic virus (TMV). The PA metabolic pathway is additionally interconnected with alternative metabolic routes involved in the formation of various signaling molecules and metabolites that are relevant in plant stress responses. PA and ethylene biosynthesis are associated through SAM, which acts as a common precursor (Fig. 2). Antagonistic effects between these compounds occur throughout leaf and flower senescence and fruit ripening (Pandey et al., 2000). PA metabolism also influences nitric oxide (NO) formation (Yamasaki and Cohen, 2006). PAs persuade the production of NO that might act as an association between PA-mediated stress responses and other stress mediators. NO acts as an intermediate signaling molecule in cytokinin, ABA, auxin, cytokinin, and ethylene signaling. H_2O_2 generated by the action of DAOs and/or PAOs is concerned in both biotic and abiotic stress signaling, as well as, in ABA-induced stomatal closure (Cona et al., 2006). Additionally, Pro levels increase in response to numerous abiotic stresses and PA catabolism is closely related to Pro accumulation in response to salt stress (Aziz et al., 1998). Interactions between stress-induced Pro and PA accumulations can reveal the fact that they share ornithine as a common precursor. In conclusion, the PA metabolism is connected to several important hormonal and metabolic pathways concerned in the development, stress responses, nitrogen assimilation, and respiratory metabolism.

5. CONCLUDING REMARKS

The PA field is complex, because PAs have manifold roles acquired during evolution and it is hard to unravel one from the other to study them in isolation. We still need to address where PAs localize, where they are necessary for their function, how they are transported from sources to sinks (and also which are the sources and the sinks), and more significantly, whether PAs are intermediary compounds in the stress protection, or have a role themselves. A detailed metabolic and signaling investigation addressing these and other fundamental questions are required to provide a broader view of the roles and mechanisms of PAs during stress. The recognition of genes underlying the differential regulation of PA levels can be achieved by traditional quantitative trait locus (QTL) mapping and cloning or by genome-wide association mapping. It must also be pointed out that a similar situation exists with respect to a surplus of other genetic manipulation approaches that have been shown to be effective in imparting short-term stress tolerance in various plant species. It is likely that the advanced high throughput techniques of genomics, transcriptomics, and proteomics, coupled with better techniques of monitoring the live plants under stress and their metabolic status (the metabolome), would provide an improved holistic picture of the consequences of upregulation or downregulation of genes likely to be involved in stress tolerance in relation to metabolites such as PAs. PAs are considered to play significant roles in protecting plant cells from stress-associated damage. So far, remarkable progress has been made in understanding the significance of PAs in stress responses. There is accumulating evidence that PA levels undergo wide-ranging changes in response to a range of abiotic stresses; and physiological, molecular, and genetic approaches have been used to identify and functionally characterize PA biosynthetic genes in various plant species. Nevertheless, many key questions remain unanswered. The causal relationship between PA accumulation and stress tolerance has not been determined, despite numerous observations of changes in PA levels in response to abiotic stresses. Furthermore, the cellular compartmentation and transportation of PAs is not well understood, although a few PA transporters have been identified, the signaling cascades linking stress responses and PA genes are still far from being well defined. In keeping with these unanswered questions, there are several promising areas of future study. The sites of PA production and actions in plant cells need to be identified, and to this end, the cellular localization of PAs and their transporters should be resolute. Moreover, the physiological and molecular mechanisms concerning the roles of PAs in stress tolerance need to be elucidated and in particular, how PAs contribute to the ROS removal through activation of antioxidant enzymes should be clearly deciphered. In conclusion, more attention should be directed toward plant responses to simultaneous, multiple abiotic stresses by PAs and to the crosstalk between abiotic and biotic stress signaling. Understanding and linking the role of microbes in PA signaling in plant stress would increase our ability to use these beneficial organisms, and also increase our understanding of biotic and abiotic stress resistance relationships in plants.

REFERENCES

Alcazar, R., Cuevas, J.C., Patron, M., Altabella, T., Tiburcio, A.F., 2006. Abscisic acid modulates polyamine metabolism under water stress in *Arabidopsis thaliana*. Physiol. Plant. 128, 448–455. https://doi.org/10.1111/j.1399-3054.2006.00780.x.

Alcazar, R., Garcia-Martinez, J.L., Cuevas, J.C., Tiburcio, A.F., Altabella, T., 2005. Overexpression of ADC2 in *Arabidopsis* induces dwarfism and late-flowering through GA deficiency. Plant J. 43, 425–436. https://doi.org/10.1111/j.1365-313X.2005.02465.x.

Alcazar, R., Altabella, T., Marco, F., Bortolotti, C., Reymond, M., Koncz, C., Carrasco, P., Tiburcio, A.F., 2010. Polyamines: molecules with regulatory functions in plant abiotic stress tolerance. Planta 231, 1237–1249. https://doi.org/10.1007/s00425-010-1130-0.

Alcazar, R., Cuevas, J.C., Planas, J., Zarza, X., Bortolotti, C., Carrasco, P., Salinas, J., Tiburcio, A.F., Altabella, T., 2011. Integration of polyamines in the cold acclimation response. Plant Sci. 180, 31–38. https://doi.org/10.1016/j.plantsci.2010.07.022.

Andronis, E.A., Moschou, P.N., Toumi, I., Roubelakis-Angelakis, K.A., 2014. Peroxisomal polyamine oxidase and NADPH-oxidase cross-talk for ROS homeostasis which affects respiration rate in *Arabidopsis thaliana*. Front. Plant Sci. 3 (5), 132. https://doi.org/10.3389/fpls.2014.00132.

Aziz, A., Martin-Tanguy, J., Larher, F., 1998. Stress-induced changes in polyamine and tyramine levels can regulate proline accumulation in tomato leaf discs treated with sodium chloride. Physiol. Plant. 104, 195–202. https://doi.org/10.1034/j.1399-3054.1998.1040207.x.

Bachrach, U., 2010. The early history of polyamine research. Plant Physiol. Biochem. 48 (7), 490–495. https://doi.org/10.1016/j.plaphy.2010.02.003.

Basu, R., Maitra, N., Ghosh, B., 1988. Salinity results in polyamine accumulation in early rice (*Oryza sativa* L.) seedlings. Aust. J. Plant Physiol. 15, 777–786.

Basu, S., Giri, R.K., Benazir, I., Kumar, S., Rajwanshi, R., Dwivedi, S.K., Kumar, G., 2017. Comprehensive physiological analyses and reactive oxygen species profiling in drought tolerant rice genotypes under salinity stress. Physiol. Mol. Biol. Plants 23 (4), 837–850. https://doi.org/10.1007/s12298-017-0477-0.

Bey, P., Danzin, C., Jung, M., 1987. Inhibition of basic amino acid decarboxylases involved in polyamine biosynthesis. In: McCann, P.P., Pegg, A.E., Sjoerdsma, A. (Eds.), Inhibition of Polyamine Metabolism. Academic Press, Orlando, pp. 1–32.

Bitonti, A.J., Carara, P.J., McCann, P.P., Bey, P., 1987. Catalytic irreversible inhibition of bacterial and plant arginine decarboxylase activities by novel substrate and product analogues. Biochem. J. 242, 69–74. https://doi.org/10.1042/bj2420069.

Borrell, A., Culianez-Macia, F.A., Altabella, T., Besford, R.T., Flores, D., Tiburcio, A.F., 1995. Arginine decarboxylase is localized in chloroplasts. Plant Physiol. 109, 771–776.

Bortolotti, C., Cordeiro, A., Alcazar, R., Borrell, A., Culianez-Macia, F.A., Tiburcio, A.F., Altabella, T., 2004. Localization of arginine decarboxylase in tobacco plants. Physiol. Plant. 120, 84–92. https://doi.org/10.1111/j.0031-9317.2004.0216.x.

Bouchereau, A., Aziz, A., Larher, F., Martin-Tanguy, J., 1999. Polyamines and environmental challenges: recent development. Plant Sci. 140, 103–125. https://doi.org/10.1016/S0168-9452(98)00218-0.

Chen, M., Chen, J., Fang, J., Guo, Z., Lu, S., 2014. Down-regulation of S-adenosylmethionine decarboxylase genes results in reduced plant length, pollen viability, and abiotic stress tolerance. Plant Cell Tissue Organ Cult. 116, 311–322. https://doi.org/10.1007/s11240-013-0405-0.

Cona, A., Rea, G., Angelini, R., Federico, R., Tavladoraki, P., 2006. Functions of amine oxidases in plant development and defence. Trends Plant Sci. 11, 80–88. https://doi.org/10.1016/j.tplants.2005.12.009.

Cuevas, J.C., Lopez-Cobollo, R., Alcazar, R., Zarza, X., Koncz, C., Altabella, T., Salinas, J., Tiburcio, A.F., Ferrando, A., 2009. Putrescine as a signal to modulate the indispensable ABA increase under cold stress. Plant Signal. Behav. 4, 219–220. https://doi.org/10.1104/pp.108.122945.

De Ronde, J.A., Laurie, R.N., Caetano, T., Greyling, M.M., Kerepesi, I., 2004. Comparative study between transgenic and nontransgenic soybean lines proved transgenic lines to be more drought tolerant. Euphytica 138, 123–132. https://doi.org/10.3389/fpls.2017.00792.

Diao, Q., Song, Y., Shi, D., Qi, H., 2017. Interaction of polyamines, abscisic acid, nitric oxide, and hydrogen peroxide under chilling stress in tomato (*Lycopersicon esculentum* Mill.) seedlings. Front. Plant Sci. 8, 203. https://doi.org/10.3389/fpls.2017.00203.

Dwivedi, S.K., Basu, S., Kumar, S., Kumar, G., Prakash, V., Kumar, S., Mishra, J.S., Bhatt, B.P., Malviya, N., Singh, G.P., Arora, A., 2017. Heat stress induced impairment of starch mobilisation regulatespollen viability and grain yield in wheat: study in Eastern Indo-Gangetic Plains. Field Crop Res. 206, 106–114. https://doi.org/10.1016/j.fcr.2017.03.006.

Flores, H.E., 1991. Changes in polyamine metabolism in response to abiotic stress. In: Slocum, R., Flores, H.E. (Eds.), The Biochemistry and Physiology of Polyamines in Plants. CRC Press, Boca Raton, FL, pp. 214–225.

Franchin, C., Fossati, T., Pasquini, E., Lingua, G., Castiglione, S., Torrigiani, P., Biondi, S., 2007. High concentrations of zinc and copper induce differential polyamine responses in micropropagated white poplar (*Populus alba*). Physiol. Plant. 130, 77–90.

Galiba, G., Kocsy, G., Kaur-Sawhney, R., Sutka, J., Galston, A.W., 1993. Chromosomal localization of osmotic and salt stress-induced differential alterations in polyamine content in wheat. Plant Sci. 92, 203–211. https://doi.org/10.1016/0168 9452(93)90207-G.

Gallardo, M., Sanchez-Calle, I., De Rueda, P.M., Matilla, A.J., 1996. Alleviation of thermo inhibition in chickpea seeds by putrescine involves the ethylene pathway. Aust. J. Plant Physiol. 23, 479–487. https://doi.org/10.1071/PP9960479.

Gill, S.S., Tuteja, N., 2010. Polyamines and abiotic stress tolerance in plants. Plant Signal. Behav. 5 (1), 26–33.

Gosal, S.S., Wani, H.S., Kang, M.S., 2009. Biotechnology and drought tolerance. J. Crop. Improv. 23, 19–54. https://doi.org/10.1080/15427520802418251.

Groppa, M.D., Benavides, M.P., 2008. Polyamines and abiotic stress: recent advances. Amino Acids 34, 35–45.

Groppa, M.D., Tomaro, M.L., Benavides, M.P., 2007. Polyamines and heavy metal stress: the antioxidant behavior of spermine in cadmium- and copper-treated wheat leaves. Biometals 20, 185–195. https://doi.org/10.1007/s10534-006-9026-y.

Gupta, K., Dey, A., Gupta, B., 2013. Plant polyamines in abiotic stress responses. Acta Physiol. Plant. 35, 2015–2036. https://doi.org/10.1007/s11738-013-1239-4.

Hanfrey, C., Sommer, S., Mayer, M.J., Burtin, D., Michael, A.J., 2001. *Arabidopsis* polyamine biosynthesis: absence of ornithine decarboxylase and the mechanism of arginine decarboxylase activity. Plant J. 27, 551–560.

Hibasami, H., Tanaka, M., Nagai, J., Ikeda, T., 1980. Dicyclohexylamine, a potent inhibitor of spermidine synthase in mammalian cells. FEBS Lett. 116 (1), 99–101.

Illingworth, C., Mayer, M.J., Elliott, K., Hanfrey, C., Walton, N.J., Michael, A.J., 2003. The diverse bacterial origins of the *Arabidopsis* polyamine biosynthetic pathway. FEBS Lett. 549, 26–30.

Kasinathan, V., Wingler, A., 2004. Effect of reduced arginine decarboxylase activity on salt tolerance and on polyamine formation during salt stress in *Arabidopsis thaliana*. Physiol. Plant. 121, 101–107. https://doi.org/10.1111/j.0031-9317.2004.00309.x.

Kovacs, Z., Simon-Sarkadi, L., 2010. Differential effects of cold, osmotic stress and abscisic acid on polyamine accumulation in wheat. Amino Acids 38, 623–631. https://doi.org/10.1007/s00726-009-0423-8.

Kramer, G.F., Wang, C.Y., 1990. Effects of chilling and temperature pre conditioning on the activity of polyamine biosynthetic enzymes in zucchini squash. J. Plant Physiol. 136, 115–122. https://doi.org/10.1016/S0176-1617(11)81624-X.

Kumar, G., Purty, R.S., Sharma, M.P., Singla-Pareek, S.L., Pareek, A., 2009. Physiological responses among *Brassica* species under salinity stress show strong correlation with transcript abundance for SOS pathway-related genes. J. Plant Physiol. 166, 507–520.

Kumar, G., Kushwaha, H.R., Panjabi-Sabharwal, V., Kumari, S., Joshi, R., Karan, R., Mittal, S., Pareek, S.L., Pareek, A., 2012a. Clustered metallothionein genes are co-regulated in rice and ectopic expression of OsMT1e-P confers multiple abiotic stress tolerance in tobacco via ROS scavenging. BMC Plant Biol. 12, 107. https://doi.org/10.1186/1471-2229-12-107.

Kumar, G., Kushwaha, H.R., Purty, R.S., Kumari, S., Singla-Pareek, S.L., Pareek, A., 2012b. Cloning, structural and expression analysis of OsSOS2 in contrasting cultivars of rice under salinity stress. Genes Genomes Genomics 6 (1), 34–41.

Kuznetsov, V.V., Shevyakova, N.I., 1997. Stress responses of tobacco cells to high temperature and salinity. Proline accumulation and phosphorylation of polypeptides. Physiol. Plant. 100, 320–326. https://doi.org/10.1111/j.1399-3054.1997.tb04789.x.

Lee, T.M., Lur, H.S., Chu, C., 1997. Role of abscisic acid in chilling tolerance of rice (*Oryza sativa* L.) seedlings: II. Modulation of free polyamine levels. Plant Sci. 126, 1–10. https://doi.org/10.1016/S0168-9452(97)00076-9.

Liu, J.H., Kitashiba, H., Wang, J., Ban, Y., Moriguchi, T., 2007. Polyamines and their ability to provide environmental stress tolerance to plants. Plant Biotechnol. J. 24, 117–126. https://doi.org/10.5511/plantbiotechnology.24.117.

Liu, M., Chen, J., Guo, Z., Lu, S., 2017. Differential responses of polyamines and antioxidants to drought in a centipedegrass mutant in comparison to its wild type plants. Front. Plant Sci. 8, 792. https://doi.org/10.3389/fpls.2017.00792.

Martin-Tanguy, J., 1985. The occurrence and possible function of hydroxylcinnamoyl acid amides in plants. Plant Growth Regul. 3, 381–399.

Moschou, P.N., Paschalidis, K.A., Delis, I.D., Andriopoulou, A.H., Lagiotis, G.D., Yakoumakis, D.I., 2008. Spermidine exodus and oxidation in the apoplast induced by abiotic stress is responsible for H_2O_2 signatures that direct tolerance responses in tobacco. Plant Cell 20, 1708–1724. https://doi.org/10.1105/tpc.108.059733.

Murkowski, A., 2001. Heat stress and spermidine: effect on chlorophyll fluorescence in tomato plants. Biol. Plant. 44, 53–57.

Nayyar, H., Chander, S., 2004. Protective effects of polyamines against oxidative stress induced by water and cold stress in chickpea. J. Agron. Crop Sci. 190, 355–365. https://doi.org/10.1111/j.1439-037X.2004.00106.x.

Ndayiragije, A., Lutts, S., 2006. Do exogenous polyamines have an impact on the response of a salt-sensitive rice cultivar to NaCl? J. Plant Physiol. 163, 506–516.

Pandey, S., Ranade, S.A., Nagar, P.K., Kumar, N., 2000. Role of polyamines and ethylene as modulators of plant senescence. J. Biosci. 25, 291–299.

Pathak, M.R., da Silva Teixeira, J.A., Wani, S.H., 2014. Polyamines in response to abiotic stress tolerance through transgenic approaches. GM Crops Food 5, 87–96.

Peng, D., Wang, X., Li, Z., Zhang, Y., Peng, Y., Li, Y., 2016. NO is involved in spermidine-induced drought tolerance in white clover via activation of antioxidant enzymes and genes. Protoplasma 253, 1243–1254. https://doi.org/10.1007/s00709-015-0880-8.

Pottosin, I., Velarde-Buendia, A.M., Bose, J., Zepeda-Jazo, I., Shabala, S., Dobrovinskaya, O., 2014a. Cross-talk between ROS and polyamines in regulation of ion transport across the plasma membrane: implications for plant adaptive responses. J. Exp. Bot. 65, 1271–1283. https://doi.org/10.1093/jxb/ert423.

Pottosin, I., Velarde-Buendia, A.M., Bose, J., Fuglsang, A.T., Shabala, S., 2014b. Polyamines cause plasma membrane depolarization, activate Ca^{2+} and modulate H^+-ATPase pump activity in pea roots. J. Exp. Bot. 65, 2463–2472. https://doi.org/10.1093/jxb/eru133.

Prakash, L., Prathapsenan, G., 1988. Effect of NaCl salinity and putrescine on shoot growth, tissue ion concentration and yield of rice (*Oryza sativa*). J. Agron. Crop Sci. 160, 325–334. https://doi.org/10.1111/j.1439-037X.1988.tb00630.x.

Racz, M., Kovacs, D., Lasztity, O., Veisz, G., Szalai, E., 1996. Effects of short-term and long-term low temperature stress on polyamine biosynthesis in wheat genotypes with varying degrees of frost tolerance. J. Plant Physiol. 148, 368–373. https://doi.org/10.1016/S0176-1617(96)80267-7.

Roy, M., Ghosh, B., 1996. Polyamines, both common and uncommon, under heat stress in rice (*Oriza sativa*) callus. Physiol. Plant. 98, 196–200. https://doi.org/10.1111/j.1399-3054.1996.tb00692.x.

Sanghera, G.S., Wani, S.H., Hussain, W., Singh, N.B., 2011. Engineering cold stress tolerance in crop plants. Curr. Genomics 12 (1), 30–43. https://doi.org/10.2174/138920211794520178.

Santa-Cruz, A., Estan, M.T., Rus, A., Bolarin, M.C., Acosta, M., 1997. Effects of NaCl and mannitol iso-osmotic stresses on the free polyamine levels in leaf discs of tomato species differing in salt tolerance. Plant Physiol. 151, 754–758. https://doi.org/10.1016/S0176-1617(97)80074-0.

Shevyakova, N.I., Rakitin, V.Y., Duong, D.B., Sadomov, N.G., Kuznetsov, V.V., 2001. Heat shock-induced cadaverine accumulation and translocation throughout the plant. Plant Sci. 161, 1125–1133. https://doi.org/10.1016/S0168-9452(01)00515-5.

Shi, H., Ye, T., Chan, Z., 2013. Comparative proteomic and physiological analyses reveal the protective effect of exogenous polyamines in the Bermuda grass (*Cynodon dactylon*) response to salt and drought stresses. J. Proteome Res. 12, 4951–4964. https://doi.org/10.1021/pr400479k.

Soudek, P., Ursu, M., Petrova, S., Vanek, T., 2016. Improving crop tolerance to heavy metal stress by polyamine application. Food Chem. 213, 223–229. https://doi.org/10.1016/j.foodchem.2016.06.087.

Szalai, G., Janda, T., Bartok, T., Paldi, E., 1997. Role of light in changes in free amino acid and polyamine contents at chilling temperature in maize (*Zea mays*). Physiol. Plant. 101, 434–438. https://doi.org/10.1111/j.1399-3054.1997.tb01018.x.

Todorova, D., Katerova, Z., Sergiev, I., Alexieva, V., 2013. Role of polyamines in alleviating salt stress. In: Ahmad, P. et al., (Ed.), In: Ecophysiology and Responses of Plants Under Salt Stress, Springer, New York, NY, pp. 355–379. https://doi.org/10.1007/978-1-4614-4747-4_13.

Urano, K., Yoshiba, Y., Nanjo, T., Igarashi, Y., Seki, M., Sekiguchi, F., 2003. Characterization of *Arabidopsis* genes involved in biosynthesis of polyamines in abiotic stress responses and developmental stages. Plant Cell Environ. 26, 1917–1926.

Verma, S., Mishra, S.N., 2005. Putrescine alleviation of growth in salt stressed *Brassica juncea* by inducing antioxidative defense system. J. Plant Physiol. 162, 669–677. https://doi.org/10.1016/j.jplph.2004.08.008.

Xing, S.G., Jun, Y.B., Hau, Z.W., Liang, L.Y., 2007. Higher accumulation of γ-aminobutyric acid induced by salt stress through stimulating the activity of diamine oxidases in *Glycine max* (L.) Merr. roots. Plant Physiol. Biochem. 45, 560–566. https://doi.org/10.1016/j.plaphy.2007.05.007.

Yamaguchi, K., Takahashi, Y., Berberich, T., Imai, A., Miyazaki, A., Takahashi, T., Michael, A., Kusano, T., 2006. The polyamine spermine protects against high salt stress in *Arabidopsis thaliana*. FEBS Lett. 580, 6783–6788. https://doi.org/10.1016/j.febslet.2006.10.078.

Yamasaki, H., Cohen, M.F., 2006. NO signal at the crossroads: polyamine-induced nitric oxide synthesis in plants. Trends Plant Sci. 11, 522–524. https://doi.org/10.1016/j.tplants.2006.09.009.

Yiu, J.C., Juang, L.D., Fang, D.Y.T., Liu, C.W., Wu, S.J., 2009. Exogenous putrescine reduces flooding-induced oxidative damage by increasing the antioxidant properties of Welsh onion. Sci. Hortic. 120, 306–314.

Zhang, M.Q., Chen, R.K., Yu, S.L., 1996. Changes of polyamine metabolism in drought-stressed sugarcane leaves and their relation to drought resistance. Acta Phys. Sin. 22, 327–732.

Zhao, H., Yang, H., 2008. Exogenous polyamines alleviate the lipid peroxidation induced by cadmium chloride stress in *Malus hupehensis* Rehd. Sci. Hortic. 116, 442–447.

FURTHER READING

Santa-Gruz, A., Perez-Alfocea, M.A., Bolarin, C., 1997. Changes in free polyamine levels induced by salt stress in leaves of cultivated and wild tomato species. Plant Physiol. 101, 341–346. https://doi.org/10.1111/j.1399-3054.1997.tb01006.x.

Yamaguchi-Shinozaki, K., Shinozaki, K., 2003. Regulatory network of gene expression in the drought and cold stress responses. Curr. Opin. Plant Biol. 6 (5), 410–417.

Yoda, H., Fujimura, K., Takahashi, H., Munemura, I., Uchimiya, H., Sano, H., 2009. Polyamines as a common source of hydrogen peroxide in host- and non-host hypersensitive response during pathogen infection. Plant Mol. Biol. 70, 103–112. https://doi.org/10.1007/s11103-009-9459-0.

CHAPTER 4

Cold Tolerance in Plants: Molecular Machinery Deciphered

Mahmood Maleki*, Mansour Ghorbanpour†
*Department of Biotechnology, Institute of Science and High Technology and Environmental Science, Graduate University of Advanced Technology, Kerman, Iran
†Department of Medicinal Plants, Faculty of Agriculture and Natural Resources, Arak University, Arak, Iran

Contents

1. INTRODUCTION

Among abiotic stresses, cold stress is a major environmental factor that limits agricultural production, causing preharvest and postharvest damage, resulting in enormous financial losses in agriculture every year (Einset et al., 2007). Cold stress also has a huge impact on the survival and geographical distribution of plants (Jan and Andrabi, 2009). An optimal temperature, or a diurnal range of temperatures, is required for the maximum rate of growth and development of plants (Fitter and Hay, 1981). When the ambient temperature deviates from optimal; physiological, biochemical, metabolic, and molecular changes occur within plants (Yadav, 2010). By applying these changes, plants will maintain their growth and development at the highest level. They also try to keep cell homeostasis under stress. This is an effort of plants to maximize growth and developmental processes and to maintain cellular homeostasis during such adverse conditions (Yadav, 2010). Under increasingly stressful conditions, the natural growth and development of plants' cellular and whole plant processes will be impaired, and plants will die under severe conditions (Yadav, 2010).

Biochemical, Physiological and Molecular Avenues for Combating Abiotic Stress in Plants
https://doi.org/10.1016/B978-0-12-813066-7.00004-8

Plants differ in their tolerance to chilling (0–15°C) and freezing (<0°C) temperatures (Jan and Andrabi, 2009). More than half of the 350,000 plant species on Earth are grown in the tropics and subtropic regions. Most of these species are damaged during storage at temperatures between 0°C and 15°C (chilling temperatures). This damage is called "chilling injury" as opposed to damage during freezing (freezing injury) (Levitt, 1980; Raison and Lyons, 1986). Plants from temperate regions are chilling tolerant, although most are not very tolerant to freezing, but can increase their freezing tolerance by being exposed to chilling, nonfreezing temperatures; a process known as cold acclimation (Levitt, 1980; Shabala, 2017), which is associated with biochemical and physiological changes (Shinozaki and Yamaguchi-Shinozaki, 1996; Thomashow, 1998; Gilmour et al., 2000). By contrast, plants of tropical and subtropical origins, including many crops such as rice, maize, and tomatoes, are sensitive to chilling stress and largely lack the capacity for cold acclimation.

The ability of plants in the seedling stage to survive low temperatures without harm, and to continue to grow, is called cold tolerance. On the contrary, in cold-sensitive species, plants, after being exposed to cold stress for a long time, suffer from external symptoms of injury that ultimately lead to plant death. Plants that show signs of obvious damage at temperatures above 15°C are very sensitive to chilling (Raison and Lyons, 1986). Chilling temperatures in temperate climates leads to a reduction in crops, or complete crop failure, due to either direct damage or delayed maturation (Lukatkin et al., 2012).

In this chapter, the impacts of cold stress on physiological processes of plants and the most important adaptation mechanisms involved in tolerance processes will be discussed in detail.

2. EFFECT OF CHILLING ON THE PHYSIOLOGICAL PROCESSES OF PLANTS

As mentioned herein, among various abiotic stresses, cold stress is one of the most important factors limiting the crop productivity and distribution of plants. But cold is not always considered a stress for plants. For some plants, chilling is essential for the passage from the vegetative phase to the reproductive phase, the process known as vernalization (Shabala, 2017). Also, low temperatures are necessary to break seed dormancy. On the other hand, chilling is necessary to induce cold stress tolerance in some species (Shabala, 2017); a phenomenon known as cold-hardening and/or cold acclimation in plants (Levitt, 1980; Shabala, 2017). However, many plants may have difficulty growing and developing in the face of cold stress. Depending on the severity of the stress, the plants will show different symptoms. Brief exposure of plants to low-temperature stress may cause transitory changes in them, and they would still survive. However, prolonged exposure to low-temperature stress causes necrosis, wilting, and chlorosis, and eventually leads to the death of plants (Adam and Murthy, 2014).

Many traits of plants are affected by cold stress. In wheat, lower temperatures of 2°C, an optimum of 8–10°C, and a maximum 20–22°C accelerate germination (Khodabandeh, 2003). However, Aslan et al. (2016) showed that the germination rate of wheat is decreased under freezing temperatures. In the case of cotton, low temperatures slow down the rate of the germination process (Marcellin, 1992).

Photosynthesis is one of the processes most sensitive to this stress (Pearce, 2001; Ruelland et al., 2009; Ploschuk et al., 2014). The cold stress, depending on the species and herb variety, has a different effect on the photosynthesis process. Qin et al. (2013) studied the effects of cold stress on photosynthesis in leaves of seven regular potato cultivars and two Hunan endemic potato species. Their results showed that the photosynthetic rate (*Pn*) was significantly reduced with decreasing environmental temperature for all the species; however, every species exhibited a different declining rate.

Ploschuk et al. (2014) showed that chilling (40 h at 4°C) and freezing (2 h at −1, −2, and −3°C) reduced up to 75%, and 100% maximum, of leaf photosynthesis (A_{max}), respectively, in *Jatropha curcas* (originated from tropical America) at 1 day after treatment. A lower-efficiency electron used for photosynthesis was detected for plants subjected to chilling and freezing stress.

C_4 plants may be affected by cold stress as well. C_4 photosynthesis, under optimal conditions, enables higher-efficiency use of light, water, and nitrogen than the C_3 form used by many crops. It is associated with the most productive terrestrial plants and crops, but is largely limited to the tropics and subtropics. It has been argued that the C_4 photosynthetic apparatus is inherently limited to warm environments. A small group of C_4 species appear to have overcome this, and in contrast to the major C_4 crop, maize, these species are able to acclimate their photosynthetic apparatus to chilling conditions (Long and Spence, 2013). Friesen and Sage (2016) showed that Rubisco activity in vitro declined in proportion to the reduction in the net CO_2 assimilation rate in chilled (12/5°C) sensitive Miscanthus hybrid plants, indicating Rubisco capacity is responsible in part for the decline in the net CO_2 assimilation rate. Pyruvate, orthophosphate dikinase activities, were also reduced by the chilling treatment when assayed at 28°C, indicating this enzyme may also contribute to the reduction in the net CO_2 assimilation rate in chilling-sensitive Miscanthus hybrids. They deduced that the carboxylation efficiency of the C_4 cycle was depressed after chilling. It should be mentioned that while Miscanthus is noted for its superior chilling tolerance relative to most C_4 species, there is substantial variation in the degree of cold tolerance between species and hybrid genotypes (Clifton-Brown and Jones, 1997; Clifton-Brown and Lewandowski, 2000; Jørgensen and Muhs, 2001; Purdy et al., 2013; Friesen et al., 2014; Głowacka et al., 2014). Miscanthus is a C_4 perennial grass being developed for bioenergy production in temperate regions where chilling events are common.

As mentioned herein, plants were sensitive to cold stress, as indicated by a reduction in the photosynthetic efficiency. However, some studies showed that the susceptibility of leaves to cold may be modified by root temperature. Paredes and Quiles (2015) showed that when the stem, but not roots, was chilled, the quantum yield of PSII and the relative electron transport rates were much lower than when the whole plant, root and stem, was chilled at 10°C. Additionally, when the whole plant was cooled, both the activity of electron donation by NADPH and ferredoxin to plastoquinone and the amount of PGR5 polypeptide, an essential component of the cyclic electron flow around PSI, increased, suggesting that in these conditions, cyclic electron flow helps protect photosystems. However, when the stem, but not the root, was cooled, cyclic electron flow did not increase, and PSII was damaged as a result of insufficient dissipation of the excess light energy.

Cold stress affects both vegetative and reproductive phases throughout the plant life cycle. In fact, cold temperatures induce flower abortion, pollen and ovule infertility, and causes breakdown of fertilization and affects seed filling, leading to low seed set and ultimately low grain yield (Thakur et al., 2010). Srinivasan et al. (1999) showed that pollen vigor (germination and tube growth) and ovule viability at low temperatures decreased in cultivars/lines of chickpeas (*Cicer arietinum* L.), but these traits were different between tolerant and sensitive cultivars. The number of ovules was not affected by cold stress in all cultivars/lines, but pollen size and viability were reduced in some of them. The reduced ovule fertilization, associated with a decline in pollen tube growth and ovule viability, was the major cause for poor seed set at low temperatures. The magnitude of effects on gamete function varied with cultivar/line and severity of stress. Clarke and Siddique (2004) demonstrated that low temperature ($<15°C$) affects both the development and function of reproductive structures in the chickpea flower. Their results showed that the function of pollen derived from chilling-sensitive plants is clearly affected most by low temperature stress, particularly the growth of the pollen tubes down the style before fertilization occurs. In contrast, pollen tubes derived from chilling tolerant plants continue to grow down the style under low-temperature stress. Although other stages of development and function were affected by low temperature, including sporogenesis, pollen germination, and the stigma, none were correlated to the phenotype of the mother plant.

Under abiotic stress such as cold, reactive oxygen species (ROS) such as $^1O_2$, H_2O_2, O_2^-, and $HO^•$, are produced (Maleki et al., 2017). ROS molecules are capable of destroying lipids, proteins, carbohydrates, and DNA (Apel and Hirt, 2004; Suzuki and Mittler, 2006). They are produced naturally in a small amount within the chloroplast, mitochondria, and peroxisomes. But under cold stress, the amount of active oxygen species increases dramatically. The main cause of ROS production in the chloroplasts is a limitation of CO_2 fixation coupled with an overreduction of the electron transport chain. In the mitochondria, overreduction of the electron transport chain is also a major mechanism of ROS production during stress (Davidson and Schiestl, 2001; Suzuki and Mittler, 2006). In contrast, in peroxisomes, when glycolate is oxidized to glyoxylic acid during photorespiration, H_2O_2 is produced (Mittler et al., 2004; Suzuki and Mittler, 2006).

3. COLD STRESS SIGNALING

The first thing that happens after a cold stress in plant cells is an increase in cytosolic Ca^{2+} as an important second messenger. This cytosolic Ca^{2+} is suggested as an important component of signal transduction and developing cold acclimation (Solanke and Sharma, 2008; Sanders et al., 2002; Dodd et al., 2006; Klimecka and Muszynska, 2007; Heidarvand and Amiri, 2010). In various studies, it has been shown a positive correlation between the accumulation of cytosolic Ca^{2+} and cold stress, as well as the expression of the cold-induced genes in many plants (Knight et al., 1991; Monroy and Dhindsa, 1995; Reddy and Reddy, 2004; Henriksson and Trewavas, 2003).

In addition to the cold stress, other nonbiotic and biotic stresses increase the cytosolic calcium for transmitting the message. It is very important that the plant can respond appropriately to stresses with changes in the expression of genes. In fact, the plant must be able to distinguish between different stresses for appropriate responses (Heidarvand and Amiri, 2010). But how do plants differentiate between the increases in cytosolic calcium induced by different stresses? One of the interesting hypotheses is that the specific characteristics of different calcium elevations ("calcium signatures") might encode specific information in plants (Allen et al., 2001; Love et al., 2004; Miwa et al., 2006; McAinsh and Pittman, 2009; Dodd et al., 2010; Short et al., 2012; Whalley and Knight, 2013).

As noted herein, each stress has a specific signature of cytosolic calcium ions. There are different calcium sensing or response elements that can recognize specific signatures induced by each stress, such as cold (Knight et al., 1996; Plieth et al., 1999; Knight and Knight, 2001; Sanders et al., 2002; Klimecka and Muszynska, 2007; Heidarvand and Amiri, 2010.). The transmission of a message resulting from the specific signature of calcium is transmitted through calcium sensors to the downstream elements, which ultimately leads to a change in the cytoskeletal rearrangements and expression of genes patterns (Sanders et al., 2002; McAinsh and Pittman, 2009).

Proteins that transmit a primary signal when cytosolic Ca^{2+} is raised are protein sensors such as calmodulin (CaM), CaM domain-containing protein kinases (CDPKs), calcineurin B-like proteins (CBLs), and CBL-interacting protein kinases (CIPKs) (Monroy et al., 1993; Solanke and Sharma, 2008).

In plants, the Ca^{2+} sensors are categorized into two classes: "sensor relay" and "responders." The first class consists of calmodulin (CaM) and calcineurin B-like proteins (CBLs) that transmit the message through conformational change and regulating gene expression. The second class relays the message to its downstream targets through effector domains such as protein kinase and phospholipase (Reddy and Reddy, 2004; Klimecka and Muszynska, 2007).

4. THE RESPONSES OF PLANTS TO COLD STRESS

When the plants are exposed to cold stress, several dysfunctions appear at cellular levels, such as membrane degradation, generation of ROS, protein denaturation, and accumulation of toxic products, and so forth (Bowers, 1994; Yuanyuan et al., 2009). Plants also try to react to this stress by a change in gene expression, modification in membrane composition, production of compatible solutes, cold shock proteins, and antioxidant enzymes, which are thought to play a role in protection of cells from freezing injury. Particularly when plants are gradually exposed to cold stress, these changes at a cellular level can cause cold stress tolerance, the process known as "cold acclimation" (Fig. 1) (Levitt, 1980; Thomashow, 1999; Shabala, 2017).

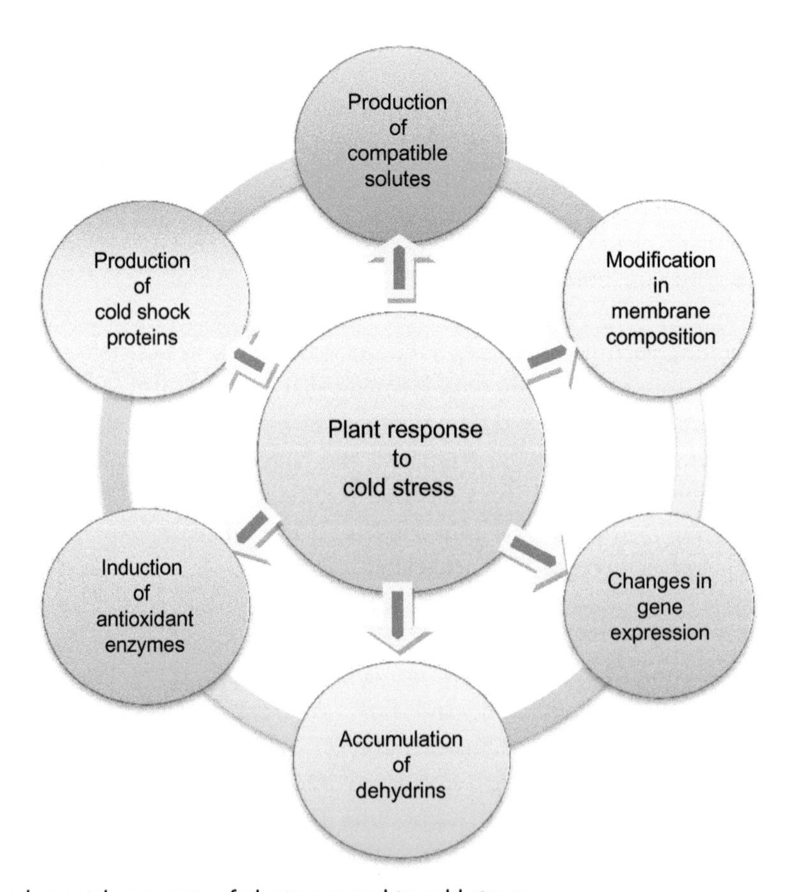

Fig. 1 Fundamental responses of plants exposed to cold stress.

4.1 Transcription Factors Involved in Cold Stress

When plants are exposed to low temperatures, they respond to cold stress with changes in gene expression. These genes encode proteins involved in cold tolerance. One of the more important genes encoded is *DREB1* (also called CBF), and DREB2 transcription factors. These transcription factors induce expression of the genes involved in tolerance to cold stress, as well as drought and salinity (Thakur and Nayyar, 2013). There are various regulatory pathways for creating cold tolerance in plants, and one of the most important of these pathways involves the expression of C-repeat binding factors (CBF1, CBF2, and CBF3), which is an important pathway for inducing cold tolerance in plants, which activates the expression of COR genes through binding to cis-elements in the promoters of COR genes (Thomashow, 1999; Chinnusamy et al., 2007). Ectopic overexpression of CBFs resulted in activation of COR genes and enhanced freezing tolerance of transgenic plants (Stockinger et al., 1997; Liu et al., 1998).

Another transcription factor, an inducer of CBF expression 1 (ICE1), is present upstream of CBF as a key regulator of cold-induced gene expression. ICE1 can bind to MYC-cis elements in the CBF3 promoter and activate the expression of CBF3. The expression of CBFF3, CBF2, and COR genes is increased with the constitutive expression of ICE1 (Chinnusamy et al., 2003, 2007), which then leads to increased cold tolerance. In fact, some research has shown that the defect in the expression of the ICE1gene leads to the nonexpression of the CBF3 gene under cold stress, which leads to hypersensitivity to chilling stress (Chinnusamy et al., 2007). Therefore, ICE1 is an important regulator of several cold responsive genes.

4.2 Modification in Membrane Composition

When plants are exposed to cold stress, they try to change the lipid composition of plasma membranes and chloroplast envelopes. These changes have been proposed to have a role in the acquisition of freezing tolerance during cold acclimation: they may prevent freeze-induced membrane damage by stabilizing the bilayer lamellar configuration (Uemura and Steponkus, 1997; Ruelland et al., 2009; Shabala, 2017). For example, cold exposure results in a decrease in the level of monogalactosyldiacylglycerol (MGDG) and an increase in the level of digalactosyldiacylglycerol (DGDG), both in the inner and outer envelopes of chloroplasts. Cold exposure also leads to a decrease in plastidic phosphatidylcholine, but only in the outer envelope. In *Solanum commersonii*, chilling exposure was correlated with an increase of phospholipids in the plasma membrane, primarily due to an increase in phosphatidylethanolamine. Such an increase in the ratio phosphatidylethanolamine to phosphatidylcholine has also been observed in spring and winter rye (Shabala, 2017).

In Arabidopsis, the proportion of diunsaturated species, such as 18:1/18:3, 18:2/18:2, and 18:2/18:3, increases in both phosphatidylcholine and phosphatidylethanolamine, and the proportion of monounsaturated species, such as 18:0/18:3 and 16:0/18:3, decreases. This seems to be a general effect of cold acclimation, observed in many other plant species (Ruelland et al., 2009; Shabala, 2017).

4.3 Production of Compatible Solutes

Carbohydrates, amino acids (proline, glycine, alanine, and serine), and polyamines are considered compatible solutions. Compatible solutions of low molecular weight molecules that are produced in large quantities during various stresses such as salinity, drought, cold, and so forth, so that the plant can withstand stress conditions.

In the case of cold stress, during the freezing period, with the initial formation of ice in the apoplastic space, the potential of water decreases, resulting in an exit of water from the cell into the extracellular compartment, causing in-cell dehydration (Shabala, 2017). To prevent dehydration of the cell, compatible solutes, such as carbohydrates, accumulate in the cell to reduce the difference in water potential between the apoplastic space and within the cell (Shabala, 2017).

In addition, the water associated with membranes is required to create the hydrophilic environment necessary to stabilize lipids in a bilayer. During freeze-induced dehydration, nonreducing sugars, such as sucrose or trehalose, can replace the lost water in creating this hydrophilic environment. Therefore, sugars may be active cryoprotectants toward membranes (Shabala, 2017).

4.4 Production of Cold Shock Proteins

Production of cold shock proteins (CSPs) was studied extensively in prokaryotes, especially in *Escherichia coli* (*E. coli*). Many studies showed that a low temperature in *E. coli* is accompanied by a spectacular accumulation of nucleic acid binding CSPs (Graumann and Marahiel, 1998; Yamanaka et al., 1998; Bae et al., 2000). CspA, the most prominent of the nine-member *E. coli* CSP family, accumulates up to 10% of total proteins during cold stress (Jiang et al., 1997). CspA has been hypothesized to prevent RNA secondary structure formation (Jiang et al., 1997), thereby enhancing translation at low temperatures (Karlson and Imai, 2003). These proteins have glycine-rich regions (Shabala, 2017).

Such proteins exist in plants and are induced by chilling temperatures. Two proteins in Arabidopsis (AtGRP2 and AtGRP2b), one protein in *Nicotiana tobacum* (NtGRP) (Kingsley and Palis, 1994) and WCSP1 in *Triticum aestivum* (Karlson et al., 2002), are found as CSPs in plants. In *Arabidopsis thaliana*, the transcript levels of *AtGRP2* increase markedly during cold stress. Root growth at 11°C of plants overexpressing GRP2 was higher than that of wild-type plants. GRP2 is capable of melting RNA secondary structures in vivo, and could be involved in RNA processing and/or translation ability under cold conditions (Kim et al., 2007; Shabala, 2017). Kim et al. (2009) demonstrated that

Arabidopsis cold shock domain protein 3 (AtCSP3), which shares a cold shock domain with bacterial CSPs, is involved in the acquisition of freezing tolerance in plants.

4.5 Roles of Dehydrins in Cold Stress Tolerance

One of the plants' responses to chilling stress is the accumulation of hydrophilic proteins predicted to form an amphipathic α-helix (Shabala, 2017). Many of the genes encoding these proteins were first characterized as being responsive to cold, drought, and abscisic acid (ABA). Therefore, many of them have been named COR (cold-responsive), LTI (low temperature-induced), RAB (responsive to abscisic acid), KIN (cold-induced), or ERD (early responsive to dehydration) (Shabala, 2017). They include dehydrins, which define group II of late embryogenesis abundant (LEA) proteins (Shabala, 2017). Dehydrins may have a role in freezing tolerance, possibly by preventing the membrane destabilization that occurs during the osmotic contraction associated with freezing (Shabala, 2017).

LEA proteins are accumulated during the late stages of seed development and are associated with the achievement of tolerance to drought stress (Liu et al., 2017), categorized into seven different groups (Battaglia et al., 2008). Dehydrins that belong to the group II LEA proteins, are considered as stress proteins involved in the formation of plants' protective reactions to dehydration (Liu et al., 2017). They can also be considered hydrophilins (Liu et al., 2017).

Although the role of dehydrins has not been fully identified, various studies have demonstrated their role in tolerance to cold stress. Hara et al. (2001) showed that the CuCOR19 protein accumulated only under cold stress, suggesting that cold is a crucial environmental cue for dehydrin production in the leaf of *Citrus unshiu*. They showed that CuCOR19 protected catalase and lactate dehydrogenase in the expression system of *E. coli* against freezing inactivation, and it was more effective than compatible solutes or BSA. Hara et al. (2003) overexpressed citrus (*Citrus unshiu* Marcov.) dehydrin in tobacco (*Nicotiana tabacum* L.) in response to chilling stress. Their results showed that dehydrin protein content was correlated with freezing tolerance in transgenics. In addition, based on their results, they suggest that dehydrin facilitates plant cold acclimation by acting as a radical-scavenging protein to protect membrane systems under cold stress. To elucidate the contribution of dehydrins (DHNs) to freezing stress tolerance in Arabidopsis, transgenic plants overexpressing multiple DHN genes were generated (Puhakainen et al., 2004). Transgenic plants exhibited improved survival when exposed to freezing stress compared with the control plants. Their results showed that dehydrins contribute to freezing stress tolerance in plants and suggests that this could be partly due to their protective effect on membranes. Overexpression of *ShDHN* (a cold-induced SK3-type dehydrin gene (*ShDHN*) isolated from wild tomato species *Solanum habrochaites*) in cultivated tomatoes increased tolerance to cold and drought stresses and improved seedling growth under salt and osmotic stresses (Liu et al., 2015). The overexpression of dehydrins *PmLEAs* (isolated from *Prunus mume*) in tobacco transgenic plants demonstrated that *PmLEAs* were involved in plant responses to cold and drought (Bao et al., 2017).

4.6 ROS Scavenging Systems

There are many reports that demonstrate the production of ROS under cold stress. To remove active oxygen species under normal and stress conditions, plants use diverse antioxidants, such as ascorbic acid and glutathione, and ROS scavenging enzymes such as superoxide dismutase (SOD), ascorbate peroxidase (APX), catalase (CAT), glutathione peroxidase (GPX), and peroxiredoxin (PrxR), thereby protecting potential cell injury and tissue dysfunction (Mittler et al., 2004; Maleki et al., 2017). Superoxide dismutase can catalyze the conversion of superoxide radicals into H_2O_2 and O_2 (Heldt, 2005). CAT, PrxR, and APX can eliminate hydrogen peroxide that has a damaging effect on many enzymes (Heldt, 2005; Steinberg, 2012).

The function of glutathione peroxidase is to reduce lipid hydroperoxides to their corresponding alcohols and to reduce free hydrogen peroxide to water, thereby protecting the membranes exposed to ROS (Steinberg, 2012).

Sofo et al. (2014) examined tolerance of wild almond species (*Prunus* spp.) to low temperatures due to their antioxidant defenses. They found the high ascorbate content and high ascorbate peroxidase activity in some almond species contributed to the decrease in H_2O_2. Generally, catalase activity increased after the cold treatments, whereas superoxide dismutase activity and $OH^{\bullet}$ levels varied markedly among the species.

The effects of cold stress were evaluated by measuring lipid peroxidation, membrane permeability, and some enzyme activities involved in the ROS scavenging system under acclimation and nonacclimation conditions in black chickpeas (*Cicer arietinum* L.), Kaka, a popular genotype planted, and accession 4322, as a landrace genotype (Nazari et al., 2012). Under nonacclimation conditions, the genotype 4322 prevented the H_2O_2 accumulation more efficiently, which led to a decrease in lipid peroxidation and membrane permeability compared with Kaka. Comparing the activities of other antioxidant enzymes, catalase was a more effective enzyme in cell protection against H_2O_2 in 4322 plants. Such a response in acclimated plants was more pronounced than in control and nonacclimated plants. The increase in guaiacol peroxidase and ascorbate peroxidase activities did not preserve cell membranes from oxidative damage in Kaka plants. It was observed that short-term acclimation can induce greater cold tolerance upon the increase of oxidative stress in chickpea plants. This was due to low levels of MDA and the electrolyte leakage index, indicating the lower lipid peroxidation and higher membrane stability under the cold stress compared with nonacclimated plants.

5. CONCLUSIONS

Currently, with the advent of new genomic study methods such as sequencing of the next generation and their bioinformatics analysis, many potential genes are identified, many of which still have not been investigated as far as their role in tolerance to cold stress. Also, information acquired from new methods that act on the transcriptome level, such as RNA-seq technology, provides very useful information about the genes involved in cold stress, to which particular attention must be paid. In addition, information obtained through the study of proteome using methods such as shotgun proteomics will also be very useful for understanding the mechanism of tolerance to cold stress. In general, the information available on the response of plants to cold stress is still limited, and requires more genomic, transcriptomics, and proteomics studies in the future, and the establishment of a logical relationship among them.

REFERENCES

Adam, S., Murthy, S.D.S., 2014. Effect of cold stress on photosynthesis of plants and possible protection mechanisms. In: Approaches to Plant Stress and their Management. Springer, India, pp. 219–226.

Allen, G.J., Chu, S.P., Harrington, C.L., Schumacher, K., Hoffmann, T., Tang, Y.Y., Grill, E., Schroeder, J.I., 2001. A defined range of guard cell calcium oscillation parameters encodes stomatal movements. Nature 411, 1053–1057.

Apel, K., Hirt, H., 2004. Reactive oxygen species: metabolism, oxidative stress, and signal transduction. Annu. Rev. Plant Biol. 55, 373–399.

Aslan, D., Ordu, B., Zencirci, N., 2016. Einkorn wheat (*Triticum monococcum* ssp. *monococcum*) tolerates cold stress better than bread wheat (*Triticum aestivum* L.) during germination. J. Field Crops Cent. Res. Inst. 25 (2), 182–192.

Bae, W., Xia, B., Inouye, M., Severinov, K., 2000. *Escherichia coli* CspA-family RNA chaperones are transcription antiterminators. Proc. Natl. Acad. Sci. U. S. A. 97 (14), 7784–7789.

Bao, F., Du, D., An, Y., Yang, W., Wang, J., Cheng, T., Zhang, Q., 2017. Overexpression of *Prunus mume* dehydrin genes in tobacco enhances tolerance to cold and drought. Front. Plant Sci. 8, 151. https://doi.org/10.3389/fpls.2017.00151.

Battaglia, M., Olvera-Carrillo, Y., Garciarrubio, A., Campos, F., Covarrubias, A.A., 2008. The enigmatic LEA proteins and other hydrophilins. Plant Physiol. 148, 6–24. https://doi.org/10.1104/pp.108.120725.

Bowers, M.C., 1994. Environmental effects of cold on plants. In: Wilkinson, R.E. (Ed.), Plant-Environment Interactions. Marcel Dekker, New York, pp. 391–411.

Chinnusamy, V., Zhu, J., Zhu, J.K., 2007. Cold stress regulation of gene expression in plants. Trends Plant Sci. 12 (10), 444–451.

Chinnusamy, V., et al., 2003. ICE1: a regulator of cold-induced transcriptome and freezing tolerance in Arabidopsis. Genes Dev. 17, 1043–1054.

Clarke, H.J., Siddique, K.H.M., 2004. Response of chickpea genotypes to low temperature stress during reproductive development. Field Crop Res. 90 (2), 323–334.

Clifton-Brown, J.C., Jones, M.B., 1997. The thermal response of leaf extension rate in genotypes of the C4 grass Miscanthus: an important factor in determining the potential productivity of different genotypes. J. Exp. Bot. 48, 1573–1581.

Clifton-Brown, J.C., Lewandowski, I., 2000. Overwintering problems of newly established Miscanthus plantations can be overcome by identifying genotypes with improved rhizome cold tolerance. New Phytol. 148, 287–294.

Davidson, J.F., Schiestl, R.H., 2001. Mitochondrial respiratory electron carriers are involved in oxidative stress during heat stress in *Saccharomyces cerevisiae*. Mol. Cell. Biol. 21, 8483–8489.

Dodd, A.N., Jakobsen, M.K., Baker, A.J., Telzerow, A., Hou, S.W., Laplaze, L., Barrot, L., Poethig, R.S., Haseloff, J., Webb, A.A.R., 2006. Time of day modulates low-temperature Ca^{2+} signals in Arabidopsis. Plant J. 48, 962–973.

Dodd, A.N., Kudla, J., Sanders, D., 2010. The language of calcium signaling. In: Merchant, S., Briggs, W.R., Ort, D. (Eds.), Annual Review of Plant Biology. In: Annual Reviews, vol. 61, pp. 593–620.

Einset, J., Winge, P., Bones, A., 2007. ROS signaling pathways in chilling stress. Plant Signal. Behav. 2 (5), 365–367.

Fitter, A.H., Hay, R.K.M., 1981. Environmental Physiology of Plants. Academic Press, New York.

Friesen, P.C., Peixoto, M.M., Busch, F.A., Johnson, D.C., Sage, R.F., 2014. Chilling and frost tolerance in Miscanthus and Saccharum genotypes bred for cool temperate climates. J. Exp. Bot. 65, 3749–3758.

Friesen, P.C., Sage, R.F., 2016. Photosynthetic responses to chilling in a chilling-tolerant and chilling-sensitive Miscanthus hybrid. Plant Cell Environ. 39 (7), 1420–1431.

Gilmour, S.J., Sebolt, A.M., Salazar, M.P., Everard, J.D., Thomashow, M.F., 2000. Overexpression of the Arabidopsis *CBF3* transcriptional activator mimics multiple biochemical changes associated with cold acclimation. Plant Physiol. 124 (4), 1854–1865.

Głowacka, K., Adhikari, S., Peng, J., Gifford, J., Juvik, J.A., Long, S.P., Sacks, E.J., 2014. Variation in chilling tolerance for photosynthesis and leaf extension growth among genotypes related to the C4 grass Miscanthus × giganteus. J. Exp. Bot. 65, 5267–5278.

Graumann, P.L., Marahiel, M.A., 1998. A superfamily of proteins that contain the cold-shock domain. Trends Biochem. Sci. 23 (8), 286–290.

Hara, M., Terashima, S., Fukaya, T., Kuboi, T., 2003. Enhancement of cold tolerance and inhibition of lipid peroxidation by citrus dehydrin in transgenic tobacco. Planta 217 (2), 290–298.

Hara, M., Terashima, S., Kuboi, T., 2001. Characterization and cryoprotective activity of cold-responsive dehydrin from *Citrus unshiu*. J. Plant Physiol. 158 (10), 1333–1339.

Heidarvand, L., Amiri, R.M., 2010. What happens in plant molecular responses to cold stress? Acta Physiol. Plant. 32 (3), 419–431.

Heldt, H.W.H., 2005. Plant Biochemistry/Hans-Walter Heldt in Cooperation with FionalHeldt. Elsevier Academic Press.

Henriksson, K.N., Trewavas, A.J., 2003. The effect of short-term low-temperature treatments on gene expression in *Arabidopsis* correlates with changes in intracellular Ca^{2+} levels. Plant Cell Environ. 26, 485–496.

Jan, N., Andrabi, K.I., 2009. Cold resistance in plants: a mystery unresolved. Electron. J. Biotechnol. 12 (3), 14–15.

Jiang, W., Hou, Y., Inouye, M., 1997. CspA, the major cold-shock protein of *Escherichia coli*, is an RNA chaperone. J. Biol. Chem. 272 (1), 196–202.

Jørgensen, U., Muhs, H.J., 2001. Miscanthus breeding and improvement. In: Jones, M.B., Walsh, M. (Eds.), Miscanthus for Energy and Fibre. James & James, London, pp. 68–85.

Karlson, D., Imai, R., 2003. Conservation of the cold shock domain protein family in plants. Plant Physiol. 131 (1), 12–15.

Karlson, D., Nakaminami, K., Toyomasu, T., Imai, R., 2002. A cold-regulated nucleic acid-binding protein of winter wheat shares a domain with bacterial cold shock proteins. J. Biol. Chem. 277 (38), 35248–35256.

Khodabandeh, N., 2003. Cereals, seventh ed. Tehran University Press, Tehran, pp. 78–111.

Kim, J.Y., Park, S.J., Jang, B., Jung, C.H., Ahn, S.J., Goh, C.H., Cho, K., Han, O., Kang, H., 2007. Functional characterization of a glycine-rich RNA-binding protein 2 in *Arabidopsis thaliana* under abiotic stress conditions. Plant J. 50, 439–451.

Kim, M.H., Sasaki, K., Imai, R., 2009. Cold shock domain protein 3 regulates freezing tolerance in *Arabidopsis thaliana*. J. Biol. Chem. 284 (35), 23454–23460.

Kingsley, P.D., Palis, J., 1994. GRP2 proteins contain both CCHC zinc fingers and a cold shock domain. Plant Cell 6 (11), 1522.

Klimecka, M., Muszynska, G., 2007. Structure and functions of plant calcium-dependent protein kinases. Acta Biochim. Pol. 54, 219–233.
Knight, H., Knight, M.R., 2001. Abiotic stress signalling pathways: specificity and cross-talk. Trends Plant Sci. 6, 262–267.
Knight, H., Trewavas, A.J., Knight, M.R., 1996. Cold calcium signaling in Arabidopsis involves two cellular pools and a change in calcium signature after cold acclimation. Plant Cell 8, 489–503.
Knight, M.R., Campbell, A.K., Smith, S.M., Trewavas, A.J., 1991. Transgenic plant aequorin reports the effects of touch and cold-shock and elicitors on cytoplasmic calcium. Nature 352, 524–526.
Levitt, J., 1980. Responses of Plants to Environmental Stresses. Vol. 1. Chilling, Freezing and High Temperatures Stresses. Academic Press, New York. 426 p.
Liu, H., Yu, C., Li, H., Ouyang, B., Wang, T., Zhang, J., Wang, X., Ye, Z., 2015. Overexpression of ShDHN, a dehydrin gene from *Solanum habrochaites* enhances tolerance to multiple abiotic stresses in tomato. Plant Sci. 231, 198–211.
Liu, Q., Kasuga, M., Sakuma, Y., Abe, H., Miura, S., Yamaguchi-Shinozaki, K., Shinozaki, K., 1998. Two transcription factors, DREB1 and DREB2, with an EREBP/AP2 DNA binding domain, separate two cellular signal transduction pathways in drought- and low temperature responsive gene expression, respectively, in Arabidopsis. Plant Cell 10, 1391–1406.
Liu, Y., Song, Q., Li, D., Yang, X., Li, D., 2017. Multifunctional roles of plant dehydrins in response to environmental stresses. Front. Plant Sci. 8, 1018.
Long, S.P., Spence, A.K., 2013. Toward cool C4 crops. Annu. Rev. Plant Biol. 64, 701–722.
Love, J., Dodd, A.N., Webb, A.A.R., 2004. Circadian and diurnal calcium oscillations encode photoperiodic information in Arabidopsis. Plant Cell 16, 956–966.
Lukatkin, A.S., Brazaityte, A., Bobinas, C., Duchovskis, P., 2012. Chilling injury in chilling-sensitive plants: a review. Agriculture 99 (2), 111–124.
Maleki, M., Ghorbanpour, M., Kariman, K., 2017. Physiological and antioxidative responses of medicinal plants exposed to heavy metals stress. Plant Gene, 11, pp. 247–254.
Marcellin, P., 1992. In: Côme, D. (Ed.), Les vegetaux et le froid. Hermann, Paris, pp. 53–105.
McAinsh, M.R., Pittman, J.K., 2009. Shaping the calcium signature. New Phytol. 181, 275–294.
Mittler, R., Vanderauwera, S., Gollery, M., Van-Breusegem, F., 2004. Reactive oxygen gene network of plants. Trends Plant Sci. 9, 490–498.
Miwa, H., Sun, J., Oldroyd, G.E., Downie, J.A., 2006. Analysis of calcium spiking using a cameleon calcium sensor reveals that nodulation gene expression is regulated by calcium spike number and the developmental status of the cell. Plant J. 48, 883–894.
Monroy, A.F., Castonguay, Y., Laberge, S., Sarhan, F., Vezina, L.P., Dhindsa, R.S., 1993. A new cold-induced alfalfa gene is associated with enhanced hardening at subzero temperature. Plant Physiol. 102, 873–879.
Monroy, A.F., Dhindsa, R.S., 1995. Low-temperature signal transduction: induction of cold acclimation-specific genes of alfalfa by calcium at 25 degrees C. Plant Cell 7, 321–331.
Nazari, M., Amiri, R.M., Mehraban, F.H., Khaneghah, H.Z., 2012. Change in antioxidant responses against oxidative damage in black chickpea following cold acclimation. Russ. J. Plant Physiol. 59 (2), 183–189.
Paredes, M., Quiles, M.J., 2015. The effects of cold stress on photosynthesis in Hibiscus plants. PLoS One. 10 (9), e0137472. https://doi.org/10.1371/journal.pone.0137472.
Pearce, R.S., 2001. Plant freezing and damage. Ann. Bot. 87, 417–424.
Plieth, C., Hansen, U.P., Knight, H., Knight, M.R., 1999. Temperature sensing by plants: the primary characteristics of signal perception and calcium response. Plant J. 18, 491–497.
Ploschuk, E.L., Bado, L.A., Salinas, M., Wassner, D.F., Windauer, L.B., Insausti, P., 2014. Photosynthesis and fluorescence responses of *Jatropha curcas* to chilling and freezing stress during early vegetative stages. Environ. Exp. Bot. 102, 18–26.
Puhakainen, T., Hess, M.W., Makela, P., Svensson, J., Heino, P., Palva, E.T., 2004. Overexpression of multiple dehydrin genes enhances tolerance to freezing stress in Arabidopsis. Plant Mol. Biol. 54 (5), 743–753.

Purdy, S.J., Maddison, A.L., Jones, L.E., Webster, R.J., Andralojc, J., Donnison, I., Clifton-Brown, J., 2013. Characterization of chilling-shock responses in four genotypes of Miscanthus reveals the superior tolerance of M. × giganteus compared with *M. sinensis* and *M. sacchariflorus*. Ann. Bot. 111, 999–1013.

Qin, Y.Z., Chen, J., Xing, Z., He, C.Z., Xiong, X.Y., 2013. Effects of low temperature stress on photosynthesis in potato leaves. J. Hunan Agric. Univ. (Nat. Sci.) 1, 7.

Raison, J.K., Lyons, J.M., 1986. Chilling injury: a plea for uniform terminology. Plant Cell Environ. 9, 685–686.

Reddy, V.S., Reddy, A.S., 2004. Proteomics of calciumsignaling components in plants. Phytochemistry 65, 1745–1776.

Ruelland, E., Vaultier, M.N., Zachowski, A., Hurry, V., 2009. Cold signalling and cold acclimation in plants. Adv. Bot. Res. 49, 35–150.

Sanders, D., Pelloux, J., Brownlee, C., Harper, J.F., 2002. Calcium at the crossroads of signaling. Plant Cell 14, S401–S417.

Shabala, S. (Ed.), 2017. Plant Stress Physiology. CABI, Boston, MA, p. 376.

Shinozaki, K., Yamaguchi-Shinozaki, K., 1996. Molecular response to drought and cold stress. Curr. Opin. Biotechnol. 7 (2), 161–167.

Short, E.F., North, K.A., Roberts, M.R., Hetherington, A.M., Shirras, A.D., McAinsh, M.R., 2012. A stress-specific calcium signature regulating an ozone-responsive gene expression network in Arabidopsis. Plant J. 71, 948–961.

Sofo, A., Rajabpoor, S., Yaghini, H., Shirani, M., Archangi, A., Sangi, S.E., Tavakoli, F., Khodai, M., Sorkheh, K., 2014. Cold-induced changes in antioxidant defenses and reactive oxygen species in eight wild almond species. Free Radicals Antioxid. 4 (1), 70.

Solanke, A.U., Sharma, A.K., 2008. Signal transduction during cold stress in plants. Physiol. Mol. Biol. Plants 14 (1–2), 69–79.

Srinivasan, A., Saxena, N.P., Johansen, C., 1999. Cold tolerance during early reproductive growth of chickpea (*Cicer arietinum* L.): genetic variation in gamete development and function. Field Crop Res. 60 (3), 209–222.

Steinberg, C.E., 2012. Stress Ecology: Environmental Stress as Ecological Driving Force and Key Player in Evolution. Springer Science & Business Media, Dordrecht, p. 480.

Stockinger, E.J., Gilmour, S.J., Thomashow, M.F., 1997. *Arabidopsis thaliana* CBF1 encodes an AP2 domain-containing transcription activator that binds to the C repeat/DRE, a cis-acting DNA regulatory element that stimulates transcription in response to low temperature and water deficit. Proc. Natl. Acad. Sci. U. S. A. 94, 1035–1040.

Suzuki, N., Mittler, R., 2006. Reactive oxygen species and temperature stresses: a delicate balance between signaling and destruction. Physiol. Plant. 126 (1), 45–51.

Thakur, P., Kumar, S., Malik, J.A., Berger, J.D., Nayyar, H., 2010. Cold stress effects on reproductive development in grain crops: an overview. Environ. Exp. Bot. 67 (3), 429–443.

Thakur, P., Nayyar, H., 2013. Facing the cold stress by plants in the changing environment: sensing, signaling, and defending mechanisms. In: Plant Acclimation to Environmental Stress. Springer, New York, pp. 29–69.

Thomashow, M.F., 1998. Role of cold-responsive genes in plant freezing tolerance. Plant Physiol. 118 (1), 1–8.

Thomashow, M.F., 1999. Plant cold acclimation: freezing tolerance genes and regulatory mechanisms. Ann. Rev. Plant Physiol. Plant Mol. Biol. 50, 571–599.

Uemura, M., Steponkus, P.L., 1997. Effect of cold acclimation on the lipid composition of the inner and outer membrane of the chloroplast envelope isolated from rye leaves. Plant Physiol. 114, 1493–1500.

Whalley, H.J., Knight, M.R., 2013. Calcium signatures are decoded by plants to give specific gene responses. New Phytol. 197 (3), 690–693.

Yadav, A.S.K., 2010. Cold stress tolerance mechanisms in plants. A review. Agron. Sustain. Dev. 30 (3), 515–527.

Yamanaka, K., Fang, L., Inouye, M., 1998. The CspA family in *Escherichia coli*: multiple gene duplication for stress adaptation. Mol. Microbiol. 27 (2), 247–255.

Yuanyuan, M., Yali, Z., Jiang, L., Hongbo, S., 2009. Roles of plant soluble sugars and their responses to plant cold stress. Afr. J. Biotechnol. 8 (10), 2004–2010.

FURTHER READING

Farooq, M., Aziz, T., Wahid, A., Lee, D.J., Siddique, K.H., 2009. Chilling tolerance in maize: agronomic and physiological approaches. Crop Pasture Sci. 60 (6), 501–516.
Kim, J.C., Lee, S.H., Cheong, Y.H., Yoo, C.M., Lee, S.I., Chun, H.J., Yun, D.J., Hong, J.C., Lee, S.Y., Lim, C.O., Cho, M.J., 2001. A novel cold-inducible zinc finger protein from soybean, SCOF-1, enhances cold tolerance in transgenic plants. Plant J. 25 (3), 247–259.

CHAPTER 5

Impact of Soil Moisture Regimes on Wilt Disease in Tomatoes: Current Understanding

Aarti Gupta[a], Dharanipathi Kamalachandran[a], Bendangchuchang Longchar, Muthappa Senthil-Kumar
National Institute of Plant Genome Research, New Delhi, India

Contents

1. INTRODUCTION

Ralstonia solanacearum is a causal agent of vascular wilt disease in more than 200 crop species, including the tomato. *R. solanacearum* is a strict soil-borne pathogen and thrives in moist soils (Van der Wolf et al., 1998). The bacterium can live for years in an infected field, and has been reported to persist for 12 months in potato fields (van Elsas et al., 2000). The sources of inoculum for agricultural fields are irrigation and surface water, weeds, infested soil, latently infected propagative plant material, and contaminated farm tools and equipment. The bacteria exhibit subterranean movement and spread from the infected plants' roots to the healthy ones (Hayward, 1991). *R. solanacearum*-caused wilt in tomato amounts to a 35%–90% yield loss under high temperatures and high moisture conditions (Singh et al., 2015). *R. solanacearum* colonizes the nutrient-poor xylem vessels, which are characterized by dead tracheary elements that have a relatively low osmotic pressure, which makes the pathogen penetration easy (Yadeta and Thomma, 2013). Vasse et al. (1995) observed that in tomatoes, the bacteria are attracted to the root wounds through an unknown mechanism, and stick to the epidermal cells' surface. The bacteria

[a] These authors contributed equally.

Biochemical, Physiological and Molecular Avenues for Combating Abiotic Stress in Plants
https://doi.org/10.1016/B978-0-12-813066-7.00005-X

then colonize the intercellular spaces in the root epidermis, followed by successive invasion of the intercellular spaces of the inner cortex, forming biofilm structures and residing there as intercellular micro-colonies (Vasse et al., 1995; Mori et al., 2016). Through the root endodermis, the *R. solanacearum* in tomatoes subsequently colonizes the vascular parenchyma. Xylem-dwelling *R. solanacearum* releases a virulence factor, extracellular polysaccharides (EPS), that increase the viscosity of the xylem fluid, block the xylem vessels, and hinder the water transport; making the plant succumb to bacterial wilt (McGarvey et al., 1999). Interestingly, under field conditions, the *R. solanacearum* incidence was drastically reduced with the decrease in soil moisture content (van Elsas et al., 2000; www.data.gov.in; Mondal et al., 2014), thus highlighting the importance of soil and plant water levels in determining the disease severity and spread. However, the interaction between the two stressors and the impact of their concurrence on the plant, specifically in tomatoes, has not been studied. In this chapter, we have reviewed the potential interaction between the *R. solanacearum* and different soil moisture regimes, and the consequent impact of the combined stress on the plant. We also present different instances of the interaction between low soil moisture and other major xylem-dwelling pathogens, such as *Xanthomonas campestris* (causes black rot and bacterial wilt), *Xylella fastidiosa* (causes bacterial leaf scorch and Pierce's disease), and *Tomato spotted wilt virus* (causes tomato spotted wilt). We have also reviewed the anatomical and physiological changes that are adopted by the tomato plants that are resistant to drought or wilt.

2. STRESS INTERACTION: AT THE JUNCTURE OF THE RHIZOSPHERE AND THE ROOTS

The soil moisture influences and interacts with the wilt pathogen in different ways. It (i) manipulates the survival of the *R. solanacearum* in soil and regulating its inoculum density, (ii) alters the pathogen infection, and (iii) modulates the pathogen progression after infection. The first two points of interaction occur in soil and at the plant's root surface in the rhizospere. The interaction described in the third point occurs *in planta* and is dependent on the soil moisture level.

2.1 Impact of Soil Moisture on Pathogen Multiplication

In a study conducted in potato fields, the *R. solanacearum* inoculum was negatively impacted by severe drought (van Elsas et al., 2000). It caused severe wilt under high moisture in the early part of the potato growing season. The coincidence of average rainfall and *R. solanacearum*-caused disease prevalence data from West Bengal, India, further supports the earlier study and reflected a negative correlation between the two (Fig. 1; Mondal et al., 2014). However, other wilt-causing pathogens, for example, *Xanthomonas campestris* pv. *musacearum* (*Xcm*, causes wilt in bananas) caused severe disease in banana plants maintained under low soil moisture as compared with normal soil moisture levels

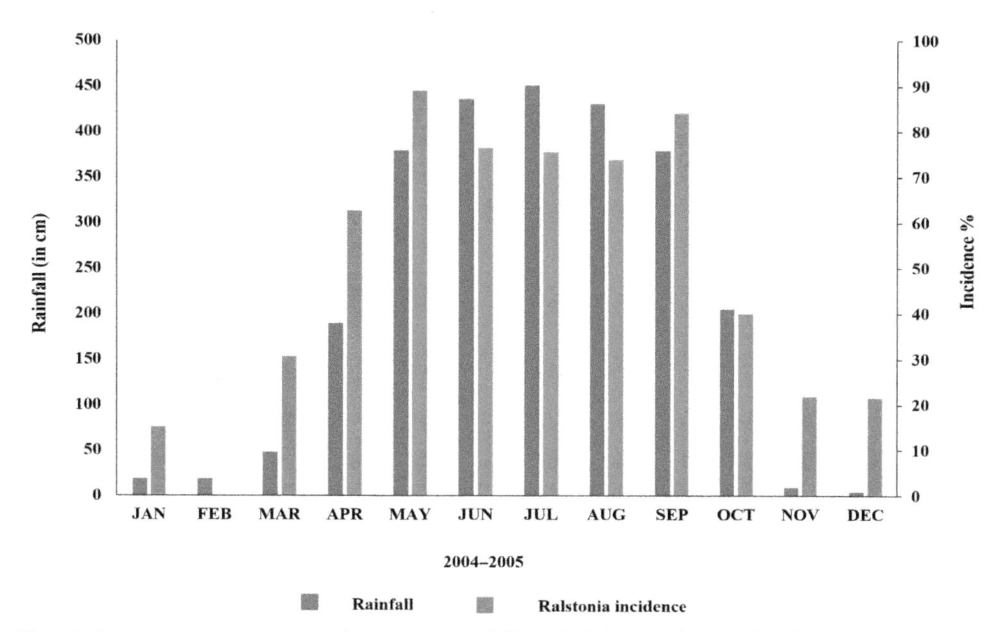

Fig. 1 Co-occurrence pattern of average rainfall and *Ralstonia* bacterial wilt incidence during 2004–2005 in West Bengal: figure displays the relationship between monthly precipitation (www.data.gov.in/) and *Ralstonia solanacearum* wilt incidence in tomatoes during 2004–2005 in West Bengal (Mondal et al., 2014). During periods of low rainfall (low soil moisture conditions), there is reduced *R. solanacearum* wilt incidence when compared with high rainfall (high soil moisture conditions) periods.

(Ochola et al., 2015). Drought stress aggravated the disease severity and advancement of *X. fastidiosa* in *Parthenocissus quinquefolia* vine (McElrone et al., 2001). Under low soil moisture conditions, *Streptomyces scabies* (causes common scabs in potatoes) multiplied to high numbers in the rhizosphere, which increased the chances for disease in potato plants (Goto, 1985). In the presented reports, soil moisture deficit effects were significant for the disease incidence and severity, whereas the resultant stress interaction between low soil moisture levels and the rhizospheric wilt pathogens varies depending upon the nature of the pathogen and the plant species.

2.2 Impact of Soil Moisture on Pathogen Infection

Despite the low soil moisture-inflicted reduction in the pathogen inoculum in the rhizosphere, the residual amount of inoculum potentially causes infection in a drought-stressed plant. *R. solanacearum* enters plant roots through small entry points, such as those generated during lateral root emergence (Vasse et al., 1995). During drought stress, the plant tends to increase the root biomass (to absorb maximum available water) by developing lateral roots. This, as such, provides an opportunity for a proximal pathogen to infect and make the plant more prone to *R. solanacearum* infection. Moreover, the concentration, composition, and the diffusion of host plant root exudates (involved in chemotaxis) may be altered under low soil moisture conditions such as to attract the bacterium to plant roots and cause infection. Thus, the low soil moisture provides the bacteria with an opportunity to infect. Moreover, the resulting stress weakens the defense systems of the tomato, allowing *R. solanacearum* to proliferate *in planta* and cause disease.

2.3 Drought and Wilt: Interactions at the Plant Interface

By influencing the host plant anatomy, the low soil moisture level can restrict the pathogen movement and the resultant disease symptoms. Under drought stress, the tomato plants develop reduced vessel size, diameter, and the number of pits leading to a compact xylem. *R. solanacearum* moves from vessel to vessel, and during its course of movement, it degenerates pit membranes (Nakaho et al., 2000). The anatomy of the xylem under such situations is one of the deciding factors of plant resistance. In a wilt-resistant tomato cultivar, the thickened pit membranes halted pathogen movement in xylem vessels, which was not observed in cultivars susceptible to *R. solanacearum* (Nakaho et al., 2000; Kim et al., 2016). Unlike the resistant tomato cultivar, a large number of the *R. solanacearum* population was observed in the vessels of both the primary and secondary xylem of a wilting-susceptible tomato cultivar. The resistant trait (to vascular wilt) was therefore associated with the thickened pits of the plant that restrained *R. solanacearum* colonization in the stem, rather than by resistance during entry through the roots (Grimault and Prior, 1993). Thus the reduced pits under low soil moisture imply reduced movement, and spread of the pathogen across the plant.

The drought stress leads to the stomatal closure, which in turn reduces the transpiration pull. Transpiration pull in the xylem vessels is one of the possible means of longitudinal movement of pathogens *in planta*. This is akin to the instance where the longitudinal spread of *R. solanacearum* in the aerial parts is inhibited in resistant cultivars (to vascular wilt) than in susceptible tomato cultivars (Grimault et al., 1994; Prior et al., 1997). Under a scenario where the low water availability restricts *R. solanacearum* movement across plant tissues, drought confers plant tolerance to wilt pathogens by restricting the systemic spread of the bacterium.

Reportedly, a low soil moisture level has been implicated in reducing the *in planta* multiplication and spread of other wilt-causing pathogens. It has been shown that the banana wilt (caused by *Xcm*) development was hastened by continued water stress, as compared with instances in which the plants were maintained under control conditions or were stressed only before inoculation. The observations in these plants suggested that both timing of infection (before or after water stress), and duration of exposure to water stress (continued after infection) are important in determining the multiplication and spread of *Xcm* in the vascular tissues of the banana (Ochola et al., 2015). Drought stress ameliorated the viral wilt disease symptoms in tomato plants by limiting the *in planta* movement of *Tomato spotted wilt virus* (Córdoba et al., 1991).

The xylem sap is a nutrient-poor medium with extremely low levels of organic and inorganic compounds (Siebrecht et al., 2003). Reportedly, the composition of xylem sap is drastically influenced by drought stress, where the nitrate, ammonium, potassium, and phosphate concentrations are reduced by about 50% (Bahrun et al., 2002; Jia and Davies, 2007). Importantly, the *R. solanacearum* wilt-resistant tomato exhibited reduced nitrate levels in comparison with the wilt-susceptible ones (Hacisalihoglu et al., 2008). This suggests an inherent dependency of the pathogen on the inorganic compounds for its pathogenicity, and may explain the negative effect of drought stress on pathogen infection.

3. PHYSIOLOGICAL CHANGES DURING STRESS INTERACTION

Based on the positive or negative interaction between the two stresses, the plant can exhibit a response additive (aggravated response), or canceled (reduced effect). In a Plant—*R. solanacearum*—drought interaction, wilting can be considered as one of the parameters to assess the net impact of the stress interaction. When wilt caused by vascular-limited pathogens occurs in drought-stressed plants, the damage may be much greater than in plants in a high soil moisture situation. Mostly, due to the xylem occlusion in the wilt infection, the plant is predisposed to drought stress (Yadeta and Thomma, 2013). Additionally, tyloses that are formed by the parenchyma cells in response to wilt pathogens, can further contribute to xylem occlusion (Fradin and Thomma, 2006; Klosterman et al., 2009; Beattie, 2011). Plants develop tyloses in the xylem vessels to inhibit pathogen movement and spread to adjoining vessels (Wallis and Truter, 1978; Rahman et al., 1999). These tyloses also block vessels and thereby inhibit water transport. In wilt-resistant tomato cultivars, tyloses blocked the colonized vessels, and although they restricted the bacterial spread, they predisposed the plant to water deficit. On the other hand in *R. solanacearum*-susceptible tomato cultivars, no tyloses were observed in the infected vessels, which failed to restrict the bacterial spread, and eventually the plant succumbed to wilt. Thus, in a wilt-resistant plant, tylose formation can make a plant more prone to drought stress (Fradin and Thomma, 2006).

Other vascular wilt pathogens also inflict dehydration in host plants. As a result of the dehydration response, the plant closes its stomata, which leads to reduced photosynthesis and decreased partitioning of photo-assimilates to the roots. The eventual reduced root growth in infected plants further predisposes a plant to low soil moisture stress (Yadeta and Thomma, 2013). *X. fastidiosa* causes an infection that prompts drought stress in alfalfa plants (Daugherty et al., 2010). Tomato plants infected with *Verticillium dahliae* (which causes Verticillium wilt) exhibited decreased leaf water potential (Ayres, 1978). The reduced shoot hydraulic conductance caused by *X. fastidiosa* infection acts additively when imposed upon by drought stress (McElrone et al., 2003). *X. fastidiosa*-infected grape plants displayed the early occurrence of embolism (the formation of air bubbles), which correlated with decreased xylem conductivity and aggravated drought stress (Pérez-Donoso et al., 2007).

These examples explain that the xylem inhabitation by wilt pathogens exposes the plant to more water-limiting conditions, such that plants become susceptible to drought stress. However, in a contrasting instance, the *Arabidopsis thaliana* infected with the xylem occluding pathogen *Verticillium longisporum* (the causal agent of vascular wilt) was tolerant to drought stress. *V. longispourum* induced the expression of the vascular-related NAC domain (*VND7*) gene in host plants and activated de novo xylem formation, which led to enhanced water storage or conductance capacity under drought stress conditions (Reusche et al., 2012).

4. CONCLUSIONS AND FUTURE PERSPECTIVES

The emergence of water limitations and increased wilt disease pressure are of serious concern to agriculture in a changing climate. In some cases of stress interaction, the plant responses under combined stressors are canceled, and some of the responses during stress interaction are neutralized by one of the dominant stressors. In the case of drought and wilt pathogens, plants display a negative interaction where the *R. solanacearum* wilt is favored by high soil moisture, and the incidence of wilt under low soil moisture is reportedly reduced. As one of the measures to control *R. solanacearum* infection in fields, the use of well drained and leveled fields is recommended. The low moisture content in the soil reduces bacterial rhizospheric inoculum, but induces lateral root development, which in turn increases the potential entry points for *R. solanacearum*. At the same time, drought stress-inflicted structural alterations, such as reduced cortical cell size, hinder *R. solanacearum* entry in the host plant. The host plant exhibits inhibited transpiration pull under low soil moisture regimes, which restricts the longitudinal movement of *R. solanacearum* in the plant. Evidence shows that both drought stress and *R. solanacearum* infection reduce the xylem vessels (Fig. 2). The production of EPS, a key virulence factor of *R. solanacearum*, is enhanced under drought conditions. Additionally, the induced tylose production in response to pathogen defense will further clog the xylem vessel and make the host plant more prone to drought stress. The low soil moisture may also intensify the wilt symptoms caused by the pathogen in drought-stressed plants. The wilt caused by *R. solanacearum* involves irreversible damage to the vasculature, but drought recovery is common. The timing and concentration of each stress, in this case, would be the determining factor in shaping the net impact in the tomato plant, which divergently affects plant fitness under combined drought stress and *R. solanacearum* infection. Under low soil moisture conditions, tomato plants reduce the vessel's size and increase thickening of pit walls, as compared with the ones under high soil moisture conditions. The more compact vasculature under low soil moisture conditions potentially restricts the *R. solanacearum* movement. Thus, as a means of predicting the outcome of plant responses to combined low soil moisture and *R. solanacearum* stress, one would need to focus on the common anatomical and physiological features modulated by these two stressors during their individual occurrences.

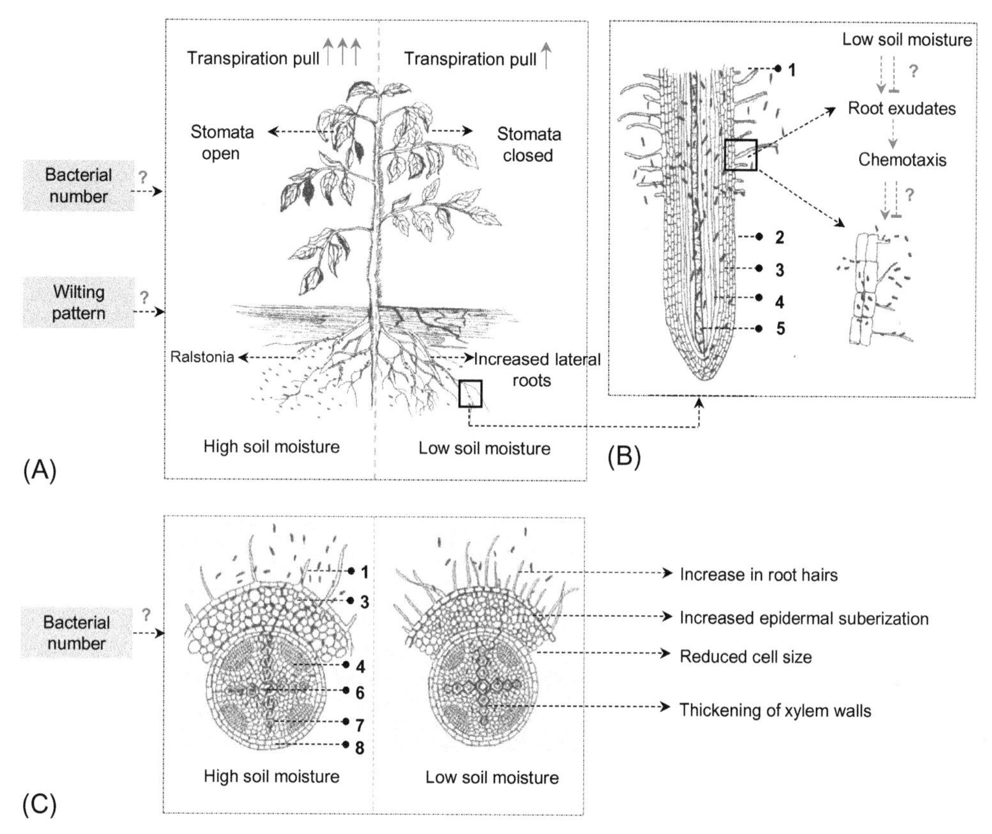

Fig. 2 Hypothetical model on effect of soil moisture on *Ralstonia solanacearum* infection in the tomato: (A) an illustration of the morpho-physiological responses of tomatoes to *R. Solanocearum* (Rs) at the above-ground and below-ground levels under high soil moisture (left side) and low soil moisture (right side) conditions. The low moisture content in the soil reduces the bacterial (Rs) population, but induces enhanced lateral root development in plants, which may increase the chances of Rs infection (through the emerging lateral root hairs). Moreover, the plants tend to close their stomata under low soil moisture, which may inhibit the transpiration pull, and hence the movement of Rs in an aerial part of the plant. (B) shows critical steps involved in Rs—root—rhizosphere interaction with a specific focus on the role of tomato root exudates and the impact of soil moisture on it. Under low soil moisture, the concentration and composition of the root exudates may change, which may regulate chemotaxis. (C) compares a transverse section of the root displaying the potential variations under different soil moisture regimes impacting Rs infection in the tomato. A plant growing under low moisture conditions has been shown to display increased root hair development, epidermal suberization, and reduced cell size, all of these posing hindrance to the Rs entry.

Continued

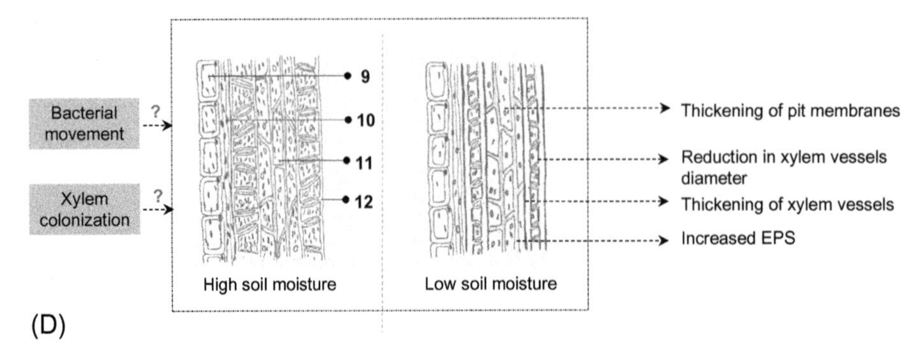

Fig. 2, cont'd (D) shows changes in the xylem tissue due to soil moisture variation, impacting Rs colonization and movement. Under low soil moisture conditions, tomato plants tend to reduce the vessel size, and increased thickening of pit walls as compared with those under high soil moisture conditions. The more compact vasculature under low soil moisture conditions potentially restricts the Rs movement. As a means of virulence, Rs tend to release exopolysaccharides (EPS) and clog the xylem vessels. Evidence from other related species has shown enhanced EPS production under drought stress. At present, the conflicting variations in the plant and the pathogen interactions under low soil moisture makes it difficult to draw any inference in the intensity of infection under low soil moisture as compared with high soil moisture. Key: 1, root hair; 2, epidermis; 3, cortex; 4, phloem; 5, xylem; 6, meta-xylem; 7, proto-xylem; 8, endodermis; 9, xylem parenchyma; 10, fibers; 11, tracheids; 12, xylem vessels. *Red upward arrows*: the intensity and direction of transpiration pull; *blunt arrows*: inhibition; *red downward arrows*: positive regulation of the process. The boxes with a question mark represent the scenarios that need to be unraveled upon study of actual combined stressors. *Rs*, *Ralstonia solanacearum*; *EPS*, exopolysaccharide.

ACKNOWLEDGMENT

MS-K Lab is supported by the DBT-innovative young biotechnologist award (BT/09/IYBA/2015/07). AG acknowledges the SERB-national post-doctoral fellowship (N-PDF/2015/000116) and BL acknowledges the DBT-RA program. The authors are thankful to DBT-eLibrary Consortium (DeLCON) for providing access to e-resources. Authors acknowledge Ms. Priyanka Ambavane for the illustrations used in Figure 2.

REFERENCES

Ayres, P.G., 1978. Water relations of diseased plants. In: Kozlowski, T.T. (Ed.), Water Deficits and Plant Growth. Academic Press, London, pp. 1–60.

Bahrun, A., Jensen, C.R., Asch, F., Mogensen, V.O., 2002. Drought-induced changes in xylem pH, ionic composition, and ABA concentration act as early signals in field-grown maize (*Zea mays* L.). J. Exp. Bot. 53, 251–263.

Beattie, G., 2011. Water relations in the interaction of foliar bacterial pathogens with plants. Annu. Rev. Phytopathol. 49, 533–555.

Córdoba, A.R., Taleisnik, E., Brunotto, M., Racca, R., 1991. Mitigation of tomato spotted wilt virus infection and symptom expression by water stress. J. Phytopathol. 133, 255–263.

Daugherty, M., Lopes, J.S., Almeida, R.P., 2010. Strain-specific alfalfa water stress induced by *Xylella fastidiosa*. Eur. J. Plant Pathol. 127, 333–340.

Fradin, E., Thomma, B., 2006. Physiology and molecular aspects of *Verticillium* wilt diseases caused by *V. dahliae* and *V. albo-atrum*. Mol. Plant Pathol. 7, 71–86.

Goto, K., 1985. The relative importance of precipitation and sugar content in potato peel for the detection of the incidence of common scab (*Streptomyces scabies*). Soil Sci. Plant Nutr. 31, 419–425.

Grimault, V., Prior, P., 1993. Bacterial wilt resistance in tomato associated with tolerance of vascular tissues to *Pseudomonas solanacearum*. Plant Pathol. 42, 589–594.

Grimault, V., Gélie, B., Lemattre, M., Prior, P., Schmit, J., 1994. Comparative histology of resistant and susceptible tomato cultivars infected by *Pseudomonas solanacearum*. Physiol. Mol. Plant Pathol. 44, 105–123.

Hacisalihoglu, G., Ji, P., Olson, S.M., Momol, M.T., 2008. Effect of *Ralstonia solanacearum* on mineral nutrients and infrared temperatures in two tomato cultivars. J. Plant Nutr. 31, 1221–1231.

Hayward, A., 1991. Biology and epidemiology of bacterial wilt caused by *Pseudomonas solanacearum*. Annu. Rev. Phytopathol. 29, 65–87.

Jia, W., Davies, W.J., 2007. Modification of leaf apoplastic pH in relation to stomatal sensitivity to root-sourced abscisic acid signals. Plant Physiol. 143, 68–77.

Kim, S.G., Hur, O., Ro, N.Y., Ko, H., Rhee, J., Sung, J.S., Lee, S.Y., Baek, H.J., 2016. Evaluation of resistance to *Ralstonia solanacearum* in tomato genetic resources at seedling stage. Plant Pathol. 32, 58–64.

Klosterman, S.J., Atallah, Z.K., Vallad, G.E., Subbarao, K.V., 2009. Diversity, pathogenicity, and management of *Verticillium* species. Annu. Rev. Phytopathol. 47, 39–62.

McElrone, A., Sherald, J., Forseth, I., 2001. Effects of water stress on symptomatology and growth of *Parthenocissus quinquefolia* infected by *Xylella fastidiosa*. Plant Dis. 85, 1160–1164.

McElrone, A., Sherald, J., Forseth, I., 2003. Interactive effects of water stress and xylem-limited bacterial infection on the water relations of a host vine. J. Exp. Bot. 54, 419–430.

McGarvey, J., Denny, T., Schell, M., 1999. Spatial-temporal and quantitative analysis of growth and EPS production by *Ralstonia solanacearum* in resistant and susceptible tomato cultivars. Phytopathology 89, 1233–1239.

Mondal, B., Bhattacharya, I., Khatua, D.K., 2014. Incidence of bacterial wilt disease in West Bengal, India. Acad. J. Agric. Res. 2, 139–146.

Mori, Y., Inoue, K., Ikeda, K., Nakayashiki, H., Higashimoto, C., Ohnishi, K., Hikichi, Y., 2016. The vascular plant-pathogenic bacterium *Ralstonia solanacearum* produces biofilms required for its virulence on the surfaces of tomato cells adjacent to intercellular spaces. Mol. Plant Pathol. 17, 890–902.

Nakaho, K., Hibino, H., Miyagawa, H., 2000. Possible mechanisms limiting movement of *Ralstonia solanacearum* in resistant tomato tissues. J. Phytopathol. 148, 181–190.

Ochola, D., Ocimati, W., Tinzaara, W., Blomme, G., Karamura, E., 2015. Effects of water stress on the development of banana *xanthomonas* wilt disease. Plant Pathol. 64, 552–558.

Pérez-Donoso, A., Greve, L., Walton, J., Shackel, K., Labavitch, J., 2007. *Xylella fastidiosa* infection and ethylene exposure result in xylem and water movement disruption in grapevine shoots. Plant Physiol. 143, 1024–1036.

Prior, P., Allen, C., Elphinstone, J., 1997. In: Bacterial wilt disease. Molecular and ecological aspects. Second International Bacterial Wilt Symposium, Gossier, Guadeloupe, France, 22–27 June. Springer, Germany, pp. 269–279.

Rahman, M., Abdullah, H., Vanhaecke, M., 1999. Histopathology of susceptible and resistant *Capsicum annuum* cultivars infected with *Ralstonia solanacearum*. J. Phytopathol. 147, 129–140.

Reusche, M., Thole, K., Janz, D., Truskina, J., Rindfleisch, S., Drubert, C., Teichmann, T., 2012. *Verticillium* infection triggers VASCULAR-RELATED NAC DOMAIN7-dependent de novo xylem formation and enhances drought tolerance in *Arabidopsis*. Plant Cell 24, 3823–3837.

Siebrecht, S., Herdel, K., Schurr, U., Tischner, R., 2003. Nutrient translocation in the xylem of poplar—diurnal variations and spatial distribution along the shoot axis. Planta 217, 783–793.

Singh, S., Gautam, R., Singh, D., Sharma, T., Sakthivel, K., Roy, S., 2015. Genetic approaches for mitigating losses caused by bacterial wilt of tomato in tropical islands. Eur. J. Plant Pathol. 143, 205–221.

Van Der Wolf, J.M., Bonants, P.J.M., Smith, J.J., Hagenaar, M., Nijhuis, E., Van Beckhoven, J.R.C.M., Saddler, G.S., Trigalet, A., Feuillade, R., Prior, P., Allen, C., 1998. Bacterial Wilt Disease: Molecular and Ecological Aspects. Springer, Berlin.

Van Elsas, J.D., Kastelein, P., van Bekkum, P., van der Wolf, J., de Vries, P., van Overbeek, L., 2000. Survival of *Ralstonia solanacearum* Biovar 2, the causative agent of potato brown rot, in field and microcosm soils in temperate climates. Phytopathology 90, 1358–1366.

Vasse, J., Frey, P., Trigalet, A., 1995. Microscopic studies of intercellular infection and protoxylem invasion of tomato roots by *Pseudomonas solanacearum*. Mol. Plant-Microbe Interact. 8, 241–251.

Wallis, F., Truter, S., 1978. Histopathology of tomato plants infected with *Pseudomonas solanacearum*, with emphasis on ultrastructure. Physiol. Plant Pathol. 13, 307–317.

Yadeta, K., Thomma, B., 2013. The xylem as battleground for plant hosts and vascular wilt pathogens. Front. Plant Sci. 4, 97.

CHAPTER 6

Field Performance of Transgenic Drought-Tolerant Crop Plants

Muhammad Sadiq, Nudrat A. Akram
Department of Botany, Government College University, Faisalabad, Pakistan

Contents

1. INTRODUCTION

Detrimental ecological conditions are a serious threat to concerned agricultural practices throughout the world. Water shortage has become a chronic factor causing the death of prevailing crops. Population explosion, as well as global climatic change, has promoted drought conditions, and thus reinforced the significance of water utility in crop husbandry systems. There is an urgent need for transgenic drought-resistant lines to meet global food challenges with limited water availability. Drought endurance is a complex metabolic phenomenon that involves activation of different transcripts. The complex genetics corresponding to desiccation are also influenced by other environmental factors (Kiriga et al., 2016).

Upon exposure to drought conditions, plants initiate a series of signaling processes at the molecular level, which regulate the corresponding genes to start transcription and translation for synthesis of various metabolites (Valliyodan and Nguyen, 2006). These pathways subsequently function to avoid stress losses such as growth reduction, accumulation of osmolytes, and limited transpiration. The interaction of a lot of genes involved in key protective roles against abiotic stresses have been identified. The foremost effect of

Biochemical, Physiological and Molecular Avenues for Combating Abiotic Stress in Plants
https://doi.org/10.1016/B978-0-12-813066-7.00006-1

drought stress is limited shoot growth and diversion of cellular solutes toward stress tolerance mechanisms. This allocation arrests plant growth, and finally, crops' economic yield. For a crop to carry almost normal growth under medium stress conditions is important for a valuable yield (Tran et al., 2007; Wan et al., 2009). Upon exposure to drought circumstances, respective genes transmit stress signals within the plant body. Accumulation of concerned compounds, and stability of membranes against reactive oxygen species (ROS) occurred. All these events potentially improve field performance of crops under drought conditions (Kiriga et al., 2016). Recently, drought-related genes have been successfully engineered in many crops, such as wheat, maize, rice, canola, and soybeans to make these crops more productive. Actual emphasis is focused on assessing target genes of drought tolerance in field-grown crops for securing more yields (Seki, 2002).

In Arabidopsis plants, several drought-related genes have been identified using microarray technology (Shinozaki and Yamaguchi-Shinozaki, 2007). These genes have been categorized into three main groups: (a) genes coding various enzymes such as osmotins, ubiquitins, chaperons, proteases, and so forth (b) regulatory genes corresponding to phosphatases, kinases, and so forth (c) genes with yet-undiscovered functions. In cereals, attempts have been made to modify drought genes at the reproductive phase to ensure grain quality (Oh et al., 2009; Xiao et al., 2009).

The cultivation of transgenic crops is particularly beneficial in less-developed areas worldwide that are facing food insufficiency as a threatening problem (Nelson et al., 2007; Zhang, 2007). National and international agriculture research communities have invested heavily in introducing genetically modified seeds, especially of drought-tolerant crops (Sinclair et al., 2004).

Genetic engineering has revolutionized the agriculture sector with better-adapted, drought-resistant cultivars; for example, water shortage-endured transgenic rice plants (Todaka et al., 2015). Similarly, biotech maize and canola (*Brassica napus*) were also introduced. The first drought-resistant study was carried out in Arabidopsis in potted soil under monitored growth conditions (Wan et al., 2009). The transgenic crops were marketed from 2006 to 2015, and beyond. The leading engineered crop, maize, was introduced in the United States in 2013. Water Efficient Maize for Africa (WEMA) introduced the first transgenic maize variety in sub-Saharan Africa in 2017, where desiccated conditions are very severe (Edmeades, 2013). Currently, more than 30 biotech crops are being cultivated in different countries utilizing 300 million acres of agricultural area (Stein and Rodriguez-Cerezo, 2009). The first generation of insect-resistant and herbicide-tolerant transgenic plants have exhibited significant commercial advantages with minimum pesticide applications, and the ability to enhance crop yield in available cultivated areas, while combating various ecological, financial, and health problems. In the coming years, the increased demand for bioengineered plants equipped with compatible quality traits will spur further exploration in the field of agriculture to address socio-economic needs of future generations (Ford et al., 2017). Due to little mutation and frequent occurrence in genomes, single nucleotide polymorphism (SNP) markers are best suited as compared with microsatellites. Marker-assisted selection (MAS) is widely used in genetic research to accelerate plant breeding mechanisms with particular traits. Drought-tolerant allelic loci engineered at the first growth phase may differ from markers at other, later stages. Moreover, MAS is a reliable method and fast growing strategy of breeding programs, which involves direct introgression of targeted genes. Recently, biotechnologists adopted SNP technology, which recognizes and analyzes intraspecific pattern variation (Telem et al., 2016). To fulfill increased global bio-energy and timber demand, researchers are exploring the trend of growing bioengineered forests on marginal lands, which will also secure native forests from deforestation (De Buck et al., 2016). The growth of transgenic crops has become a progressive adaptive strategy in recent times. Due to hardships of stable GM production, and gene transfer technology and related issues, it has been delayed for commercial applications (Bakshi and Dewan, 2013). Comprehensive and precise security recommendations are required for transgenic crops raised for food production. Specifically, overexpression of regulatory genes and plant metabolic activities may cause unintentional losses (Nam et al., 2014). This chapter describes a brief account of transgenic crop plants at the molecular level, their field performance, and major issues and benefits of GM plants.

2. CROP DEVELOPMENT AND RESPONSE AGAINST DROUGHT STRESS

Drought is the most deleterious abiotic stress factor that is causing serious losses in the agricultural sector globally. Crop yield improvements, as well as the increased productivity under soil water deficiency, are the immediate issues that must be addressed to fulfill the food demands of a rapidly increasing world population (Yang et al., 2010). Plants respond to moisture deficiency morphologically and physiologically by initiating a series of different signaling cascades. Reduced leaf area, leaf wilting, senescence, and impaired growth patterns are some of the physiological responses to drought. Even under extreme conditions, stress hormones, that is, abscisic acid (ABA)-induced stomatal closure, ultimately hamper transpiration. All these adaptations are used to conserve cellular water contents (Kiriga et al., 2016). Soil water deficits reduce stomatal conductance, growth, root proliferation, photosynthetic pigments, protein contents, and impaired photosynthetic pigments, and alter hormone distribution and increased activities of antioxidant enzymes to cope unfavorably with osmotic changes (Praba et al., 2009). Moreover, drought stress influences yield parameters such as pollen sterility, reduced grains, and production of ABA in wheat spikes (Ji et al., 2010). Under soil dehydration, stomatal closure-reduced leaf CO_2 levels that altered carbon assimilation in photorespiration (Demirevska et al., 2009) and induced oxidative stress slacken photosynthetic electron flow and photosystem II activity (Farooq et al., 2009). Under oxidative stress, plants develop complex system of nonenzymatic and enzymatic antioxidants.

The nonenzymatic antioxidants (ascorbic acid, glutathione, and carotenoids) retain photosynthetic apparatus integrity, whereas enzymatic antioxidants (catalase, peroxidase, and superoxide dismutase) scavenge harmful oxidative free radicals. Physiologically, plants respond to stress conditions by accumulating various osmolytes such as proline, mannitol, glycinebetaine, phenolics, trehalose, tocopherols, and inorganic K^+ ions. These osmolytes not only retain a cellular hydrated state, but also preserve macromolecule structures by scavenging ROS during desiccated conditions (Chaves et al., 2003).

The drought-responsive genes can be identified by exposing genetically modified crops to stresses. The general dehydration avoidance approach is utilized to overcome agricultural drought and retain crop yield before drought accentuation. The lowland, rain-fed areas are drought avoiders for rice crops. The drought-resistant genotypes are better adapted under water stress to provide a better harvest. Studies targeted rice plants during moisture deficits to monitor their productivity in drought-prone, rain-fed areas. The implementation of gene technology is similar to plant breeding approaches with the aim to compensate crop yield under stressful conditions. It is therefore important to monitor plant biomass production and yield parameters resulting from gene manipulation (Hervé and Serraj, 2009).

3. DROUGHT ENDURANCE AND CROP ACHIEVEMENTS

Endurance means rehabilitation of ecosystem dynamics after stress conditions. It is the actual use of biological sources without damage. This technique is used frequently in cultivation for compensating for losses from drought (O'farrell et al., 2009). Significant research has been done on GM plants with an emphasis on drought-tolerant genes under severe stress conditions. On the contrary, the drought-avoidance method is used to cultivate crops before agricultural drought occurs, such as with rice in lowland rain-fed areas; it is cultivated before the dehydration period. The drought-tolerant cultivars produced a higher grain yield, as they retained osmotic levels at pre- and postantithesis stages under drought conditions. Reasoning irrigation systems are based on an accurate assessment of crop requirements. In citrus, it is estimated based on canopy cover (Villalobos et al., 2013). Sometimes, withholding irrigation is used as a strategy to increase fruit quality (Ballester et al., 2011). Farmers use this technique to raise their crop/tree yield. Precise estimates are necessary to maintain proper irrigation during water-deficit field conditions (Naor, 2006).

The water requirements of plants vary according to changes in soil mediums (Joshi et al., 2016). The mechanism of drought tolerance in plants is a positive approach to meet the demands of the world's increasing population (Todaka et al., 2012). In this regard, the metabolic reactions that are sensitive to stress signals can initiate actual adaptive processes in transgenic plants against stress conditions (Sanchez et al., 2011). The stress response of plants is correlated with growth stages and water-use efficiency, as the flowering stage in many plant species is susceptible to soil dehydration (Farooq et al., 2017). Having sessile positions, plants have developed different physiological, molecular, and biochemical mechanisms against soil desiccation. Stomatal closure, leaf rolling, chlorophyll reduction, enhanced root length, and wax deposition are some visible symptoms of low soil-moisture contents in plants (Khazaei et al., 2013). Leaf rolling is a primary reaction to osmotic stress that helps plants avoid heat losses and transpiration, and thus improve water-use efficiency (Singh and Laxmi, 2015). The species that retain osmotic potential under water shortages are drought-tolerant (Joshi and Karan, 2013).The soil water contents support the root growth pattern. The soil osmotic stress reduces primary root growth, and induces growth of lateral branches (Ji et al., 2014). Under abiotic stresses, plants accumulate various osmoprotectants, antioxidants, and reactive oxygen species (ROS). These osmoprotectants confer the harms of drought stress by stabilizing biological membranes and enzyme activities (Singh and Laxmi, 2015). At the genetic level, plants activate the corresponding genes and regulatory pathways under stress conditions (Buchanan et al., 2015). It is necessary to detect stress regulatory pathways to indicate important regulators that could be utilized to breed or transgene stress-resistant plants. By employing the latest omics tools, progress has been made in defining the explanation of concerned stress-associated pathways during stress responses. In different molecular studies, few related genes of transcriptional networks have been characterized (Todaka et al., 2015). These signaling routes indicate signal perception, transduction, and activation of relevant metabolic processes (Pérez-Clemente et al., 2013). During this process, membrane-bounded sensors or receptors perceive the stress signals. The receiver molecules stimulate intracellular messengers such as inositol phosphate, Ca^{++}, cyclic nucleotides, ROS, nitric oxide, and sugars that subsequently transduce the signals (Bhargava and Sawant, 2013). During these transduction pathways, regulation of phosphorylation and dephosphorylation of proteins is carried out. The role of calcium-dependent protein kinases (CDPKs) and mitogen-activated protein kinases (MAPKs) are important in stress signaling pathways under osmotic stress (Huang et al., 2012). Last, protein phosphatases or kinases activate or suppress the transcription factors (TFs) and directly downstream the genes during interaction with cis elements in their promoter region (Danquah et al., 2014). Moreover, at the transcriptional level, TFs are regulated by other upstream factors, and after several alterations such as sumoylation and ubiquitination, a complex regulatory system expressed stress genes by catalyzing various cellular reactions (Mizoi et al., 2013).

4. MAJOR DROUGHT-TOLERANT TRANSGENIC PLANTS

4.1 Genes Related to Osmoprotectants

Various techniques have been applied to generate stress-resistant plants using plant breeding and classical genetic approaches. One method to produce stress-tolerant plants exhibiting better performance under abiotic stresses is the genetic approach. This method is partially applicable due to varying patterns of precipitation and the complex polygenic nature of stress genes. The transfer of stress-tolerant genes from their wild ancestors to their current crop plants has had limited success. Gene manipulation is carried out using classical genetic approaches to produce drought-tolerant plants. Recent important achievements have resulted from current engineering strategies to improve plant stress tolerance. However, genetically complex mechanisms are susceptible to potential threats and harmful side effects that have consequently made this approach very difficult to apply (Table 1).

Genetic engineering has opened new horizons for biosynthesis of compatible compounds into transgenic plants that have better traits related to stress endurance. Overproduction of these solutes has made them more osmotolerant. Among these solutes, glycinebetaine and proline are the most important. The Δ 1-pyrroline-5-carboxylate (P5C) and glutamic-γ-semialdehyde (GSA) pathways are used for proline biosynthesis. The glutamate is catalyzed into P5C by P5C synthase and then reduced to proline by action of P5C reductase. In transgenic rice and tobacco plants, P5Cs overexpressed, and more proline is produced, which scavenges free radicals and produces better growth results under water deficiency. Similarly, betaine provides a better response in plants against water stress. Under abiotic stress, in chloroplast, choline is catalyzed into betaine aldehyde by choline mono-oxygenase (CMO). The enzyme aldehyde dehydrogenase converts betaine aldehyde into glycinebetaine (GB). Some important cash crops such

Table 1 Genetically modified (GMO) crops/vegetables cultivars

Name of crop/vegetable	Cultivar/variety	Distinguished characteristics for adoptions	References
Cotton	Bt cotton	Insect resistant	Manjunath (2004)
Banana	Rasthali (AAB)	Expression of surface antigen of the hepatitis B virus	Kumar et al. (2005)
Potato	TH110-51	Expression of lymphotoxin-beta protein	Tacket (2007)
Tomato	Zhongshu	positive role in tolerance to drought stress	Lu et al. (2010)
Rice	Golden Rice	Dietary Golden Rice in vitamin A	Tang et al. (2009)
Maize	Bt maize	Production of insecticidal Cry proteins	Lang and Otto (2010)
Corn	Corn, MON810	Insecticide resistant	Frizzas et al. (2014)
Tomato	Rio Grande, Moneymaker and Roma	Cold stress tolerance	Shah et al. (2016)

as tomatoes, potatoes, and rice do not produce GB naturally; genetically engineered plants show better performance against water stress. Transgenic cultivars of Brassica, tobacco, and Arabidopsis have overexpressed (codA/cox) choline oxidase genes, and have elevated GB levels under stress conditions. The overexpressed *N*-methyltransferase gene in Arabidopsis produces higher GB contents in leaves, roots, and flowers during exposure to stress conditions. The other engineered plants and cyanobacteria also have better stress adaptation. Like other solutes, plants also produce polyamines under drought stress. In Arabidopsis, spermidine synthase is overexpressed to make plants more tolerant against multiple stresses, including drought. Many other compatible solutes such as sugar alcohols, including sorbitol, myo-inositol, trehalose, and mannitol overexpressed in target-transformed plants. In genetically modified tobacco plants, cDNA expressed to perform myo-inositol, 1-methyltransferase (IMT1), which aggregates dononitol levels, and this solute protects photosynthetic apparatus and ameliorates the effects of salt and dehydration. Transgenic rice plants with elevated levels of trehalose showed more resistance against abiotic stresses and retained a higher K^+/Na^+ ratio compulsory for cellular processes. The expression of stress-activated enzymes such as alfalfa aldose aldehyde reductase in engineered tobacco plants made them more tolerant against oxidative stress. Ion homeostasis is also another crucial adaptation in plants against salinity and dehydration. Various proton pumps such as H^+-PPase, V-ATPase, and P-ATPase played a significant role in this matter. The engineered Aradiposis plants genetically modified with the *AVP1* gene, overexpressed for H^+-PPase pump activity, made plants more resistant to drought and salt than untreated plants. Later embryogenesis abundant protein (LEA protein) aggregates in plants in response to dehydration. Integral expression of ABF3 in Arabidopsis activates genes *RAB18* and *RD29*, which induces more levels of LEA-type proteins, and these transformed plants prove to be more osmotic tolerant at the seedling stage. Overexpression of *HAV1* in transgene rice and *3LEA* protein genes in barley showed more resistance against drought and saline stress than wild cultivars (Lisar et al., 2012).

4.2 Genes Related to Ions/Mineral Nutrients

The intracellular and transcellular pathways are used for transport of water and ions in plants. The plasma membrane-bounded aquaporin proteins transport small molecules, water, gas, and glycerol across the membranes and play a role in water homeostasis. Active transport is another method of mineral movement across cellular compartments (Li et al., 2010). Engineered plants such as Arabidopsis have upregulated gene overexpression of *AVP1* in tomatoes (Park et al., 2007), vacuolar-H^+-PPase in tobacco, Na^+/H^+ antiporter in wheat and TVP1, and H^+-PPase, *TNHX1* in Arabidopsis (Li et al., 2008). All these transgenic plants have displayed more salinity and drought tolerance with better growth and well-developed root systems. Transgenic plants are also equipped with modified aquaporins (Miyazawa et al., 2008). In apoplastic pathways, aquaporins transfer water and minerals across membranes (Jang et al., 2007). Different aquaporins have various roles, depending on the nature of the prevailing stressor. In transgenic rice, *RWC3*, and Arabidopsis, *Rd28*genes were overexpressed under dehydration, while *AtPIP1* and *NtQP1* remained unaltered during water deficit adversaries. Moreover, in certain aquaporins genes, *AtPIP1b* diminished resistance capability in some plants, whereas *Brassica napusBnPIP1*, *Vicia fabaPIP1*, *Brassica junceaBjPIP1*, *Panax ginsengPgTIP1* promoted drought tolerance (Cui et al., 2008). In isoforms of aquaporins linked with various physiological processes, the plants respond differently according to aquaporins, and upregulation or downregulation under drought conditions (Peng et al., 2006). The upregulation of aquaporins have also been applied during heavy metal stress (Zhang et al., 2008).

The biotechnological approaches are used to improve endurance of crop plants to continue to grow and give yield under mineral-limited environments. Nitrogen is vital for plant growth. Nitrates are necessary for regulation of several plant processes (Wang et al., 2007). Nitrate transporters such as *NRT1.2* and *NRT2.1* are not responsible for N absorption, but also behave as sensor molecules for low rhizospheric N concentration.

The signaling transporter *NR 1.1* caused accumulation of nitrate-rich patches in soil and promoted root proliferation in coordination with *MADS-boxTF* and *ANR1* genes (Remans et al., 2006). Both these genes regulate the growth of root primordials and tips. Transporter genes have been engineered to modify nitrogen uptake efficiency in several plants. For example, in leaves of lettuce, a bacterial gene, asparagine synthase, overexpressed to alter N status. These transgenic lettuce plants showed 1.3-fold higher vegetative growth (leaf number and leaf area) for 35 days of postgermination. However, the modified plants showed earlier flowering, which was not economically beneficial for leaf vegetables. Similar findings were found in transgenic lotus (Vincent et al., 1997) and oil seed rape (Seiffert et al., 2004). These GM plants retained two times more aspartate and asparagines than water treated with higher levels of glutamine (Curtis et al., 1999). In phosphorus deficient soils, the P uptake or transporter genes may have high or low affinity with P absorption from soil. In barley, upregulation of high affinity P transport genes has no profound effect on P uptake or on plant growth while transgenic rice culture has demonstrated an increased uptake of P. In addition, homologous, high-affinity P transporter genes overexpressed in rice showed twofold P uptakes as compared with normally irrigated rice plants, and they gave more tillars (Seo et al., 2008). Later investigation revealed complex interaction between plant P transporters and vesicular arbuscular mycorrhizae (Glassop et al., 2007). Plants alter the rhizopheric sphere with specific exudates to improve the uptake of phosphorus. This exudation phenomenon was observed in transgenic rice (Park et al., 2007) and Arabidopsis (Xiao et al., 2009). The phytase gene has ecotopic regulation in roots of potatoes (Hong et al., 2008). Iron is also an important bioelement in plants and animals. A transgenic line of rice has more quantities of phytosiderophores (PS), which is more tolerant to Fe-limited soils. Other achievements in the approach are transgenic barley with nicotianamine transferase genes (Takahashi et al., 2001), and engineered rice in saline soils with upregulation of Fe chelate reductase genes (Ishimaru et al., 2007).

4.3 Genes Related to Antioxidants

Abiotic stresses generate various reactive oxygen species (ROS) such as $O_2^{\bullet -}$, $HO^{\bullet}$, $1O_2$, and H_2O_2. These species cause lipid peroxidation, disrupt membrane integrity, and damage DNA and proteins (Miller et al., 2010). Researchers are in search of concerned genes responsible for producing stress tolerance. The enzymatic antioxidants belong to the family aldehyde dehydrogenases (ALDHs), which transform aldehyde to relevant acids, which detoxify acetaldehydes (Nakazono et al., 2000). McKersie et al. (1996) reported that in alfalfa plants, the gene *Mn-superoxide dismutase* (*Mn SOD3.1*) regulated the catalysis of O_2^- into H_2O_2. Under field study, in tomatoes, the regulation of the *cAPX* gene enhanced heat resistance (Wang et al., 2006). The rice became more salt tolerant by expression of the *katE* gene (Nagamiya et al., 2007). The activation of various antioxidant enzymes exhibited a promising strategy against stress adversaries. The combined expression of CAT genes and glutathione *S*-transferase (GST) in transgenic rice plants at the vegetative stage improved stress tolerance during oxidative stress. The combined interaction of three antioxidant enzymes, APX, copper zinc superoxide dismutase (CuZnSOD), and dehydroascorbater-eductase (DHAR) caused more stress tolerance in tobacco plants (Lee et al., 2007). The superoxide dismutase encoding gene was introduced in engineered potatoes, alfalfa, and rice (Wang et al., 2005). The overexpression of glyoxalate I and II genes was reported by Viveros et al. (2013) in transformed tobacco. The transformants have reduced lipid peroxidation and H_2O_2 levels under saline stress, while the wild types have decreased chlorophyll *a* and *b*. Hill et al. (2013) investigated that overexpression of *TomloxD* promoted lipoxygenase activity and higher levels of endogenous jasmonic acid and antioxidants enzymes in transgenic tomatoes, and found them to be more tolerant to biotic and abiotic stresses. Shan et al. (2007) described that transgenic tobacco plants with *GhDREB1* from cotton were more tolerant to chilling stress as compared with wild types. In cereal crops, especially in wheat and rice, *DREB* genes are specifically important for transformation against abiotic stresses (Chen and Murata, 2008). Cui et al. (2011) found significant overexpression of *OsDREB2A* in transgenic rice against drought and saline stress, and Qin et al. (2007) observed overexpression of *CaMV35S* and *ZmDREB2A* in transgenic maize. In rice, the gene *SNAC1* expressed during the reproductive phase and made it tolerant against severe drought/saline stress (Hu et al., 2006). Under severe water deficiency treatment in transgenic rice, the expression of *OsWRKY11* under the promoter factor *HSP101* made them more tolerant than untransformed plants. Oh et al. (2009) also described several transformants other than *DREB*.

4.4 Genes Related to Plant Hormones

Plant hormones are substances produced in minor quantities that regulate various metabolic processes in plants (Vob et al., 2014). In response to stresses, they regulate and reprogram the cellular genetic metabolic machinery. They also coordinate internal and external stimuli (Kazan, 2015). Abscisic acid (ABA) has been recognized as a stress hormone. It retains seed dormancy, inhibitors of plant growth, stomatal movement, and leaf and fruit abscission. For biotechnologists, transformation of phytohormones could be a better choice to improve plants commercially and nutritionally. The first response of plants against moisture stress is the accumulation of ABA transcripts in roots (Thompson et al., 2007), which is conducted to shoot via xylem, and as a result, stomata is closed and plant development is slowed. It may also produce in the leaf (Wilkinson and Davies, 2010). In Arabidopsis, many genes are characterized that are responsible for ABA metabolism and downstream signaling (Cutler et al., 2010). However, in cereal crops such as rice, a key enzyme *LOS5/ABA3* is transcribed for water stress tolerance. In transgenic rice plants, the LOS5 overexpressed in drought conditions and prominently enhanced yield components in field trials (Xiao et al., 2009). In transcribed tomatoes, the upregulation of *LeNCED1* caused up-levels of ABA, but remained nonresponsive to drought tolerance (Thompson et al., 2007).

Cytokinine (Ck) has antagonistic effects compared with ABA. It has decreased levels in soil moisture deficiency conditions. However, its higher levels promote growth, even in water deficiency conditions with proline contents, and it ultimately minimized the abscission. Handling of CK transcripts proved beneficial in transgenic plants. The isopentenyltransferase (*IPT*) transcript has been engineered in many plants. The type of promoter to drive *IPT* regulation caused variance against drought tolerance. In transformed tobacco, the *IPT* gene under promoter *SARK* expressed, which consequently increased photosynthetic performance, dehydration tolerance, and yield (Rivero et al., 2010). Similarly, expression of transcription of *PSAK:IPT* in transgenic rice increased grain yield in drought tolerance conditions. In transgenic cassava, the expression of *IPT* under field conditions decreased leaf abscission and more drought tolerance than wild cultivars (Zhang et al., 2010).

Jasmonic acid (JA) plays a key role in plant defense and development. The overtranscription of *AtJMT* in transgenic rice enhances a higher concentration of JA levels in panicles (Kim et al., 2009). In response to ecological cues, the gibberellic acid (GA) hormone plays a role in hormonal interactions. The mutual interaction of gibberellic acid, ethylene, cytokinins, and ABA is important for stress tolerance of plants (Krouk et al., 2011). Against responses of abiotic stresses, auxin plays its necessary role in signaling. In transgenic rice, the gene *OsPIN3t*'s expression is involved in polar auxin transport, root growth abnormalities in the seedling stage, and tolerance against drought. The application of 20% PEG induces stress, and β-glucuronidase activity significantly improves naphthaleneacetic acid. It is therefore inferred that transcription of the OsPIN3t gene is responsible for auxin conduction and water deficit responses.

5. FIELD PERFORMANCE OF TRANSGENIC PLANTS

Many authors do substantial research on genetically modified plants to focus on water-deficit-associated genes. This strategy has little scope for cereal plants (Sinclair et al., 2004). Drought avoidance is a general approach to maintaining crop productivity before the onset of unfavorable dehydrated conditions. For example, rice in lowland rain-fed areas adopts a desiccation avoidance strategy. Some cultivars of crops are drought tolerant; they produce a higher grain yield under adverse stress conditions. Most research is focused on rice cultivars in drought-stressed rainy areas. Like plant breeding programs, the primary goal is engineering gene technologies within cereal crops to improve the yield of potential crop plants within a target environment. Farmers are interested in drought-tolerant crops with high yield. Therefore, it is important to monitor GM plants for biomass production and yield potential (Hervé and Serraj, 2009).

Because of osmotic stress, the biosynthesis of various osmoprotectants has been undertaken, and their degradation has been suppressed. The GM plants with a potent concentration of osmolyte proved to be more drought-tolerant. It is noteworthy that all compatible solutes cannot accumulate in all tolerant plants. For example, all higher halophytes and xerophytes plants cannot accumulate glycinebetaine. Similarly, the drought-tolerant wild watermelon cultivar contains citrulline in its leaves.

Various genetic techniques have been deployed to generate stress-resistant plants. One crucial approach is to select cultivars that produce a higher yield in a dehydrated environment. It is a partially successful approach due to the complex polygenic machinery of osmotic tolerance. The gene manipulation technique has also had limited success. In this approach, the resistant genes from wild cultivars are transferred to cereal crops. The modified plants have the ability to maintain cellular processes and structures under adversaries. The current applied genetic strategies include transfer of one or several genes leading to stress pathways. Despite beneficial achievements, there are still potentially harmful side effects, which made this technology extremely risky.

Genetic engineering has produced stress-tolerant transgenic plants equipped with overexpressed novel pathways for biosynthesis of several compatible solutes. Such compatible solutes may be glycinebetaine, proline, polyamines, sugar alcohols, and late embryogenesis abundant (LEA) proteins in droughted plants. Recently, modified transformants have been produced possessing avoidance and tolerant characters. Rivero et al. (2007) manipulated a gene of leaf senescence. The phenomenon of accelerated leaf senescence in crop plants reduced biomass and yield, and ultimately led to plant death. An investigation was conducted into tobacco plants after suppression of water deficiency as a tool to accelerate drought. The transgenic plants, after drought recovery, resumed their photosynthetic rate, growth, and leaf turgor, while wild cultivars did not. The transgenic plants had significantly higher water-use efficiency than the nontransformants ones. Although Bt crops are widely used in integrated pest control programs, some serious threats are linked to their cultivation due to perseverance of larvicidal Cry proteins in agri-ecosystems. The persistence of Cry proteins affects the soil microbial activity (Zhang et al., 2017a,b). The *Leymus chinensis*-derived expression of S-adenosylmethionine decarboxylase gene (*LcSAMDC1*) in transgenic Arabidopsis made it salt- and cold-tolerant. The overexpression of γ-tocopherol methyl transferase (γ-tmt) in transformant *Codonopsis lanceolata* caused higher levels of polyphenolic compounds and α-tocopherol, which enhanced antimicrobial processes of leaf and root secretions (Ghimire et al., 2017). In GM sweet potatoes, the excessive expression of the *IbCBF3* gene caused more cold and drought resistance compared with water-treated plants (Jin et al., 2017). In Arabidopsis, the regulation of the *FtMYB9* gene activates the stress-responsive pathways during drought and salinity stress. So the gene *FtMYB9* can be incorporated in other plants for genetic breeding (Gao et al., 2017). In rice plants, Park et al. (2017) reported that, as a result of abiotic stress in paddy fields, the higher expression of the glutathione synthetase gene (*OsGS*) produced glutathione, which made rice plants more tolerant. Despite different ecological stresses in *poplar trichocarpa*, the introgression of *Populus trichocarpa* nucleoside diphosphate kinase (*PtNDPK2*) was beneficial in improving plant biomass (Zhang et al., 2017a,b).

6. MAJOR PROBLEMS UNDER FIELD CONDITIONS

The mismatching of genotypes and phenotypes of transgenic crops relevant to their environment are important issues (Lemaux, 2009). In agriculture, there has been concern that transgenic crops may harm nontarget organisms and thus affect biodiversity; or that farming of GM plants may raise a new progeny of pathogens and pests. On the other hand, transformed crops may enhance sustainable farming. For example, herbicide-resistant transgenic sugar beets caused less harm than the conventional crop to the field operations (Bennet et al., 2004). According to Brookes and Barfoot (2012), less pesticide spray was required for transgenic crops, thus decreasing the impact on the environment. Moreover, transgenic crops reduced the emission of greenhouse gases from the cropping fields.

Some activists warn that farmers' dependence on climate-tolerant GM crops may become preferable to nontransformants, leading to the jeopardy of food security and threats to biodiversity. On the contrary, many plant biotechnologists view new crops as a gift for small farms in stressful areas; giving them the opportunity for a larger harvest and a greater share in the market economy. Farre et al. (2010) viewed that resource-poor farmers may not get benefits from engineered plants without overcoming technical and political barriers. Some potential hazards are produced during genetic manipulation, which can create ecological risks in an agroecosystem. GM plants have hazardous effects on soil microflora, which plays basic role in residue decomposition and the biochemical nutrient cycle. For example, Bt cotton produced Bt toxin in pollen, which deposited on vegetation of nearby fields, causing harm to nontarget organisms that feed these plants. The GM plants affected microflora that reduce soil fertility (Giovannetti et al., 2005).

An updated assessment framework may be required for regular approval of stress-prone transgenic crops (Ortiz et al., 2007). For example, manipulation of genes in plants' genetic setup has a cascading effect on different genetic pathways. The GM plants may differ from first-generation GM plants. A new framework will be needed to evaluate the influence of regular GM genes on the environment and human health. For example, new phenotypes of GM technology against stress tolerance may increase competition with wild varieties, which would result in an escape of target crops to other areas.

7. SOCIAL AND ECONOMIC BENEFITS OF GENETICALLY MODIFIED PLANTS

Genetically modified crops have a great deal of potential. Many environmental benefits can be achieved upon cultivating GM crops. In developing countries, the new cultivation technologies have improved the economic status of those living in rural areas (Fan et al., 2005). The same expectations are associated with GM crops (FAO, 2004). Genetically modified crops not only fulfill nutritional demands, but also provide health benefits. Genetically modified crops aim to produce oil crops with a better fatty acids profile, staple foods with more proteins, and food with elevated levels of antioxidants and loaded with improved dietary fiber (Newell-McGloughlin, 2008). Transgenic golden rice has transcripts for carotenogenes in grains without altering agronomic features. The latest cultivars of golden rice are engineered with transgenes of maize. This version has both cereal qualities. Moreover, the engineered golden rice with β-carotene is a source of vitamin A for children. In China, the intake of 100–150 g of cooked golden rice provides 60% of vitamin A required by 6–8 year-old-children. Similarly, a maize variety was bred genetically that has folate, β-carotene, and ascorbate in its endosperm. This attempt could be advantageous for the people of Latin America and Africa, where maize is used as a staple food. These engineering techniques have improved the quality of seed plants with a better amino acid profile, with reduced risks of myocardial infections (Haslam et al., 2013). Until now, 10 such transgenes have been engineered. A clear majority of people around world benefitted pharmaceutically from these transformed plants. The new transformed plants are engineered to produce more levels of amino acids, vitamins, and Omega-3 fatty acids.

Globally, many research agencies work to monitor the development of transformed plants in agro-industries to assure the biosafety of GM organisms. However, it is crucial to set up an international coordinated network for safe use of this GM technology within the next several years (Singh et al., 2006). Mostly, GM plants have a gene-lock trait for one growing season. Growers need to purchase quality seeds from ethical agribusiness companies. This is a very costly, and thus unaffordable, practice for farmers of poor countries each year.

Despite the benefits, GM technology has some significant issues. Public opposition has become more widespread in Europe, and reservations have also reached other countries through public media (Paarlberg, 2008). The core threats correspond to health, ecological hazards, and adverse social associations, as in the USA and Europe, where many children suffered from life-threatening allergies by consuming GM foods.

8. CONCLUSIONS AND PROSPECTS

The current scenario focuses on the transformation of important crops for improved drought endurance. The literature cites that genes are not only vulnerable to drought stress, but also to cold and salinity stresses. It is of the utmost importance that we produce crops that are more heat- and drought-tolerant. Information about stress-tolerant pathways is still limited. It is time to identify the signaling networks missing in this literature that can improve our understanding of crosstalk between signaling networks. Engineered plants may have still-unknown effects on animals, uncultivated lands, and even on ecosystems. It is better to perform trials of these novel technologies for overexpressed stress-tolerant genes and their random combinations in crops before commercial release. This will lead to sovereignty of these genes, which is best suited to climatic conditions, thus enhancing gene diversity. Plant transgenic techniques explore new genetic pathways in molecular physiology against abiotic stresses. For developing new versions of GM crops, further field trials are required to assess their growth and tolerance during their entire life cycles.

REFERENCES

Bakshi, S., Dewan, D., 2013. Status of transgenic cereal crops: a review. Clon. Transgen. 3, 1–13.

Ballester, C., Castel, J., Intrigliolo, D.S., Castel, J.R., 2011. Response of Clementina de Nules citrus trees to summer deficit irrigation. Yield components and fruitcomposition. Agric. Water Manag. 98 (6), 1027–1032.

Bennet, R., Phipps, R., Strange, A., Grey, P., 2004. Environmental and human health impacts of growing genetically modified herbicide-tolerant sugar beet: a life-cycle assessment. Plant Biotechnol. J. 2, 273–278.

Bhargava, S., Sawant, K., 2013. Drought stress adaptation: metabolic adjustment and regulation of gene expression. Plant Breed. 132, 21–32.

Brookes, G., Barfoot, P., 2012. Global impact of biotech crops: environmental effects, 1996–2010. GM Crops Food 3 (2), 129–137.

Buchanan, B.B., Gruissem, W., Jones, R.L., 2015. Biochemistry and Molecular Biology of Plants, second ed. John Wiley & Sons, Somerset, NJ.

Chaves, M.M., Maroco, J.P., Pereira, J.S., 2003. Understanding plant response to drought. Funct. Plant Biol. 30, 239–264.

Chen, T., Murata, N., 2008. Glycinebetaine: an effective protectant against abiotic stress in plants. Trends Plant Sci. 13, 499–505.

Cui, M., Zhang, W.J., Zhang, Q., Xu, Z.Q., Zhu, Z.G., Duan, F., Wu, R., 2011. Induced overexpression of the transcription factor *OsDREB2A* improves drought tolerance in rice. Plant Physiol. Biochem. 49, 1384–1391.

Cui, X.H., Hao, F.S., Chen, H., Chen, J., Wang, X.C., 2008. Expression of the *Vicia fabaVfPIP1* gene in *Arabidopsis thaliana* plants improves their drought resistance. J. Plant Res. 121 (2), 207–214.

Curtis, I.S., Power, J.B., de Laat, A.M.M., Caboche, M., Davey, M.R., 1999. Expression of a chimeric nitrate reductase gene in transgenic lettuce reduces nitrate in leaves. Plant Cell Rep. 18, 889–896.

Cutler, S.R., Rodriguez, P.L., Finkelstein, R.R., Abrams, S.R., 2010. Abscisic acid: emergence of a core signaling network. Annu. Rev. Plant Biol. 61, 651–679.

Danquah, A., de Zelicourt, A., Colcombet, J., Hirt, H., 2014. The role of ABA and MAPK signaling pathways in plant abiotic stress responses. Biotechnol. Adv. 32, 40–52.

De Buck, S., Ingelbrecht, I., Heijde, M., Van Montagu, M., 2016. Innovative Farming and Forestry Across the Emerging World: The Role of Genetically Modified Crops and Trees. International Industrial Biotechnology Network (IIBN), Gent, Belgium.

Demirevska, K., Zasheva, D., Dimitrov, R., Simova-Stoilova, L., Stamenova, M., Feller, U., 2009. Drought stress effects on Rubisco in wheat: changes in the Rubisco large subunit. Acta Physiol. Plant. 31, 1129–1138.

Edmeades, G.O., 2013. Progress in Achieving and Delivering Drought Tolerance in Maize—An Update. Ithaca, NY, ISAAA.

Fan, S., Chan-Kang, C., Qian, K., Krishnaiah, K., 2005. National and international agricultural research and rural poverty: the case of rice research in India and China. Agric. Econ. 33, 369–379.

FAO, 2004. The State of Food and Agriculture 2003–04; Agricultural Biotechnology: Meeting the Needs of the Poor? FAO, Rome.

Farooq, M., Gogoi, N., Barthakur, S., Baroowa, B., Bharadwaj, N., Alghamdi, S.S., Siddique, K.H.M., 2017. Drought stress in grain legumes during reproduction and grain filling. J. Agron. Crop Sci. 203, 81–102.

Farooq, M., Wahid, A., Kobayashi, N., Fujita, D., Basra, S.M.A., 2009. Plant drought stress: effects, mechanisms and management. Agron. Sustain. Dev. 29, 185–212.

Farre, G., Ramessar, K., Twyman, R.M., Capelli, T., Christou, P., 2010. The humanitarian impact of plant biotechnology: recent breakthroughs vs bottlenecks for adoption. Curr. Opin. Plant Biol. 13, 219–225.

Ford, K., Bailey, A.M., Foster, G.D., 2017. Beneficial Uses of Genetically Modified Crops. John Wiley & Sons Ltd., Chichester. https://doi.org/10.1002/9780470015902.a0023735

Frizzas, M.R., SilveiraNeto, S., Oliveira, C.M.D., Omoto, C., 2014. Genetically modified corn on fall armyworm and earwig populations under field conditions. Ciênc. Rural 44 (2), 203–209.

Gao, F., Zhou, J., Deng, R.Y., Zhao, H.X., Li, C.L., Chen, H., Suzuki, T., Park, S.U., Wu, Q., 2017. Overexpression of a tartary buckwheat R2R3-MYB transcription factor gene, FtMYB9, enhances tolerance to drought and salt stresses in transgenic Arabidopsis. J. Plant Physiol. 214, 81–90.

Ghimire, B.K., Seong, E.S., Yu, C.Y., Kim, S.H., Chung, I.M., 2017. Evaluation of phenolic compounds and antimicrobial activities in transgenic *Codonopsislanceolata* plants via overexpression of the γ-tocopherolmethyltransferase (γ-tmt) gene. S. Afr. J. Bot. 109, 25–33.

Giovannetti, M., Sbrana, C., Turrini, A., 2005. The impact of genetically modified crops on soil microbial communities. Rev. Biol. 98 (3), 393–417.

Glassop, D., Godwin, R.M., Smith, S.E., Smith, F.W., 2007. Rice phosphate transporters associated with phosphate uptake in rice roots colonised with arbuscular mycorrhizal fungi. Can. J. Bot. 85, 644–651.

Haslam, R.P., Ruiz-Lopez, N., Eastmond, P., Moloney, M., Sayanova, O., Napier, J.A., 2013. The modification of plant oil composition via metabolic engineering-better nutrition by design. Plant Biotechnol. J. 11, 157–168.

Hervé, P., Serraj, R., 2009. Gene technology and drought: a simple solution for a complex trait? Afr. J. Biotechnol. 8 (9), 1740–1749.

Hill, C.B., Taylor, J.D., Edwards, J., Mather, D., Bacic, A., Langridge, P., 2013. Whole-genome mapping of agronomic and metabolic traits to identify novel quantitative trait lociin bread tomato grown in a water-limited environment. Plant Physiol. 162, 1266–1281.

Hong, Y.F., Liu, C.Y., Cheng, K.J., Hour, A.L., Chan, M.T., Tseng, T.H., Chen, K.Y., Shaw, J.F., Yu, S.M., 2008. The sweet potato sporamin promoter confers high-level phytase expression and improves organic phosphorus acquisition and tuber yield of transgenic potato. Plant Mol. Biol. 67, 347–361.

Hu, H., Dai, M., Yao, J., Xiao, B., Li, X., Zhang, Q., Xiong, L., 2006. Overexpressing a NAM, ATAF, and CUC (NAC) transcription factor enhances drought resistance and salt tolerance in rice. Proc. Natl. Acad. Sci. U. S. A. 103, 12987–12992.

Huang, G.T., Ma, S.L., Bai, L.P., Zhang, L., Ma, H., Jia, P., Liu, H., Zhong, M., Guo, Z.F., 2012. Signal transduction during cold, salt, and drought stresses in plants. Mol. Biol. Rep. 39, 969–987.

Ishimaru, Y., Kim, S., Tsukamoto, T., Oki, H., Kobayashi, T., Watanabe, S., Matsuhashi, S., Takahashi, M., Nakanishi, H., Mori, S., Nishizawa, N.K., 2007. Mutational reconstructed ferric chelate reductase confers enhanced tolerance in rice to iron deficiency in calcareous soil. Proc. Natl. Acad. Sci. U. S. A. 104, 7373–7378.

Jang, J.Y., Lee, S.H., Rhee, J.Y., Chung, G.C., Ahn, S.J., Kang, H., 2007. Transgenic Arabidopsis and tobacco plants overexpressing an aquaporin respond differently to various abiotic stresses. Mol. Biotechnol. 40, 280–292.

Ji, H., Liu, L., Li, K., Xie, Q., Wang, Z., Zhao, X., Li, X., 2014. PEG-mediated osmotic stresses induce premature differentiation of the root apical meristem and outgrowth of lateral roots in wheat. J. Exp. Bot. 65, 4863–4872.

Ji, X., Shiran, B., Wan, J., Lewis, D.C., Jenkins, C.L.D., Condon, A.G., Richard, R.A., Dolferus, R., 2010. Importance of pre-anthesis anther sink strength for maintenance of grain number during reproductive stage water stress in wheat. Plant Cell Environ. 33 (6), 926–942.

Jin, R., Kim, B.H., Ji, C.Y., Kim, H.S., Li, H.M., Ma, D.F., Kwak, S.-S., 2017. Overexpressing *IbCBF3* increases low temperature and drought stress tolerance in transgenic sweet potato. Plant Physiol. Biochem. https://doi.org/10.1016/j.plaphy.2017.06.002.

Joshi, R., Karan, R., 2013. Physiological, biochemical and molecular mechanisms of drought tolerance in plants. In: Gaur, R.K., Sharma, P. (Eds.), Molecular Approaches in Plant Abiotic Stress. CRC Press, Boca Raton, FL, pp. 209–231.

Joshi, R., Singh, B., Bohra, A., Chinnusamy, V., 2016. Salt stress signalling pathways: specificity and cross talk. In: Wani, S.H., Hossain, M.A. (Eds.), Managing Salinity Tolerance in Plants: Molecular and Genomic Perspectives. CRC Press, Boca Raton, FL, pp. 51–78.

Kazan, K., 2015. Diverse roles of jasmonates and ethylene in abiotic stress tolerance. Trends Plant Sci. 20, 219–229.

Khazaei, H., Street, K., Bari, A., Mackay, M., Stoddard, F.L., 2013. The FIGS (Focused Identification of Germplasm Strategy) approach identifies traits related to drought adaptation in *Vicia faba* genetic resources. PLoS One. 8(5), e63107.

Kim, E.H., Kim, Y.S., Park, S.H., Koo, Y.J., Do Choi, Y., Chung, Y.Y., Kim, J.K., 2009. Methyl jasmonate reduces grain yield by mediating stress signals to alter spikelet development in rice. Plant Physiol. 149 (4), 1751–1760.

Kiriga, W.J., Yu, Q., Bill, R., 2016. Breeding and genetic engineering of drought-resistant crops. Int. J. Agric. Crop Sci. 9 (1), 7–12.

Krouk, G., Ruffel, S., Gutiérrez, R.A., Gojon, A., Crawford, N.M., Coruzzi, G.M., Lacombe, B., 2011. A framework integrating plant growth with hormones and nutrients. Trends Plant Sci. 16 (4), 178–182.

Kumar, G.B.S., Ganapathi, T.R., Revathi, C.J., Srinivas, L., Bapat, V.A., 2005. Expression of hepatitis B surface antigen in transgenic banana plants. Planta 222 (3), 484–493.

Lang, A., Otto, M., 2010. A synthesis of laboratory and field studies on the effects of transgenic *Bacillus thuringiensis* (Bt) maize on non-target Lepidoptera. Entomol. Exp. Appl. 135 (2), 121–134.

Lee, Y.P., Kim, S.H., Bang, J.W., Lee, H.S., Kwak, S.S., Kwon, S.Y., 2007. Enhanced tolerance to oxidative stress in transgenic tobacco plants expressing three antioxidant enzymes in chloroplasts. Plant Cell Rep. 26 (5), 591–598.

Lemaux, P.G., 2009. Genetically engineered plants and foods: a scientist's analysis of the issues. Part II. Annu. Rev. Plant Biol. 60, 511–559.

Li, B., Wei, A., Song, C., Li, N., Zhang, J., 2008. Heterologous expression of the *TsVP* gene improves the drought resistance of maize. Plant Biotechnol. J. 6, 146–159.

Li, X.J., Yang, M.F., Chen, H., Qu, L.Q., Chen, F., Shen, S.H., 2010. Abscisic acid pretreatment enhances salt tolerance of rice seedlings: proteomic evidence. Biochim. Biophys. Acta Protein Proteomics 1804 (4), 929–940.

Lisar, S.Y., Rahman, I.M., Hossain, M.M., Motafakkerazad, R., 2012. Water Stress in Plants: Causes, Effects and Responses. INTECH Open Access Publisher, Rijeka, Croatia.

Lu, C., Li, Y., Chen, A., Li, L., Zuo, J., Tian, H., Zhu, B., 2010. LeERF1 improves tolerance to drought stress in tomato (*Lycopersicon esculentum*) and activates downstream stress-responsive genes. Afr. J. Biotechnol. 9 (38), 6294–6300.

Manjunath, T.M., 2004. Bt Cotton in India: The Technology Wins as the Controversy Wanes. http://www.monsanto.co.uk/news/ukshowlib.html?wid=8478.

McKersie, B.D., Bowley, S.R., Harjanto, E., Leprince, O., 1996. Water-deficit tolerance and field performance of transgenic alfalfa over expressing superoxide dismutase. Plant Physiol. 111, 1177–1181.

Miller, G., Suzuki, N., Ciftci-Yilmaz, S., Mittler, R., 2010. Reactive oxygen species homeostasis and signalling during drought and salinity stresses. Funct. Plant Biol. 33, 453–467.

Miyazawa, S.I., Yoshimura, S., Shinzaki, Y., Maeshima, M., Miyake, C., 2008. Deactivation of aquaporins decreases internal conductance to CO_2 diffusion in tobacco leaves grown under long-term drought. Funct. Plant Biol. 35, 556–564.

Mizoi, J., Ohori, T., Moriwaki, T., Kidokoro, S., Todaka, D., Maruyama, K., Kusakabe, K., Osakabe, Y., Shinozaki, K., Yamaguchi-Shinozaki, K., 2013. *GmDREB2A*; 2, a canonical dehydration-responsive element-bindingprotein2-type transcription factorin soybean, is post translationally regulated and mediates dehydration-responsive element-dependent gene expression. Plant Physiol. 161, 346–361.

Nagamiya, K., Motohashi, T., Nakao, K., Prodhan, S., Hattori, E., Hirose, S., Ozawa, K., Ohkawa, Y., Takabe, T., Takabe, T., Komamine, A., 2007. Enhancement of salt tolerance in transgenic rice expressing an *Escherichia coli* catalase gene, *kat E*. Plant Biotechnol. Rep. 1, 49–55.

Nakazono, M., Tsuji, H., Li, Y., Saisho, D., Arimura, S., Tsutsumi, N., Hirai, A., 2000. Expression of a gene encoding mitochondrial aldehyde dehydrogenase in rice increases under submerged conditions. Plant Physiol. 124, 587–598.

Nam, K.H., Kim, D.Y., Shin, H.J., Nam, K.J., An, J.H., Pack, I.S., Kim, C.G., 2014. Drought stress-induced compositional changes in tolerant transgenic rice and its wild type. Food Chem. 153, 145–150.

Naor, A., 2006. Irrigation scheduling and evaluation of tree water status in deciduousorchards. Hortic. Rev. 32, 111–165.

Nelson, D.E., Repetti, P.P., Adams, T.R., Creelman, R.A., Wu, J., Warner, D.C., Anstrom, D.C., Bensen, R.J., Castiglioni, P.P., Donnarummo, M.G., Hinchey, B.S., Kumimoto, R.W., Maszle, D.R., Canales, R.D., Krolikowski, K.A., Dotson, S.B., Gutterson, N., Ratcliffe, O.J., Heard, J.E., 2007. Plant nuclear factor *Y (NF-Y) B* subunits confer drought tolerance and lead to improved corn yields on water-limited acres. Proc. Natl. Acad. Sci. U. S. A. 104 (42), 16450–16455.

Newell-McGloughlin, M., 2008. Nutritionally improved agricultural crops. Plant Physiol. 147, 939–953.

O'farrell, P.J., Anderson, P.M.L., Milton, S.J., Dean, W.R.J., 2009. Human response and adaptation to drought in the arid zone: lessons from southern Africa. S. Afr. J. Sci. 105 (1–2), 34–39.

Oh, S.J., Kim, Y.S., Kwon, C.W., Park, H.K., Jeong, J.S., Kim, J.K., 2009. Overexpression of the transcription factor *AP37* in rice improves grain yield under drought conditions. Plant Physiol. 150 (3), 1368–1379.

Ortiz, R., Iwanaga, M., Reynolds, M.P., Wu, H., Crouch, J.H., 2007. Overview on crop genetic engineering for drought-prone environments. J. SAT Agric. Res. 4, 1–30.

Paarlberg, R.L., 2008. Starved for Science: How Biotechnology is Being Kept out of Africa. Harvard University Press, Cambridge, MA.

Park, E.J., Jeknic, Z., Pino, M.T., Murata, N., Chen, T.H., 2007. Glycinebetaine accumulation in chloroplasts is more effective than that in cytosol in protecting transgenic tomato plants against abiotic stress. Plant Cell Environ. 30, 994–1005.

Park, S.I., Kim, Y.S., Kim, J.J., Mok, J.E., Kim, Y.H., Park, H.M., Yoon, H.S., 2017. Improved stress tolerance and productivity in transgenic rice plants constitutively expressing the *Oryza sativa* glutathione synthetase *OsGS* under paddy field conditions. J. Plant Physiol. 215, 39–47.

Peng, L.X., Gu, L.K., Zheng, C.C., Li, D.Q., Shu, H.R., 2006. Expression of *MaMAPK* gene in seedlings of *Malus* L. under water stress. Acta Biochim. Biophys. Sin. 38, 281–286.

Pérez-Clemente, R.M., Vives, V., Zandalinas, S.I., López-Climent, M.F., Muñoz, V., Gómez-Cadenas, A., 2013. Biotechnological approaches to study plant responses to stress. Biomed. Res. Int. 2013, 654120. https://doi.org/10.1155/2013/654120.

Praba, M.L., Cairns, J.E., Babu, R.C., Lafitte, H.R., 2009. Identification of physiological traits underlying cultivar differences in drought tolerance in rice and wheat. J. Agron. Crop Sci. 195, 30–46.

Qin, F., Kakimoto, M., Sakuma, Y., Maruyama, K., Osakabe, Y., Tran, L.P., Shinozaki, K., Yamaguchi-Shinozaki, K., 2007. Regulation and functional analysis of ZmDREB2A in response to drought and heat stress in *Zea mays* L. Plant J. 50, 54–69.

Remans, T., Nacry, P., Pervent, M., Filleur, S., Diatloff, E., Mounier, E., Tillard, P., Forde, B.G., Gojon, A., 2006. The Arabidopsis NRT1.1 transporter participates in the signaling pathway triggering root colonization of nitrate-rich patches. Proc. Natl. Acad. Sci. U. S. A. 103, 19206–19211.

Rivero, R.M., Gimeno, J., Van Deynze, A., Walia, H., Blumwald, E., 2010. Enhanced cytokinin synthesis in tobacco plants expressing *PSARK::IPT* prevents the degradation of photosynthetic protein complexes during drought. Plant Cell Physiol. 51, 1929–1941.

Rivero, R.M., Kojima, M., Gepstein, A., Sakakibara, H., Mittler, R., Gepstein, S., Blumwald, E., 2007. Delayed leaf senescence induces extreme drought tolerance in a flowering plant. Proc. Natl. Acad. Sci. U. S. A. 104, 19631–19636.

Sanchez, D.H., Pieckenstain, F.L., Szymanski, J., Erban, A., Bromke, M., Hannah, M.A., Kraemer, U., Kopka, J., Udvard, M.K., 2011. Comparative functional genomics of salt stress in related model and cultivated plants identifies and overcomes limitations to translational genomics. PLoS One. 6. e17094.

Seiffert, B., Zhou, Z.W., Wallbraun, M., Lohaus, G., Mollers, C., 2004. Expression of a bacterial asparagine synthetase gene in oilseed rape (*Brassica napus*) and its effect on traits related to nitrogen efficiency. Physiol. Plant. 121, 656–665.

Seki, M., 2002. Monitoring the expression profiles of 7000 Arabidopsis genes under drought, cold and high-salinity stresses using full-length cDNA microarray. Plant J. 31, 279–292.

Seo, H.M., Jung, Y., Song, S., Kim, Y., Kwon, T., Kim, D.H., Jeung, S.J., Yi, Y.B., Yi, G., Nam, M.H., Nam, J., 2008. Increased expression of *OsPT1*, a high-affinity phosphate transporter, enhances phosphate acquisition in rice. Biotechnol. Lett. 30, 1833–1838.

Shah, S.H., Ali, S., Hussain, Z., Jan, S.A., Din, J.U., Ali, G.M., 2016. Genetic improvement of tomato (*Solanum lycopersicum*) with *AtDREB1A* gene for cold stress tolerance using optimized Agrobacterium-mediated transformation system. Int. J. Agric. Biol. 18, 471–482.

Shan, D.P., Huang, J.G., Yang, Y.T., Guo, Y.H., Wu, C.A., Yang, G.D., Gao, Z., Zheng, C.C., 2007. Cotton *GhDREB1* increases plant tolerance to low temperature and is negatively regulated by gibberellic acid. New Phytol. 176, 70–81.

Shinozaki, K., Yamaguchi-Shinozaki, K., 2007. Gene networks involved in drought stress response and tolerance. J. Exp. Bot. 58, 221–227.

Sinclair, T.R., Purcell, L.C., Sneller, C.H., 2004. Crop transformation and the challenge to increase yield potential. Trends Plant Sci. 9 (2), 70–75.

Singh, D., Laxmi, A., 2015. Transcriptional regulation of drought response: a tortuous network of transcriptional factors. Front. Plant Sci. 6, 1–12.

Singh, O.V., Ghai, S., Paul, D., Jain, R.K., 2006. Genetically modified crops: success, safety assessment, and public concern. Appl. Microbiol. Biotechnol. 71 (5), 598–607.

Stein, A.J., Rodríguez-Cerezo, E., 2009. The Global Pipeline of New GM Crops: Implications of Asynchronous Approval for International Trade. European Commission, Joint Research Centre, Institute for Prospective Technology Studies, Sevilla.

Tacket, C.O., 2007. Plant-based vaccines against diarrheal diseases. Trans. Am. Clin. Climatol. Assoc. 118, 79–87.

Takahashi, M., Nakanishi, H., Kawasaki, S., Nishizawa, N.K., Mori, S., 2001. Enhanced tolerance of rice to low iron availability in alkaline soils using barley nicotianamine aminotransferase genes. Nat. Biotechnol. 19, 466–469.

Tang, G., Qin, J., Dolnikowski, G.G., Russell, R.M., Grusak, M.A., 2009. Golden Rice is an effective source of vitamin A. Am. J. Clin. Nutr. 89 (6), 1776–1783.

Telem, R.S., Wani, S.H., Singh, N.B., Sadhukhan, R., Mandal, N., 2016. Single nucleotide polymorphism (SNP) marker for abiotic stress tolerance in crop plants. In: Advances in Plant Breeding Strategies: Agronomic, Abiotic and Biotic Stress Traits. Springer International Publishing, Switzerland, pp. 327–343.

Thompson, A.J., Andrews, J., Mulholland, B.J., McKee, J.M.T., Hilton, H.W., Horridge, J.S., Farquhar, G.D., Smeeton, R.C., Smillie, I.R.A., Black, C.R., Taylor, I.B., 2007. Overproduction of abscisic acid in tomato increases transpiration efficiency and root hydraulic conductivity and influences leaf expansion. Plant Physiol. 143, 1905–1917.

Todaka, D., Nakashima, K., Shinozaki, K., Yamaguchi-Shinozaki, K., 2012. Toward understanding transcriptional regulatory networks in abiotic stress responses and tolerance in rice. Rice 5 (6), 1–9.

Todaka, D., Shinozaki, K., Yamaguchi-Shinozaki, K., 2015. Recent advances in the dissection of drought-stress regulatory networks and strategies for development of drought-tolerant transgenic rice plants. Front. Plant Sci. 6 (84), 1–20.

Tran, L.S., Nakashima, K., Shinozaki, K., Yamaguchi-Shinozaki, K., 2007. Plant gene networks in osmotic stress response: from genes to regulatory networks. Methods Enzymol. 428, 109–128.

Valliyodan, B., Nguyen, H.T., 2006. Understanding regulatory networks and engineering for enhanced drought tolerance in plants. Curr. Opin. Plant Biol. 9, 189–195.

Villalobos, F.J., Testi, L., Orgaz, F., Garcia-Tejera, O., Lopez-Bernal, A., Gonzalez-Dugo, V., Ballester-Lurbe, C., Castel, J.R., Alarcon-Cabañero, J.J., Nicolas-Nicolas, E., Girona, J., Marsal, J., Fereres, E., 2013. Modelling canopy conductance andtranspiration of fruit trees in Mediterranean areas: a simplified approach. Agric. For. Meteorol. 171, 93–103.

Vincent, R., Fraisier, V., Chaillou, S., Limami, M.A., Deleens, E., Phillipson, B., Douat, C., Boutin, J.P., Hirel, B., 1997. Overexpression of a soybean gene encoding cytosolic glutamine synthetase in shoots of transgenic *Lotus corniculatus* L. plants triggers changes in ammonium assimilation and plant development. Planta 201, 424–433.

Viveros, A.M.F., Inostroza-Blancheteau, C., Timmermann, T., González, M., Arce- Johnson, P., 2013. Overexpression of *GlyI* and *GlyII* genes in transgenic tomato (*Solanumlycopersicum* Mill.) plants confers salt tolerance by decreasing oxidative stress. Mol. Biol. Rep. 40, 3281–3290.

Vob, U., Bishopp, A., Farcot, E., Bennett, M.J., 2014. Modelling hormonal response and development. Trends Plant Sci. 19, 311–319.

Wan, J., Griffiths, R., Ying, J., McCourt, P., Huang, Y., 2009. Development of drought-tolerant canola (*Brassica napus* L.) through genetic modulation of ABA-mediated stomatal responses. Crop Sci. 49, 1539–1554.

Wang, R.C., Xing, X.J., Crawford, N., 2007. Nitrite acts as a transcriptome signal at micromolar concentrations in Arabidopsis roots. Plant Physiol. 145, 1735–1745.

Wang, Y., Wisniewski, M., Meilan, R., Cui, M., Fuchigami, L., 2006. Transgenic tomato (*Lycopersicon esculentum*) overexpressing *cAPX* exhibits enhanced tolerance to UV-B and heat stress. J. Appl. Hortic. 8, 87–90.

Wang, Y.J., Hao, Y.J., Zhang, Z.G., Chen, T., Zhang, J.S., Chen, S.Y., 2005. Isolation of trehalose-6-phosphate phosphatase gene from tobacco and its functional analysis in yeast cells. J. Plant Physiol. 162, 215–223.

Wilkinson, S., Davies, W.J., 2010. Drought, ozone, ABA and ethylene: new insights from cell to plant to community. Plant Cell Environ. 33, 510–525.

Xiao, B.Z., Chen, X., Xiang, C.B., Tang, N., Zhang, Q.F., Xiong, L.Z., 2009. Evaluation of seven function-known candidate genes for their effects on improving drought resistance of transgenic rice under field conditions. Mol. Plant 2, 73–83.

Yang, S., Vanderbeld, B., Wan, J., Huang, Y., 2010. Narrowing down the targets: towards successful genetic engineering of drought-tolerant crops. Mol. Plant 3 (3), 469–490.

Zhang, J., Movahedi, A., Sang, M., Wei, Z., Xu, J., Wang, X., Wu, X., Wang, M., Yin, T., Zhuge, Q., 2017a. Functional analyses of *NDPK2* in *Populus trichocarpa* and overexpression of *PtNDPK2* enhances growth and tolerance to abiotic stresses in transgenic poplar. Plant Physiol. Biochem. 117, 61–74.

Zhang, P., Wang, W.Q., Zhang, G.L., Kaminek, M., Dobrev, P., Xu, J., Gruissem, W., 2010. Senescence-inducible expression of isopentenyltransferase extends leaf life, increases drought stress resistance and alters cytokinin metabolism in cassava. J. Integr. Plant Biol. 52, 653–669.

Zhang, Q., 2007. Strategies for developing green super rice. Proc. Natl. Acad. Sci. U. S. A. 104 (42), 16402–16409.

Zhang, Y.J., Xie, M., Li, Q., Zhang, X.L., Zhang, Z.R., 2017b. Monitoring changes in the action bacterial field communities present in the rhizosphere soil of a transgenic cotton producing Cry1Ab/Ac proteins. Crop Prot. 91, 1–7.

Zhang, Y.Y., Li, Y., Gao, T., Zhu, H., Wang, D.J., Zhang, H.W., Guo, H.S., 2008. Arabidopsis *SDIR1* enhances drought tolerance in crop plants. Biosci. Biotechnol. Biochem. 72 (8), 2251–2254.

CHAPTER 7

DNA Helicase-Mediated Abiotic Stress Tolerance in Plants

Maryam Sarwat*, Narendra Tuteja†
*Amity Institute of Pharmacy, Amity University, Noida, India
†Plant Molecular Biology, International Centre for Genetic Engineering and Biotechnology (ICGEB), New Delhi, India

Contents

1. INTRODUCTION

Helicases are the molecular motors that convert nucleic acids from double-stranded into single-stranded. They function by breaking the hydrogen bonds in nucleic acids, as well as by disrupting the non-covalent bonds between complementary strands (Vashisht and Tuteja, 2006). Helicases use the energy of nucleoside 5′-triphosphates, usually ATP, for these processes (Kornberg and Baker, 1991; Matson et al., 1994; Tuteja and Tuteja, 1996).

The helicases are classified broadly by their nucleic acid preferences; DNA helicases unwind DNA, and RNA helicases unwind RNA, and some helicases can also unwind DNA:RNA hybrids. DNA helicases, also known as the "DNA unwinding enzyme," were first isolated from *E. coli* in 1976 (Abdel-Monem et al., 1976). To date, hundreds of DNA helicases have been isolated from various organisms, for example, bacteria, bacteriophages, viruses, yeasts, *Drosophila*, frogs, mice, cows, humans, and plants (Ilyina et al., 1992; Lohman, 1992; Matson et al., 1994; Tuteja and Tuteja, 1996). Thirteen different DNA helicases have been isolated from *E. coli* (Lohman, 1992; Matson et al., 1994). Several have been isolated from humans, calf thymuses, yeasts, and viruses (Matson et al., 1994; Borowiec, 1996; Tuteja and Tuteja, 1996). In 1978, the first plant DNA helicase

Biochemical, Physiological and Molecular Avenues for Combating Abiotic Stress in Plants
https://doi.org/10.1016/B978-0-12-813066-7.00007-3

was isolated from a lily (Hotta and Stern, 1978). Throughout the years, progress in the research of plant helicases has been very slow.

1.1 Properties of Helicases

The three important characteristics of helicases are: (i) they bind to nucleic acids, (ii) they bind to NTP/dNTP and hydrolyze them, and (iii) they cause unwinding of nucleic acids (Hall and Matson, 1999). Therefore, they possess two enzymatic activities; DNA- or RNA-dependent ATPase, and NTP- (or dNTP)-dependent strand displacement.

The helicases differ in their biochemical properties, substrate specificities, and the NTPs that can act as cofactors. They also differ in their affinity for single-stranded versus double-stranded nucleic acids, as well as the processivity of the enzyme in the unwinding reaction. They further differ in the directionality of unwinding. The helicases are often found in multimeric subunits of two or six.

2. GENOMICS

There has been a debate over how many helicases are required for the proper functioning and viability of a particular organism. Researchers have determined that out of the 11 helicases found in *E. coli*, only 1 (dnaB) is essential, as its absence is lethal to the organism. When the single gene is mutated, it causes severe phenotypes, such as UV sensitivity or slow growth. When more than one gene is mutated, it can be lethal. For example, *rep* and *uvrD* double mutants lose their viability.

3. STRUCTURE OF THE HELICASE

Bioinformatic analyses revealed seven short conserved helicase motifs named I, Ia, II, III, IV, V, and VI (Gorbalenya et al., 1989, 1998; Gorbalenya and Koonin, 1993; Tanner and Linder, 2001).

Based on the similarities and differences of these motifs, helicases have been classified into three super-families: SF1, SF2, and SF3; with a few exceptions, such as the DnaB proteins in bacteria. The high conservation in helicase sequences suggests that the helicases are evolved from a common ancestor. The seven helicase motifs of SF1 and SF2 are present in 200–700 amino acids, called core regions. In between these conserved motifs, there are stretches of amino acids that have low sequence conservation, but are highly conserved in length. The conserved regions are responsible for helicase activity, and the divergent regions are responsible for individual protein functions. Motifs I and II are homologous to Walker boxes A and B, and should play a role in the binding of nucleotide cofactors, as they appear in many NTP-binding proteins. In motif I, the highly conserved amino acids 'GxGKS/T' interact with the carbon 6 atom of the sugar and phosphates of the nucleotide co-factor. Motif II has a conserved aspartate residue that should interact with Mg^{2+} ion for catalysis. The conserved glutamine residue of motif II should coordinate the reacting water during ATP hydrolysis. Because of these two residues, the two different helicase families are named DExx-box, or DEAD-box helicases. Any mutation in motifs I or II cause impaired ATPase and DNA-unwinding activities in enzymes. It is now known that the DEAD-box helicase family contains more than 55 members in *Arabidopsis*. In yeast, a Q motif has been identified upstream of motif I. It is thought to regulate ATP binding and hydrolysis in eIF4a (Tanner et al., 2003).

The enzyme kinetic studies and binding data have proposed two models for the DNA unwinding mechanism of helicases, the "active rolling model" (SenGupta and Borowiec, 1992; Wong and Lohman, 1992; Lohman, 1992) and the "inchworm model" (Yarranton and Gefter, 1979; Lohman, 1992; Lohman and Bjornson, 1996). Both models indicated that the enzyme has different states that are modulated by co-factor binding and hydrolysis. It depends on which state is kinetically favored to bind the single-stranded or double-stranded DNA. The active rolling model emphasizes the two subunits of the enzyme, whereas the inchworm model indicates that only a monomer is required to unwind the DNA. The former model also says that, in a cycle, the number of base pairs (bp) unwound is equal to the number of nucleotides bound. However, in the latter they can be different.

Other helicase motifs are specific to particular helicases and play an important role in DNA binding, and other activities imperative for enzyme catalysis. Amino acid substitutions in motif III can impair the enzyme from its strand displacement activity, while the NTPase activity remains intact.

4. IDENTIFICATION OF HELICASES

The research on helicases has shown two major methods for their identification:

(a) through the presence of helicase activity, and

(b) through the presence of seven characteristic and well-conserved motifs in the amino acid sequence.

Sometimes, a protein contains all the seven helicase motifs, and yet does not possess the biochemical activity of a helicase, and vice versa. This is the drawback of identifying helicases bioinformatically. One of the best examples is the eukaryotic SWI2/SNF2 protein, which contains seven motifs, and is a potent DNA-dependent ATPase, but is not a DNA unwinding enzyme. Its function, however, is to reorganize, or remodel, chromatin. Such proteins are called helicase pretenders. Through bioinformatics analysis, these helicase pretenders are grouped into phylogenetically identifiable families based on their amino acid sequence alignments. Seventeen SWI2/SNF2 homologues have been identified in *Saccharomyces cerevisiae*. They participate in diverse functions such as transcriptional regulation and DNA repair and recombination. For these functions, the chromatin structure should be changed, so that it is accessible to other protein factors. This shows the functional similarity between real helicases and imposters, as both provide accessibility; thus converting a closed duplex into two open single strands so that the DNA can act on other protein factors.

In the case of plants, helicases are present in all the three organelles: the nucleus, mitochondrion, and chloroplast, which contain their own genomes (Gagliardi et al., 1999; Pham et al., 2000; Tuteja et al., 1996, 2001a). This suggests their involvement in the growth and development of the plants.

Most of the helicases belong to one of the three classes:

- chromatin-remodeling proteins,
- RNA helicases,
- or DNA helicases.

5. DNA HELICASES

The general functions of DNA helicases include DNA replication, DNA repair and recombination, chromosome segregation, and transcription initiation. Many aspects of DNA helicases have yet to be studied. Some of these are the structural biology of the different helicases, and the molecular mechanisms that specify their cellular functions. Now that we have complete, analyzed sequences of the genomes of major model organisms, our next venture should be to investigate the field of helicase genomics. Putative helicases are recognized by the presence of the seven motifs, or by proof of their helicase activity through biochemical assays.

5.1 Plant DNA Helicases

We have far less information about plant DNA helicases. The first plant DNA helicase identified from a lily was partially purified by Hotta and Stern (1978). The research group of Tuteja et al. (1996) has identified and purified the first plant DNA helicase. Later, one of the group members (Pham et al., 2000) cloned the first plant DNA helicase encoding biochemically active protein.

5.1.1 Functions of Plant DNA Helicases

There was a need for a systematic study of the functions of plant DNA helicases. There is indirect evidence shedding light on their roles in many different pathways. Herein, we have described the functions of biochemically active helicases, and the possible functions of some of the plant DNA helicases. Hotta and Stern (1978) have suggested the possible role of lily DNA helicases in plant DNA recombination, as it is prominent during the meiotic prophase of plants. Studies on chloroplast DNA revealed (Tuteja and Tewari, 1999) that replication begins by the formation of two D-loops, or displacement loops, that start at oriA and oriB, and expand toward each other. DNA helicases help in unwinding the DNA at these points and forming a Cairns replicative forked structure.

These structures expand bi-directionally, and form two daughter molecules (Tuteja and Tewari, 1999). The chloroplast DNA helicase II works in a replication fork; Tuteja and Phan (1998) suggested its role in chloroplast DNA replication. The mutants lacking a particular helicase cause defects in the biochemical, cellular, and genetic levels, among others. However, the mechanism behind it is still unknown. The common properties of all DNA helicases are their ATPase and strand displacement enzymatic activities. Researchers are finding out the substrate specificities of the helicases and their interacting proteins, or protein complexes. They are of the opinion that it will shed light on the mechanistic aspect of a particular helicase. The interactions between a helicase and other proteins can change the preferences of a particular substrate functioning in a biochemical pathway. Some of the plant DNA helicases and their functions are discussed as follows.

5.1.2 Pea DNA Helicase 45

Pham et al. (2000) identified a DNA helicase from *Pisum sativum* (PDH45) that is homologous to eIF-4A. It conducts multiple activities, suggesting its involvement in DNA and RNA metabolism in the initiation of translation, in addition to maintaining the basic activities of the cell. Its transcript was reported to be upregulated in pea plants in response to high salinity, cold stress, abscisic acid (ABA), dehydration, and early wounding. The first direct evidence that overexpression of PDH45 confers salinity stress tolerance without yield loss has also been reported by the group led by Dr. Tuteja. The promoter analysis of the PDH45 gene showed cis-regulatory elements that might act as binding sites for RNA polymerase and transcription factors, and control the regulation of gene expression. His group has also reported certain stress-responsive cis-regulatory elements that were believed to regulate the expression of PDH45 under abiotic stress conditions (Tajrishi and Tuteja, 2011).

The PDH45, when introduced in chili, showed an improved response to a wide variety of simulated abiotic stresses and enhanced expression of several stress-responsive genes. The survival and recovery of transgenic plants was significantly higher, and the plants exhibited increased growth and productivity with improved water use efficiency (Shivakumara et al., 2017). Augustine et al. (2015a) over-expressed the PDH45 gene in sugarcane, and reported that the transgenic plants exhibit significantly higher cell membrane thermostability, transgene expression, relative water content, gas exchange parameters, chlorophyll content, and photosynthetic efficiency under soil moisture stress, compared with the wild type (WT). The transgenic plants showed higher germination ability and better chlorophyll retention than the WT under salinity stress. They also showed upregulation of DREB2-induced downstream stress-related genes.

Gene pyramiding has proven to be a useful approach to express more than one good character in a single plant. When PDH45 was expressed along with the *Pennisetum glaucum* heat-shock factor (*PgHSF4*) and the Alfalfa zinc finger 1 (*Alfin1*) gene, which is a root growth-associated transcription factor, higher growth and productivity was observed under drought stress conditions. The transgenic lines also exhibited higher tolerance to ethrel-induced senescence and methyl viologen-induced oxidative stress. The stress-responsive genes such as heat-shock proteins (*HSPs*), RING box protein-1 (*RBX1*), aldose reductase, late embryogenesis abundant-5 (*LEA5*) and proline-rich protein-2 (*PRP2*), showed increased expression under stress conditions (Ramu et al., 2016). It has already been reported that the *DREB2* gene from *Erianthus arundinaceus* (*EaDREB2*) overexpression causes enhanced tolerance to drought and salinityin sugarcane. When gene pyramiding is done, taking this gene, along with PDH45, all the physiological, molecular, and morphological parameters showed that the transgenic plants exhibited a greater level of salinity tolerance than the single-gene transgenics (Augustine et al., 2015b).

A homologue of the PDH45 was isolated from *Medicago sativa* (alfalfa), referred to as *M. sativa* helicase 1 (MH1). When the onion epidermis was transiently transfected by the 35S::MH1-GFP construct, it showed that MH1 was localized in the nucleus of the roots, stems, and leaves of alfalfa. Its transcript level was found to be increased under mannitol, NaCl, methyl viologen, and ABA stress. When the MH1 was expressed in *Arabidopsis*, the transgenic plants showed improved seed germination and plant growth under drought, salt, and oxidative stress conditions. The capacity for osmotic adjustment, superoxide dismutase, and ascorbate peroxidase activities and proline content were also elevated in these plants (Luo et al., 2009).

Tuteja et al. (2001a) identified a plant nuclear DNA helicase (PDH65) that was localized within the dense fibrillar component of pea nucleoli in the regions around the rDNA transcription sites. Studies suggest that PDH65 may be involved both in rDNA transcription, and in the early stages of pre-rRNA processing.

5.1.3 Ku Protein

Another DNA helicase "Ku" has been reported to play an important role in DSB repair in the NHEJ pathway. It has also been involved inV(D)J recombination in differentiating lymphocytes, DNA replication, transcription regulation, regulation of heat shock-induced responses, regulation of structure of telomeric termini, and also in G2 and M phases of the cell cycle (Mimori et al., 1981; Tuteja and Tuteja, 2000). When the DNA-damaging agents (bleomycin and methylmethane sulphonate) are applied to plants, the transcription of Ku genes (*AtKu70* and *AtKu80*) was up-regulated. It suggests the importance of plant Ku proteins for DSB repair, probably through the NHEJ pathway (Tamura et al., 2002). Other reports show a lack of Ku70 causes dramatic deregulation of telomere length control, thus indicating Ku's role in telomeric strand maintenance in plants.

In *Arabidopsis*, the *KU70/80* genes are found to be ubiquitously expressed, and their products are reported to form stable heterodimers in vitro (Riha et al., 2002). The T-DNA insertion mutants at *KU70* have no growth or developmental defects, but the mutant seedlings exhibited hypersensitivity to gamma-irradiation-induced double-strand breaks.

These mutants are also reported to be hypersensitive to methyl methanosulphonate during seed germination. They behave normally during seedling development, thus indicating that the plants require NHEJ differentially during their course of development. In the *Arabidopsis* plants lacking Ku70, telomeric length control is deregulated (Riha et al., 2002). We already know that the telomerase enzyme is responsible for telomere elongation. When the telomerase, or telomere capping proteins become dysfunctional, it leads to end-to-end chromosome fusions. The Ku protein seems to prevent this dimerization, as the Tu 70/80 heterodimer is found to be present in telomeres. In the case of mammals, the Tu 70/80 heterodimer is shown to prevent fusion of telomeres. Research on Ku70 mutant plants, and in plants deficient in both Ku70 and the catalytic subunit of telomerase (TERT), has shown extended terminal 3 overhang (Riha and Shippen, 2003). This indicates that Ku is an integral part of the maintenance of the telomeric C-rich strand. In Ku70, TERT double mutant-increased telomere shortening was reported, which is consistent with C-strand maintenance. In the terminal plants, where telomeres are critically shortened, the anaphase bridges do not require Ku for the formation of end-to-end chromosome fusions in the *Arabidopsis* plants, which are deficient in the telomerase enzyme. However, the situation is quite different in mammals, where Ku deficiency leads to chromosome fusions. All these findings indicate that Ku 70 is imperative for the maintenance of the telomeric C-rich strand in higher eukaryotes (Riha and Shippen, 2003).

5.1.4 Nucleolin

Nucleolin is a well-known DNA and RNA helicase in humans. Its helicase activity in plants is being explored now. Its role has been directly or indirectly shown in many metabolic processes, such as ribosome biogenesis, cell proliferation and growth, cytoplasmic-nucleolar transport of ribosomal components, replication, signal transduction, and many more (Bogre et al., 1996; Tuteja and Tuteja, 1998). Bogre et al. (1996) reported the induction of the plant homologue of the human nucleolin gene *nucMs1* in roots and other meristematic cells of the plant. Being tightly linked to cell division and cell proliferation, its role in the growth and development of plants is imperative. The nucleolin gene (*nucMs1*) and the cyclingene (*cycMs4*) of *M. sativa* showed simultaneous expression in the G1 phase before the onset of DNA synthesis (Bogre et al., 1996).

The nucleolin gene is used as a marker for proliferation events during the course of flower development. Tong et al. (1997) showed pea nucleolin to be light (phytochrome) regulated.

5.1.5 XPB and XPD

Nucleotide excision repair (NER) is an important component of DNA repair machinery in mammals. One of its constituents is TFIIH, which consists of six to nine subunits. XPB (ERCC3) and XPD (ERCC2) DNA helicases are part of this complex, and play an important role in NER (Tuteja and Tuteja, 2001). The NER system is not well characterized in plants (Tuteja et al., 2001b). Ribeiro et al. (1998) cloned the plant homologue of XPB from *Arabidopsis* (*araXPB*). It gives molecular evidence of the conservation of the NER pathway. Xu et al. (1998) cloned the homologue of human *ERCCI* from a lily. When the araXPB protein is compared with its yeast and human homologues, it's found to be 50% identical. Moreover, 70% of their amino acids were found to be conserved. The bioinformatics analysis of plant XPB revealed all the functional domains, including the nuclear localization signal, DNA-binding domain, and helicase motifs present; thus predicting its possible role in plant NER machinery (Ribeiro et al., 1998). To date, no one has shown the DNA unwinding activity of plant XPB. However, research on *AtXPB1*, the *XPB/RAD25* homologue gene from *Arabidopsis thaliana*, has suggested its possible role in DNA repair and plant development (Costa et al., 2001). In another report, the bioinformatic study of rice *XPB* (*OsXPB2*) revealed cis-elements, accounting for various abiotic stresses (salt, dehydration, or cold). It contains CACG, GTAACG, CACGTG, and CGTCA CCGCCGCGCT cis-acting elements, which are reported to be salt, dehydration, cold, MeJA, or ABA responsiveness, respectively. When theOsXPB2::GUS chimeric construct was analyzed in a transient assay in tobacco leaves, it showed hormone-induced (Auxin, ABA, or MeJA) GUS expression/activity in the promoter-reporter assay (Raikwar et al., 2015).

5.1.6 Mini-Chromosome Maintenance Proteins

The MCM proteins also known as mini-chromosome maintenance proteins were first identified in yeast mutants. They are involved in the initiation and elongation of DNA during eukaryotic DNA replication. The human MCM protein complex (MCM4/6/7) is known for its role as a DNA helicase (Ishimi, 1997). The *MCM* genes are reported in plants such as *Arabidopsis*, maize, tobacco, peas, and rice (Ivanova et al., 1994; Sabelli et al., 1996, 1999; Springer et al., 1995, 2000; Tuteja et al., 2011). They are highly expressed in dividing tissues such as the shoot apex and root tips, localized in the nucleus and cytosol. They also play an important role in DNA replication in plants, megagametophyte, and embryo development. Six *MCM* coding genes were identified from peas and *Arabidopsis* that belong to six distinct classes of MCM protein in higher plants. Their function also seems to be conserved among the eukaryotes.

A single subunit MCM6 from peas has been shown to contain helicase and ATPase activities in vitro. The transcript level was found to be upregulated in pea plants in response to high salinity and cold stress. The transcript level was unaltered with ABA, drought, and heat stress. Overexpression of the single subunit MCM6 causes salinity stress tolerance without yield loss. The promoter of the pea MCM6 single subunit has also been studied and found to contain stress responsive elements which might play a role in its performance under abiotic stress conditions (Dang et al., 2011).

5.1.7 SUV3

The *SUV3* (suppressor of *Var 3*) gene products act as both DNA and RNA helicases. It also exhibits ATPase activity. It is localized in mitochondria, and is a subunit of the degradosome complex involved in the regulation of RNA surveillance and turnover, and ATPase activities. When SUV3 is overexpressed in rice plants, it confers salinity tolerance and drought stress. The rice SUV3 transgenic lines showed lesser lipid peroxidation, electrolyte leakage, and H_2O_2 production, along with higher activities of antioxidant enzymes under salinity stress, as compared with WT, vector control plants. The possible mechanism could be that it functions by improving photosynthesis and antioxidant machinery in transgenic rice under these stressful conditions (Tuteja et al., 2013).

Another report from the group showed the role of *OsSUV3* in cadmium and zinc stress. The transcript level of *OsSUV3* is induced under these stresses, and overexpression of *OsSUV3* confers the metal stress tolerance in transgenic IR64 rice plants (Sahoo and Tuteja, 2014).

5.2 DNA Helicases and Abiotic Stress Tolerance

The DEAD-box proteins are known as RNA helicases, but some of them play a role as DNA helicases as well (Aubourg et al., 1999; Tuteja, 2000). The cDNA microarray analysis revealed a DEAD-box helicase gene (accession number AB050574) in *Arabidopsis* (Seki et al., 2001). This gene is also induced by cold-stress, suggesting its dual role as a plant helicase and in cold stress tolerance. Gong et al. (2002) reported two chilling and freezing stress-inducible DEAD-box helicase genes in *Arabidopsis*. Their mutants showed impairment in the cold-regulated expression of CBF and their downstream target genes. The study suggested the role of these plant DEAD-box helicases not only in positive regulation of CBF genes, but also in chilling resistance. Chamot et al. (1999) reported a DEAD-box helicase gene (*crhC*) from cynobacteria that expressed specifically only under cold-shock conditions. The mechanism of action of these helicases is not clear. One of the hypotheses is that they might act by removing cold-stabilized secondary structures in cold-shock mRNAs, thus improving the cold-induced blockage, and thus inducing the translation initiation under cold-shock conditions (Thieringer et al., 1998).

Few other functions of DEAD-box helicases are reported by researchers. Jacobsen et al. (1999) showed unregulated cell division in the *Arabidopsis* floral meristems caused by a disrupted putative helicase/RNA III gene. The *VDL* gene in tobacco, responsible for variegated and distorted leaves, also acts as a DEAD-box helicase in plastids. They played an important role in chloroplast differentiation and plant morphogenesis (Wang et al., 2000). Likewise, Li et al. (2001) have demonstrated the role of *RNA helicase I* of *Vigna radiata* in the viability of its seeds.

The SDE3 gene of *Arabidopsis* that codes for a putative helicase is also a post-transcriptional regulator that acts through gene silencing (Dalmay et al., 2001).

Bourc'his and Bestor (2002) reported the DDM1for gene maintenance and genomic methylation as a homologue of the SW12/SNF2 family of helicases in *Arabidopsis*.

The HUA and HEN (HUA ENHANCER) proteins play important roles in flower and vegetative development in *Arabidopsis* (Chen and Meyerowitz, 1999; Jack, 2002; Li et al., 2001; Chen et al., 2002; Cheng et al., 2003), other than HEN being a putative DExH box helicase. *AGAMOUS (AG)* is the Arabidopsis floral organ identity gene. As described herein, *HUA1*, *HUA2*, *HEN1*, and *HEN2* genes encode nuclear proteins that have important roles in RNA metabolism. These genes also function redundantly as components of the AG pathway of floral development.

Lal et al. (2003) have shown the possible role of plant DNA helicases in a new class of transposon through a DNA helicase-bearing transposable element maize genome. These helitrons are autonomous, and codes for a 5′ to 3′ DNA helicase. They transpose by a rolling-circle mode of DNA replication, and are called rolling-circle transposons (Kapitonov and Jurka, 2001).

6. CONCLUSION

In conclusion, DNA helicases play important roles in plant physiology, apart from their well-documented roles of DNA unwinding and ATP hydrolysis. In this chapter, we have highlighted some of these diversified roles. These proteins are involved in phytohormone signaling pathways, signal transduction pathways, and as stress responsive proteins. Various studies have shown the modulation of their gene transcript in response to salt, dehydration, hormones, and cold and heat treatments. These studies imply that the genes might perform crucial functions directly involved in cellular response to specific abiotic stresses, suggesting that they are components of general stress response mechanisms. Their protein functions are further characterized by overexpression in the same plant, or ectopic expression in a different plant species. The gene pyramiding approach has also been undertaken for some helicases to develop plants, which plays a potent role in multiple physiological processes. Only a small number of DNA helicases have been functionally validated, so there's a need for exploitation of these valuable proteins in other plant systems as well.

REFERENCES

Abdel-Monem, M., Durwald, H., Hoffmann-Berling, H., 1976. Enzymic unwinding of DNA II chain separation by an ATP dependent DNA unwinding enzyme. Eur. J. Biochem. 65, 441–449.

Aubourg, S., Kreis, M., Lecharny, A., 1999. The DEAD box RNA helicase family in *Arabidopsis thaliana*. Nucleic Acids Res. 27, 628–636.

Augustine, S.M., Ashwin Narayan, J., Syamaladevi, D.P., Appunu, C., Chakravarthi, M., Ravichandran, V., Tuteja, N., Subramonian, N., 2015a. Introduction of pea DNA helicase 45 into sugarcane (*Saccharum*spp. hybrid) enhances cell membrane thermostability and upregulation of stress-responsive genes leads to abiotic stress tolerance. Mol. Biotechnol. 57, 475–488.

Augustine, S.M., Ashwin Narayan, J., Syamaladevi, D.P., Appunu, C., Chakravarthi, M., Ravichandran, V., Tuteja, N., Subramonian, N., 2015b. Overexpression of EaDREB2 and pyramiding of EaDREB2 with the pea DNA helicase gene (PDH45) enhance drought and salinity tolerance in sugarcane (*Saccharum* spp. hybrid). Plant Cell Rep. 34, 247–263.

Bogre, L., Jonak, C., Mink, M., et al., 1996. Developmental and cell cycle regulation of alfalfa nucMs1, a plant homolog of the yeast Nsr1 and mammalian nucleolin. Plant Cell 8, 417–428.

Borowiec, J.A., 1996. DNA helicases. In: De Pamphilis, M.L. (Ed.), DNA Replication in Eukaryotic Cells. Cold Spring Harbor Laboratory Press, Cold Spring Harbor, pp. 545–574.

Bourc'his, D., Bestor, T.H., 2002. Helicase homologues maintain cytosine methylation in plants and mammals. BioEssays 2, 297–299.

Chamot, D., Magee, W.C., Yu, E., Owttrim, G.W., 1999. A cold shock induced cyanobacterial RNA helicase. J. Bacteriol. 181, 1728–1732.

Chen, X., Meyerowitz, E.M., 1999. HUA1 and HUA2 are two members of the floral homeotic AGAMOUS pathway. Mol. Cell 3, 349–360.

Chen, X., Liu, J., Cheng, Y., Jia, D., 2002. HEN1 functions pleiotropically in *Arabidopsis* development and acts in C function in the flower. Development 129, 1085–1094.

Cheng, Y., Kato, N., Wang, W., Li, J., Chen, X., 2003. Two RNA binding proteins, HEN4 and HUA1, act in the processing of AGAMOUS pre-mRNA in *Arabidopsis thaliana*. Dev. Cell 4, 53–66.

Costa, R.M., Morgante, P.G., Berra, C.M., Nakabashi, M., Bruneau, D., Bouchez, D., Sweder, K.S., Van Sluys, M.A., Menck, C.F., 2001. The participation of AtXPB1, the XPB/RAD25 homologue gene from *Arabidopsis thaliana*, in DNA repair and plant development. Plant J. 28, 385–395.

Dalmay, T., Horsefield, R., Braunstein, T.H., Baulcombe, D.C., 2001. SDE3 encodes an RNA helicase required for post-transcriptional gene silencing in *Arabidopsis*. EMBO J. 20, 2069–2078.

Dang, H.Q., Tran, N.Q., Tuteja, R., Tuteja, N., 2011. Promoter of a salinity and cold stress-induced MCM6 DNA helicase from pea. Plant Signal. Behav. 6, 1006–1008.

Gagliardi, D., Kuhn, J., Spadinger, U., Brennicke, A., Leaver, C.J., Binder, S., 1999. An RNA helicase (AtSUV3) is present in *Arabidopsis thaliana* mitochondria. FEBS Lett. 458, 337–342.

Gong, Z., Lee, H., Xiong, L., Jagendorf, A., Stevenson, B., Zhu, J.K., 2002. RNA helicase-like protein as an early regulator of transcription factors for plant chilling and freezing tolerance. Proc. Natl. Acad. Sci. U. S. A. 99, 11507–11512.

Gorbalenya, A.E., Koonin, E.V., 1993. Helicases: amino acids sequence comparisons and structure–function relationship. Curr. Opin. Struct. Biol. 3, 419–429.

Gorbalenya, A.E., Koonin, E.V., Donchenko, A.P., Blinov, V.M., 1989. Two related super families of putative helicases involved in replication, recombination, repair and expression DNA and RNA genomes. Nucleic Acids Res. 17, 4713–4730.

Gorbalenya, A.E., Koonin, E.V., Donchenko, A.P., Blinov, V.M., 1998. A conserved NTP-motif in putative helicases. Nature 333, 22.

Hall, M.C., Matson, S.W., 1999. Helicase motifs: the engine that powers DNA unwinding. Mol. Microbiol. 34, 867–877.

Hotta, Y., Stern, H., 1978. DNA unwinding protein from meiotic cells of *Lilium*. Biochemistry 17, 1872–1880.

Ilyina, T., Gorbalenya, A.E., Koonin, E.V., 1992. Organization and evolution of bacterial and bacteriophage primase-helicase systems. J. Mol. Evol. 34, 351–357.

Ishimi, Y., 1997. A DNA helicase activity is associated with an MCM4, -6, and -7 protein complex. J. Biol. Chem. 272, 24508–24513.

Ivanova, M.I., Todorov, I.T., Atanassova, L., DeWitte, W., Onckelen, H.A.V., 1994. Co-localization of cytokinins with proteins related to cell proliferation in developing somatic embryos of *Dactylis glomerata* L. J. Exp. Bot. 45, 1009–1017.

Jack, T., 2002. New members of the floral organ identity AGAMOUS pathway. Trends Plant Sci. 7, 286–287.

Jacobsen, S.E., Running, M.P., Meyerowitz, E.M., 1999. Disruption of an RNA helicase/RNAse III gene in *Arabidopsis* causes unregulated cell division in floral meristems. Development 126, 5231–5243.

Kapitonov, V.V., Jurka, J., 2001. Rolling-circle transposons in eukaryotes. Proc. Natl. Acad. Sci. U. S. A. 98, 8714–8719.

Kornberg, A., Baker, T.A., 1991. DNA Replication, Second ed. Freeman and Co., New York.
Lal, S.K., Giroux, M.J., Brendel, V., Vallejos, C.E., Hannah, L.C., 2003. The maize genome contains a Helitron insertion. Plant Cell 15, 381–391.
Li, J., Jia, D., Chen, X., 2001. HUA1, a regulator of stamen and carpel identities in *Arabidopsis*, codes for a nuclear RNA binding protein. Plant Cell 13, 2269–2281.
Lohman, T.M., 1992. *Escherichia coli* DNA helicases: mechanisms of DNA unwinding. Mol. Microbiol. 6, 5–14.
Lohman, T.M., Bjornson, K.P., 1996. Mechanism of helicase-catalyzed DNA unwinding. Annu. Rev. Biochem. 65, 169–214.
Luo, Y., Liu, Y.B., Dong, Y.X., Gao, X.Q., Zhang, X.S., 2009. Expression of a putative alfalfa helicase increases tolerance to abiotic stress in *Arabidopsis* by enhancing the capacities for ROS scavenging and osmotic adjustment. J. Plant Physiol. 166, 385–394.
Matson, S.W., Bean, D.W., George, J.W., 1994. DNA helicases: enzymes with essential roles in all aspects of DNA metabolism. BioEssays 16, 13–22.
Mimori, T., Akizuki, M., Yamagata, H., Irada, S., Yoshida, S., Homma, M., 1981. Characterization of a high molecular weight acidic nuclear protein recognised by autoantibodies from patients with polymyositis-schleroderma overlap. J. Clin. Invest. 68, 611–620.
Pham, X.H., Reddy, M.K., Ehtesham, N.Z., Matta, B., Tuteja, N., 2000. A DNA helicase from *Pisum sativum* is homologous to translation initiation factor and stimulates topoisomerase I activity. Plant J. 24, 219–229.
Raikwar, S., Srivastava, V.K., Gill, S.S., Tuteja, R., Tuteja, N., 2015. Emerging importance of helicases in plant stress tolerance: characterization of *Oryza sativa* repair helicase XPB2 promoter and its functional validation in tobacco under multiple stresses. Front. Plant Sci. 6, 1094.
Ramu, V.S., Swetha, T.N., Sheela, S.H., Babitha, C.K., Rohini, S., Reddy, M.K., Tuteja, N., Reddy, C.P., Prasad, T.G., Udayakumar, M., 2016. Simultaneous expression of regulatory genes associated with specific drought-adaptive traits improves drought adaptation in peanut. Plant Biotechnol. J. 14, 1008–1020.
Ribeiro, D.T., Machdo, C.R., Costa, R.M.A., Praekelt, U.M., Van-sluys, M.A., Menck, C.F.M., 1998. Cloning of a cDNA from *Arabidopsis thaliana* homologous to the human XPB gene. Gene 208, 207–213.
Riha, K., Shippen, D.E., 2003. Ku is required for telomeric C-rich strand maintenance but not for end-to-end chromosome fusions in *Arabidopsis*. Proc. Natl. Acad. Sci. U. S. A. 100, 611–615.
Riha, K., Watson, J.M., Parkey, J., Shippen, D.E., 2002. Telomere length deregulation and enhanced sensitivity to genotoxic stress in *Arabidopsis* mutants deficient in Ku70. EMBO J. 21, 2819–2826.
Sabelli, P.A., Burges, S.R., Kush, A.K., Yong, M.R., Shewry, P.R., 1996. cDNA cloning and characterization of a maize homologue of the MCM proteins required for the initiation of DNA replication. Mol. Gen. Genet. 252, 125–136.
Sabelli, P.A., Parker, J.S., Barlow, P.W., 1999. cDNA and promoter sequences for MCM3 homologues from maize, and protein localization in cycling cells. J. Exp. Bot. 50, 1315–1322.
Sahoo, R.K., Tuteja, N., 2014. OsSUV3 functions in cadmium and zinc stress tolerance in rice (*Oryza sativa* L. cv IR64). Plant Signal. Behav. 9. e27389.
Seki, M., Narusaka, M., Abe, H., Kasuga, M., Yamaguchi-Shinozaki, K., Carninci, P., Hayashizaki, Y., Shinozaki, K., 2001. Monitoring the expression pattern of 1300 *Arabidopsis* genes under drought and cold stresses by using a full-length cDNA microarray. Plant Cell 13, 61–72.
SenGupta, D., Borowiec, J.A., 1992. Strand-specific recognition of a synthetic DNA replication fork by the SV40 large tumor antigen. Science 256, 1656–1661.
Shivakumara, T.N., Sreevathsa, R., Dash, P.K., Sheshshayee, M.S., Papolu, P.K., Rao, U., Tuteja, N., UdayaKumar, M., 2017. Overexpression of pea DNA helicase 45 (PDH45) imparts tolerance to multiple abiotic stresses in chili (*Capsicum annuum* L.). Sci. Rep. 7, 2760.
Springer, P.S., McCombie, W.R., Sundaresan, V., Martienseen, R.A., 1995. Gene trap tagging of PROLIFERA, an essential MCM 2-3-5-like gene in *Arabidopsis*. Science 268, 877–880.
Springer, P.S., Holding, D.R., Groover, A., Yordan, C., Martienssen, R.A., 2000. The essential Mcm7 protein PROLIFERA is localized to the nucleus of dividing cells during the G1 phase and is required maternally for early *Arabidopsis* development. Development 127, 1815–1822.
Tajrishi, M.M., Tuteja, N., 2011. Isolation and in silico analysis of promoter of a high salinity stress-regulated pea DNA helicase 45. Plant Signal. Behav. 6, 1447–1450.

Tamura, K., Adachi, Y., Chiba, K., Oguchi, K., Takahashi, H., 2002. Identification of Ku70 and Ku80 homologues in *Arabidopsis thaliana*: evidence for a role in repair of double-strand breaks. Plant J. 29, 771–781.

Tanner, N.K., Linder, P., 2001. DExD/H-box RNA helicases: from generic motors to specific dissociation functions. Mol. Cell 8, 251–262.

Tanner, N.K., Cordin, O., Banroques, J., Doe're, M., Linder, P., 2003. The Q motif: a newly identified motif in DEAD box helicases may regulate ATP binding and hydrolysis. Mol. Cell 11, 127–138.

Thieringer, H.A., Jones, P.G., Inouye, M., 1998. Cold shock and adaptation. BioEssays 20, 49–57.

Tong, C.-G., Reichler, S., Blumenthal, S., Balk, J., Hsieh, H.-L., Roux, S.J., 1997. Light regulation of the abundance of mRNA encoding a nucleolin-like protein localized in the nucleoli of pea nuclei. Plant Physiol. 114, 643–652.

Tuteja, N., 2000. Plant cell and viral helicases: essential enzymes for nucleic acid transactions. Crit. Rev. Plant Sci. 19, 449–478.

Tuteja, N., Phan, T.N., 1998. A chloroplast DNA helicase II from pea that prefers fork-like replication structures. Plant Physiol. 118, 1029–1038.

Tuteja, N., Tewari, K.K., 1999. Molecular biology of chloroplast genome. In: Singhal, G.S., Renger, G., Sopory, S.K., Irrgang, K.D., Govindjee, (Eds.), Concepts in Photobiology: Photosynthesis and Photomorphogenesis. Narosa Publishing House/Kluwer Academic Press, New Delhi/Dordrecht, pp. 691–738.

Tuteja, N., Tuteja, R., 1996. DNA helicases: the long unwinding road. Nat. Genet. 13, 11–12.

Tuteja, R., Tuteja, N., 1998. Nucleolin: a multifunctional major nucleolar phosphoprotein. Crit. Rev. Biochem. Mol. Biol. 33, 407–436.

Tuteja, R., Tuteja, N., 2000. Ku autoantigen: a multifunctional DNA binding protein. Crit. Rev. Biochem. Mol. Biol. 35, 1–33.

Tuteja, N., Tuteja, R., 2001. Unraveling DNA repair in human: molecular mechanisms and consequences of repair defect. Crit. Rev. Biochem. Mol. Biol. 36, 261–290.

Tuteja, N., Phan, T.N., Tewari, K.K., 1996. Purification and characterization of a DNA helicase from pea chloroplast that translocates in the 30-to-50 direction. Eur. J. Biochem. 238, 54–63.

Tuteja, N., Beven, A.F., Shaw, P.J., Tuteja, R., 2001a. A pea homologue of human DNA helicase I is localized within the dense fibrillar component of the nucleolus and stimulated by phosphorylation with CK2 and cdc2 protein kinases. Plant J. 25, 9–17.

Tuteja, N., Singh, M.B., Misra, M.K., Bhalla, P.L., Tuteja, R., 2001b. Molecular mechanisms of DNA damage and repair: progress in plants. Crit. Rev. Biochem. Mol. Biol. 36, 337–397.

Tuteja, N., Tran, N.Q., Dang, H.Q., Tuteja, R., 2011. Plant MCM proteins: role in DNA replication and beyond. Plant Mol. Biol. 77, 537–545.

Tuteja, N., Sahoo, R.K., Garg, B., Tuteja, R., 2013. OsSUV3 dual helicase functions in salinity stress tolerance by maintaining photosynthesis and antioxidant machinery in rice (*Oryza sativa* L. cv. IR64). Plant J. 76, 115–127.

Vashisht, A.A., Tuteja, N., 2006. Stress responsive DEAD-box helicases: a new pathway to engineer plant stress tolerance. J. Photochem. Photobiol. B 84, 150–160.

Wang, Y., Duby, G., Purnelle, B., Boutry, M., 2000. Tobacco VDL gene encodes a plastid DEAD-box RNA helicase and is involved in chloroplast differentiation and plant morphogenesis. Plant Cell 12, 2129–2142.

Wong, I., Lohman, T.M., 1992. Allosteric effects of nucleotide cofactors on *Escherichia coli* Rep helicase-DNA binding. Science 256, 350–355.

Xu, H., Swoboda, I., Bhalla, P.L., Sijbers, A.M., Zhao, C., Ong, E.K., Hoeijmakers, J.H.J., Singh, M.B., 1998. Plant homologue of human excision repair gene ERCC1 points to conservation of DNA repair mechanisms. Plant J. 13, 823–829.

Yarranton, G.T., Gefter, M.L., 1979. Enzyme catalyzed DNA unwinding: studies on *Escherichia coli* rep protein. Proc. Natl. Acad. Sci. U. S. A. 76, 1658–1662.

FURTHER READING

Sahoo, R.K., Ansari, M.W., Tuteja, R., Tuteja, N., 2015. Salt tolerant SUV3 overexpressing transgenic rice plants conserve physicochemical properties and microbial communities of rhizosphere. Chemosphere 119, 1040–1047.

CHAPTER 8

RNAi Technology: The Role in Development of Abiotic Stress-Tolerant Crops

Tushar Khare*, **Varsha Shriram**†, **Vinay Kumar***,‡
*Department of Biotechnology, Modern College of Arts, Science and Commerce (Savitribai Phule Pune University), Pune, India
†Department of Botany, Prof. Ramkrishna More College (Savitribai Phule Pune University), Pune, India
‡Department of Environmental Science, Savitribai Phule Pune University, Pune, India

Contents

1. RNAi-BASED TECHNOLOGY: AN EMERGING NOVEL APPROACH

A challenge in the 21st century is the production of an adequate amount food to meet the population's food demand, regardless of the drop in quantity, as well the quality, of available arable land and water resources, and progressively changing weather patterns related to climate change. Over the past several decades, increasingly harmful environmental conditions have been responsible for observable crop loss throughout the world (Mickelbart et al., 2015). The decline in yields of major crops including rice, wheat, and maize is projected according to the analysis generated by various integrated climate change and crop production models (Iizumi et al., 2013; Rosenzweig et al., 2014). Along with improvement in germplasm development and agronomic practices, susceptibility of crops to environmental variability has also risen, owing to higher sowing densities and

Biochemical, Physiological and Molecular Avenues for Combating Abiotic Stress in Plants
https://doi.org/10.1016/B978-0-12-813066-7.00008-5

increasing competition for water and nutrients (Lobell et al., 2014). Hence, to satisfy global food demands, implementation of resilient crop genotypes is an urgent necessity (Mickelbart et al., 2015). The abiotic stress environment is one of the major factors responsible for a reduction in crop yields. Crops are routinely exposed to a number of abiotic stress factors instantaneously under the field conditions presenting the unique stress responses, which cannot be characterized by studying different stress factors individually (Zandalinas et al., 2017). To bypass these hurdles, breeding methods supplemented with modern biotechnological tools consisting of genomics, proteomics, and transcriptomics are essential to develop stress-tolerant and high-yielding crop varieties (Mittler and Blumwald, 2010; Tester and Langridge, 2010). The precision of certain biotechnological methods that are more effective than conventional breeding has contributed significantly to rapid crop improvement by offering a comprehensive array of novel genes and traits that can be successfully inserted into elite crops to improve yield as well as to confer resistance to various stresses, including abiotic factors (Jagtap et al., 2011). Although the advanced biotechnological tools are providing a well-defined and promising method of crop improvement, there are some concerns regarding the use of such engineered crops in contemporary agricultural practices; mainly related to biosafety guidelines and the acceptability of crops bearing genes from organisms other than the plants. Before releasing genetically modified crops, facts, including those related to erosion of biodiversity and ecological disturbance, must be considered with respect to the potential risks in the future. Therefore, developing transgenic crops requires further time, cost, and expertise. Hence, there is an absolute requirement for novel strategies and safe methods of resilient crop development that would be substantially more acceptable to the general public (Jagtap et al., 2011).

Concerning all these factors, RNA silencing, or RNA interference (RNAi) technology, has become a technology of choice by molecular biologists and molecular breeders for crop improvement. RNAi is a molecular mechanism leading the posttranscriptional gene silencing (PTGS) elicited by double-stranded RNAs (dsRNAs) to inhibit specific gene expression (Younis et al., 2014). It is a form of sequence-specific gene regulation mediated by dsRNA for prevention of translational/transcriptional repression. Since the discovery of this technology, the precision, efficiency, stability, and general advantages over antisense technology have made RNAi an efficient tool for not only crop improvement, but also many other molecular engineering-based modifications (Saurabh et al., 2014). The technique has proven effective for molecular technology-based improvement in terms of bio-fortification and bio-elimination. RNAi technology has been successfully employed to achieve the modifications of numerous desired traits, including nutritional fortifications, allergen/toxic content reduction, morphological amendments, male sterility altering, secondary metabolite enhancement, and boosted defense against various biotic/abiotic stresses in many plants (Saurabh et al., 2014).

2. RNAi: BRIEF HISTORY AND BASIC MECHANISM

The enlivening phenomenon of RNAi encompasses dsRNA-mediated, gene-specific expression prevention by instigating a sequence of specifically targeted mRNA degradation in cytoplasm. Various organisms showing the presence of this mechanism have been studied.

2.1 History

The term "RNAi" was first created by Craig Mello and coworkers to define the unknown contrivance through which the exogenously-delivered sense-antisense RNAs efficiently silenced the gene expression in *Caenorhabditis elegans* (Rocheleau et al., 1997). Further structural and functional elucidation of RNAi and its delivery revealed the fact that integration of target-gene-specific dsRNA in *C. elegans* blocked the amassing of endogenous RNA transcripts more effectively than injecting sense/antisense strands independently (Fire et al., 1998). The effectiveness of a very small quantity of injected dsRNA in gene silencing led to the speculation about the amplification module of the interference process (Koch and Kogel, 2014). Before these studies, induced PTGS upon transgene introduction in plants and fungi was observed, which was manifested by decreased accumulation of transgene transcripts, or if transgenes showed homology with endogenous genes; a decrease in transcript for both the genes. The latter occurrence was termed "co-suppression" in fungi and plants (Hammond et al., 2001). The investigation regarding the association between endogenously generated dsRNAs and PTGS was done by Hamilton and Baulcombe (1999). The tomato and tobacco lines transformed with endogenous or foreign genes, or those diseased with Potato Virus X, provided the foresight about the instigation of gene silencing and accumulation of short (~25 nt) antisense RNAs, whose sequences were complementary with target transcripts (Hamilton and Baulcombe, 1999). The relativity of this phenomenon in animals, plants, and fungi was further demonstrated by the presence of proteins involved in PTGS and RNAi (Fagard et al., 2000). Since then, the capability of short antisense RNAs to regulate gene expression was pronounced in varied groups of organisms, including plants (Napoli et al., 1990). The RNAi was first witnessed in plants by Napoli et al. (1990), where enhanced production of anthocyanin pigments was achieved in *Petunia hybrid* L. by introducing the chalcone synthase gene (*CHS A*). The unexpected phenomenon of co-suppression was observed as white or chimeric flowers, which were produced by transgenic plants, instead of dark purple flowers, because of silencing of the endogenous homolog of the gene. The occurrence of RNAi is conserved in a wide range of organisms, designated as PTGS in plants, quelling in fungi (Romano and Macino, 1992), and RNA interference in animals (Fire et al., 1998). Owing to the potential of the gene expression alterations by RNAi, the technique has been extensively used for differential studies in mammalian cells, *Drosophila*, nematodes, algae, moss, and plants, and in some prokaryotes as well, which incorporate types of small noncoding RNAs such as small interfering RNAs (siRNAs) and microRNAs (miRNAs) (Koch and Kogel, 2014).

2.2 Basic Mechanism

RNAi is depicted as conserved evolutionary defense machinery comprising dsRNAs that can target cellular and viral mRNAs. In a typical biological route, small RNA molecules interfere with mRNA transcript translation, which ultimately suppresses the expression of target genes. The involved small noncoding RNAs are the products formed from the cleavage of dsRNAs, generally denoted as miRNA (microRNAs) and siRNAs (small interfering RNAs). The cleavage of dsRNA is mediated by ribonuclease, known as the Dicer-like enzyme (DICER) (Pare and Hobman, 2007). These small noncoding RNAs, along with RNA-induced silencing complex (RISC) (Redfern et al., 2013; Wilson and Doudna, 2013), Argonaute (AGO) (Ender and Gunter, 2010; Riley et al., 2012), and other effector proteins lead to the RNAi. The generated ds-siRNAs are then unwound by an ATP-activated RISC complex. Loss of sense strands from ds-siRNA is achieved through RNA helicases, and the simultaneously remaining antisense siRNA strand is integrated inside the nuclease containing the RISC complex (Kusaba, 2004). The antisense RNA-RISC complex then aims homologous transcripts by complementary base pairing and cleaves mRNA/inhibits translation, hence indirectly blocking protein synthesis (Bartel, 2004). The siRNA molecules are synthesized from dsRNAs. The endogenous sources of dsRNAs include dsRNAs produced through natural cis-antisense, gene-paired encoded mRNAs, dsRNAs generated via heterochromatin and DNA repeats, and as a product of miRNA-directed cleavage of ssRNAs (Chinnusamy et al., 2007). On the other hand, biosynthesis of miRNA is mediated through single-stranded primary miRNA (pri-miRNA) transcripts, which are transcribed from MIR genes through the action of DNA polymerases II. This procedure is then followed by formation of improperly paired stem-loop-shaped precursor RNAs (pre-miRNA) (Chinnusamy et al., 2007). This hairpin-shaped precursor is then transformed into a miRNA-miRNA* duplex in which the guide strand is miRNA, whereas miRNA* is degraded by the Dicer-like 1 protein (DCL1), HYPONASTIC LEAVES 1 (HYL1); a dsRNA binding protein, and SERRATE (SE); a zinc finger protein. The process, if then followed by methylation at 3′, ends by HEN1 (HUA ENHANCER 1).

The premature or mature RNAs are then exported to the cytoplasm from the nucleus by the HASTY exporting protein. The processing of miRNAs continues further by combining with the RISC complex to target complementary transcripts and subsequent protein synthesis inhibition (Voinnet, 2009; Khraiwesh et al., 2010, 2012).

The small RNAs hence play a central role in the RNAi-mediated gene silencing in plants. Different individual approaches merging traditional cloning, in silico prediction tools, and high-throughput sequencing of small RNA libraries have recognized some classes of small RNAs that can be categorized with respect to their size and functions. These classes include miRNAs, ra-siRNAs (repeat associated small interfering RNAs), nat-siRNAs (natural antisense transcript-derived small interfering RNAs), ta-siRNAs (trans-acting small interfering RNAs), hc-siRNAs (heterochromatic small interfering RNAs), secondary transitive siRNAs, primary siRNAs, lsiRNAs (long small interfering RNAs) (Khraiwesh et al., 2012), piRNAs (PIWI-interacting RNAs), qiRNAs (QDE-2-interacting RNA) and svRNAs (small vault RNAs) with different biochemical origins (Aalto and Pasquinelli, 2012). Briefly, the initial biogenesis of siRNA and miRNA differs from their respective dsRNA precursors. Afterward, both the small RNA species are produced by Dicer-mediated cleavage of dsRNA. Last, both the small RNA molecules alongside the RISC and AGO and other effector proteins lead to gene silencing (Saurabh et al., 2014). The simplified diagrammatic summery of the overall RNAi mechanism is provided in Fig. 1.

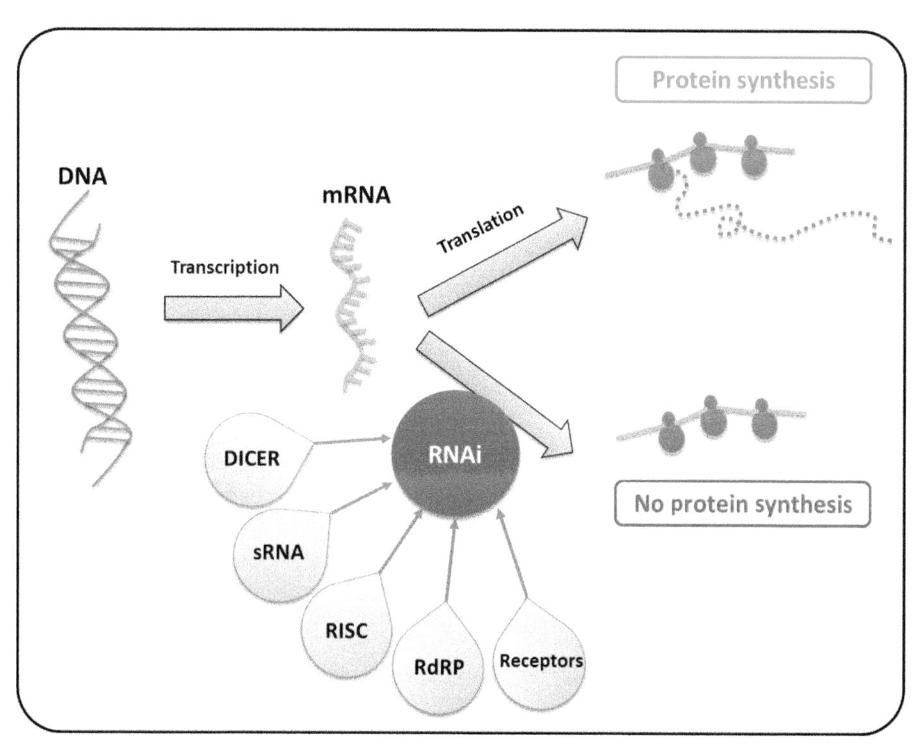

Fig. 1 Simplified procedure of RNAi (*DNA*, deoxyribonucleic acid; *mRNA*, messenger RNA; *DICER*, Dicer-LIKE enzyme; *sRNA*, small RNA; *RISC*, RNA-induced silencing complex; *RdRP*, RNA dependent RNA polymerase).

2.3 Phases of RNAi Mediated mRNA Degradation

The elucidated complex procedure of RNAi-mediated gene silencing has provided a four-phased automatous model, which includes (Fig. 2) the induction phase, completion phase, multiplication phase, and degeneration phase (Bhardwaj et al., 2014). In the initial induction phase, siRNAs are manufactured by processing dsRNAs, which are then amalgamated into a large ribo-nucleo-protein complex. The enzyme responsible for initiation of the RNAi process is from the RNase III family, with specificity for dsRNAs (Bernstein et al., 2001). In the completion phase, the large ribo-nucleo-protein effector complexes interfere with the expression pattern of genes by utilizing small RNA strands showing complementarity with target mRNAs, leading to its degradation (Agrawal et al., 2003). Small dsRNAs are produced during the process via Dicer-mediated cleavage that is elicited by dsRNA-precursors further dissociating into competent single strands. These strands later on act as guides for RISCs. However, very few dsRNA molecules are necessary to denature incessantly, producing target mRNA for a longer time. During the process, RdRP plays crucial role in amplification of RNAi (Lipardi et al., 2001; Sijen et al., 2001). Both ssRNAs and dsRNAs play a role as a template for copying RdRP. Followed by this new full length, dsRNAs are manufactured instantaneously and cleaved further. In the last degeneration phase, RISC formation takes place via binding between dsRNAs and RNAi-specific protein complexes. The complex is activated in the presence of ATP to allow the RISC complex to perform the downstream RNAi reaction (Bhardwaj et al., 2014).

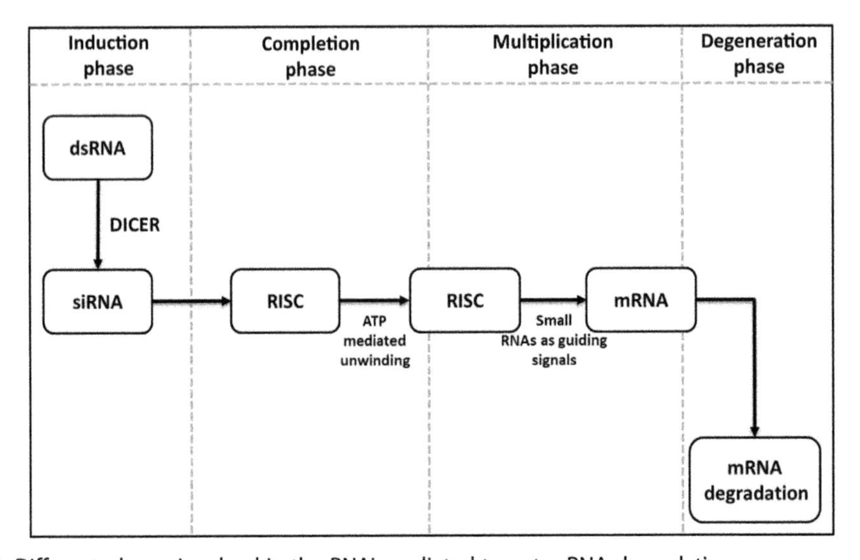

Fig. 2 Different phases involved in the RNAi mediated target-mRNA degradation.

3. INVOLVEMENT OF RNAi IN ABIOTIC STRESS RESPONSES

Abiotic environmental factors including salinity, drought, temperature, UV-radiation, heavy metals, and nutrient deprivation are the prime reasons for depletion in the productivity of many major crops throughout the world (Shriram et al., 2016). Numerous advancements in the analytical tools for biochemical, molecular, genomic, proteomic, and overall metabolic analysis are now allowing better understanding of the complex regulatory network of stress-mediated responses. Using such tools, the involvement of RNAi (small RNA species and their respective targets) in abiotic stress responses in many crops is well documented (Khraiwesh et al., 2012; Sun, 2012; Sunkar et al., 2012; Shriram et al., 2016; Kumar et al., 2017).

3.1 Salinity

The progressively emergent threat of salinity is a factor in qualitative and quantitative crop loss worldwide (Sunkar et al., 2012; Khare et al., 2015; Kumar and Khare, 2014, 2016; Joshi et al., 2016; Kumar et al., 2015, 2016). The moderate salinity level affects the crop yield negatively; whereas severe salt levels are cause for the low survival rate of the plants. Depending on the salt tolerance potential of the plant (glycophytes: salt sensitive; halophytes: salt tolerant), differential behavior of the small RNAs is observed in many plants under the course of salinity. During stress conditions, expression levels of miRNAs and other related genes are altered (Shriram et al., 2016). Salt stress responsiveness of miRNAs is recognized in many plant species, including numerous crops. The salinity responsive miRNAs are identified in *Arabidipsis thaliana* (Barciszewska-Pacak et al., 2015), *Zea mays* (Fu et al., 2017), *Gossypium hirsutum* (Gao et al., 2016), *Raphanus sativus* (Sun et al., 2015), *Populus tomentosa* (Ren et al., 2013), *Cicer arietinum* (Kohli et al., 2014), *Triticum aestivum* (Gupta et al., 2014a), *Oryza sativa* (Mittal et al., 2016), and so forth. Some of the important miRNA families showing altered expressional behavior under saline conditions include *miR156*, *miR159*, *miR168*, *miR169*, *miR393*, and *miR398*.

3.2 Drought

Because of the extent of negative effects on crops, drought stress, or water deficits, are becoming important factors in plant abiotic stress research. In response to drought, numerous genes, along with their targets, have been analyzed using genome-wide expression studies (Kruszka et al., 2012). Some of the miRNA species have been identified as conserved entities across the plant species with drought-mediated regulatory mechanisms (Ferdous et al., 2015). The miRNA expression profiles in response to drought are documented in *Sorghum bicolor* (Hamza et al., 2016), *Gossypium hirsutum* (Wang et al., 2013), *Oryza rufipogon* (Zhang et al., 2016), *Solanum tuberosum* (Zhang et al., 2014), *Triticum turgidum* (Giusti et al., 2017), *Hordeum vulgare* (Ferdous et al., 2017), *Cucumis sativus* (C. Li et al., 2016a), *Panicum virgatum* (Xie et al., 2014), and

Elettaria cardamomum (Anjali et al., 2017). Interestingly, some of the drought-responsive elements have shown their presence in the promoter region of miRNA genes, which differentially express under drought conditions (Kruszka et al., 2012).

3.3 Temperature Variations (Cold and Heat)

Geographical conditions and seasonal deviations in temperature are responsible for stunted growth and reduced productivity of crops. In response to inconsistent temperature variations, plants alter their gene expression patterns at posttranscriptional levels (Shriram et al., 2016). Several temperature-responsive miRNA species have been identified in plants (Cao et al., 2014). *Panicum virgatum* (Hivrale et al., 2016), *Oryza sativa* (Li et al., 2015; Liu et al., 2017; Mangrauthia et al., 2017), and *Triticum aestivum* (Kumar et al., 2015) have deciphered the heat responsive alterations in different miRNA species; whereas chilling-responsive miRNAs have been characterized in *Glycine* max (Xu et al., 2016), *Zea mays* (Li et al., 2016b), and *Solanum habrochaites* (Cao et al., 2014).

3.4 Heavy Metals

Industrial and anthropogenic activities have led to extensive heavy metal pollution in vast agricultural lands. Heavy metal pollution is a prime cause of abiotic stress-environments, and the consequent hazardous effects on physiological and metabolic processes that deleteriously affect the growth, survival, development, and overall crop productivity (Gupta et al., 2014b; Shriram et al., 2016). The essential heavy metals (copper, iron, zinc, and manganese) are toxic at higher concentrations; whereas nonessential heavy metals (cadmium, aluminum, and mercury) are toxic even at low concentrations (Gielen et al., 2012). Because heavy metals' interaction with crops ultimately result in different heavy metals in the human food chain, investigation into this phenomenon is crucial. Recent studies have incorporated investigations that have established relationships between heavy metals and their molecular targets, such as small RNAs. Documentation of heavy-metal-responsive miRNAs is observable, such as cadmium-responsive miRNAs in *Brassica napus* (Huang et al., 2010), *Oryza sativa* (Ding et al., 2011), and *Raphanus sativus* (Xu et al., 2013); mercury-responsive miRNAs in *Medicago truncatula* (Zhou et al., 2012); manganese-responsive miRNAs in *Phaselous vulgaris* (Valdés-López et al., 2010); aluminum-responsive miRNAs in *Glycine soja* (Zeng et al., 2012); and arsenic-responsive miRNAs in *Oryza sativa* (Yu et al., 2012).

Apart from these abiotic factors, factors such as ultra-violet radiation, nutrient deficiencies, and numerous biotic factors have been studied for their small RNA-mediated mechanisms in many plants. The outcome of these investigations provides insight regarding the regulatory role of many of the miRNAs and their interconnected network (Fig. 3), which has proven helpful for the development of small RNA-facilitated crop improvement strategies.

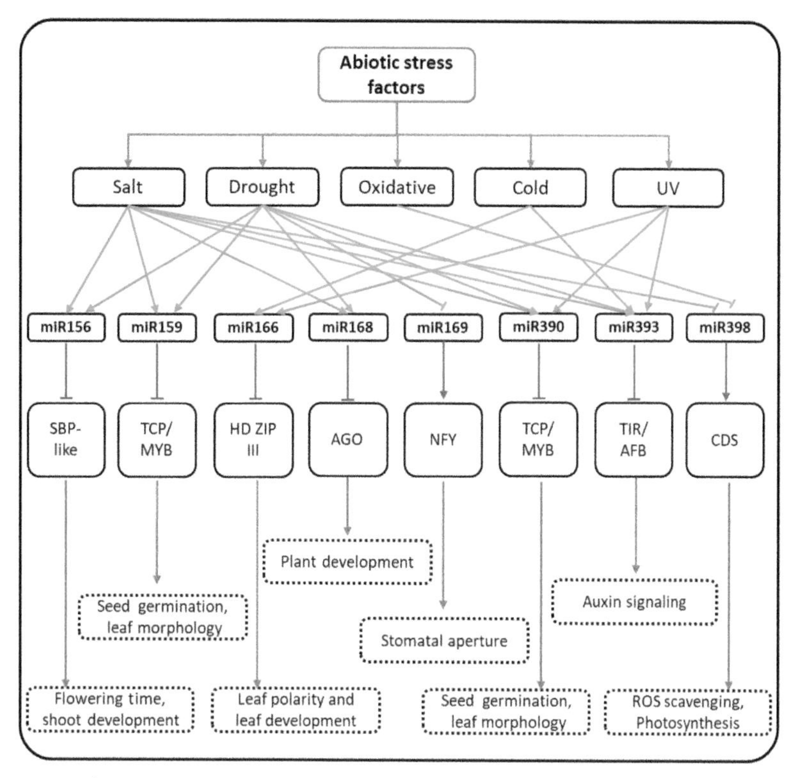

Fig. 3 Summary of abiotic stress responsive important miRNAs and their identified/predicted targets.

4. UTILIZATION OF RNAi FOR CROP IMPROVEMENT

Manipulation for quality traits in crops can now be accomplished by RNAi. The progressively emergent RNAi technique can be employed for crop improvement by following defined steps. The main events in the procedure include identification and characterization of the target gene, development of the vectors (RNAi constructs), raising transgenic plant lines, and screening and assessment of manipulated traits (Saurabh et al., 2014). The RNAi constructs are developed in such a manner that they will express self-complementary sequences homologous to the target sequences, as hairpin RNA. The addition of spacer sequences between complementary sequences elicits efficiency. Modifying the RNAi construct as per the required gene silencing pattern is a crucial step. Along with all the precautions, correct selection of vectors, promoters, markers, and mode of transformation are the factors deciding the efficiency of RNAi (Saurabh et al., 2014). By utilizing appropriate strategies, many successful demonstrations of amended abiotic stress tolerance in a wide variety of crops are now emerging worldwide.

Novel regulatory genes are prime targets for crop upgrading. One of such target is a receptor for activated C-kinase 1 (*RACK1*). The *RACK1* is a highly conserved protein scaffold with versatility in its functions, which has proven to be imperative in plant growth and development. Transgenic *Oryza sativa* lines were raised by Li et al. (2009) bearing the silenced expression of *RACK1* gene expression, which was achieved by RNAi. The study explicated the role of *RACK1* in drought responses in the crop. The expression of RACK1 in transgenic lines was inhibited to one half, and the transgenic lines showed higher drought tolerance than their wild counterparts. The *RACK1*-mediated negative regulation of the redox system in drought stress was correlated with lower lipid peroxidation levels and higher superoxide dismutase activities in the transgenic lines (Li et al., 2009).

Squalene synthase (*SQS*) is a type of farnesyl-diphosphate farnesyltransferase that plays a catalytic role in sterol biosynthesis. Early reports have already suggested its role in the plant defense system (Manavalan et al., 2012). RNAi-mediated disruption of the *SQS* synthase gene in rice by maize *SQS* has demonstrated incremental drought tolerance at the vegetative as well as reproductive plant growth phases. In 32 days of water stress, transgenics showed delayed wilting, higher degrees of soil water, and improvement in recovery. At the reproductive stage, plants also showed reduced stomatal conductance, a successively lower rate of water loss, and improved relative water content in leaves; which can be positively correlated to improved drought tolerance (Manavalan et al., 2012).

The miR319 family is a conserved miRNA family among the diverse plant species essential regulators in plant development. Yang et al. (2013) reported the expression pattern analysis and functional characterization of the two miR319 family members in rice, namely *Osa-MIR319a* and *Osa-MIR319b*. Overexpression of these genes in transgenic lines showed improved cold tolerance in rice. Apart from this, the target genes of miR319, namely *OsPCF5* and *OsPCF8,* were genetically down-regulated using the RNAi technique. The RNAi plants also exhibited enhanced cold tolerance (Yang et al., 2013).

The involvement of *Arabidopsis* β-subunit of farnesyltransferase (*ERA1*) in the plants' sensitivity to abscisic acid and drought has been reported by Wang et al. (2009). The *AtHPR1*, from the *Arabidopsis* single-gene family encodes hydroxypyruvate reductase, which converts hydroxypyruvate to glycerate during photorespiration, which is then transported from peroxisomes to chloroplasts for the Calvin cycle. The *AtHPR1* expression is drought inducible. By using these interconnected genetic links, conditional and precise down-regulation of farnesyltransferase in *Brassica napus* by the *AtHPR1* promoter driving an RNAi construct caused yield protection to drought stress in the field (Wang et al., 2009).

The ubiquitin ligase gene has been utilized for RNAi in *Oryza sativa* to augment drought tolerance. Rice knockdown of a RING finger E3 ligase gene-*OsDSG1* leads to enhanced drought tolerance (Park et al., 2010). Similarly, silencing of *OsDIS1* (for *Oryza sativa* drought-induced SINA protein 1), a C3HC4 RING finger E3 ligase by RNAi enhanced drought tolerance (Kamthan et al., 2015). In *Medicago sativa,* the importance of the miR156 wasin drought environment was tested by Arshad et al. (2017).

The overexpression of miR156 in the transgenic plants resulted in significant improvement against drought stress as the transgenic line showed higher survival, reduced water loss, and higher stomatal conductance. The miR156 encodes for *SPL13* (Squamosa Promoter Binding Protein-LIKE), which is important in stress responses. The *SPL13*-RNAi transgenic plants were also assessed for their drought responses. The *SPL13*-RNAi line exhibited enhanced root length, increased stomatal conductance, higher chlorophyll content and photosynthetic assimilation, and reduced water loss; and hence performed better in drought conditions compared with their nontransformed wild lines (Arshad et al., 2017).

Apart from the exact gene silencing approach for a target trait, many reports have been published in recent years where the overexpression of the specific small RNA molecule is the prime strategy for achieving the desired expression level of the gene to achieve abiotic stress tolerance in targeted plant species. Table 1 summarizes the examples in which the RNAi technique is employed for development of abiotic stress tolerant crop lines.

Table 1 Utilization of RNAi technology with specific target genes for development of abiotic stress resistant crops

Target gene/small RNA molecule	Crop used	Enhanced characters	Refs.
Receptors for activated C-kinase 1 (*RACK1*)	*Oryza sativa*	Higher drought tolerance, lower lipid peroxidation levels and higher superoxide dismutase activities	Li et al. (2009)
Squalene synthase (*SQS*)	*Oryza sativa*	Delayed wilting, conserved higher degrees of soil water and improvement in recovery, lower rate of water lose and improved relative water content, improved drought tolerance	Manavalan et al. (2012)
OsPCF5 and *OsPCF8*	*Oryza sativa*	Enhanced tolerance against cold	Yang et al. (2013)
Farnesyltransferase (*FTase*)	*Brassica napus*	Enhanced tolerance against drought	Wang et al. (2009)
OsDSG1 (RING finger E3 ligase gene)	*Oryza sativa*	Enhanced tolerance against drought	Park et al. (2010)
SPL13 (Squamosa Promoter Binding Protein-LIKE)	*Medicago sativa*	Enhanced root length, increased stomatal conductance, higher chlorophyll content and photosynthetic assimilation, reduced water loss; enhanced drought tolerance	Arshad et al. (2017)

5. CONCLUSION: PROS AND CONS OF RNAi, AND THE FUTURE

The abiotic stress environment is a major cause of qualitative and quantitative losses in agricultural industries, which will definitely lead to the inability of agro-industries to fulfill the needs of the progressively growing population of the world. Hence, development of crop varieties performing with the most efficiency in abiotic environments is the prime aim of plant breeders. By taking into the account the success of RNAi-based programs, RNAi-grounded research has demonstrated its potential in crop improvement to overcome the problem. Rigorous research has been conducted to explore the small RNA molecules and their respective targets involved in abiotic stress mechanisms, which is generating milestones for agricultural improvement, thereby refining our way of life. The application of RNAi-mediated gene silencing has generated many successful transgenic crop examples with developed tolerance against abiotic stress factors. This strategy has some definite advantages over other molecular approaches. Advantages include sequence specificity, targeting more than a single gene at a time, control over the extent of gene silencing at required stages and tissues. Absence of transgene protein expression also ensures the absence of the extra metabolic load on transgenic plants. Most importantly, RNAi pose minimal or lesser biosafety issues. However, there are some drawbacks regarding this approach. Though the approach is sequence specific, there are chances of off-target effects leading to unwanted traits in developed transgenic lines. Generation of unintended secondary effects is also a possibility sometimes when the approach aims toward resistance against multiple factors at once. Besides, delivery methods for RNA molecules are rate- and efficiency-determining steps of the technique. Therefore, even though the technique has acquired success in many aspects of crop improvement, there are still significant challenges in developing and commercializing the RNAi technology for its best and most efficient utilization. Hence, it can be concluded that RNAi technology would serve as an important approach for crop improvement, which will contribute greatly to crop productivity. There are numerous prospects for the uses of RNAi in crop breeding for its enhancement, such as stress tolerance and enhanced nutritional levels. Thus, there is an enormous potential in RNAi-based technologies for improving the agricultural yield substantially.

ACKNOWLEDGMENTS

The research in VK's lab is supported through the Science and Engineering Research Board (SERB), Department of Science and Technology (DST), and Government of India funds (grant number SR/FT/LS-93/2011 and EMR/2016/003896). TK acknowledges the Senior Research Fellowship from SERB under the project EMR/2016/003896. The authors acknowledge the use of facilities created under the DST-FIST program and Star College Scheme of Department of Biotechnology (DBT), Government of India implemented at Modern College, Ganeshkhind, Pune.

REFERENCES

Aalto, A.P., Pasquinelli, A.E., 2012. Small non-coding RNAs mount a silent revolution in gene expression. Curr. Opin. Cell Biol. 24, 333–340. https://doi.org/10.1016/j.ceb.2012.03.006.

Agrawal, N., Dasaradhi, P.V.N., Mohmmed, A., Malhotra, P., Bhatnagar, R.K., Mukherjee, S.K., 2003. RNA interference: biology, mechanism, and applications. Microbiol. Mol. Biol. Rev. 67, 657–685. https://doi.org/10.1128/MMBR.67.4.657-685.2003.

Anjali, N., Nadiya, F., Thomas, J., Sabu, K.K., 2017. Discovery of MicroRNAs in cardamom (*Elettariacardamomum* Maton) under drought stress. Dataset Pap. Sci. 2017, 1–4. https://doi.org/10.1155/2017/9507485.

Arshad, M., Feyissa, B.A., Amyot, L., Aung, B., Hannoufa, A., 2017. MicroRNA156 improves drought stress tolerance in alfalfa (*Medicagosativa*) by silencing SPL13. Plant Sci. 258, 122–136. https://doi.org/10.1016/j.plantsci.2017.01.018.

Barciszewska-Pacak, M., Milanowska, K., Knop, K., Bielewicz, D., Nuc, P., Plewka, P., Pacak, A.M., Vazquez, F., Karlowski, W., Jarmolowski, A., Szweykowska-Kulinska, Z., 2015. *Arabidopsis* microRNA expression regulation in a wide range of abiotic stress responses. Front. Plant Sci. 6, 410. https://doi.org/10.3389/fpls.2015.00410.

Bartel, D.P., 2004. MicroRNAs: genomics, biogenesis, mechanism, and function. Cell 116, 281–297.

Bernstein, E., Caudy, A.A., Hammond, S.M., Hannon, G.J., 2001. Role for a bidentate ribonuclease in the initiation step of RNA interference. Nature 409, 363–366. https://doi.org/10.1038/35053110.

Bhardwaj, R., Ohri, P., Kaur, R., Rattan, A., Kapoor, D., Bali, S., Kaur, P., Khajuria, A., Singh, R., 2014. Gene Silencing: A Novel Cellular Defense Mechanism Improving Plant Productivity Under Environmental Stresses. Academic Press, Elsevier, San Diego. https://doi.org/10.1016/B978-0-12-800876-8.00010-2.

Cao, X., Wu, Z., Jiang, F., Zhou, R., Yang, Z., 2014. Identification of chilling stress-responsive tomato microRNAs and their target genes by high-throughput sequencing and degradome analysis. BMC Genomics 15, 1130. https://doi.org/10.1186/1471-2164-15-1130.

Chinnusamy, V., Zhu, J., Zhou, T., Zhu, J.-K., 2007. Small RNAs: big role in abiotic stress tolerance of plants. In: Advances in Molecular Breeding Toward Drought and Salt Tolerant Crops. Springer, Dordrecht, pp. 223–260. https://doi.org/10.1007/978-1-4020-5578-2_10.

Ding, Y., Chen, Z., Zhu, C., 2011. Microarray-based analysis of cadmium-responsive microRNAs in rice (*Oryza sativa*). J. Exp. Bot. 62, 3563–3573. https://doi.org/10.1093/jxb/err046.

Ender, C., Gunter, M., 2010. Argonaute proteins at a glance. J. Cell Sci. 123, 1819–1823. https://doi.org/10.1242/jcs.055210.

Fagard, M., Boutet, S., Morel, J.B., Bellini, C., Vaucheret, H., 2000. AGO1, QDE-2, and RDE-1 are related proteins required for post-transcriptional gene silencing in plants, quelling in fungi, and RNA interference in animals. Proc. Natl. Acad. Sci. USA 97, 11650–11654. https://doi.org/10.1073/pnas.200217597.

Ferdous, J., Hussain, S.S., Shi, B.J., 2015. Role of microRNAs in plant drought tolerance. Plant Biotechnol. J. 13, 293–305. https://doi.org/10.1111/pbi.12318.

Ferdous, J., Sanchez-Ferrero, J.C., Langridge, P., Milne, L., Chowdhury, J., Brien, C., Tricker, P.J., 2017. Differential expression of microRNAs and potential targets under drought stress in barley. Plant Cell Environ. 40, 11–24. https://doi.org/10.1111/pce.12764.

Fire, A., Xu, S., Montgomery, M.K., Kostas, S.A., Driver, S.E., Mello, C.C., 1998. Potent and specific genetic interference by double-stranded RNA in *Caenorhabditis elegans*. Nature 391, 806–811. https://doi.org/10.1038/35888.

Fu, R., Zhang, M., Zhao, Y., He, X., Ding, C., Wang, S., Feng, Y., Song, X., Li, P., Wang, B., 2017. Identification of salt tolerance-related microRNAs and their targets in maize (*Zea mays* L.) using high-throughput sequencing and degradome analysis. Front. Plant Sci. 8, 1–13. https://doi.org/10.3389/fpls.2017.00864.

Gao, S., Yang, L., Zeng, H.Q., Zhou, Z.S., Yang, Z.M., Li, H., Sun, D., Xie, F., Zhang, B., 2016. A cotton miRNA is involved in regulation of plant response to salt stress. Sci. Rep. 6, 19736. https://doi.org/10.1038/srep19736.

Gielen, H., Remans, T., Vangronsveld, J., Cuypers, A., 2012. MicroRNAs in metal stress: specific roles or secondary responses? Int. J. Mol. Sci. 13, 15826–15847. https://doi.org/10.3390/ijms131215826.

Giusti, L., Mica, E., Bertolini, E., De Leonardis, A.M., Faccioli, P., Cattivelli, L., Crosatti, C., 2017. MicroRNAs differentially modulated in response to heat and drought stress in durum wheat cultivars with contrasting water use efficiency. Funct. Integr. Genomics 17, 293–309. https://doi.org/10.1007/s10142-016-0527-7.

Gupta, O.P., Meena, N.L., Sharma, I., Sharma, P., 2014a. Differential regulation of microRNAs in response to osmotic, salt and cold stresses in wheat. Mol. Biol. Rep. 41, 4623–4629. https://doi.org/10.1007/s11033-014-3333-0.

Gupta, O.P., Sharma, P., Gupta, R.K., Sharma, I., 2014b. MicroRNA mediated regulation of metal toxicity in plants: present status and future perspectives. Plant Mol. Biol. 84, 1–18. https://doi.org/10.1007/s11103-013-0120-6.

Hamilton, A.J., Baulcombe, D.C., 1999. A species of small antisense RNA in posttranscriptional gene silencing in plants. Science 286, 950–952. https://doi.org/10.1126/science.286.5441.950.

Hammond, S.M., Caudy, A.A., Hannon, G.J., 2001. Post-transcriptional gene silencing by double-stranded RNA. Nat. Rev. Genet. 2 (2), 110–119. https://doi.org/10.1038/35052556.

Hamza, N.B., Sharma, N., Tripathi, A., Sanan-Mishra, N., 2016. MicroRNA expression profiles in response to drought stress in *Sorghum bicolor*. Gene Expr. Patterns 20, 88–98. https://doi.org/10.1016/j.gep.2016.01.001.

Hivrale, V., Zheng, Y., Puli, C.O.R., Jagadeeswaran, G., Gowdu, K., Kakani, V.G., Barakat, A., Sunkar, R., 2016. Characterization of drought- and heat-responsive microRNAs in switchgrass. Plant Sci. 242, 214–223. https://doi.org/10.1016/j.plantsci.2015.07.018.

Huang, S.Q., Xiang, A.L., Che, L.L., Chen, S., Li, H., Song, J.B., Yang, Z.M., 2010. A set of miRNAs from *Brassica napus* in response to sulphate deficiency and cadmium stress. Plant Biotechnol. J. 8, 887–899. https://doi.org/10.1111/j.1467-7652.2010.00517.x.

Iizumi, T., Sakuma, H., Yokozawa, M., Luo, J.-J., Challinor, A.J., Brown, M.E., Sakurai, G., Yamagata, T., 2013. Prediction of seasonal climate-induced variations in global food production. Nat. Clim. Chang. 3, 904–908. https://doi.org/10.1038/nclimate1945.

Jagtap, U.B., Gurav, R.G., Bapat, V.A., 2011. Role of RNA interference in plant improvement. Naturwissenschaften 98, 473–492. https://doi.org/10.1007/s00114-011-0798-8.

Joshi, R., Singh, B., Bohra, A., Chinnusamy, V., 2016. Salt stress signalling pathways: specificity and cross-talk. In: Wani, S.H., Hossain, M.A. (Eds.), Managing Salinity Tolerance in Plants: Molecular and Genomic Perspectives. CRC Press, Boca Raton, FL, pp. 51–78.

Kamthan, A., Chaudhuri, A., Kamthan, M., Datta, A., 2015. Small RNAs in plants: recent development and application for crop improvement. Front. Plant Sci. 6208. https://doi.org/10.3389/fpls.2015.00208.

Khare, T., Kumar, V., Kishor, P.B.K., 2015. Na^+ and Cl^- ions show additive effects under NaCl stress on induction of oxidative stress and the responsive antioxidative defense in rice. Protoplasma 252, 1149–1165. https://doi.org/10.1007/s00709-014-0749-2.

Khraiwesh, B., Arif, M.A., Seumel, G.I., Ossowski, S., Weigel, D., Reski, R., Frank, W., 2010. Transcriptional control of gene expression by MicroRNAs. Cell 140, 111–122. https://doi.org/10.1016/j.cell.2009.12.023.

Khraiwesh, B., Zhu, J.-K., Zhu, J., 2012. Role of miRNAs and siRNAs in biotic and abiotic stress responses of plants. Biochim. Biophys. Acta, Gene Regul. Mech. 1819, 137–148. https://doi.org/10.1016/j.bbagrm.2011.05.001.

Koch, A., Kogel, K.H., 2014. New wind in the sails: improving the agronomic value of crop plants through RNAi-mediated gene silencing. Plant Biotechnol. J 12, 821–831. https://doi.org/10.1111/pbi.12226.

Kohli, D., Joshi, G., Deokar, A.A., Bhardwaj, A.R., Agarwal, M., Katiyar-Agarwal, S., Srinivasan, R., Jain, P.K., 2014. Identification and characterization of wilt and salt stress-responsive microRNAs in chickpea through high-throughput sequencing. PLoS One 9, e108851. https://doi.org/10.1371/journal.pone.0108851.

Kruszka, K., Pieczynski, M., Windels, D., Bielewicz, D., Jarmolowski, A., Szweykowska-Kulinska, Z., Vazquez, F., 2012. Role of microRNAs and other sRNAs of plants in their changing environments. J. Plant Physiol. 169, 1664–1672. https://doi.org/10.1016/j.jplph.2012.03.009.

Kumar, V., Khare, T., 2014. Individual and additive effects of Na^+ and Cl^- ions on rice under salinity stress. Arch. Agron. Soil Sci. 61, 381–395. https://doi.org/10.1080/03650340.2014.936400.

Kumar, V., Khare, T., 2016. Differential growth and yield responses of salt-tolerant and susceptible rice cultivars to individual (Na^+ and Cl^-) and additive stress effects of NaCl. Acta Physiol. Plant. 38, 170. https://doi.org/10.1007/s11738-016-2191-x.

Kumar, R.R., Pathak, H., Sharma, S.K., Kala, Y.K., Nirjal, M.K., Singh, G.P., Goswami, S., Rai, R.D., 2015. Novel and conserved heat-responsive microRNAs in wheat (*Triticum aestivum* L.). Funct. Integr. Genomics 15, 323–348. https://doi.org/10.1007/s10142-014-0421-0.

Kumar, V., Wani, S.H., Sah, S.K., Khare, T., Shriram, V., 2016. Engineering phytohormones for abiotic stress tolerance in crop plants. In: Ahammed, G.J., Yu, J. (Eds.), Plant Hormones Under Challenging Environmental Factors. Springer, Dordrecht, pp. 247–266. https://doi.org/10.1007/978-94-017 7758-2_10.

Kumar, V., Khare, T., Shriram, V., Wani, S.H., 2017. Plant small RNAs: the essential epigenetic regulators of gene expression for salinity stress responses and tolerance. Plant Cell Rep. 37, 61–75. https://doi.org/10.1007/s00299-017-2210-4.

Kusaba, M., 2004. RNA interference in crop plants. Curr. Opin. Biotechnol 15, 139–143. https://doi.org/10.1016/j.copbio.2004.02.004.

Li, D.-H., Liu, H., Yang, Y.-l., Zhen, P.-p., Liang, J.-S., 2009. Down-regulated expression of RACK1 gene by RNA interference enhances drought tolerance in rice. Rice Sci. 16, 14–20. https://doi.org/10.1016/S1672-6308(08)60051-7.

Li, J., Wu, L.-Q., Zheng, W.-Y., Wang, R.-F., Yang, L.-X., 2015. Genome-wide identification of microRNAs responsive to high temperature in rice (*Oryza sativa*) by high-throughput deep sequencing. J. Agron. Crop Sci. 201, 379–388. https://doi.org/10.1111/jac.12114.

Li, S.P., Dong, H.X., Yang, G., Wu, Y., Su, S.Z., Shan, X.H., Liu, H.K., Han, J.Y., Liu, J.B., Yuan, Y.P., 2016a. Identification of microRNAs involved in chilling response of maize by high-throughput sequencing. Biol. Plant. 60, 251–260. https://doi.org/10.1007/s10535-016-0590-x.

Li, C., Li, Y., Bai, L., He, C., Yu, X., 2016b. Dynamic expression of miRNAs and their targets in the response to drought stress of grafted cucumber seedlings. Hortic. Plant J. 2, 41–49. https://doi.org/10.1016/j.hpj.2016.02.002.

Lipardi, C., Wei, Q., Paterson, B.M., 2001. RNAi as random degradative PCR: siRNA primers convert mRNA into dsRNAs that are degraded to generate new siRNAs. Cell 107, 297–307. https://doi.org/10.1016/S0092-8674(01)00537-2.

Liu, Q., Yang, T., Yu, T., Zhang, S., Mao, X., Zhao, J., Wang, X., Dong, J., Liu, B., 2017. Integrating small RNA sequencing with QTL mapping for identification of miRNAs and their target genes associated with heat tolerance at the flowering stage in rice. Front. Plant Sci. 8, 43. https://doi.org/10.3389/fpls.2017.00043.

Lobell, D.B., Roberts, M.J., Schlenker, W., Braun, N., Little, B.B., Rejesus, R.M., Hammer, G.L., 2014. Greater sensitivity to drought accompanies maize yield increase in the U.S. Midwest. Science 344, 516–519. https://doi.org/10.1126/science.1251423.

Manavalan, L.P., Chen, X., Clarke, J., Salmeron, J., Nguyen, H.T., 2012. RNAi-mediated disruption of squalene synthase improves drought tolerance and yield in rice. J. Exp. Bot. 63, 163–175. https://doi.org/10.1093/jxb/err258.

Mangrauthia, S.K., Bhogireddy, S., Agarwal, S., Prasanth, V.V., Voleti, S.R., Neelamraju, S., Subrahmanyam, D., 2017. Genome-wide changes in microRNA expression during short and prolonged heat stress and recovery in contrasting rice cultivars. J. Exp. Bot. 68, 2399–2412. https://doi.org/10.1093/jxb/erx111.

Mickelbart, M.V., Hasegawa, P.M., Bailey-Serres, J., 2015. Genetic mechanisms of abiotic stress tolerance that translate to crop yield stability. Nat. Rev. Genet. 16, 237–251. https://doi.org/10.1038/nrg3901.

Mittal, D., Sharma, N., Sharma, V., Sopory, S.K., Sanan-Mishra, N., 2016. Role of microRNAs in rice plant under salt stress. Ann. Appl. Biol. 168, 2–18. https://doi.org/10.1111/aab.12241.

Mittler, R., Blumwald, E., 2010. Genetic engineering for modern agriculture: challenges and perspectives. Annu. Rev. Plant Biol. 61, 443–462. https://doi.org/10.1146/annurev-arplant-042809-112116.

Napoli, C., Lemieux, C., Jorgensen, R., 1990. Introduction of a chimeric chalcone synthase gene into petunia results in reversible co-suppression of homologous genes in trans. Plant Cell 2, 279–289. https://doi.org/10.1105/tpc.2.4.279.

Pare, J.M., Hobman, T.C., 2007. Dicer: structure, function and role in RNA-dependent gene-silencing pathways. In: Industrial Enzymes: Structure, Function and Applications. Springer, Dordrecht, pp. 421–438. https://doi.org/10.1007/1-4020-5377-0_24.

Park, G.G., Park, J.J., Yoon, J., Yu, S.N., An, G., 2010. A RING finger E3 ligase gene, *Oryza sativa* delayed seed germination 1 (*OsDSG1*), controls seed germination and stress responses in rice. Plant Mol. Biol. 74, 467–478. https://doi.org/10.1007/s11103-010-9687-3.

Redfern, A.D., Colley, S.M., Beveridge, D.J., Ikeda, N., Epis, M.R., Li, X., Foulds, C.E., Stuart, L.M., Barker, A., Russell, V.J., Ramsay, K., Kobelke, S.J., Li, X., Hatchell, E.C., Payne, C., Giles, K.M., Messineo, A., Gatignol, A., Lanz, R.B., O'Malley, B.W., Leedman, P.J., 2013. RNA-induced silencing complex (RISC) proteins PACT, TRBP, and Dicer are SRA binding nuclear receptor coregulators. Proc. Natl. Acad. Sci. 110, 6536–6541. https://doi.org/10.1073/pnas.1301620110.

Ren, Y., Chen, L., Zhang, Y., Kang, X., Zhang, Z., Wang, Y., 2013. Identification and characterization of salt-responsive microRNAs in *Populus tomentosa* by high-throughput sequencing. Biochimie 95, 743–750. https://doi.org/10.1016/j.biochi.2012.10.025.

Riley, K.J., Yario, T.A., Steitz, J.A., 2012. Association of Argonaute proteins and microRNAs can occur after cell lysis. RNA 18, 1581–1585. https://doi.org/10.1261/rna.034934.112.

Rocheleau, C.E., Downs, W.D., Lin, R., Wittmann, C., Bei, Y., Cha, Y.H., Ali, M., Priess, J.R., Mello, C.C., 1997. Wnt signaling and an APC-related gene specify endoderm in early *C. elegans* embryos. Cell 90, 707–716. https://doi.org/10.1016/S0092-8674(00)80531-0.

Romano, N., Macino, G., 1992. Quelling: transient inactivation of gene expression in Neurospora crassa by transformation with homologous sequences. Mol. Microbiol. 6, 3343–3353. https://doi.org/10.1111/j.1365-2958.1992.tb02202.x.

Rosenzweig, C., Elliott, J., Deryng, D., Ruane, A.C., Müller, C., Arneth, A., Boote, K.J., Folberth, C., Glotter, M., Khabarov, N., Neumann, K., Piontek, F., Pugh, T.A.M., Schmid, E., Stehfest, E., Yang, H., Jones, J.W., 2014. Assessing agricultural risks of climate change in the 21st century in a global gridded crop model intercomparison. Proc. Natl. Acad. Sci. USA 111, 3268–3273. https://doi.org/10.1073/pnas.1222463110.

Saurabh, S., Vidyarthi, A.S., Prasad, D., 2014. RNA interference: concept to reality in crop improvement. Planta 239, 543–564. https://doi.org/10.1007/s00425-013-2019-5.

Shriram, V., Kumar, V., Devarumath, R.M., Khare, T.S., Wani, S.H., 2016. MicroRNAs as potential targets for abiotic stress tolerance in plants. Front. Plant Sci. 7, 8173389–8173817. https://doi.org/10.3389/fpls.2016.00817.

Sijen, T., Fleenor, J., Simmer, F., Thijssen, K.L., Parrish, S., Timmons, L., Plasterk, R.H.A., Fire, A., 2001. On the role of RNA amplification in dsRNA-triggered gene silencing. Cell 107, 465–476. https://doi.org/10.1016/S0092-8674(01)00576-1.

Sun, G., 2012. MicroRNAs and their diverse functions in plants. Plant Mol. Biol. 80, 17–36. https://doi.org/10.1007/s11103-011-9817-6.

Sun, X., Xu, L., Wang, Y., Yu, R., Zhu, X., Luo, X., Gong, Y., Wang, R., Limera, C., Zhang, K., Liu, L., 2015. Identification of novel and salt-responsive miRNAs to explore miRNA-mediated regulatory network of salt stress response in radish (*Raphanus sativus* L.). BMC Genomics. 16, 197. https://doi.org/10.1186/s12864-015-1416-5.

Sunkar, R., Li, Y.F., Jagadeeswaran, G., 2012. Functions of microRNAs in plant stress responses. Trends Plant Sci. 17, 196–203. https://doi.org/10.1016/j.tplants.2012.01.010.

Tester, M., Langridge, P., 2010. Breeding technologies to increase crop production in a changing world. Science 327 (80), 818–822. https://doi.org/10.1126/science.1183700.

Valdés-López, O., Yang, S.S., Aparicio-Fabre, R., Graham, P.H., Reyes, J.L., Vance, C.P., Hernández, G., 2010. MicroRNA expression profile in common bean (*Phaseolus vulgaris*) under nutrient deficiency stresses and manganese toxicity. New Phytol. 187, 805–818. https://doi.org/10.1111/j.1469-8137.2010.03320.x.

Voinnet, O., 2009. Origin, biogenesis, and activity of plant microRNAs. Cell 136, 669–687. https://doi.org/10.1016/j.cell.2009.01.046.

Wang, Y., Beaith, M., Chalifoux, M., Ying, J., Uchacz, T., Sarvas, C., Griffiths, R., Kuzma, M., Wan, J., Huang, Y., 2009. Shoot-specific down-regulation of protein farnesyltransferase (α-subunit) for yield protection against drought in Canola. Mol. Plant 2, 191–200. https://doi.org/10.1093/mp/ssn088.

Wang, M., Wang, Q., Zhang, B., 2013. Response of miRNAs and their targets to salt and drought stresses in cotton (*Gossypium hirsutum* L.). Gene 530, 26–32. https://doi.org/10.1016/j.gene.2013.08.009.

Wilson, R.C., Doudna, J.A., 2013. Molecular mechanisms of RNA interference. Annu. Rev. Biophys. 42, 217–239. https://doi.org/10.1146/annurev-biophys-083012-130404.

Xie, F., Stewart, C.N., Taki, F.A., He, Q., Liu, H., Zhang, B., 2014. High-throughput deep sequencing shows that microRNAs play important roles in switchgrass responses to drought and salinity stress. Plant Biotechnol. J. 12, 354–366. https://doi.org/10.1111/pbi.12142.

Xu, L., Wang, Y., Zhai, L., Xu, Y., Wang, L., Zhu, X., Gong, Y., Yu, R., Limera, C., Liu, L., 2013. Genome-wide identification and characterization of cadmium-responsive microRNAs and their target genes in radish (*Raphanus sativus* L.) roots. J. Exp. Bot. 64, 4271–4287. https://doi.org/10.1093/jxb/ert240.

Xu, S., Liu, N., Mao, W., Hu, Q., Wang, G., Gong, Y., 2016. Identification of chilling-responsive microRNAs and their targets in vegetable soybean (*Glycine max* L.). Sci. Rep. 6, 26619. https://doi.org/10.1038/srep26619.

Yang, C., Li, D., Mao, D., Liu, X., Ji, C., Li, X., Zhao, X., Cheng, Z., Chen, C., Zhu, L., 2013. Overexpression of microRNA319 impacts leaf morphogenesis and leads to enhanced cold tolerance in rice (*Oryza sativa* L.). Plant Cell Environ. 36, 2207–2218. https://doi.org/10.1111/pce.12130.

Younis, A., Siddique, M.I., Kim, C.K., Lim, K.B., 2014. RNA interference (RNAi) induced gene silencing: a promising approach of hi-tech plant breeding. Int. J. Biol. Sci. 10, 1150–1158. https://doi.org/10.7150/ijbs.10452.

Yu, L.J., Luo, Y.F., Liao, B., Xie, L.J., Chen, L., Xiao, S., Li, J.T., Hu, S.N., Shu, W.S., 2012. Comparative transcriptome analysis of transporters, phytohormone and lipid metabolism pathways in response to arsenic stress in rice (*Oryza sativa*). New Phytol. 195, 97–112. https://doi.org/10.1111/j.1469-8137.2012.04154.x.

Zandalinas, S.I., Mittler, R., Balfagón, D., Arbona, V., Gómez-Cadenas, A., 2017. Plant adaptations to the combination of drought and high temperatures. Physiol. Plant 162, 2–12. https://doi.org/10.1111/ppl.12540.

Zeng, Q.-Y., Yang, C.-Y., Ma, Q.-B., Li, X.-P., Dong, W.-W., Nian, H., 2012. Identification of wild soybean miRNAs and their target genes responsive to aluminum stress. BMC Plant Biol. 12, 182. https://doi.org/10.1186/1471-2229-12-182.

Zhang, N., Yang, J., Wang, Z., Wen, Y., Wang, J., He, W., Liu, B., Si, H., Wang, D., 2014. Identification of novel and conserved microRNAs related to drought stress in potato by deep sequencing, in: Margis, R. (Ed.), PLoS One. Public Library of Science, 9 p. e95489. https://doi.org/10.1371/journal.pone.0095489

Zhang, F., Luo, X., Zhou, Y., Xie, J., 2016. Genome-wide identification of conserved microRNA and their response to drought stress in Dongxiang wild rice (*Oryza rufipogon* Griff.). Biotechnol. Lett. 38, 711–721. https://doi.org/10.1007/s10529-015-2012-0.

Zhou, Z.S., Zeng, H.Q., Liu, Z.P., Yang, Z.M., 2012. Genome-wide identification of *Medicago truncatula* microRNAs and their targets reveals their differential regulation by heavy metal. Plant Cell Environ. 35, 86–99. https://doi.org/10.1111/j.1365-3040.2011.02418.x.

CHAPTER 9

Genome-Wide Association Studies (GWAS) for Abiotic Stress Tolerance in Plants

Surekha Challa, Nageswara R.R. Neelapu
Department of Biochemistry and Bioinformatics, Gandhi Institute of Technology and Management (GITAM), Deemed-to-be-University, Visakhapatnam, India

Contents

1. INTRODUCTION

Plants are continuously exposed to combinations of various biotic stresses (bacteria, fungi, insects, viruses, and weeds) and abiotic stresses (metals, nutrients, photo-oxidation or light, salt, temperature, water, and wind) (Khraiwesh et al., 2012; Neelapu et al., 2015; Surekha et al., 2015). Plants use carrier/ion channel/transporter proteins, compatible organic solutes, chaperones, hormones, late embryogenesis abundant proteins, protein kinases, proteins related to phospholipid metabolism, proteinases that remove denatured proteins, water channel proteins, and transcription factors to develop tolerance to abiotic stresses. Plants also enhance antioxidant production to combat abiotic stresses (Neelapu et al., 2015; Surekha et al., 2015; Table 1). Although the preceding candidate genes are used in transgenic plants for developing tolerance to abiotic stresses, there is a limitation with the use of candidate genes, as abiotic stresses in plants are controlled by multiple traits.

Biochemical, Physiological and Molecular Avenues for Combating Abiotic Stress in Plants
https://doi.org/10.1016/B978-0-12-813066-7.00009-7

Table 1 Biomolecules generated in response to abiotic stress in plants

Biomolecules	Role	References
Compatible organic solutes		
Proline, trehalose, sucrose, polyols, quaternary ammonium compounds such as glycine betaine, prolinebetaine, alaninebetaine, hydroxyprolinebetaine, pipecolatebetaine, and choline *O*-sulfate diamines, triamines, tetraamines, and polyamines	Overproduction of different types of compatible organic solutes	Rhodes and Hanson (1993) and Surekha et al. (2014)
Water channel proteins		
Aquaporins	Reduce membrane injury, improve ion distribution, and maintain osmotic balance	Surekha et al. (2015)
Ion channel/transporters/carrier proteins		
Proton pumps, antiporters (Na^+/H^+ antiporters), or transporters	Ion uptake or exchange of ions, transport or compartmentalization are the mechanisms to prevent abiotic tolerance	Ratner and Jacoby (1976), Niu et al. (1993), Shi et al. (2003), and Suneetha et al. (2016)
Enhancing antioxidant production		
Antioxidants—ascorbate, glutathione, α-tocopherol, lutein, morin, β-carotene, quercetin, kaempferol, xanthophylls, and catechins. Antioxidant enzymes—superoxide dismutase (SOD), catalase (CAT), ascorbate peroxidase (POD), glutathione peroxidase, and enzymes of ascorbate-glutathione cycle	Scavenging/detoxifying the reactive oxygen species (ROS) by increasing the expression of antioxidant enzymes and decreasing ROS and subsequently abiotic stress	Ashraf (2009)
Late embryogenesis abundant (LEA) proteins		
COR (cold regulated), ERD (early responsive to dehydration), KIN (cold inducible), RAB [responsive to abscisic acid (ABA)], and RD (responsive to dehydration) genes encode LEA proteins	LEA proteins are expressed in vegetative tissues of plants to impart dehydration tolerance	Shinozaki and Yamaguchi-Shinozaki (2000) and Zhu (2002)

Table 1 Biomolecules generated in response to abiotic stress in plants—cont'd

Biomolecules	Role	References
Chaperones		
Hsp70 family, chaperonins (Hsp60), Hsp90 family, Hsp100/Clp family, sHsp family	Chaperones stabilize membranes and proteins and assist in protein refolding under stress conditions	Wang et al. (2003)
Proteinases that remove denatured proteins		
Dehydrins, chaperones, proteases	Maintenance of functional conformation of a protein	Vierstra (1996) and Feller (2004)
Hormones		
Phytohormones such as ABA, auxin, brassinosteroids (BRs), cytokinins (CKs), gibberellic acid (GA), jasmonates (JAs), salicylic acid (SA), and triazoles (TRs)	Phytohormones play an important role by regulating plant responses to stresses	Surekha et al. (2015)
Transcription factors		
ABA-responsive element (ABRE), MYC (myelocytomatosis oncogene)/MYB (myeloblastosis oncogene), CBF (C-repeat binding factor)/DREBs (dehydration-responsive element binding protein, and NAC (NAM, ATAF and CUC) are the transcription factors	Phytohormones play an important role by regulating plant responses to stresses	Lata et al. (2011)
Protein kinases		
Inducible protein kinase Esi47, ABA-induced ICK1, cyclin-dependent protein kinases (CDKs), mitogen-activated protein kinase kinase (MKK)	Protein kinases hormonal signaling, inhibition of cell division, and intra- and extracellular signaling in plants have been identified in plants	Shen et al. (2001), Zhu (2001), Wang et al. (2000), and Oh et al. (2014)
Protein-related phospholipid metabolism		
Phosphoinositides, phosphatidic acid	Activates phospholipids like phosphoinositides, phosphatidic acid to mediate early signaling events	Wang et al. (2007)

Approaches used to study and understand multiple traits in plants are quantitative trait loci (QTL) mapping and genome-wide association studies (GWAS). QTL mapping is a statistical method that helps in linking two types of information or data, that is, complex phenotypes (traits with measurements) with genotypes (specific regions of chromosomes) (Falconer and Mackay, 1996; Kearsey, 1998; Lynch and Walsh, 1998). Molecular markers such as single nucleotide polymorphisms (SNPs) and amplified fragment length polymorphism (AFLPs) are used for mapping QTLs, and then they are correlated with observed phenotypic data. Several studies on plant species reported usage of QTL mapping to study multiple traits linked to abiotic stresses (Fan et al., 2015; Mora et al., 2016; Gahlaut et al., 2017). The advantage with QTL mapping is that variants for a trait can be mapped in F_2 recombinant inbred line (RIL) populations. The same advantage showed limitations with QTL mapping, where allelic diversity between parents at the F_2 cross can be assayed, and the mapping is limited only to RIL populations (Korte and Farlow, 2013). The drawbacks of QTL mapping are overcome by GWAS. GWAS is hypothesis-free; yet capable of generating hypotheses. At the same time, GWAS can identify more variants associated with traits. GWAS identifies or studies the correlation between the genetic variants/traits/phenotypes in a population of any organism based on SNPs in the sequence data. GWAS explores the complete genome, in contrast to other approaches that exactly investigate a minor amount of prespecified chromosomal areas.

2. GENOME-WIDE ASSOCIATION STUDY, DESIGN, AND ANALYSIS

Genome-wide association studies, or whole genome association studies, investigate a genome-wide set of genetic variants in different varieties to see if any variant is associated with a trait (Manolio, 2010). GWAS usually emphasize associations between SNPs and traits, for example, DNA of plant varieties is compared with different phenotypes for a particular trait. GWAS implementation is based on designs, genotyping technologies, and statistical concepts for analysis, replication, interpretation, and follow-up of association results (Bush and Moore, 2012).

2.1 GWAS Design

A standard approach or design is required for an effective investigation. GWAS design is usually case versus control, or quantitative design, which compares the DNA of a variety of plants with varying phenotypes for a particular trait (Bush and Moore, 2012). If the study focuses on a trait, it can be case versus control; whereas if the study focuses on the measurement of a trait, it can be a quantitative design. In circumstances of case versus control, the variety may be a case, and a similar variety without traits that can be controlled, or variety with different phenotypes for a particular trait. In general, we classify all varieties either into a case or a control. In some instances, the influence of genetic variations in the study can easily measure a quantitative trait that is a part of a quantitative design (Bush and Moore, 2012). The study will be effective, and depend on the implementation of GWAS design to capture SNP data.

2.2 Capturing SNPs

SNPs are single base-pair changes (mutations) in the DNA and are the modern units of genetic variation in any genome of an organism. More than a million SNPs can be captured using genotyping technologies such as customized, chip-based microarray technology (Kumar et al., 2015), Illumina (San Diego, California), Affymetrix (Santa Clara, California) (Distefano and Taverna, 2011), or other next-generation sequencing technologies (Neelapu and Surekha, 2016). These captured SNPs can be analyzed or associated with phenotypes or traits based on GWAS design.

2.3 GWAS Analysis

Linkage (group of alleles/genes) of the hereditary unit for traits in a population is passed on from parent to progeny. Changes in the linkage are due to mutations/breakage of linkage/SNPs and can be measured as linkage disequilibrium (LD). The concept behind GWAS is to assess SNPs in the genome-influencing phenotype. SNPs influencing phenotypes are assessed in two ways: either by direct association, or indirect association (Bush and Moore, 2012). When the SNP-influencing phenotype is directly genotyped and is statistically associated with the trait it is known as direct association; whereas, if the SNP is not directly genotyped, but is statistically associated to the phenotype, it is referred to as an indirect association (Bush and Moore, 2012). The correlation between allele and SNP in a population is performed for single-locus or multilocus analysis (Padmavathi et al., 2003; Devi et al., 2006, 2007). If an association between each SNP and phenotype is examined independently, then it is a single-locus statistic test. If the association between each SNP and phenotype/trait is examined, then it is multilocus analysis (Padmavathi et al., 2003; Devi et al., 2006, 2007; Bush and Moore, 2012).

When the GWAS analysis is for quantitative design, analysis for traits will be carried out using logistic regression. When GWAS analysis is for case versus control design, analysis for traits will be carried out using a generalized linear model (GLM) and analysis of variance (ANOVA). The null hypothesis or assumptions of GLM, ANOVA, and logistic regression are: (1) the trait is normally distributed; (2) the trait variance within each group is the same; (3) the groups are independent (Bush and Moore, 2012). Although there are advantages with GWAS in associating SNPs with phenotypes or traits, some limitations also exist. SNPs are most likely to be associated with traits during GWAS analysis due to population stratification. Generally, genetic analysis is a measure of population substructure (Bush and Moore, 2012). To prevent population stratification, population structures can be studied and subjected to principal component analysis (PCA) to minimize effects in the data (Bush and Moore, 2012).

2.4 Replication of GWAS

A replication results in generalization of the association between SNPs and traits, and the genetic effect is relevant to multiple populations. Replication studies must be carried out

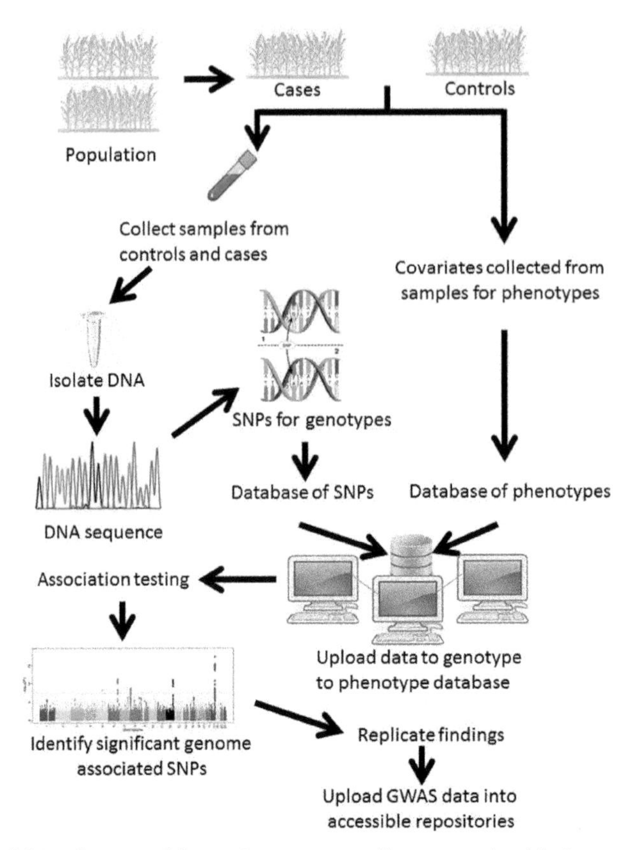

Fig. 1 GWAS model implemented by various case studies to study abiotic stresses like drought, salinity, and temperature stress. *(Modified and reprinted with permission, https://www.nature.com/articles/nrcardio.2010.53).*

with independent datasets derived from similar samples, as with GWAS, in order to endorse the outcome of the population (Bush and Moore, 2012). If the results are consistent across case studies and replication studies, then the association of SNP with phenotypes or traits can be endorsed (Fig. 1).

2.5 Metaanalysis of GWAS

Numerous GWAS case studies have been published since the first successful publication in 2005. Generally, scientific publications may agree or disagree on a common hypothesis. Metaanalysis is used to resolve this ambiguity in publications. Metaanalysis is a powerful statistical procedure that combines GWAS data from multiple studies published to examine the same hypothesis and to identify the common outcome of combined studies (Bush and Moore, 2012). Before considering the data for metaanalysis, the data from combined studies should be imputated, as the data is not directly genotyped for SNPs in the study and is composed of data that is obtained from multiple genome sequence platforms (Bush and Moore, 2012).

There are several advantages with this approach; for example, the outcomes of the study can be generalized to a greater population. Methods can improve the precision and accuracy of the estimates. Discrepancies of findings among the GWAS reports can be enumerated and studied. A summary of the estimates in a GWAS study can be used in the metaanalysis to test the hypothesis, and so forth (Bush and Moore, 2012).

3. APPLICATIONS OF GWAS FOR ABIOTIC STRESS TOLERANCE IN PLANTS

GWAS has a wide range of applications in plants, which includes studying biotic and abiotic stress, flowering traits, yield, and other agronomic traits, and so forth. GWAS have been used to study drought tolerance (Verslues et al., 2013; Thoen et al., 2017; Qin et al., 2016), salt tolerance (Kumar et al., 2015; Shi et al., 2017; Thoen et al., 2017; Wan et al., 2017), thermal tolerance (Lafarge et al., 2017; Thoen et al., 2017; Chopra et al., 2017; Chen et al., 2017), boron toxicity (de Abreu Neto et al. 2017), yield and other agronomic traits (Begum et al., 2015), and flowering traits (Wang et al., 2017) (Table 2). Further, this section will review implementation of GWAS to study abiotic stresses such as drought, salt, and temperature.

Table 2 Recent genome-wide association study (GWAS) conducted on different plant species to understand and identify QTL's responsible for various abiotic and biotic stresses

S. No	Plant	Stress	Study	Outcome	References
1.	*Arabidopsis thaliana*	Drought stress	Conducted a genome-wide association study of low water potential-induced proline accumulation	GWAS identifies genes connecting pro accumulation to cellular metabolic and redox status	Verslues et al. (2013)
		Abiotic and biotic stress	Conducted a genome-wide association study for various biotic (salt, drought, osmotic) biotic stress (parasitic plant, nematode, whitefly, Aphid, Thrips, caterpillar, fungus)	Revealed genetic architecture of plant stress resistance for multitraits in *Arabidopsis thaliana*	Thoen et al. (2017)

Continued

Table 2 Recent genome-wide association study (GWAS) conducted on different plant species to understand and identify QTL's responsible for various abiotic and biotic stresses—cont'd

S. No	Plant	Stress	Study	Outcome	References
2.	*Oryza sativa*	Salt stress	Implemented a genome-wide association study and mapping of salinity tolerance	Identified novel and major QTL Saltol and other stress-related genomic regions known to control salinity tolerance at seedling stage in rice	Kumar et al. (2015)
		Temperature stress	Performed a genome-wide association study for thermal and other stresses in rice	Identified possible spikelet sterility (SPKST) candidate genes in genomic regions associated with anthesis and for several secondary traits potentially affecting panicle micro-climate and thus the fertilization process	Lafarge et al. (2017)
		Salt stress	Conducted a genome-wide association study of salt tolerance at the seed germination stage in rice	Identified 11 loci for salt tolerance in rice	Shi et al. (2017)
3.	*Aegilops tauschii*	Drought stress	Performed a genome-wide association study of drought-related resistance traits	Identified candidate genes that might control drought resistance traits, including genes encoding enzymes, storage proteins, and drought-induced proteins	Qin et al. (2016)

Table 2 Recent genome-wide association study (GWAS) conducted on different plant species to understand and identify QTL's responsible for various abiotic and biotic stresses—cont'd

S. No	Plant	Stress	Study	Outcome	References
4.	*Brassica napus* L.	Salt stress	Conducted a genome-wide association study to identify salt tolerance-related QTL	Identified 38 possible candidate genes in genomic regions associated with salt tolerance. These genes include transcription factors, aquaporins, transporters, and enzymes	Wan et al. (2017)
5.	*Sorghum bicolor*	Cold and heat stress	Employed a genome-wide association study of diverse Sorghum under thermal stress	Identified potential gene networks in *Sorghum* responsible for adaptation to thermal stresses like cold and heat stress during plant developmental stages	Chopra et al. (2017)
		Temperature stress	Employed genome-wide association study for studying heat tolerance during vegetative growth stages in a *Sorghum*	Identified gene candidates either directly or indirectly linked to a biological pathway involved in plant stress responses (heat stress response)	Chen et al. (2017)

3.1 GWAS Identifies Genes Associated With Drought Tolerance

Several groups have employed GWAS on different plant species to identify drought response genes. The case studies include pro-accumulation of drought response genes under low water potential (Verslues et al., 2013), response to collective stresses (Thoen et al., 2017), and understanding drought resistance genes in *Aegilops tauschii* (Qin et al., 2016).

Verslues et al. (2013) used powerful and promising methods from GWAS-guided-reverse genetics to identify new effector genes related to drought response. Verslues et al. (2013) investigated the accumulation of proline in *Arabidopsis thaliana* when a drought response was induced under low water potential. The prioritized genomic datasets inducing pro-accumulation under low water potential from natural accessions and publicly available SNPs were considered by the group for analysis. Further, phenotypic datasets of low water potential induced pro-accumulation were also prioritized. Subsequently, the associations between the SNPs of the genomic and phenotypic datasets are analyzed and reported candidate regions controlling pro-accumulation. Additionally, reverse genetics identified pro effector genes such as thioredoxins and Universal Stress Protein A domain proteins, mitochondrial protease LON1, ribosomal protein RPL24A, protein phosphatase 2A subunit A3, a MADS box protein, and a nucleoside triphosphate hydrolase involved in pro accumulation. This study provided information on pro-effectors in pro-accumulation of molecules in response to drought stress.

Plants are continuously exposed to combinations of stresses, and have the capability to cope with these stresses. Plants employ a wide range of protective mechanisms influencing multiple traits in response to combined stress. Thoen et al. (2017) implemented a comprehensive approach to investigate and understand the mechanisms influencing multiple traits in response to collective stress. Thoen et al. (2017) studied plant responses to 11 single stresses and their combinations in 350 *A. thaliana* varieties using a multitrait mixed model (MTMM). Drought was one among the 11 single stresses that was studied. GWAS analysis was performed for 214,000 SNPs and tailored multitrait mixed models. The 30 most significant SNPs were identified for MTMM. The analyses recognized QTLs with distinct and related responses to stresses. The analysis also observed larger QTL sizes for a combination of stresses than for single stresses. Plants seem to utilize wide range of protective mechanisms and share similar phytohormonal signaling pathways to various stress responses. This study provided information on genes and phytohormonal signaling pathways that are distinct and similar in response to mixed stresses.

The species *A. tauschii* is known for its superior quality, that is, in holding treasure of numerous resistance genes related to abiotic stresses (Ashraf et al., 2009). An understanding of the genetic architecture of *A. tauschii* is required for breeders to develop drought-resistant cultivars. Qin et al. (2016) assessed 13 traits in 373 *A. tauschii* varieties under simulated drought stress. The genomic datasets with 7185 SNP markers were selected for performing GWAS with all phenotypic characteristics. Qin et al. (2016) applied general linear models (GLM) and mixed linear models (MLM) to identify significant SNPs associated with all phenotypic characteristics. GLM and MLM identified 208 and 28 significant SNPs, respectively, associated with traits; whereas 25 significant SNPs were revealed with both models. Further, a search on SNPs and their genes in several public databases revealed a large number of genes belonging to categories such as enzymes, storage proteins, and drought-induced proteins related to drought resistance that can be used for plant breeding. The study provided information on drought response genes in *A. tauschii*.

The preceding case studies provided information on pro-effectors in pro-accumulation of molecules (Verslues et al., 2013), genes, and phytohormonal signaling pathways that are distinct and similar (Thoen et al., 2017) enzymes, storage proteins, and drought-induced proteins (Qin et al., 2016) in response to drought stress. This knowledge can be used for breeding plants to develop drought-tolerant varieties.

3.2 GWAS Identifies Genes Associated With Salt Tolerance

Several case studies were reported by various groups on usage of GWAS to identify salt-tolerant genes. GWAS was utilized to study different plant species, such as rice (Kumar et al., 2015; Shi et al., 2017), *A. thaliana* (Thoen et al., 2017), and rapeseed (Wan et al., 2017) for sustainable agriculture. GWAS was applied to identify complex QTLs controlling salinity tolerance in plants. The potential of GWAS was harnessed to reveal the genes controlling salinity stress tolerance in rice (Kumar et al., 2015; Shi et al., 2017). Kumar et al. (2015) used infinium high-throughput assay, which is based on customized gene-based SNP arrays to genotype 220 rice varieties. The gene-based SNP array was designed for 6000 SNPs containing stress-responsive genes. Phenotypic datasets include 12 different traits for stress. The association between 12 traits and 6000 SNPs were analyzed. Twenty significant SNPs were identified to be associated with Na^+/K^+ ratio; whereas 44 SNPs were observed to be associated with other stresses. Further, QTLs containing salt tolerance genes/alleles were reported and mapped in the genomic regions on chromosomes 1, 4, 6, and 7. Saltol, a major QTL on chromosome 1 in rice, is known for its association with salinity stress tolerance at the seedling stage. Kumar et al. (2015) also reported association among Saltol, Na^+/K^+ ratio, and salinity stress tolerance in rice at the reproductive stage. This study provided information on salinity-tolerant QTLs at the reproductive stage, which are spread across genomic regions of chromosomes 1, 4, 6, and 7.

Shi et al. (2017) investigated 6,361,920 SNPs from 478 diverse rice varieties for seven seed germination-related traits under control and salt-stress conditions. The mixed linear model was applied to analyze the association between SNPs and traits. Twenty two significant SNPs spread across 11 loci were reported to be associated with salt tolerance. Of these seven, loci were reported on chromosomes 1, 5, 6, 11, and 12. Two strong associations for salt tolerance were identified on chromosomes 1 and 2. The genomic region on chromosome 1 is of $\sim$ 13.3 kb and is known for controlling the Na^+:K^+ ratio, total Na^+ uptake, and total K^+ concentration. The genomic region on chromosome 2 is of $\sim$ 164.2 kb, which harbors two genes, OsNRT2.1 and OsNRT2.2, related to the nitrate transporter family. This study provided information on salinity-tolerant QTLs at the reproductive stage, which are spread across genomic regions of chromosomes 1, 5, 6, 11, and 12. This valuable information on salt tolerance in rice may assist in plant breeding.

GWAS was utilized to delve into salt tolerance to identify salt tolerance genes at the seedling stage. Wan et al. (2017) investigated salt tolerance in rapeseed (*Brassica napus* L.) at the seedling stage using GWAS. Wan et al. (2017) assessed the genomic datasets from 368 *B. napus* cultivars using a *Brassica* 60 K Illumina Infinium SNP array. Salt tolerance-related traits were included as the phenotypic datasets. GWAS assessed the association between SNPs and salt tolerance-related traits. Wan et al. (2017) reported 25 QTLs based on 75 SNPs spread across 14 chromosomes. These 25 QTLs based on 75 SNPs were associated with four salt tolerance-related traits. Wan et al. (2017) also identified 38 genes that include aquaporins, enzymes, transcription factors, and transporters that are associated with salt tolerance. This valuable information on salt tolerance in rapeseed at seedling stages may assist in plant breeding.

The preceding case studies provided information on salinity-tolerant QTLs on chromosomes 1, 4, 5, 6, 7, 11, and 12 in rice, which control Na^+:K^+ ratio, total Na^+ uptake, and total K^+ concentration at both the seedling and reproductive stages (Kumar et al., 2015; Shi et al., 2017). Another case study provided information on 25 QTLs in rapeseed, which include aquaporins, enzymes, transcription factors, and transporters at the seedling stage (Wan et al., 2017). This knowledge on salt tolerance at both the seedling and reproductive stages can be used for breeding of plants to develop drought-tolerant varieties.

3.3 GWAS Identifies Genes Associated With Thermal Tolerance

Change in the microclimate around the plant influences crop growth and yield. Temperature is one such factor that influences the microclimate, and thereby, fertilization and productivity of the crop. Plants were known to be resilient to temperature stress for their ultimate survival. Therefore, understanding the adaptive mechanism of the major crops to temperature fluctuations at different stages of the plant will be of prime importance to plant breeders. Studies explaining the association between phenotype temperature stress and the genome would help in breeding thermo-tolerant varieties. GWAS was employed to study thermal tolerance to identify thermal-tolerance genes in rice (Lafarge et al., 2017), *A. thaliana* (Thoen et al., 2017), and *Sorghum* (Chopra et al., 2017; Chen et al., 2017) for sustainable agriculture.

Lafarge et al. (2017) explored 167 rice varieties for traits such as spikelet sterility (SPKST), and secondary traits (panicle micro-climate, fertilization process) to study the effect of temperature sensitivity during anthesis. The genomic datasets were prioritized by genotyping an average density of one marker per 29 kb. Lafarge et al. (2017) performed GWAS using three methods of single-marker regression, haplotype regression, and simultaneous fitting of all markers. Lafarge et al. (2017) reported that significant association exists between 14 loci and SPKST. The loci were reported with functions such as cell division, gametophyte development, heat shock proteins, regulating plant response, sensing abiotic stresses, and wall-associated kinases. Associations were also detected for secondary traits. Analysis of loci revealed SPKST-favorable alleles within 3000 rice genomes. Further, the N22 variety, and some Indian and Taiwanese varieties are reported to be heat tolerant. This study provides valuable information for breeding of heat-tolerant varieties.

Chopra et al. (2017) investigated the *Sorghum* varieties and identified responses to heat and cold stresses, and thermo-tolerant genes for understanding thermal tolerance. The genomic datasets were prioritized for SNPs in *Sorghum* varieties and the phenotypic datasets were prioritized for traits related to heat and cold stresses. Chopra et al. (2017) performed GWAS analysis between the genotypes and phenotypic datasets. Chopra et al. (2017) reported huge variations under cold and heat stress for *Sorghum* varieties at the seedling stage. Thirty SNPs of genes such as regulators of anthocyanin expression and soluble carbohydrate metabolism were strongly associated with traits of cold stress at the seedling stage. Twelve SNPs of genes such as sugar metabolism, and ion transport pathways were significantly associated with heat stress at the seedling stage. Chopra et al. (2017) analyzed a network of genes for both stresses, and identified complex gene interactions that are significantly associated with SNPs. Among them, genes were upregulated during cold stress in a moderately tolerant variety when compared with the sensitive variety. Expression and validation of four genes identified Sb06g025040, a basic-helix-loop-helix (bHLH) transcription factor that is involved in purple color pigmentation of leaves. This study provided information on genes and their networks controlling thermal stress (cold and heat) adaptations. Further, work on these findings may help in understanding plant response to thermal fluctuations and resilencing to thermal stress at plant developmental stages for developing thermo-tolerant varieties.

Chen et al. (2017) explored *Sorghum bicolor* (L.) Moench at the vegetative stage for heat stress traits such as leaf firing (LF) and leaf blotching (LB). Chen et al. (2017) performed GWAS to study the association between SNPs of the genotype and heat stress traits LF and LB. Chen et al. (2017) reported nine SNPs significantly associated with LF; whereas five SNPs were associated with LB. Further, genes in the vicinity of SNPs were probed and 14 genes were identified. These genes were required for plants in response to abiotic stress. The study provided information of leaf traits in response to heat stress in the vegetative stage of *Sorghum,* which can be used in breeding enhanced heat tolerance varieties.

The preceding case studies provided valuable information that the following genes related to cell division, gametophyte development, heat shock proteins, regulating plant response, sensing abiotic stresses, wall associated kinases, regulators of anthocyanin expression, soluble carbohydrate metabolism, sugar metabolism, and ion transport pathways. This knowledge can be used for breeding of plants to develop thermo-tolerant varieties.

4. CONCLUSION

In conclusion, GWAS is a powerful tool for studying multiple traits in response to stresses such as drought, salt, and temperature. GWAS on various plants/crops such as *A. thaliana, O. sativa, A. tauschii, B. napus* L., and *S. bicolor* have identified novel QTLs/gene candidates or genes responsible for abiotic stress. This valuable information on abiotic stress can be used for plant breeding and in designing future crops.

ACKNOWLEDGMENTS

We would like to thank GITAM, Visakhapatnam, India for providing the facility and support.

REFERENCES

Ashraf, M., 2009. Biotechnological approach of improving plant salt tolerance using antioxidants as markers. Biotechnol. Adv. 27, 84–93.

Ashraf, M., Ozturk, M., Athar, H.R., 2009. Salinity and Water Stress: Improving Crop Efficiency, first ed. Springer, Berlin.

Begum, H., Spindel, J.E., Lalusin, A., Borromeo, T., Gregorio, G., Hernandez, J., Virk, P., Collard, B., McCouch, S.R., 2015. Genome-wide association mapping for yield and other agronomic traits in an elite breeding population of tropical rice (*Oryza sativa*). PLoS One 10 (3), e0119873. https://doi.org/10.1371/journal.pone.0119873.

Bush, W.S., Moore, J.H., 2012. Genome-wide association studies. PLoS Comput. Biol. 8 (12), e1002822. https://doi.org/10.1371/journal.pcbi.1002822.

Chen, J., Chopra, R., Hayes, C., Morris, G., Marla, S., Burke, J., Xin, Z., Burow, G., 2017. Genome-wide association study of developing leaves' heat tolerance during vegetative growth stages in a *Sorghum* association panel. Plant Genome 10 (2), 1–15. https://doi.org/10.3835/plantgenome2016.09.0091.

Chopra, R., Burow, G., Burke, J.J., Gladman, N., Xin, Z., 2017. Genome-wide association analysis of seedling traits in diverse *Sorghum* germplasm under thermal stress. BMC Plant Biol. 17 (1), 12. https://doi.org/10.1186/s12870-016-0966-2.

de Abreu Neto, J.B., Hurtado-Perez, M.C., Wimmer, M.A., Frei, M., 2017. Genetic factors underlying boron toxicity tolerance in rice: genome-wide association study and transcriptomic analysis. J. Exp. Bot. 68 (3), 687–700.

Devi, K.U., Reineke, A., Reddy, N.N.R., Rao, C.U.M., Padmavathi, J., 2006. Genetic diversity, reproductive biology and speciation in the entomopathogenic fungus *Beauveria bassiana* (Balsamo) Vuillemin. Genome 49 (5), 495–504.

Devi, K.U., Reineke, A., Rao, C.U.M., Reddy, N.N.R., Khan, A.P.A., 2007. AFLP and single-strand confirmation polymorphism studies of recombination in the entomopathogenic fungus *Nomuraea rileyi*. Mycol. Res. 111, 716–725.

Distefano, J.K., Taverna, D.M., 2011. Technological issues and experimental design of gene association studies. Methods Mol. Biol. 700, 3–16.

Falconer, D.S., Mackay, T.F.C., 1996. Introduction to Quantitative Genetics, fourth ed. Longmans Green, Harlow.

Fan, Y., Shabala, S., Ma, Y., Xu, R., Zhou, M., 2015. Using QTL mapping to investigate the relationships between abiotic stress tolerance (drought and salinity) and agronomic and physiological traits. BMC Genomics 16 (1), 43. https://doi.org/10.1186/s12864-015-1243-8.

Feller, U., 2004. Proteolysis. In: Nooden, L.D. (Ed.), Plant Cell Death Processes. Elsevier, Amsterdam, pp. 107–123.

Gahlaut, V., Jaiswal, V., Tyagi, B.S., Singh, G., Sareen, S., Balyan, H.S., Gupta, P.K., 2017. QTL mapping for nine drought-responsive agronomic traits in bread wheat under irrigated and rain-fed environments. PLoS One 12 (8), e0182857. https://doi.org/10.1371/journal.pone.0182857.

Kearsey, M.J., 1998. The principles of QTL analysis (a minimal mathematics approach). J. Exp. Bot. 49, 1619–1623.

Khraiwesh, B., Zhu, J.K., Zhu, J., 2012. Role of miRNAs and siRNAs in biotic and abiotic stress responses of plants. Biochim. Biophys. Acta 1819 (2), 137–148.

Korte, A., Farlow, A., 2013. The advantages and limitations of trait analysis with GWAS: a review. Plant Methods 9, 29. https://doi.org/10.1186/1746-4811-9-29.

Kumar, V., Singh, A., Mithra, S.V., Krishnamurthy, S.L., Parida, S.K., Jain, S., Tiwari, K.K., Kumar, P., Rao, A.R., Sharma, S.K., Khurana, J.P., Singh, N.K., Mohapatra, T., 2015. Genome-wide association mapping of salinity tolerance in rice (*Oryza sativa*). DNA Res. 22 (2), 133–145.

Lafarge, T., Bueno, C., Frouin, J., Jacquin, L., Courtois, B., Ahmadi, N., 2017. Genome-wide association analysis for heat tolerance at flowering detected a large set of genes involved in adaptation to thermal and other stresses. PLoS One 12 (2), e0171254. https://doi.org/10.1371/journal.pone.0171254.

Lata, C., Yadav, A., Prasad, M., 2011. Role of plant transcription factors in abiotic stress tolerance. In: Shanker, A. (Ed.), Abiotic Stress Response in Plants—Physiological, Biochemical and Genetic Perspectives. InTech, Rijeka.

Lynch, M., Walsh, B., 1998. Genetics and Analysis of Quantitative Traits. Sinauer Associates, Sunderland.

Manolio, T.A., 2010. Genomewide association studies and assessment of the risk of disease. N. Engl. J. Med. 363 (2), 166–176.

Mora, F., Quitral, Y.A., Matus, I., Russell, J., Waugh, R., Del Pozo, A., 2016. SNP-based QTL mapping of 15 complex traits in barley under rain-fed and well-watered conditions by a mixed modeling approach. Front. Plant Sci. 7(909). https://doi.org/10.3389/fpls.2016.00909.

Neelapu, N.R.R., Surekha, C., 2016. Next-generation sequencing and metagenomics. In: Wong, K.-C. (Ed.), Computational Biology and Bioinformatics: Gene Regulation. CRC Press, Boca Raton, FL, pp. 331–351.

Neelapu, N.R.R., Deepak, K.G.K., Surekha, C., 2015. Transgenic plants for higher antioxidant contents and salt stress tolerance. In: Wani, S.H., Hossain, M.A. (Eds.), Managing Salt Tolerance in Plants: Molecular and Genomic Perspectives. CRC Press, Boca Raton, FL, pp. 391–406.

Niu, X., Bressan, R.A., Hasegawa, P.M., 1993. Halophytes up-regulate plasma membrane Hþ-ATPase gene more rapidly than glycophytes in response to salt stress. Plant Physiol. 102, 130.

Oh, S.K., Jang, H.A., Lee, S.S., Cho, H.S., Lee, D.H., Choi, D., Kwon, S.Y., 2014. *Cucumber* Pti1-L is a cytoplasmic protein kinase involved in defense responses and salt tolerance. J. Plant Physiol. 171 (10), 817–822.

Padmavathi, J., Uma Devi, K., Rao, C.U.M., Reddy, N.N.R., 2003. Telomere fingerprinting for assessing chromosome number, isolating typing and recombination in the entomopathogen *Beauveria bassiana*. Mycol. Res. 107 (5), 572–580.

Qin, P., Lin, Y., Hu, Y., Liu, K., Mao, S., Li, Z., Wang, J., Liu, Y., Wei, Y., Zheng, Y., 2016. Genome-wide association study of drought-related resistance traits in *Aegilops tauschii*. Genet. Mol. Biol. 39 (3), 398–407.

Ratner, A., Jacoby, B., 1976. Effect of Kþ, its counter anion, and pH on sodium efflux from barley root tips. J. Exp. Bot. 27, 843–850.

Rhodes, D., Hanson, A.D., 1993. Quaternary ammonium and tertiary sulfonium compounds in higher-plants. Annu. Rev. Plant Physiol. Plant Mol. Biol. 44, 357–384.

Shen, W., Gómez-Cadenas, A., Routly, E.L., Ho, T.H., Simmonds, J.A., Gulick, P.J., 2001. The salt stress-inducible protein kinase gene, *Esi47*, from the salt-tolerant wheatgrass *Lophopyrum elongatum* is involved in plant hormone signaling. Plant Physiol. 125 (3), 1429–1441.

Shi, H., Wu, S.J., Zhu, J.K., 2003. Overexpression of a plasma membrane Na^+/H^+ antiporter improves salt tolerance in *Arabidopsis*. Nat. Biotechnol. 21, 81–85.

Shi, Y., Gao, L., Wu, Z., Zhang, X., Wang, M., Zhang, C., Zhang, F., Zhou, Y., Li, Z., 2017. Genome-wide association study of salt tolerance at the seed germination stage in rice. BMC Plant Biol. 17 (1), 92. https://doi.org/10.1186/s12870-017-1044-0.

Shinozaki, K., Yamaguchi-Shinozaki, K., 2000. Molecular response to dehydration and low temperature: differences and cross-talk between two stress signaling pathways. Curr. Opin. Plant Biol. 3, 217–223.

Suneetha, G., Neelapu, N.R.R., Surekha, C., 2016. Plant vacuolar proton pyrophosphatases (VPPases): structure, function and mode of action. IJRSR 7 (6), 12148–12152.

Surekha, C., Nirmala Kumari, K., Aruna, L.V., Suneetha, G., Arundhati, A., Kavi Kishor, P.B., 2014. Expression of the *Vigna aconitifolia* P5csf129a gene in transgenic pigeonpea enhances proline acculumation and salt tolerance. Plant Cell Tissue Organ Cult. 116, 27–36.

Surekha, C., Aruna, L.V., Hossain, M.A., Wani, S.H., Neelapu, N.R.R., 2015. Present status and future prospects of transgenic approaches for salt tolerance in plants/crop plants. In: Wani, S.H., Hossain, M.A. (Eds.), Managing Salt Tolerance in Plants: Molecular and Genomic Perspectives. CRC Press, Boca Raton, FL, pp. 329–352.

Thoen, M.P., Davila Olivas, N.H., Kloth, K.J., Coolen, S., Huang, P.P., Aarts, M.G., Bac-Molenaar, J.A., Bakker, J., Bouwmeester, H.J., Broekgaarden, C., Bucher, J., Busscher-Lange, J., Cheng, X., Fradin, E.F., Jongsma, M.A., Julkowska, M.M., Keurentjes, J.J., Ligterink, W., Pieterse, C.M., Ruyter-Spira, C., Smant, G., Testerink, C., Usadel, B., van Loon, J.J., van Pelt, J.A., van Schaik, C.C., van Wees, S.C., Visser, R.G., Voorrips, R., Vosman, B., Vreugdenhil, D., Warmerdam, S., Wiegers, G.L., van Heerwaarden, J., Kruijer, W., van Eeuwijk, F.A., Dicke, M., 2017. Genetic architecture of plant stress resistance: multi-trait genome-wide association mapping. New Phytol. 213 (3), 1346–1362.

Verslues, P.E., Lasky, J.R., Juenger, T.E., Liu, T.W., Kumar, M.N., 2013. Genome-wide association mapping combined with reverse genetics identifies new effectors of low water potential-induced proline accumulation in *Arabidopsis*. Plant Physiol. 164 (1), 144–159.

Vierstra, R., 1996. Proteolysis in plants: mechanisms and functions. Plant Mol. Biol. 32, 275–302.

Wan, H., Chen, L., Guo, J., Li, Q., Wen, J., Yi, B., Ma, C., Tu, J., Fu, T., Shen, J., 2017. Genome-wide association study reveals the genetic architecture underlying salt tolerance-related traits in rapeseed (*Brassica napus* L.). Front. Plant Sci. 8, 593. https://doi.org/10.3389/fpls.2017.00593.

Wang, H., Zhou, Y., Gilmer, S., Whitwill, S., Fowke, L.C., 2000. Expression of the plant cyclin-dependent kinase inhibitor ICK1 affects cell division, plant growth and morphology. Plant J. 24 (5), 613–623.

Wang, W., Vinocur, B., Altman, A., 2003. Plant responses to drought, salinity and extreme temperatures: towards genetic engineering for stress tolerance. Planta 218, 1–14.

Wang, X., Zhang, W., Li, W., Mishra, G., 2007. Phospholipid signaling in plant response to drought and salt stress. In: Jenks, E., Matthew, A., Hasegawa, E., Paul, M., Jain, E., Mohan, S. (Eds.), Advances in Molecular Breeding Toward Drought and Salt Tolerant Crops. Springer, Dordrecht, pp. 183–192.

Wang, H., Xu, C., Liu, X., Guo, Z., Xu, X., Wang, S., Xie, C., Li, W.X., Zou, C., Xu, Y., 2017. Development of a multiple-hybrid population for genome-wide association studies: theoretical consideration and genetic mapping of flowering traits in maize. Sci. Rep. 7, 40239. https://doi.org/10.1038/srep40239.

Zhu, J.K., 2001. Plant salt tolerance. Trends Plant Sci. 6 (2), 66–71.

Zhu, J.K., 2002. Salt and drought stress signal transduction in plants. Annu. Rev. Plant Biol. 53, 247–273.

FURTHER READING

Wani, S.H., Dutta, T., Neelapu, N.R.R., Surekha, C., 2017. Transgenic approaches to enhance salt and drought tolerance in plants. Plant Gene 11, 219–231. https://doi.org/10.1016/j.plgene.2017.05.006.

Yamamoto, T., Kuboki, Y., Lin, S.Y., Sasaki, T., Yano, M., 1998. Fine mapping of quantitative trait loci Hd-1, Hd-2 and Hd-3, controlling heading date of rice, as single Mendelian factors. Theor. Appl. Genet. 97, 37–44.

Yamamoto, T., Lin, H., Sasaki, T., Yano, M., 2000. Identification of heading date quantitative trait locus Hd6 and characterization of its epistatic interactions with Hd2 in rice using advanced backcross progeny. Genetics 154, 885–891.

Yano, M., Harushima, Y., Nagamura, Y., Kurata, N., Minobe, Y., Sasaki, T., 1997. Identification of quantitative trait loci controlling heading date in rice using a high-density linkage map. Theor. Appl. Genet. 95, 1025–1032.

CHAPTER 10

Targeting the Redox Regulatory Mechanisms for Abiotic Stress Tolerance in Crops

Punam Kundu*, Ritu Gill*, Shruti Ahlawat†, Naser A. Anjum‡, Krishna K. Sharma†, Abid A. Ansari§, Mirza Hasanuzzaman¶, Akula Ramakrishna||, Narsingh Chauhan#, Narendra Tuteja, Sarvajeet S. Gill***

*Centre for Biotechnology, Maharshi Dayanand University, Rohtak, India
†Department of Microbiology, Maharshi Dayanand University, Rohtak, India
‡Department of Botany, Aligarh Muslim University, Aligarh, India
§Department of Biology, Faculty of Science, University of Tabuk, Tabuk, Saudi Arabia
¶Department of Agronomy, Faculty of Agriculture, Sher-e-Bangla Agricultural University, Dhaka, Bangladesh
||Monsanto Crop Breeding Station, Bangalore, India
#Department of Biochemistry, Maharshi Dayanand University, Rohtak, India
**Plant Molecular Biology Group, International Centre for Genetic Engineering and Biotechnology (ICGEB), New Delhi, India

Contents

Biochemical, Physiological and Molecular Avenues for Combating Abiotic Stress in Plants
https://doi.org/10.1016/B978-0-12-813066-7.00010-3

1. INTRODUCTION

Various environmental stresses adversely affect the sustainability of crop yields worldwide (Horie and Schroeder, 2004). Plants, due to their sessile nature, cannot avoid these stresses by moving from one place to another (Hasanuzzaman et al., 2013). In many agricultural areas, environmental factors such as temperature, drought, and salinity result in severe loss of crop yields every year (Shrivastav and Kumar, 2014). Plants exhibit various responses when exposed to such stresses, and these responses may occur at molecular, cellular, or whole plant levels (Suzuki et al., 2000). Plants' ability to tolerate these stresses is determined by multiple physiological and biochemical mechanisms. These mechanisms include osmolyte biosynthesis, ion homeostasis, reactive oxygen species (ROS) scavenging systems, and so forth (Gill and Tuteja, 2010; Hossain and Dietz, 2016). During stress, phytohormone biosynthesis also fluctuates; for example, in drought conditions, ABA biosynthesis increases and plays a role in the closure of stomata, and thereby avoids/minimizes transpiration (Jarzyniak and Jasinski, 2014). To combat the various threats coming from the outside environment, plants have developed a broad range of mechanisms or strategies. Redox regulatory mechanisms to counteract abiotic stress also include the synthesis of ROS at a high level (Gill and Tuteja, 2010; Mendoza, 2011). Synthesis of ROS at higher concentrations is also called respiratory, or oxidative, burst. ROS were initially recognized as toxic side products of aerobic metabolism in a favorable condition, and these toxic products are removed by defense machinery of the plant, such as antioxidants (Sewelam et al., 2016). When the ROS level surpasses in the cell, then it is said to be under oxidative stress. Responses of plants to abiotic stress either lead to destructive change, or develop a tolerance index (Das et al., 2014). In unfavorable environmental conditions, different signaling pathways come together, and start a signaling network. In the past decade it became apparent that ROS plays an important role in plant signaling, and controls various processes such as growth, development, and programmed cell death. Whether ROS act as messengers in signaling, or damage the system, depend on the equilibrium between its synthesis and removal (Sharma et al., 2012). In plants, highly efficient scavenging mechanisms have developed that enabled plant cells to overcome ROS toxic products. ROS have the potential to induce cellular damage by the oxidation of biomolecules. These species lead to lipid peroxidation, oxidation of proteins, and inactivation of enzymes (Sharma et al., 2012). ROS also alter gene expression when they act as a signaling molecule. The importance of ROS in signaling has been demonstrated in many studies. What the effects of ROS will be, and whether its production is beneficiary or harmful, depends on their production site, how much is produced, and the activity of the antioxidant defense system of the plant itself (Circu and Aw, 2010).

By targeting the activity of ROS, which act on intracellular key signaling molecules such as the MAP kinases pathway, GPCR, and phytohormone signaling, we can develop stress tolerance in crops. A better understanding of the roles of ROS in plant signaling would help to improve plant adaptability under abiotic stress (Choudhury et al., 2013). ROS interaction with other signaling molecule such as calcium, MAPK, G-protein plant hormones ABA, and transcription factors demonstrate its importance at the transcription level (Sewelam et al., 2016).

2. REACTIVE OXYGEN SPECIES: CHEMICAL BEHAVIOR, HISTORY, AND PRODUCTION SITES

Our planet has abundant free oxygen in the atmosphere, which distinguishes it from all other planets in the solar system. Molecular oxygen is generated by photosynthesis, the biological process in which water molecules are split in the presence of solar energy, and oxygenic photosynthesis by lower organisms, which leads the evolution of eukaryotes in genetic complexity (Sellers et al., 2003). In 1954, Gerschman stated that free radicals are the cause behind most of the damaging effects of oxygen (Gershman et al., 1954). Decreased antioxidant defense, or increased partial pressure of oxygen contributes equally toward cell and tissue damage, keeping in mind that oxygen toxicity is a steady process. Similarly, Elstner showed that ROS were formed in the photosynthetic electron transport chain (ETC) in plants (Río, 2015) (Table 1). In aerobic organisms, molecular oxygen, due to its chemical behavior, forms a reactive oxygen species (ROS) species. ROS are generated when excited electrons fall on oxygen (Gutowski and Kowalczyk, 2013). These are highly reactive oxygen-containing molecules that are either produced due to incomplete reduction of molecular oxygen, namely hydrogen peroxide (H_2O_2) and hydroxyl radicals ($OH^{\bullet}$), superoxide radical anion ($O_2^{\bullet -}$), or another form, such as ozone (O_3), singlet oxygen ($^1O_2$) (Gill and Tuteja, 2010).

2.1 Chemical Behavior of ROS

The reduction of molecular oxygen (O_2) to the superoxide radical anion ($O_2^{\bullet -}$) is the precursor of most other reactive oxygen species:

$$O_2 + e^- \rightarrow O_2^{\bullet -}$$

Dismutation of superoxide produces (H_2O_2)

$$2H^+ + O_2^{\bullet -} + O_2^{\bullet -} \rightarrow H_2O_2 + O_2$$

Hydrogen peroxide either generates the hydroxyl radical ($^{\bullet}OH$), or is fully reduced to water

Table 1 List of important landmarks in history related to reactive oxygen species (ROS)

S. no.	Inventor	Year of discovery	Country of origin	Discovery
1.	Rebeca Gerschman	1954	New York (USA)	Free radicals are the cause behind most of damaging effects of oxygen. Decreased antioxidant defense or increased partial pressure of oxygen contributes equally toward cell and tissue damage with a concluding remark that O_2 toxicity is a steady process.
2.	Joe McCord and Irwin Fridovich	1969	Durham (USA)	Discovered an enzyme superoxide dismutase (SOD). They found that SOD catalyzes superoxide radicals into molecular oxygen and hydrogen peroxide.
3.	Bernard M Babior and coworkers	1973, 2004	Boston (USA)	They found that WBCs (leukocytes) generates superoxide radicals as a result of "respiratory burst" which are utilized as potent bactericidal agent to kill engulfed bacteria during phagocytosis. Later on they discovered NADPH oxidase in phagocytes which generates superoxide radicals.
4.	Helmut Sies and Britton Chance	1970	–	Both of them in collaborative work identified H_2O_2 as a normal aerobic metabolite and devised a method for quantification H_2O_2 in cells.
5.	Helmut Sies	1985	Germany	Gave the concept of "oxidative stress."
6.	Erich F Elstner	1990, 1994	Germany	Investigated relationship among phytohormone ethylene and ROS. He was first to show the role of ethylene and ROS in signal transduction in both plant and animals. He also showed that ROS were formed in photosynthetic electron transport chain (ETC) in plants.
7.	Kozi Asada	1973, 1974	Japan	He demonstrated that molecular oxygen can be reduced to superoxide in chloroplast in response to illumination. He for the first time demonstrated the presence of superoxide dismutase in chloroplast and purified and characterized the SOD also.
8.	Kozi Asada, Christine H Foyer and Barry Halliwell	1976	Japan, London (UK)	They discovered ascorbate glutathione cycle also called Foyer-Halliwell-Asada cycle, a mechanism for H_2O_2 metabolism in chloroplasts.
9.	Alain Puppo	1982	France	He studied the role of ROS in symbiosis among N_2-fixing bacteria (rhizobia) and legumes.
10.	Fewson and Nicholas	1960	–	He observed that plant uses NO.
11.	L. Klepper	1979	–	He showed that plant emits NO.
12.	Ya'acov Leshem, Tomoya Noritake	1996	Israel, Japan	They showed the role of ROS in senescence and plant immunity. Leshem published the first book on NO in plants.
13.	Dr. Lorenzo Lamattina	1997	Argentina	He demonstrated the effect of NO in chlorophyll preservation in the plants infected by a fungus.

$$H_2O_2 + e^- \rightarrow HO^- + OH^\bullet \text{ (Partial reduction)}$$

$$2\,H_2O_2 \rightarrow 2\,H_2O + O_2 \text{ (Complete reduction)}$$

Molecular oxygen in its ground state does not show reactivity, but in its radical form, it is highly reactive, due to the presence of unpaired single electrons in an excited state, while nonradical forms of ROS are quite toxic.

2.2 Site of ROS Production

In plants subjected to abiotic stresses, generation of the ROS within various intracellular compartments is one of the earliest responses. An increase in production of ROS either leads to metabolic disorders, oxidative damage, or plays an important role in signaling. Most important sites to initiate the response via ROS are mitochondria, chloroplasts, cell walls, peroxisomes, apoplast endoplasmic reticulum, and plasma membranes (Sharma et al., 2012) (Fig. 1).

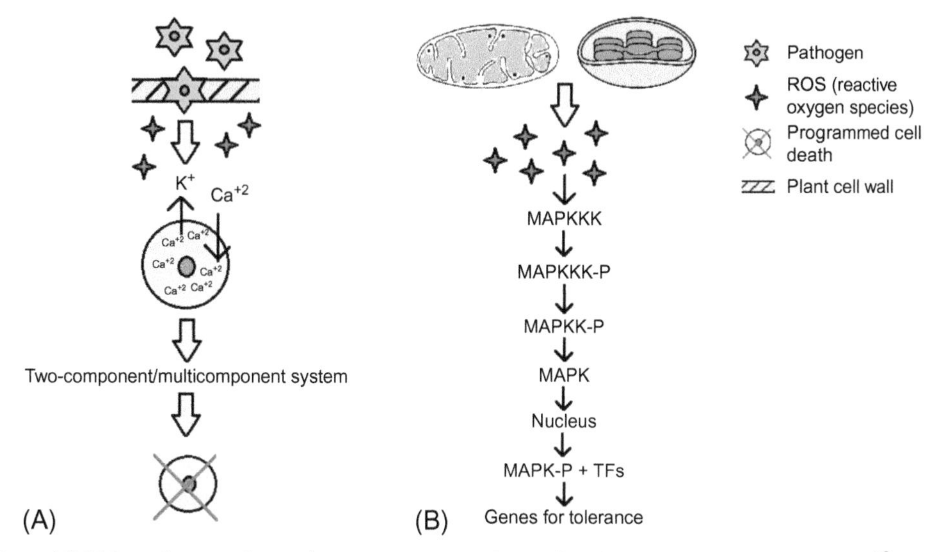

Fig. 1 (A) ROS production after pathogen penetrates plant cell wall. Cationic channels such as Ca^{+2} and K^+ sense produced ROS converting signal into physiological responses such as increased intracellular Ca^{+2} due to Ca^{+2} influx and K^+ efflux. Increased intracellular Ca^{+2} activate two-/multicomponent system inducing programmed cell death (PCD). (B) ROS production from chloroplast and mitochondria activates MAPKKK (MAPKK kinase) which upon activation phosphorylates and activates MAPKK (MAPK kinase). Phosphorylated MAPKK activates MAPK (MAP kinase), which upon activation enters nucleus and get phosphorylated such that it can bind to various transcription factors (TFs) regulating expression of various genes involved in tolerance against stress.

2.2.1 Mitochondria

During unfavorable conditions, ROS are accumulated into cells through various mechanisms (Fig. 1). In plant cells, oxidative phosphorylation for production of ATP is the crucial role of mitochondria (Das et al., 2015). Compared with other eukaryotes, in plants, mitochondria play a crucial role in many aspects, such as in metabolic processes of leaves, where it develops complex enzymatic machinery for the activation of a number of biochemical and metabolic pathways, like decarboxylation of glycine and synthesis of vitamins (Mittler, 2002). In plant cells, mitochondria are the major source of ROS production because of the site of the electron transport system (ETS). However, the relative contribution of mitochondria to ROS production in plants in normal conditions is quite low, which elevates under the stressed cellular environment. Stress-induced accumulation of reactive oxygen species (ROS) and alterations in the cellular energy state can cause oxidation of cellular components. In oxidative phosphorylation, electrons are released from NADH and $FADH_2$ to O_2, forming H_2O. Electrons are transferred via enzyme complexes named complex I, II, and III that perform their function in the presence of cofactors. Complexes I and III of mitochondrial ETC are the very well-known sites of $O_2^{\bullet-}$ production. Complex I of the respiratory chain is involved in the direct reduction of oxygen to $O_2^{\bullet-}$ (Arora et al., 2002). In general, electrons flow in a substantial order from complex I and II to complex III. However, when NAD^+-linked substrates for complex I are limited, electron transport can occur in the reverse direction from complex II to complex I that lead to the production of ROS, and this process is regulated by ATP hydrolysis (Turrens, 2003). Reactivity of $O_2^{\bullet-}$ increases in an aqueous solution, but this can be further reduced by dismutation, and its conversion into H_2O_2 takes place (Quan et al., 2008). Up to 5% of mitochondrial O_2 accumulated in the form of H_2O_2. H_2O_2 shows reactivity with metallic ions such as copper and iron to produce highly toxic $OH^{\bullet}$ that causes peroxidation of lipids present in the membrane of mitochondria, which in turn leads to the formation of cytotoxic lipid aldehydes, alkenals, and hydroxyalkenals, such as 4-hydroxy-2-nonenal and malondialdehyde (Rhoads et al., 2006). Once these products are formed, they react with cellular biomolecular machinery such as proteins and nucleic acids, and cause cellular damage. Some enzymes occurring in the mitochondrial matrix are also responsible for the production of ROS such as aconitase, while others have the ability to capture electrons from ETC. $O_2^{\bullet-}$ is the primary ROS formed, and then it is converted quickly either by the SOD or APX into the relatively stable H_2O_2 that is finally converted to an extremely active hydroxyl radical ($OH^{\bullet}$) in the Fenton reaction (Andreyev et al., 2005; Rasmusson et al., 2008). Under stresses, generally there are respiratory fluctuations in the plants that trigger a signal that downregulates other signals for adaptation toward these stresses, which is a process by which their cellular metabolism is reprogrammed so it can reinstate the balance of their cellular energy at the transcription level (Bosch et al., 2013).

2.2.2 Chloroplast

In oxygenic photosynthesis, light energy is captured and converted into ATP where H_2O acts as an electron donor, and evolution of oxygen takes place. This is also referred to as photolysis of water. In thylakoids, light-induced electrons transport from water to $NADP^+$, and oxygen is evolved. Transport of electrons from the donor to acceptor involves the collaboration of two reaction centers PSI (P700) and PSII (P680) (Fig. 1). The electron flow begins from PSII by absorbing light energy, then it transfers e^- to the next acceptor, pheophytin, then transfers it to Q_A, and Q_A to Q_B, and likewise PS1 to the A_0 to A_1 Fe-S center, and then finally to ferredoxin. It means there are a number of steps in this electron transport chain in which chances of e^- leakage increases, and electron transportation up to the final acceptor is reduced. In the case of stressful conditions, due to lower supply of NADP, there is leakage of electrons from ferredoxin to O_2, which is reduced into $O_2^{\bullet -}$, and then it is spontaneously dismutated to H_2O_2 on the external stromal surface. This is called the Mehler reaction (Sharma and Dubey, 2005).

$$2O_2 + 2Fd_{red} - \rightarrow 2O_2^{\bullet -} + 2Fd_{ox}$$

Additionally, rubisco (ribulose-1,5 bisphosphate carboxylase-oxygenase), a chloroplastic enzyme that assimilates atmospheric CO_2 in normal conditions, as a substrate to form two molecules of 3-carbon compound D-phosphoglycerate (3PGAL), also exhibits oxygenase activity that allows it to react with oxygen to produce phosphoglycolate, and thereby also initiates the process known as photorespiration (Reumann and Weber, 2006). In addition, thylakoid pigments involve photodynamic energy transfer to ground-state triplet O_2, which leads to the formation of a highly reactive singlet O_2.

2.2.3 Peroxisome

Peroxisomes are dynamic subcellular organelles that lack DNA and ribosomes, and they are ubiquitous in nature. These are metabolically active organelles, and a major source for the production of reactive oxygen species, because here, essentially, oxidative metabolism of biomolecules takes place (Demarquoy and Le Borgne, 2015) (Fig. 1). Production of three different types of ROS has been reported in peroxisomes; namely hydrogen peroxide (H_2O_2), singlet oxygen, and superoxide radical anion ($O_2^{\bullet -}$). Peroxisomes contain hydrogen peroxide-producing oxidase and H_2O_2-degrading catalase (Corpas et al., 2001). In peroxisome, there are many oxidative metabolic processes which includes fatty acid β-oxidation, photorespiration, nucleic acid, polyamine catabolism, and so forth, that are responsible for the production of ROS species (Rio et al., 2006). In peroxisomes, production and scavenging of H_2O_2 and O_2^- takes place simultaneously, which allows it to regulate dynamic changes in ROS levels that further play a role in signaling, because peroxisomal ROS participate in more complex signaling networks involving calcium, hormones, and redox homeostasis, which finally determines the response of plants to outer environments that can be adaptable or lead to plant-programmed cell death (Rio et al., 2006). In plants, peroxisomes contain amorphous inclusion that has catalase enzymatic content that helps in the scavenging of H_2O_2. Catalytic enzymatic content includes ascorbate peroxidase (APX), glutathione peroxidase (GPX), and catalase (CAT). Both these enzymes are able to scavenge H_2O_2 with different mechanisms. APX activity is contrary to CAT, requires an ascorbate and coenzyme glutathione (GSH) regeneration system, and then converts H_2O_2 into H_2O; it directly converts H_2O_2 into H_2O and $1/2O_2$. In peroxisomes, β-oxidation is primarily committed to the breakdown of fatty acids and subsequent use of acetyl-CoA as a carbon and energy source for growth. β-Oxidation was also shown to be involved in the endogenous turnover of intermediates leaking out of the fatty acid biosynthetic pathway. The β-oxidation of fatty acids, and the metabolism of nitrogen compounds, are common metabolic functions of peroxisomes in most living beings. In plants, peroxisomes differentiate into several location-specific and metabolically specialized variant glyoxysomes that predominantly contain the enzymes required for β oxidation of fatty acids to convert storage oil into carbohydrates.

2.2.4 Apoplast

Apoplast is the space outside the plasma membrane that allows free movement of material. This route plays a crucial role, and allows diffusion of a solvent across the tissue or organ. The response to different environmental and biotic stimuli in apoplast can be an oxidative or respiratory burst (Fig. 1). Oxidative bursts include several classes of enzymes, including cell wall peroxidases (Bindschedler et al., 2006). Another enzyme, plasma membrane NADPH oxidases, are commonly known as respiratory burst enzymes and respiratory burst oxidase homologs (Rboh), and are transmembrane flavoproteins that oxidize cytoplasmic NADPH and translocate electrons across plasma membranes and reduce extracellular ambient (triplet) oxygen to yield $O_2^{\cdot-}$ in the cell wall (Shapiguzov et al., 2012). These types of ROS are charged ions, unable to passively cross the lipid bilayer because of the shorter lifecycle, and leftovers in the apoplast do not have the ability to convey a long-distance signal, where it finds hydrogen ions and rapidly converts into another species, H_2O_2, either automatically or in a reaction catalyzed by the SOD. The functions of the plant NADPH oxidases, other than stress responses, include its role in translocation of ions, and intriguingly, also in long-distance ROS signaling (Sagi and Fluhr, 2006). In the plants, heat stress and increased salinity results in an RbohD-dependent systemic increase of the oxidative break and open signals that are triggered by intracellular Ca^{2+} at the affected site. Signaling is expanded by accumulation of ROS through apoplast and by symplastic signals, one of which might be ROS production in chloroplasts, which influences other organelles by fluctuation. This ROS "wave" travels across the plant at a constant rate (Miller et al., 2009). Unlike superoxide, the H_2O_2 molecule is relatively stable under physiological conditions, and is similar to a molecule of water. The dipole moment of H_2O_2, similar to that of H_2O, limits passive diffusion of H_2O_2 through biological membranes. Aquaporins, a ubiquitous family of channel proteins to transport apoplastic H_2O_2, has undergone an extensive expansion in vascular plants. In *Arabidopsis*, aquaporin for H_2O_2 transport has been shown. However, the role of H_2O_2 transport during the oxidative burst is not clear, and is supposed to react with, and modify, compounds located in the apoplast.

2.2.5 Other Cellular Sites

The plasma membrane, a barrier between living cells and their environments, plays an important role in signaling related to the changes in the surrounding environment. As a component of the plasma membrane, the ion channel plays an important role in transducing signaling from the outer environment (Apel and Hirt, 2004). The physical state of the membrane is also very sensitive and dynamic according to condition; for example, due to heat stress membrane fluidity change. H_2O_2 responds to this change, leading to the activation of the synthesis of small heat shock proteins. In plant membrane-bound NADPH oxidase (NOX), also called respiratory burst oxidase homologue (RBOH) has the capability to transfer electrons through the chain from intracellular NADPH across the plasma membrane to molecular oxygen in the apoplast site and generate $O_2^{\bullet-}$, which can be converted to H_2O_2 through different mechanisms (Hossain and Dietz, 2016). The plant Rboh protein has two main apparatus: membrane-bound respiratory burst oxidase homologue (Rboh), and its regulator Rac, present in the cytoplasm (Zhang et al., 2010). Integral plasma membrane protein is composed of transmembrane domains and multipasses with helical structures. Transmembrane domains contain His residues. These residues are required to bind two heme groups, C-terminal FAD and NADPH hydrophilic domains, and two N-terminal calcium-binding (EF-hand) motifs, and some phosphorylation target sites (Aguirre et al., 2005). Regulatory components of Rboh involved in phosphorylation, such as calcium-dependent protein kinases (CDPKs), a Ser/Thr protein kinase that includes a Ca^{2+}-binding calmodulin-like domain), Ca^{2+}/CaM-dependent protein kinase (CaMK). Posttranslational protein modifications are also mediated by signaling in plants such as *Arabidopsis*, especially by ROS. These modifications include phosphatidic acid binding and S-nitrosylation of Rboh homologues and other antioxidant elements that provide a clear interrelationship between ROS and reactive nitrogen species for intracellular processes. Rboh is involved in many plant processes, including cell growth, plant development, stomatal closure, pollen tube growth, symbiotic interactions, abiotic stress, and pathogen responses. However, the number of Rboh isozymes, which are differentially expressed, suggests a certain grade of specialization for each one.

3. OVERPRODUCTION OF ROS SPECIES IN STRESSFUL ENVIRONMENTS

Reactive oxygen species are generated during the metabolic processes and electron transport in both chloroplast and mitochondria. Furthermore, abiotic and biotic stresses can also induce the production and accumulation of ROS (Apel and Hirt, 2004). Several sources for production of ROS have been reported in plant cells via photosynthetic and respiratory electron transport chains, NADPH oxidase, photorespiration, and cell wall-bound peroxidases (Mittler, 2002). Chloroplasts can produce ROS at the level of photosystem (PSI and PSII). Under stress conditions, the absorbed light energy exceeds the capacity of photosynthetic machinery. Therefore, various ROS are formed, including singlet oxygen (PSII) and superoxide radicals (PSI and PSII) (Asada, 2006; Schmitt et al., 2014). Under various biotic and abiotic stresses, the production of ROS in plant cells is drastically increased, which eventually disturbs the balance of $OH^{\bullet}$, $O_2^{\bullet-}$, and H_2O_2 (Gill and Tuteja, 2010). The effects of ROS production due to various environmental stresses (Fig. 1) such as drought, salinity, cold, metal toxicity, and pathogen attack are discussed as follows.

3.1 Drought

One of the inevitable consequences of drought stress is enhanced ROS production in the different cellular compartments, namely in the chloroplasts, the peroxisomes, and the mitochondria. This enhanced ROS production is, however, kept under tight control by a versatile and cooperative antioxidant system that modulates intracellular ROS concentration and sets the redox-status of the cell. Unlike biotic stress, where an oxidative burst is part of a defense response that frequently triggers programmed cell death (PCD), the role of ROS production and control during drought stress is yet to be resolved. However, as stated by Dat and collaborators (Dat et al., 2000), ROS seems to have a dual effect under abiotic stress conditions that depends on their overall cellular amount. If kept at relatively low levels, they are likely to function as components of a stress-signaling pathway, triggering stress defense/acclimation responses. However, increased ROS under stress conditions functions as an alarm that triggers defense responses by specific signal transduction pathways (e.g., H_2O_2 as secondary messenger). ROS signaling is linked to abscisic acid (ABA), Ca^{2+} fluxes, and sugar sensing under drought stress. Nevertheless, if drought stress is prolonged to a certain extent, ROS production will overwhelm the scavenging action of the antioxidant system, resulting in extensive cellular damage and death (Carvalho, 2008). Under drought stress, ROS production is enhanced in several ways. Inhibition of CO_2 assimilation, coupled with the changes in photosystems activities and photosynthetic transport capacity under drought stress resulting in accelerated production of ROS via the Mehler reaction in chloroplast (Asada, 1999). Further, CO_2 fixation is limited due to the closure of stomata, which in turn leads to reduced $NADP^+$ regeneration through the Calvin cycle. Due to the lack of an electron acceptor, overreduction of the photosynthetic ETC occurs that leads to a higher leakage of electrons to O_2 by the Mehler reaction. More than 50% leakage of photosynthetic electrons to the Mehler reaction has been reported in drought-stressed wheat plants, as compared with unstressed plants (Biehler and Fock, 1996). Photosynthetic activity is inhibited in plant tissues due to an imbalance between light-capture and its utilization under drought stress (Foyer and Noctor, 2000). Dissipation of excess light energy in the PSII core and antenna leads to generation of ROS, which are potentially dangerous under drought stress conditions (Foyer and Harbinson, 1994). Under drought stress, the photorespiratory pathway is also enhanced, especially when RUBP oxygenation is maximal due to limitations in CO_2 fixation (Noctor et al., 2002). Noctor et al. (2002) have estimated that photorespiration is likely to account for more than 70% of total H_2O_2 production under drought stress conditions.

3.2 Salinity

Salinity stress results in an excessive generation of ROS (Tanou et al., 2009a,b). High salt concentrations lead to overproduction of the ROS—$O_2^{\bullet -}$, $OH^{\bullet}$, H_2O_2, and $^1O_2$ by impairment of the cellular electron transport within different subcellular compartments, such as chloroplasts and mitochondria, as well as from induction of metabolic pathways such as photorespiration. Salt stress can lead to stomatal closure, which reduces CO_2 availability in the leaves, and inhibits carbon fixation, which in turn causes exposure of chloroplasts to excessive excitation energy and reduction of photosynthetic electron transport systems, leading to enhanced generation of ROS and induced oxidative stress. A low chloroplastic CO_2/O_2 ratio also favors photorespiration, leading to increased production of ROS, such as H_2O_2 (Sharma et al., 2012). Elevated CO_2 reduces the oxidative stress caused by salinity, which eventually reduces ROS generation. It also maintains redox homeostasis as a consequence of higher assimilation rates and lower photorespiration (Perez-Lopez et al., 2009). Interestingly, salinity-induced ROS disrupts normal metabolism through lipid peroxidation, denaturing proteins, and nucleic acids in several plant species (Tanou et al., 2009a,b). Earlier, differential genomic and proteomic screenings carried out in *Physcomitrella patens* plants suggested that they responded to salinity stress by upregulating expression of several genes responsible for an antioxidant defense mechanism (Wang et al., 2009a). Therefore, the antioxidative system may play a crucial role in protecting cells from oxidative damage following exposure to salinity stress in plants. There is also a report on differential responses of the antioxidative defense system and the salt tolerance in genotypes of *Oryza sativa* L. (Mishra et al., 2013). Higher status of antioxidants (AsA and GSH) and a coordinated higher activity of the stress enzymes (SOD, CAT, GPX, APX, and GR) have been suggested to be the major factors for salt tolerance in Indica rice seedlings (Mishra et al., 2013).

3.3 Cold/Chilling

Chilling stress is an important abiotic/environmental factor limiting growth and productivity of crop plants. It causes oxidative damage, and changes in signaling molecules (NO, SA, and ABA) and NR activity in maize plants. Chilling leads to the overproduction of

ROS by exacerbating the imbalance between light absorption and light use by inhibiting Calvin-Benson cycle activity (Logan et al., 2006), enhancing photosynthetic electron flux to O_2, and causing overreduction of respiratory ETC (Hu et al., 2008). Nitric oxide (NO) improves the chilling tolerance of maize by regulating the biochemical mechanisms of the chilling response via apoplastic antioxidative enzymes in leaves (Esim and Atici, 2014). Chilling stress also causes significant reductions in *rbc*L and *rbc*S transcripts, RUBISCO content, and initial RUBISCO activity, leading to higher electron flux to O_2 (Zhou et al., 2006). H_2O_2 accumulation in chloroplast was negatively correlated with the initial RUBISCO activity and photosynthetic rate (Zhou et al., 2006). Chilling-induced oxidative stress, evident by increased accumulation of ROS, including H_2O_2 and $O_2^{\bullet -}$, lipid peroxidation, and protein oxidation, is a significant factor in relation to chilling injury in plants (Prasad, 1997). In maize seedlings, protein carbonyl content increases after exposure to chilling temperatures, which indicates oxidative damage (Prasad, 1997). Responses to chilling-induced oxidative stress include alteration in activities of enzymes of antioxidant defense systems. The activities of antioxidative enzymes APX, MDHAR, DHAR, GR, and SOD increase during chilling periods in different plants, for example, maize and strawberry leaves (Fryer et al., 1998). Lipoxygenase activity, as well as lipid peroxidation, was increased in maize leaves during low temperatures, suggesting that lipoxygenase-mediated peroxidation of membrane lipids contributes to the oxidative damage occurring in chill-stressed maize leaves (Fryer et al., 1998). However, if the duration of chilling stress is too long, the defense system may not remove overproduced ROS effectively, which may result in severe damage, or even death (Gill and Tuteja, 2010). Nonenzymatic antioxidants (AsA, GSH, carotenoids, and α-tocopherol) also play an important role in cold response. Under cold stress conditions, low-molecular weight antioxidants, especially that of reduced AsA, have been suggested to be an important component in plant cell defense (Radyuk et al., 2009). Many comparative studies using chilling-tolerant and sensitive genotypes have shown greater antioxidant capacity in chilling-tolerant species compared with sensitive ones (Sharma et al., 2012).

3.4 Metal/Metalloid Toxicity

The increasing levels of both redox-active metals and redox-inactive metals into the environment affects plant growth and metabolism vis-a-vis severe losses in crop yields. One of the consequences of the presence of toxic metals within plant tissues is the formation of ROS, which can be initiated directly or indirectly by the metals and cause oxidative damage to different cell constituents (Sharma et al., 2012). Under metal stress conditions, net photosynthesis decreases due to damage to photosynthetic metabolism, which also includes electron transport (Vinit-Dunand et al.,

2002). For example, copper has been shown to negatively affect components of both the light reactions (e.g., PSII, thylakoid membrane structure, and chlorophyll content). These alterations in photosynthetic metabolism lead to overproduction of ROS such as $OH^{\bullet}$, $O_2^{\bullet -}$, and H_2O_2 (Vinit-Dunand et al., 2002). Redox-active metals, such as iron, copper, and chromium, undergo redox cycling, producing ROS; whereas redox-inactive metals, such as lead, cadmium, mercury, and others, deplete the cell's major antioxidants, particularly thiol-containing antioxidants and enzymes (Sharma et al., 2012). The increased activity of antioxidative enzymes in metal-stressed plants appears to serve as an important component of the antioxidant defense mechanism of plants to combat metal-induced oxidative injury. Several studies have indicated that exposure of plants to high levels of heavy metals induces ROS, either directly or indirectly, by influencing metabolic processes. GSH participates in the control of H_2O_2 levels of plant cells (Foyer and Noctor, 2000). Further, the change in the ratio of its reduced (GSH) to oxidized (GSSG) formed during degradation of H_2O_2 is important in certain redox signaling pathways. It has been suggested that the GSH/GSSG ratio, an indicative of the cellular redox balance, may be responsible in ROS signaling.

4. DUAL BEHAVIOR OF ROS

In plants, ROS are either produced due to electron leakage to molecular oxygen from electron transport activities in the chloroplast, mitochondria, and plasma membrane; or as a byproduct of metabolic pathways in various subcellular compartments (Sharma et al., 2012) (Table 2). Chloroplast is the primary source for ROS generation in plants (Wang and Song, 2008). ROS, especially the hydroxyl radicals ($OH^{\bullet}$) and singlet oxygen ($^1O_2$), are powerful oxidizing species such that they can react with almost any component of the cell, causing severe damage to lipids, proteins, and nucleic acids. This can be defined as "oxidative stress." To prevent oxidative stress, plants harbor a large number of enzymatic and nonenzymatic antioxidants, which scavenge surplus oxidants protecting plant biomolecules from damage (Foyer and Shigeoka, 2011). However, the concept of "oxidative stress" has been reapplied with "oxidative signaling" or "redox signaling" in plant biology, suggesting that ROS production, which was previously considered an extremely harmful process, is now a part of the signaling network that plants use for their development, and in response to environmental challenges. Progression of powerful antioxidant systems has enabled plants to conquer ROS toxicity, and to utilize ROS as signal transducers in a number of signaling pathways involved in growth, development, and plant metabolism, in response to biotic and environmental stimuli and programmed cell death. This represents the "double role of ROS" (Río, 2015).

Table 2 List of different ROS produced in plants along with their site of generation

S. no.	Oxidants (ROS)	Organelle	Producers	Effect	Antioxidant
1.	H_2O_2 (hydrogen peroxide)	Peroxisomes	Flavin-containing enzymes glycolate oxidase and acyl-CoA oxidase, which are part of photorespiratory and fatty acid β-oxidation pathways, respectively.	It oxidizes cysteine (—SH) or methionine residues (—SCH_3), and cause enzyme inactivation by oxidizing their thiol groups, for example Calvin cycle enzymes, Cu/Zn-SOD, and Fe-SOD, oxidizes protein kinases, phosphatases, and transcription factors having thiolate residue. At high concentration cause programmed cell death.	CAT (catalases), peroxidises, flavonoids and SOD (superoxide dismutase).
		Peroxisomes, chloroplast, and mitochondria	Oxidative and electron transport reactions.		
		Cell wall (apoplast)	NADPH oxidase and extracellular heme-containing Class III peroxidises.		
2.	$O_2^{\bullet -}$ (super oxide anion)	Plasma membrane	NADPH oxidase (NOX).	It is the primary ROS formed in cell which initiates a cascade of reaction to generate secondary ROS.	SOD (superoxide dismutase).
		Peroxisomes, chloroplast and mitochondria	Oxidative and electron transport reactions.		
3.	$^1O_2$ (singlet oxygen)	Chloroplasts	Photodynamic reactions in reaction center of photosystem II (PS II). Limited CO_2 availability during environmental stress-drought and salinity.	It directly oxidizes nearly all biomolecules, modifies nucleic acid via selective reaction with deoxyguanosine, cause light-induced loss of PSII activity.	β-Carotene and α-tocopherol.
4.	$^{\bullet}OH$ (hydroxyl radical)	–	Fenton reaction and Haber-Weiss cycle. Heme oxygenases (HO), cytochrome P450, superoxide reductases, bleomycin and some PS II proteins.	Primary cause of oxidative damage. It damages proteins, nucleic acids as well as lipids during oxidative damage and involved in programmed cell death.	Specific $^{\bullet}OH$ scavengers or antioxidants not found. Flavonoids and proline.
5.	NO (nitric oxide)	Peroxisomes	Xanthine oxidoreductase, cytochrome P450, peroxidase, few hemeproteins.	NO in combination with other ROS is involved in various processes such as cell death and disease resistance in plants.	Class 1 nonsymbiotic hemoglobin.

4.1 ROS Communication With Other Signaling Molecules

ROS are important metabolites that take part in metabolism, growth, and morphogenesis of plant cells.

4.1.1 MAP Kinases and ROS

MAP kinases play an essential role in transducing different types of signals that are further involved in regulation of several important biological processes to control and regulate chromosomal and extra-chromosomal gene expression. Many of the basic processes that are involved in cell wall growth and dynamics require a highly oxidizing environment (Jalmi and Sinha, 2015). Different types of stresses are responsible for MAPK cascade activity, and no single stress is enough. For example, in *Arabidopsis*, more than one type of stressor, such as O_3, H_2O_2, Ethylene, ABA, and JA involved in AtMPK6 activation further activates others downstream and plays a role in embryonic development. Under abiotic stress, such as salinity and drought, MAPK expression increases in plants. In a study on *Arabidopsis* it has been shown that the MEKK1 (a MAPKKK) mRNA accumulated as a result of abiotic stresses, including high salinity (Teige et al., 2004). In rice transgenic plants, overproduction of OsMAPK5 increased tolerance, while its suppression led to hypersensitivity to various stresses. Scavenging of ROS and its neutralization is very necessary to protect the cell in order to undergo programmed cell death, which is achieved by antioxidants and scavengers such as catalase, glutathione, and superoxide dismutase. Plants defeat oxidative stress with the production of scavenger enzymes such as catalase and hydrogen peroxide (Sinha et al., 2011). MAPK signaling to ROS is a common response of stresses, and it induces activation and participation in the production of antioxidants and scavengers. The redox status of a cell determines the type of MAP kinases activated and the fate of the cell.

4.1.2 G-Proteins and ROS

G- Proteins are heterodimeric proteins that switch a number of signals inside the cell, and they are made up of three subunits and transmit signals from outside stimuli to the inside of the cell. In general, it binds to GPCR (G-protein coupled receptor) at the cell surface; for example, the ABA receptor is a GPCR that is present at the surface of the guard cell, and is responsible for stomatal movement (i.e., closing and opening of the stomata). These cells are present on the lower surfaces of leaves. In response to salinity stress, in ROS mediated stomatal movement, ABA starts accumulating in the cell, and this accumulation directs the alteration at the expression level, and stomatal closure limits water loss through transpiration (Shinozaki and Shinozaki, 2007). So under abiotic stress signals, transduction pathways become operative because of membrane receptors. Salinity stress is one of the major pressures that a plant faces, resulting in osmotic stress in the cells. Ionic stress triggered by an intracellular increase in concentration of Na^+ and Cl^- ions due to uptake of their salt by cell. It disturbs the sodium and potassium ratio inside the cell, hence causing ionic imbalance (Blumwald et al., 2000). Nuclear or chromosomal gene encoding proteins are used as the machinery for expression of the plastid gene and this shifting is associated with redox potential. Plant signaling networks have a high ability to balance the effects of disturbance in neighboring nodes and a number of other signaling pathways during stress.

4.1.3 NADPH Oxidases

In plants, NADPH oxidases, an enzyme complex which generates $O_2^{\bullet-}$ free radical that further converted into H_2O_2, have two EF hand-like domains, and their activity depends on the presence of calcium ion, thus acting as a calcium sensor. Here, in the presence of calcium ions, production of ROS takes place, likewise in the presence of the ROS, and the intracellular calcium ion concentration increases, which activates other downstream processes. A plant cell plasma membrane has channels, a number of ionophores, transmembrane proteins, and ionic sensors that allow the signal either in the form of an ion that allows it to enter from outside, or sense and perceive a signal from biotic and abiotic stimuli. In *Arabidopsis thaliana*, NADPH oxidases are present in plasma membranes and require phosphorylation of its serine residues present at the N terminal. CDPK phosphorylates its N-terminal in the presence of calcium ions. For activation, oxidases require calcium for binding with EF hand domains, and also for phosphorylation (Ogasawara et al., 2008).

The production of NO (or "nitric oxide," sometimes considered a reactive oxygen species) also depends on calcium ions. ROS-induced fluctuations in Ca^{+2} concentrations are essential signals for many biological responses, including stomatal movement, adaptation in stressful conditions, and numerous types of programmed cell death. To overcome the different type of stresses, plants maintain the balance at different levels that can be molecular, cellular, tissue, organ, system, anatomical, or morphological (Atkinson and Urwin, 2012). There is a tight link between cellular metabolism and ROS signaling. The ROS level is dynamic in terms of production and scavenging, and disturbance in this equilibrium accompanies several different signaling events (Nishimura and Dangl, 2010). When a reporter gene is subjected to the control of a rapid ROS-response promoter in plants, it demonstrates a rapid raise in ROS levels due to this one cell communicating to another. ROS signals propagate like a wave throughout the tissue and organs, and lead to the production of a cascade of signal transduction (Mittler et al., 2011). In chloroplast and mitochondria, phosphatases activity requires oxidation of the cysteine residues that is initiated by ROS. The H_2O_2 signal and its response are wide and nonspecific, and bear the hallmark for signaling activation and transduction. Mitochondrial electron transport chains in animals are upstream from the tyrosine-protein kinase *lyn*, which is further upstream from another tyrosine protein kinase, *syk*, which controls other downstream processes by signaling and regulating transcription, translation, and cell division (Patterson et al., 2015). ROS controls the hormone signaling pathway, leading to an increase in the level of SAR, SA, JA, and ethylene. The cytosolic concentration of calcium increases in response to ROS $O_2^{\bullet-}$, which triggers the influx of calcium actions of SA, leading to rapid production of ROS. An increase in $[Ca^{2+}]$cyt was checked by experiments using chemiluminescent probes specific for $O_2^{\bullet-}$, in tobacco (Kawano et al., 1998).

4.2 ROS Role in Stress Signaling

In plants, ROS are produced by direct excitation energy transferred from chlorophyll forming $^1O_2$, or by reduction of univalent oxygen at PS I in the Mehler reaction, leading to oxidative injuries, which compels the plants to initiate redox homeostatic mechanisms to manage accumulating ROS. ROS play a dual role based on their accumulation levels within the cell. At high intracellular concentrations, they cause extensive cell injury and cell death; whereas, at low or moderate concentrations, they act as signaling molecules or inducers under various environmental stresses in the plants for programmed secondary metabolisms, cell wall differentiation, and activation of MAPKs (Mitogen-activated protein kinases) arching toward tolerance for stress (Baxter et al., 2014). In plants, the ROS signal is sensed, transduced, and translated into an appropriate cellular response by redox-sensitive proteins, protein phosphorylation, calcium mobilization, and gene expression (Sharma et al., 2012). A balanced intracellular redox is important for an organism, be it plants or animals. A slight shift in this redox homeostasis induces the expression of specific gene-encoding antioxidant proteins, oxidative scavengers, protein phosporylation cascades, and compatible molecules. Excessive ROS cease mitochondrial electron transport, generating ATP scarcity (Kotchoni and Gachomo, 2006). ROS act as secondary messengers in ABA (abscisic acid) pathways in guard cells (Wang and Song, 2008). ABA is known to induce many stress-responsive genes (Ahmad et al., 2010). ABA-inferred ROS mediates stomatal closure to stop water loss by opening calcium channels in the plasma membrane (Song et al., 2014). In addition to ROS's role in stomatal closure, it has a role in root gravitropism, dormancy alleviation, defense gene induction in the tomato plants due to wounding, PCD (programmed cell death), osmotic stress, and heavy metal and low temperature signal transduction pathways (Sharma et al., 2012). NO, along with ROS, plays a role in defensive mechanisms against pathogens in plants (Río, 2015). Whenever a pathogen penetrates a plant cell wall, a number of events follows: cross-linking of cell wall proteins and phenolic compounds, calcium-pectate gel formation, glycoproteins aggregation, accumulation of callose-containing papilles, and ROS production arresting infection in the affected area. Cation channels such as Ca^{2+} and K^+ sense ROS signals and convert them into physiological responses such as Ca^{+2} influx, cytosolic Ca^{2+} elevation, and K^+ efflux. Elevated Ca^{2+} activates many two- and multicomponent signaling systems, causing genetic and metabolic adjustments.

K^+ efflux via ROS-activated K^+ channels activates cytosolic cell death proteases and endonucleases, resulting in programmed cell death (PCD). PCD regulates various aspects of growth and development, such as removing infected, damaged, or dead cells during environmental stresses or pathogenic attacks (Río, 2015). Other ROS sensors include histidine kinases, ROS-sensitive phosphitases, and redox-sensitive transcription factors. Reversible oxidation of S-containing amino acids behave as regulatory redox switches in plant cells (Kotchoni and Gachomo, 2006; Demidchik, 2014). The presence of NO and GSNO (S-nitrosoglutathione) in plant tissues and the formation of $ONOO^-$ results in covalent posttranslational modifications (PTMs) such as S-nitrosylation and the nitration of proteins in plants under natural, as well as stress conditions (Corpas and Barroso, 2013). S-nitrosylation controls key signaling molecules such as H_2O_2. Various gene families implicated in providing protection against pathogen attacks were found to be upregulated in response to exogenous application of ROS (H_2O_2), including at least 152 genes in *A. thaliana* (Kotchoni and Gachomo, 2006).

The signaling importance of ROS is that even a single fluctuation in this leads to change of the cellular homeostasis and metabolism of a particular cell. ROS are formed instantly after the onset of the stress and are very reactive; they can react with membrane lipids, carbohydrates, proteins, and DNA. ROS such as H_2O_2 can pass through the biological membranes, channel like aquaporins, and develop responses. Ion channels such as aquaporin are membrane proteins that allow hydrophilic molecules to pass through. In *Arabidopsis*, it was found that when plant cells are exposed to high concentrations of H_2O_2, the plasma membrane intrinsic protein PIP2.1 was found to enhance the toxicity of H_2O_2. Under oxidative stress, polyunsaturated fatty acids (PUFAs) are attacked by different ROSs, especially $^1O_2$ and $OH^•$, which causes production of lipid hydroperoxides, leading to a decrease of membrane fluidity (Gill and Tuteja, 2010). That leads to another response by changing the membrane's physical state, which could activate downstream signaling intermediates. Evidence implies that ROSs, especially H_2O_2, are active signaling molecules, and their regular addition into the cell through redox sensing leads to a number of different kinds of cellular responses (Kovalchuk and Dutta Gupta, 2010). ROS, in high concentration, results in cellular damage or even cell death of highly sensitive cells (Gill and Tuteja, 2010); whereas lower concentrations of ROS develop signals, controlling various functions in the plant. ROS such as H_2O_2, O_2 are the major members of many signaling pathways the activate the gene so that acclimation, stress tolerance, and other defense responses can develop in a plant cell. For example, ethylene, ABA, and salicylic acid are produced under stress, and ROS and the antioxidant molecule glutathione are important contributors to maintaining the redox state of the plant cell. In a plant cell, signals develops in a single compartment and spread through its network in the whole plant; for example, under light intensity, or heat or cold stresses, there is an increase in ROS levels in the chloroplast or peroxisome, which is then transferred to the entire plant. ROS signaling is primarily used as a general signal to activate the cellular signaling network and specificity of the signaling conveyed by other signals that function together. Other signals could be biomolecules such as small peptides, lipids, cell wall fragments, hormones, and others. The phytohormone abscisic acid (ABA) responds to drought stress by stomatal closure, leading to the reduction of transpirational water loss. In guard cells of stomata, ABA signaling involves redox regulation and reactive oxygen species (ROS) involving hydrogen peroxide (H_2O_2) that serve as a key mediator of ABA activation of Ca^{2+} channels (Murata et al., 2001; Kwak et al., 2003). Increasing H_2O_2 concentration activates Ca^{2+} channels and evokes a guard cell [Ca^{2+}]cyt increase. Here, NAD(P)H oxidases in plasma membranes are responsible for ABA-induced ROS production in guard cells. ROS production by apoplastic enzymes are also involved in induction of stomatal closure (Hossain et al., 2013). At the molecular level, the changes in the expression of stress-related genes occur to overcome the problems faced by the cell (Grativol et al., 2012; Shinozaki and Shinozaki, 2007). Important mechanisms that initiate responses to abiotic stress include the accumulation of osmolyte and biochemical solutes. These include chemically diverse metabolites, such as proline, polyamines, glycine betaine, γ-amino-*N*-butyric acid (GABA), raffinose, trehalose, sucrose, and polyols (sorbitol, myo-inositol).

Plant resistance to pathogens consists of systems for signal perception and transduction. The insight of pathogen attacks initiates production of various molecules such as ROS, protein kinases, and transcription factors, which activates downstream defense responses. In the plants, signaling is regulated by protein phosphorylation and protein dephosphorylation involving ROS (H_2O_2) induced mitogen-activated protein kinase (MAPK) cascades (Hancock et al., 2000). The MAPK-signaling pathways have three components from the protein kinase family: (1) MAPK, (2) MAPK kinase (MAPKK), (3) MAPKK kinase (MAPKKK). MAPKKK because phosphorylated in response to accumulating ROS, causing the activation of MAPKK. Upon activation, the MAPKK becomes phosphorylated and activates MAPK. The activated MAPK enters the nucleus, where it becomes phosphorylated and binds to the transcription factors to cause subsequent expression of the targeted genes enhancing plant tolerance to numerous stresses. In *A. thaliana*, H_2O_2 induces ANP (nicotinamaprotein kinase), a class of MAPK kinase-kinases (MAPKKKs) (Kotchoni and Gachomo, 2006) (Fig. 2).

4.3 Elevated ROS and Oxidative Damage to Biomolecules

As detailed herein, at low concentrations, ROS as signaling molecules activate hormones as abscisic acid and play a role in stomatal movement (Murata et al., 2001). Auxin plays a role in root gravitropism; jasmonic acid in lignin biosynthesis; and GA in seed germination, programmed cell death, salicylic acid hypersensitivity, and osmotic response. This process takes place in an inducible manner; for example, a constitutive

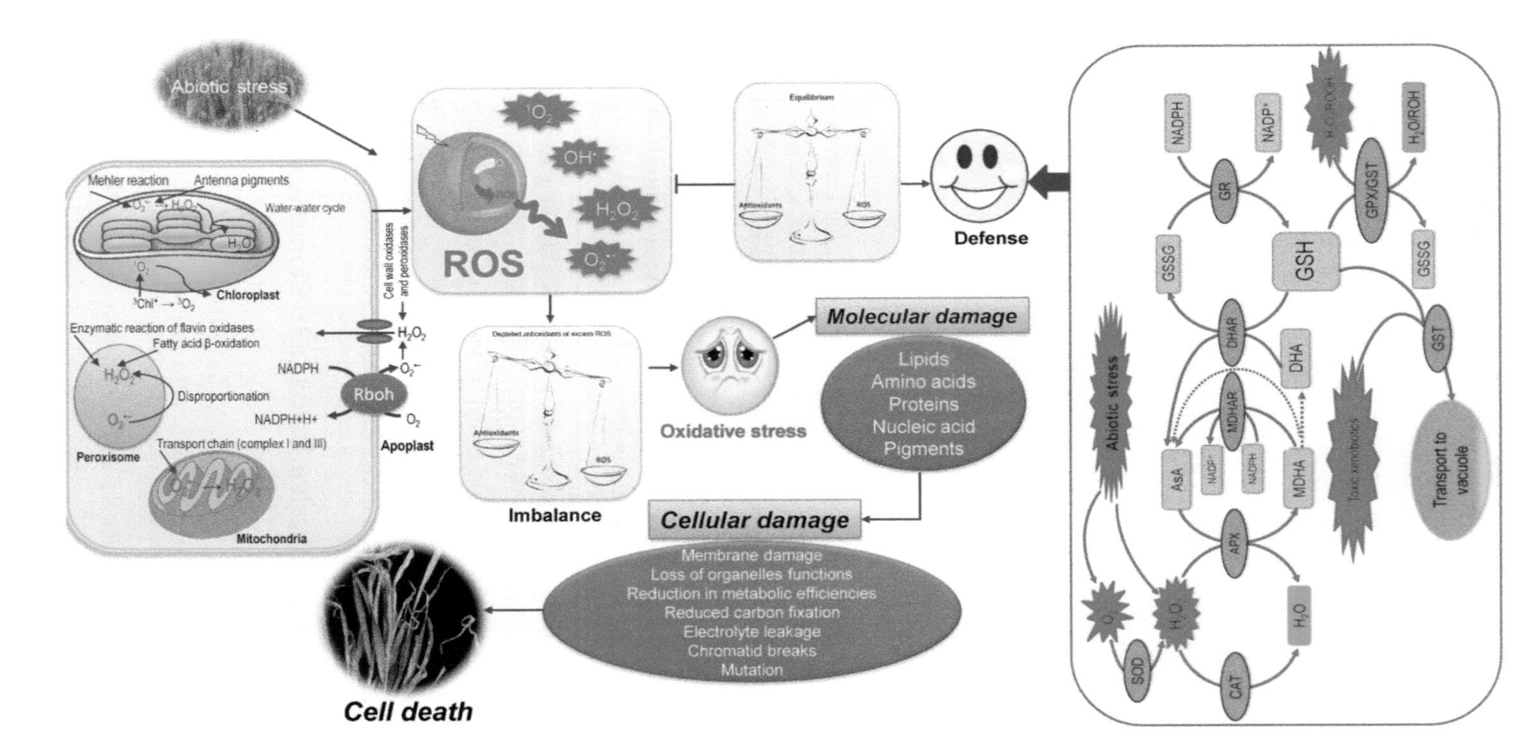

Fig. 2 Reactive oxygen generation, oxidative stress and antioxidant defense in plants under abiotic stress conditions.

rise in concentration of H_2O_2 does not happen because of the ABA-induced stomatal closure. In the dormant seeds of barley, GA and ROS are low, but ABA is high, causing dormancy. Whereas ROS levels and GA result in seed germination. ROS production and its scavenging must be rigorously controlled to keep away cells from oxidative stress, when its production level exceeds from scavenging the cell said to be under stress. It happens when plants face certain abiotic stress conditions such as drought, salinity, heat, heavy metal, and biotic conditions such as pathogen attacks; but if elevated levels of ROS cross a defined threshold level, causing oxidative damage of biomolecules such as proteins, lipids, and DNA, the resulting intrinsic property of the cell influences membrane fluidity change, ion transport gets affected, inhibition of protein synthesis happens, and the cell is ultimately committed to programmed cell death.

4.3.1 Lipid Peroxidation and Protein Oxidation

Lipids and proteins are considered as major targets of oxidative modifications in abiotic stressed plants (Anjum et al., 2016b). Cell membranes are made up of phospholipids that contain long unsaturated fatty acid chains. These polyunsaturated fatty acid chains are very sensitive to ROS. Lipid peroxidation begins with the interaction of $^{\bullet}OH$ or $^1O_2$ with double bond of PUFA (polyunsaturated fatty acids), resulting in a chain reaction such that functional lipids are converted into toxic aldehydes and ketones (Ahmad et al., 2010). Lipid peroxidation involves three stages: initiation, propagation, and termination. Initiation involves hydrogen atom abstraction out of lipid molecules by hydroxyl, alkoxyl, peroxyl radicals, and peroxynitrite forming—$^{\bullet}CH$, or lipid radical ($L^{\bullet}$). $L^{\bullet}$ activates O_2 forming lipidperoxyl radicals ($LOO^{\bullet}$), which abstracts H^+ from neighboring fatty acids generating lipid hydroperoxide (LOOH) and another lipid radical ($L^{\bullet}$). LOOH begins the propagation phase by undergoing "reductive cleavage" by the involvement of transition metals forming lipid alkoxyl radicals ($LO^{\bullet}$), which further abstracts H^+ from neighboring fatty acids. Another method of lipid peroxidation is by direct reaction of double bonds with singlet oxygen from the PSII reaction center, forming LOOH (Das and Roychoudhury, 2014; Demidchik, 2014; Poirier et al., 2001; Svingen et al., 1979).

$$OH^{\bullet} + RH \rightarrow H_2O + R^{\bullet}$$

$$RO^{\bullet} + RH \rightarrow ROH + R^{\bullet}$$

$$R^{\bullet} + O_2 \rightarrow RO_2^{\bullet}$$

$$RO_2^{\bullet} + RH \rightarrow ROH + R^{\bullet}$$

Lipid peroxidation products include aldehydic secondary products such as HNE (4-hydroxy-2-nonenal), MDA (malondialdehyde), acrolein and 4-hydroxy-2-hexenal, which act as biomarkers of oxidative stress (Demidchik, 2014). In an oxidative environment, intracellular protein and the proteins present in plasma membranes show a number of modifications, such as the formation of the disulfide bond, S-nitrosylation, glutathionylation, and carbonylation that lead to site-specific amino acid modification, and breakdown of peptide chains that force a protein to go under the process of proteolysis. Carbonylated protein concentration was found to increase when a tissue was injured in response to oxidative stress (Moller and Kristensen, 2004). Proteinogenic amino acid-oxidation by ROS results in the loss of protein-mediated functions and toxic protein aggregation. Most of the protein modifications are irreversible, except sulfur-containing amino acids such Cys and Met. Different results can be expected from oxidation of different amino acids in a protein (Demidchik, 2014). Most common protein oxidation is carbonylation, that is, the insertion of a carbonyl group, after Cys and Met modifications. Carbonylation is defined as the formation of ketones and aldehydes due to oxidation of side chains of lysine, threonine, proline, and arginine and their detection using 2,4-dinitrophenylhydrazine. It involves more input of energy, more amino acids, and induces significantly huge changes in protein structure and function. Another phenomenon known as "secondary protein carbonylation" is defined as a reaction due to the addition of aldehydes products as a result of lipid peroxidation, which modifies a protein's enzymatic and physiological activities. Plant stresses such as drought, salinity, and cadmium toxicity induce protein carbonylation, which is considered to be irreversible; however, recent evidence suggests that it can be reversed for some TFs (transcription factors) suggesting its role as a novel ROS signaling mechanism (Demidchik, 2014). In *Arabidopsis thaliana*, proteomic studies suggested that the nitrosative stress causes S-nitrosylation by peroxynitrite resulting in severe loss in protein function (Demidchik, 2014).

4.3.2 Nucleic Acid

Being genetic material, any damage in nucleic acid can cause malfunctioning all over the cell because it will code defective protein or complete inactivation of the protein. ROS cause oxidative damage of all types of DNA, including nuclear, chloroplastic, and mitochondrial DNA. Plants exposed to various environmental stresses have shown that sugar and base moieties of DNA are susceptible to oxidation by ROS (Gill and Tuteja, 2010). Notably, oxidation of DNA is extensively studied in animals due to its role in carcinogenesis, but due to this missing role, it was not studied extensively in plants. Despite this, it may be a reason for aging of seed stocks, and at times, crop plant death. It includes three types of lesions: chemically modified bases, mismatched bases, and double-strand breaks. Hydroxyl radicals are the major damaging factor for nucleic acids, they reacts with nucleic acid both by adding H^+ to the double bond of bases and removing H^+ from C—H bonds of 2′-deoxyribose and $—CH_3$ of thymine. End products of $^{\bullet}OH$ induced DNA/RNA oxidation include 8-oxo-7,8-dihydroguanine (8-oxoG) and 2,6-diamino-4-hydroxy-5-formamidopyrimidine (FapyG). Plants have repair systems to prevent DNA oxidation, which consists of base and nucleotide replacement and direct repair of damage. Antioxidant defense is important for protection against ROS-induced DNA damage. Protection is also provided by (1) substances that decrease transition metal catalytic activity, including phytochelatines, metollothioneines, pectins, and other structural proteins; (2) forming protein isoforms and lipids that are less sensitive to oxidation; (3) living tissues using dead cell layers, which can rapidly die in response to programmed ROS-induced mechanisms providing shield against infection or stressors; (4) enhanced mycorrhization providing protection against heavy metals; (5) activating biosynthesis systems for reparation of damaged components (Gill and Tuteja, 2010).

4.3.3 Carbohydrates

Carbohydrates are the most abundant biomolecules in the plants; however, they are the least studied biomolecule in terms of oxidative stress. They play various roles in plants, such as shaping and supporting plant cells, regulation of osmotic pressure, storage of reduced forms of carbon, acting as nonenzymatic antioxidants, and many others. ROS induced damage to carbohydrates is probably very harmful for plants (Gill and Tuteja, 2010).

5. DEFENCE SYSTEM AGAINST ROS: THE ROLE OF ANTIOXIDANTS

Elevation in the generation of ROS is one of the common consequences in crops/plants exposed to varied abiotic stress factors. Elevated and/or nonmetabolized ROS are bound to bring impairments in cellular organelles and redox homeostasis (Gill and Tuteja, 2010; Anjum et al., 2012, 2016a,b). Crops/plants possess antioxidant defense systems to counteract ROS-accrued anomalies and maintain cellular redox homeostasis. Components of crop/plant antioxidant defense systems can be grouped into enzymatic and nonenzymatic antioxidants. This section aims to briefly discuss major enzymatic (APX, CAT, GR, GST, GPX, SOD) and nonenzymatic (AsA and GSH) antioxidants and highlight their modulation and role in crops/plants under abiotic stresses such as salinity, drought, metals/metalloids, and temperature extremes (Fig. 1).

5.1 Ascorbate Peroxidases

Ascorbate peroxidases (APX; EC 1.11.1.11) exhibit a higher affinity for peroxides (mainly H_2O_2) and utilize ascorbate (AsA) as a specific electron donor in order to reduce H_2O_2 into H_2O in organelles, including chloroplasts, cytosol, mitochondria, and peroxisomes. In fact, the scavenging of H_2O_2 by APX is the first step of the AsA-GSH cycle, and plays an essential role in scavenging ROS and protecting cells in higher plants (reviewed by Anjum et al., 2016b).

$$H_2O_2 + 2AsA \rightarrow 2H_2O + 2MDHA\ (\rightarrow DHA)$$

Information related to enzymatic properties, distribution, location, function mechanisms, and molecular characteristics is widely available (Zhang et al., 2013). APX is multigenic in nature, and at least five different isoforms, including mitochondrial (mAPX), thylakoid (tAPX) and glyoxisome membrane forms (gmAPX), as well as chloroplast stromal soluble forms (sAPX) and cytosolic forms (cAPX) belonging to the APX family (Noctor and Foyer, 1998; Caverzan et al., 2012). A great deal of literature is available on the role of APX in plants exposed to different abiotic stress factors (reviewed by Anjum et al., 2016b).

5.1.1 Drought and Salinity

Increased APX activity and the amount detected by western blotting, enzyme activity assays, and biophoton emission techniques, protected *Glycine max* against drought stress impacts (Kausar et al., 2012). In transgenic tobacco (*Nicotiana tabacum*, cv. Wisconsin), Zarei et al. (2012) reported enhanced APX activity, where the highest activity was observed in 10% and 20% of the PEG treatment. There existed a relationship between drought and APX in tree species such as almond (*Prunus* spp.) under drought stress (Sofo et al., 2015). Overexpression of cytosolic APX (cyt*apx*) in transgenic tobacco (*Nicotiana tabacum* cv. *Xanthi*) protected this plant against drought-caused damages. In drought-exposed rice seedlings, the loss of function in *OsAPX2* was reported to affect the growth and development of rice seedlings, and resulted in semidwarf seedlings, yellow-green leaves, leaf lesion mimic, and seed sterility (Zhang et al., 2013). Notably, the authors observed a lower APX activity and a drought-sensitivity in *Osapx2* mutants; whereas, overexpression of *OsAPX2* was found to increase APX activity and enhance drought tolerance. An elevated drought (and also salinity)-tolerance was reported in transgenic tobacco overexpressing the peroxisomal ascorbate peroxidase (SbpAPX) gene cloned from halophyte *Salicornia brachiate* (Singh et al., 2014).

Mutants deficient in cytosolic APXs were reported to exhibit their susceptibility to the salinity-caused oxidative damage (Bonifacio et al., 2011; Mittova et al., 2015). Transgenic plum plants expressing the cyt*apx* (cytosolic APX) genes exhibited their enhanced tolerance to salt stress (100 mm NaCl) (Diaz-Vivancos et al., 2013). In a similar study, transgenic *Arabidopsis thaliana*, the APX (*PutAPX*) gene of *Puccinellia tenuiflora* (a perennial wild grass able to grow in extreme saline-alkali soil environments) was reported to tolerate 150 and 175 mM NaCl (Guan et al., 2015). Rice APx1/2s mutants (double mutants for cytosolic APXs) exhibited an altered redox homeostasis (Bonifacio et al., 2011).

5.1.2 Metals/Metalloids

Plant species types, and doses and types of metals/metalloids (hereafter called "metals"), and also an exposure period significantly modulated the activity of APX in plants. Extensive literature is available on the Cd toxicity in plants, and the modulation and role therein of various components of the antioxidant defense system. Cd exposure caused an increase in APX activity in a number of crop/plant species, including *Brassica juncea*

(Mobin and Khan, 2007), *Zea mays* (Krantev et al., 2008), *Triticum aestivum* (Khan et al., 2007), *Vigna mungo* (Singh et al., 2008), and *B. campestris* (Anjum et al., 2008), *V. radiata* (Anjum et al., 2011), and *B. napus* (Hasanuzzaman et al., 2012b). However, APX activity was also reported to decrease in plants such as *Hordeum vulgare* (Hegedüs et al., 2001); *B. juncea*, *B. napus*, and *B. campestris* (Nouairi et al., 2009); and *Cucumis sativus* (Zhang et al., 2003). As mentioned herein, plant organs and the exposure period can differentially modulate APX activity in metal/Cd-exposed plants. To this end, low Cd levels were reported to increase APX activity in *Glycine max* roots and nodules; however, APX activity decreased with a high Cd concentration (Balestrasse et al., 2001). In another report, a short-term (10h) Cd stress significantly increased APX activity; whereas, a decrease in APX activity increased in the Cd-exposure period (Schutzendubel et al., 2002). *T. aestivum* root tissues exhibited decreased APX activity with the increasing level of Ni concentration, while activity increased significantly in shoot tissues (Gajewska and Sklodowska, 2008). Corroborated with the changes at transcript levels of cAPX mRNA, an exposure time- and genotype-dependent response pattern for APX was reported in Cd stress (5 μM Cd)-exposed *H. vulgare* genotypes Weisuobuzhi and Dong 17 (Chen et al., 2010). Cd-tolerant cultivar Pusa 9531 of *Vigna radiata* showed a 324.8% increase in APX activity, while PS 16 showed a 186.5% increase in APX activity at 100 $mg\,Cd\,kg^{-1}$ (Anjum et al., 2011). The authors also observed a lower Cd-induced oxidative stress in terms of lesser increments in H_2O_2 in the tolerant cv. Pusa 9531 (vs. PS 16). All three cultivars of *B. napus* exhibited a profound increase in APX activity with 0.75 mM $CdCl_2$, while the activity decreased at a higher dose of Cd (1.5 and 2.25 mM $CdCl_2$) (Touiserkani and Haddad, 2012). In Cd-exposed *B. napus* seedlings, APX activity increased by 39% and 43%, at 0.5 and 1.0 mM $CdCl_2$, respectively (vs. control (Hasanuzzaman et al., 2012b). Another naturally occurring element, arsenic (As), is not essential for plant growth and can be accumulated in plants up to toxic levels. About 1.5–3 $mg\,kg^{-1}$ As can be found in the Earth's crust (Drewniak and Sklodowska, 2013). A linear increase in APX activity was noted in the leaves of *B. juncea* grown under As exposure (5 and 25 μM As) (Khan et al., 2009). In another report, activities of total APX as well as chl-APX were found higher in As^{3+}-treated seedlings of As sensitive rice genotypes Malviya-36 and Pant-12 grown under 25 and 50 μM As_2O_3 (vs. controls) (Mishra et al., 2011). Hasanuzzaman and Fujita (2013) reported 24% and 34% increase in APX activity (vs. control) in *T. aestivum* seedlings exposed to 0.25 and 0.5 mM As, respectively.

Reports are also available on the Cu-mediated modulation in APX activity in plants (Khatun et al., 2008; Posmyk et al., 2009). In *B. oleracea* seedlings, Cu (0.5 mM) exposure caused little or no change in APX activity (Posmyk et al., 2009). In contrast, the authors observed a linear increase in the APX activity with 2.5 mM of Cu stress. Greater APX activity in leaves and AsA regeneration were advocated to protect the Indian ginseng (*Withania somnifera*) plant against $CuSO_4$ (0–200 μM) stress (Khatun et al., 2008). The authors also corroborated these results with a higher induction in the six APX isoforms under Cu exposure (Khatun et al., 2008). In another report, lower Cu levels increased APX activity in *M. sativa*. Fe was reported to trigger a rapid induction of APX gene expression in *B. napus* (Vansuyt et al., 1997). A significant elevation in leaf APX activity was observed in *T. aestivum* seedlings treated with 100 and 300 μM Fe; whereas 500 μM Fe treatment resulted in about a 28% decrease in APX activity (Li et al., 2012). In *Myracrodruon urundeuva* leaves, 50 and 80 mg Zn kg^{-1} caused induction in APX activity; whereas Zn doses greater than 80 mg kg^{-1} decreased its activity (Gomes et al., 2013). Wang et al. (2009a,b) observed a decrease in APX activity with Ni exposure (100 mg Ni kg^{-1} soil) in *Zea mays*. However, a marked increase in APX activity was noted in *B. juncea* with increasing levels of Ni (0.0, 0.2, 0.4, and 0.6 mM, 60 d) (Kanwar et al., 2012). In another study, Li et al. (2012) found a close association of APX activity with H_2O_2 production, where an enhanced APX activity somehow diminished the oxidative stress.

5.2 Monodehydroascorbate Reductase and Dehydroascorbate Reductase

The AsA-GSH cycle has another important enzyme monodehydroascorbate reductase that catalyzes the conversion from the oxidized state of monodehydroascorbate (MDHA) to the reduced state of AsA using NADH or NADPH as an electron donor (Zhang et al., 2013). Another related enzyme is dehydroascorbate reductase that is a key component of the AsA recycling system, and regenerates AsA from the oxidized state of DHA to the reduced state of ascorbate (AsA) with GSH as the electron donor (Martínez and Araya, 2010). In fact, MDHA is rapidly reduced by MDHAR; whereas the former is then reduced to AsA by the action of DHAR (Chen et al., 2003). Because it is involved in AsA regeneration, MDHAR plays an important role in maintaining the antioxidant properties of AsA. DHAR regulates the cellular AsA redox state and allows the plant to recycle DHA, thereby recapturing AsA before it is lost, and thus rapidly regenerating (Martínez and Araya, 2010).

Following are the catalytic reactions mediated by MDHAR and DHAR:

$$2\text{MDHA} + \text{NADPH} \xrightarrow{\text{MDHAR}} 2\ \text{Ascorbate} + \text{NADP}^+$$

$$\text{DHA} + \text{GSH} \xrightarrow{\text{DHAR}} \text{Ascorbate} + \text{GSSG}$$

Extensive reports are available on the MDAHR and DHAR-mediated control of plant oxidative stress tolerance and acclimation (Mittova et al., 2003; Hasanuzzaman et al., 2010, 2011a,b; Hasanuzzaman and Fujita, 2011, 2013). Modulation and the role of MDHAR and DHAR in metal/metalloid-stressed plants is briefly discussed as follows.

5.2.1 MDHAR

The response of MDAHR was reported to depend on the genotypes of plants as well as the dose and duration of metal exposure. Cd-exposed *Pinus sylvestris* and poplar hybrids (*Populus × canescens*) exhibited, respectively, increased and decreased MDHAR activity (Schützendübel et al., 2001). In *T. aestivum* roots, Paradiso et al. (2008) reported unchanged MDHAR activity upon 3 d of Cd exposure; whereas 7 d of Cd-exposure increased its activity with a concomitant increase in AsA levels. MDHAR activity considerably decreased in *Ceratophyllum demersum* exposed to Cd (10 μM $CdCl_2$, 7 d) (Aravind and Prasad, 2005). A short-term exposure to Cd did not change MDHAR activity in *T. durum* roots; whereas prolonged Cd treatment (7 d) significantly increased MDAHR activity (Paradiso et al., 2008). Tobacco bright yellow-2 cells exposed to Cd (100 μM) exhibited a marked decrease in MDHAR activity (Islam et al., 2009). *Brassica juncea* plants exposed to 10, 30, 50, and 100 μM of Cd for 5 d in a hydroponic culture were analyzed by Markovska et al. (2009). The activity of MDHAR increases upon exposure to any level of Cd in dose-dependent manners. The activity was highest at 30 and 50 μM Cd. However, this upregulation could not increase the level of AsA, and thus the seedlings were affected by Cd-induced oxidative stress. In a recent study, Hasanuzzaman et al. (2012b) observed that Cd stress significantly reduced the activity of MDHAR in 12 d old *B. napus* seedlings. Treatment with 0.5 and 1.0 mM $CdCl_2$ for 48 h, reduced the activity of MDHAR by 16% and 32%, respectively, over control. However, when the seedlings were supplemented with Se (50 and 100 μM Se Na_2SeO_4), the activities markedly increased, which caused enhanced regeneration of AsA and provided better protection to Cd-induced oxidative stress (Hasanuzzaman et al., 2012b). Short-term treatment of *H. vulgare* roots with Cd (30 or 60 μM $CdCl_2$, 1–6 h) caused differential activities of MDHAR in a dose-dependent manner (Zelinova et al., 2013). A transient decrease of MDHAR activity was observed at the zone immediately behind the root apex containing a meristematic and elongation zone. However, in the second root segment containing the beginning of a differentiation zone, the activity of MDHAR increased after Cd treatment (Zelinova et al., 2013). In *Vigna radiata*, Cd exposure (25–100 mg Cd kg^{-1} soil) caused a significant increase in MDHAR activity in a dose-dependent manner (Anjum et al., 2011). However, the responses were different in two genotypes varying in tolerance to Cd. Upon exposure to a higher level of Cd (100 mg Cd kg^{-1} soil), Cd-tolerant cv. Pusa 9531 showed relatively higher activity (24.2% increase over control) than Cd-susceptible cv. PS 16 (18.5% increase over control). This helped in effective regeneration of AsA, and partially alleviates the oxidative stress (Anjum et al., 2011). MDHAR activity depended on the period of exposure to As^{3+} in *Oryza sativa* (Mishra et al., 2011). Five days of As^{3+} exposure declined the activity of MDHAR, whereas a concomitant increase was observed in MDHAR activity during 10–20 d of As^{3+} treatment (25 and 50 μM) (Mishra et al., 2011). No change in MDHAR activity was reported in *T. aestivum* seedlings exposed to As (0.25 and 0.5 mM $Na_2HAsO_4{\cdot}7H_2O$) (Hasanuzzaman and Fujita,

2013). In *Typha latifolia*, MDHAR activity responded differently to Pb stress, where 100 μM Pb increased MDHAR activity by 250% and a higher dose of Pb (250 μM) dropped the activity of MDHAR (vs. control) (Lyubenova and Schröder, 2011). Additionally, MDHAR activity was found to increase in *O. sativa* seedlings subjected to Ni stress, suggesting that Ni activates the AsA regenerating system in order to maintain the elevated level of AsA (Maheshwari and Dubey, 2009). In *B. juncea*, increasing levels of Ni (0.0, 0.2, 0.4, and 0.6 mM, 60 d) caused a marked increase in MDHAR activity (Kanwar et al., 2012).

5.2.2 DHAR

The activity of DHAR in metals/metalloids-exposed plants also depended on metal/metalloid types, concentration, plant type/organs, and exposure period. In poplar (*Populus* × *canescens*) roots, Cd (50 μM) caused significant decreases in DHAR activity after 15 h of treatment; whereas increasing the duration of Cd exposure caused an increase in DHAR activity (Schutzendubel et al., 2002). A reduction in DHAR activity was noted in *Ceratophyllum demersum* exposed to Cd (10 μM $CdCl_2$) (Aravind and Prasad, 2005). The authors revealed a close association of these decreases with increasing oxidative stress. In the roots of Cd-exposed *T. durum*, Balestrasse et al. (2001) observed a higher increase in DHAR activity at an earlier stage (3 d) versus a later stage (7 d). Among 4 doses of Cd (10, 30, 50, and 100 μM), only 10 μM Cd caused increases in DHAR activity (41.6% over control); whereas the other Cd levels (30, 50, and 100 μM Cd) resulted in significant decreases in the activity of DHAR in *B. juncea* (Markovska et al., 2009). The authors noticed a concomitant decrease in AsA content in Cd-stressed *B. juncea* seedlings. A marked decrease in DHAR activity was also observed by Islam et al. (2009) in tobacco bright yellow-2 cells exposed to Cd (100 μM). Plant genotypes exhibiting their differential sensitivity to metal/metalloid toxicity also showed a differential DHAR activity. The Cd-tolerant *B. juncea* cultivar Pusa Jai Kisan exhibited increased activity of DHAR at both 25 and 50 μM Cd (Iqbal et al., 2010). The increases in DHAR activity were corroborated with a higher AsA content in *B. juncea* cv. Pusa Jai Kisan. However, no change in DHAR activity was noted in the Cd-sensitive SS2 *B. juncea* cultivar. In another report, a differential activity of DHAR in response to Cd stress (25–100 mg Cd kg^{-1} soil) was observed in *Vigna radiata* genotypes with varying levels of tolerance to Cd stress (Anjum et al., 2011). In response to higher levels of Cd (100 mg Cd kg^{-1} soil), the authors noted a relatively higher activity of DHAR (46% increase over control) in Cd-tolerant cv. Pusa 9531 than Cd-susceptible cv. PS 16 (29.6% increase over control). In *B. napus* seedlings, Hasanuzzaman et al. (2012b) reported a decreased activity of DHAR (24% and 43%) upon exposure to 0.5 and 1.0 mM $CdCl_2$, respectively. The authors also revealed a decrease in AsA content in Cd-stressed seedlings. On the other hand, Aksoy and Dinle (2012) did not observe any changes in DHAR activities in Cd-treated [0.5 mM Cd $(NO_3)_2$, 5 d)] *Glycine max* seedlings. A short-term exposure of *H. vulagre* to Cd resulted in the marked change in DHAR activity in its roots (Zelinova et al., 2013). Despite insignificant changes in DHAR activity in *Cucurbita pepo* roots under Al [50 μM $Al_2(SO_4)_3$] exposure, Dipierro et al. (2005) argued that this slight increase in the activity contributed to maintaining the AsA system in the reduced state. In a recent study, Mishra et al. (2011) observed differential responses in terms of DHAR activity in two cultivars of *O. sativa* (Malviya-36 and Pant-12) exposed to As^{3+} for variable time durations. Overexpression of DHAR (but not of MDHAR) was advocated to play a role in the maintenance of a high AsA level and subsequent Al tolerance in transgenic *Nicotiana tabacum* plants overexpressing *Arabidopsis* cytosolic MDHAR or DHAR (Yin et al., 2010). In *O. sativa* shoots and roots, an Ni-mediated increase in DHAR activity was suggested to act as an activator of the AsA regenerating system that maintains the elevated level of AsA required for cellular redox homeostasis (Maheshwari and Dubey, 2009). At a lower dose of Ni (0.2 and 0.4 mM, 60 d), no changes in DHAR activity were observed in *B. juncea*; whereas higher doses of Ni (0.5–0.6 mM) significantly increased DHAR activity (Kanwar et al., 2012). A marked decrease in DHAR activity was argued to limit the regeneration of AsA in chloroplast in *Z. mays* plants exposed to Ni (100 mg Ni kg^{-1} soil) (Wang et al., 2009a,b). Plant age was also reported to modulate DHAR activity in plants exposed metal types. To this end, young *Helianthus annuus* sunflowers showed a more pronounced activity of DHAR versus adult plants under metal toxicity (Cd and Zn) (Nehnevajova et al., 2012). In seedlings of *O. sativa* (Malviya-36 and Pant-12) cultivars exposed to As^{3+}, Mishra et al. (2011) reported a declined activity of DHAR during 5–10 days of growth. However, a growth period of 15–20 days with As^{3+} showed increased DHAR activity compared with controls (Mishra et al., 2011). Along with a decrease in AsA concentration, a 33% and 30% decrease in DHAR activity was noted in *T. aestivum* seedlings exposed to 0.25 and 0.5 mM As (Hasanuzzaman and Fujita, 2013).

5.3 Catalase

Catalase (CAT; EC 1.11.1.6), a tetrameric, heme-containing enzyme mainly occurs in all differentiated plant peroxisomes and catalyzes the dismutation reaction without requiring any reductant (Su et al., 2014). Varied abiotic stress factors can differentially modulate CAT activity in plants.

5.3.1 Metals/Metalloids

Inconsistent activity of CAT has been reported in plants exposed to metal(oids), and was found to be dependent on plant species, tissue type, plant age/developmental stage, and metal type, and the concentration and duration of exposure (reviewed by Anjum et al., 2016b). A great deal of literature is available on Cd-mediated modulation of CAT activity in plants, where Cd was evidenced to increase (Hasanuzzaman et al., 2012b; Hsu and Kao

2004; Mobin and Khan, 2007; Khan et al., 2007; Agami and Mohamed 2013) or decrease (Balestrasse et al., 2001; Agami and Mohamed, 2013) the activity of CAT. Decreased CAT activity (by 45%) was reported in Mn (100 μM $MnCl_2$) (Saidi et al., 2014). Cd (50 μg L^{-1}) was reported to significantly modulate the mRNA expression of the CAT2 gene, and not the CAT1 gene in *Suaeda salsa* (Cong et al., 2013). Azpilicueta et al. (2007) reported an enhanced CAT3 (mitochondrial) transcript level in the leaf disks of Cd (300 and 500 μM $CdCl_2$)-exposed *Helianthus annuus*. Al-accrued enhancements in the expression of CAT cDNA were noted in *Capsicum annuum* (Kwon and An, 2001). Hg (20–40 μM $HgCl_2$) has also been reported to induce the CAT3 gene in *Arabidoposis thaliana* Columbia wild type (Heidenreich et al., 2001). A twofold increase in the transcript level of *PgCAT1* was reported in 50 μM Cu-exposed *Panax ginseng* (Purev et al., 2010). In Pb-exposed tomatoes (*Solanum lycopersicum*), Kabir and Wang (2011) reported a high transcript level of *SlCAT1* and *SlCAT2*. An Mn-mediated increase in the specific activity of CAT and also upregulation in the expression of gene encoding CAT (CAT, TDF no. 103-2) were reported by Zhou et al. (2013). In soybeans exposed to Cd (100 μM $CdCl_2$), Hossain et al. (2012) observed a higher accumulation of CAT protein and enhanced expression of molecular chaperones. El-Beltagi et al. (2010) reported an increase in the density with increasing Cd concentration in roots and leaves of *Raphanus sativus*. However, Cu failed to bring any change in the pattern of CAT isozymes in *Solanum nigrum* (Fidalgo et al., 2013). *ScCAT1*, a novel peroxisomal catalase gene from sugarcane that is localized in the plasma membrane and cytoplasm, is induced in response to Cu and Cd exposures, and exhibits tolerance (Su et al., 2014). A transient overexpression of *ScCAT1* in the leaves of *Nicotiana benthamina* induced a hypersensitive reaction response and cell death (Su et al., 2014). In leaves of Cd-exposed poplar (*Populus deltoides* × *Populus nigra*) cuttings, increased activity of two isoforms of CAT alleles (CAT1 and CAT2) was detected (Zhang et al., 2014). Verma and Dubey (2003) reported two CAT isoforms (with Rf 0.15 and 0.29) in the roots, whereas three isoforms (with Rf 0.15, 0.29, and 0.42) were found in the shoots at 500 μM Pb.

5.3.2 Drought, Salinity, and Other Abiotic Stresses

In dry and arid regions, two major stresses, namely drought and salinity, are most common, causing reduced plant growth and productivity (Pinheiro and Chaves, 2011). In general, an increase in CAT activity has been positively correlated to the degree of drought experienced by plants (Mittler et al., 2011; Pinheiro and Chaves, 2011). Drought tolerance in *Panicum sumatrense* was related with an increased activity of antioxidant enzymes, including CAT (Ajithkumar and Panneerselvam, 2014). The role of CAT in the oxidative protection against increased cellular leaf H_2O_2 content was reported in drought-tolerant *Jatropha curcas* (Silva et al., 2012). CAT activity in C_3 and C_4 plants may exhibit a differential response. To this end, in drought-exposed seedlings of *Cleome spinosa* (C_3) and *Cleome gynandra* (C_4), Uzilday et al. (2012) correlated CAT1 gene expression in *C. spinosa* with CAT activity, but the authors did not find this correlation of CAT1 gene expression in *C. gynandra* at 10 d. Notably in *C. spinosa*, the antioxidant defence system was insufficient to suppress the increasing ROS production under stress conditions. On the other hand, induction of CAT in *C. gynandra*, although lower as compared with *C. spinosa*, was able to cope with ROS formation under drought stress. In another study, in three drought-exposed Australian bread wheat cultivars [(*Triticum aestivum* L.; Kukri (drought-intolerant), Excalibur (drought-tolerant), and RAC875 (drought-tolerant)], Ford et al. (2011) reported the significance of increased CAT isoforms in oxidative stress metabolism and ROS scavenging capacity.

Increased salinization of arable land has been expected to have devastating global effects (Grover et al., 2011; Vanderauwera et al., 2012). High soil salinity can cause water deficit and ion toxicity in many plant species, and can also become a severe constraint on crop growth and productivity in many regions. In a salt-sensitive rice variety pretreated with two exogenous osmoprotectants (proline and threalose), Nounjan et al. (2012) reportedly increased the CAT activity and transcript upregulation of genes encoding CatC, and correlated these responses with a significant reduction in H_2O_2. Sorkheh et al. (2012) observed salt stress-mediated significant induction in CAT activity in various wild almond species. CAT played a key role in salt stress acclimation induced by H_2O_2 pretreatment in maize (Gondim et al., 2012). Catalase (CAT2) can play a key role in protecting the plant genome against photorespiratory-dependent H_2O_2-induced DNA damage (Vanderauwera et al., 2011). Low temperature affects crop production and quality, and causes oxidative stress in plants. In addition to other antioxidant enzymes, CAT was reported to play a key role in the process of cold acclimation through their effect on the redox state of the cell in pepper (*Capsicum annum*) (Airaki et al., 2012). Sugarcane varieties resistant to the smut pathogen *Sporisorium scitamineum* exhibited a higher CAT activity when compared with susceptible varieties (Su et al., 2014). Additionally, a significant induction of ScCAT1 was also reported by these authors in smut pathogen-resistant *S. scitamineum*.

5.4 Glutathione Reductase

Glutathione reductase (GR) is a flavoprotein oxidoreductase, and is localized in chloroplast (70%–80%), mitochondria, and cytosol. GR is associated with the family of flavoenzymes in which the catalyzation of NADPH-dependent reduction of glutathione disulfide (GSSG) to glutathione (GSH) occurs. Thus, GR maintains an appropriate GSH/GSSG ratio, and reduced GSH levels in cells (reviewed by Gill et al., 2013). GR is differentially modulated in different plant species exposed to metal—metalloids, salinity, drought, and other various abiotic stresses (reviewed by Gill et al., 2013).

5.4.1 Metals/Metalloids

A differential modulation of GR activity was reported in plants exposed to various metals/metalloids, where the role of the levels of the metals supplied, plant species, the age of the plant, and duration of the treatment were observed (reviewed by Anjum et al., 2012). *B. juncea* plants treated with As significantly increased the activity of GR, as well as the contents of GSH, which provided an indication regarding the capacity of mustard plants to detoxify the low As level (Khan et al., 2009; Garg and Singla, 2011). On the contrary, Kertulis-Tartara et al. (2009) did not find any change in GR activity in As-treated fronds of *P. vittata*. In *O. sativa* seedlings, the upregulation of GR substantiated that As accumulation generated oxidative stress (Shri et al., 2009). In *B. napus* seedlings, Hasanuzzaman et al. (2012b) observed a profound increase in GR activities in response to 0.5 mM $CdCl_2$ (66 % higher), whereas at 1.0 mM $CdCl_2$, the activity significantly decreased by 24%. When the seedlings were pretreated with exogenous Se they showed enhanced activities compared with the Cd-stressed seedlings without Se supplementation. This increased GR activity, coupled with enhanced GSH content and their redox stress, conferred more tolerance to Cd-induced oxidative stress to the plants. Recently, the activity of GR in *T. aestivum* seedlings increased (by 31%) at 0.25 mM As, while no change was observed at 0.5 mM (Hasanuzzaman and Fujita, 2013). The authors reported no significant GR activity increase for protection of *T. aestivum* seedlings against As-induced damages.

5.4.2 Drought, Salinity, and Other Abiotic Stresses

The upregulation of the antioxidant defense system is one of the adaptive mechanisms of plants under drought, and has indicated an important role of GR in this process. Increased GR activity was observed in *Oryza sativa* (Sharma and Dubey, 2005), *B. juncea* L. (Alam et al., 2013), *Cleome gynandra* and *C. spinosa* (Uzilday et al., 2012), and *Arachis hypogea* (Solanki and Sarangi, 2014), and these provided tolerance to drought. However, the effect of drought on the pattern of GR activity is very complex, and in most cases, it is difficult to analyze with respect to drought tolerance/susceptibility of cultivars. Sharma and Dubey (2005) observed the increase in GR activity in rice seedlings up to a certain period (10 d), and then the activities declined. In *Ctenanthe setosa*, the activity of GR in leaves was increased up to 32 d of drought and then decreased. However, in root and petiole, the GR activity increased in time-dependent manners (Terzi and Kadioglu, 2006). Selote and Khanna-Chopra (2010) demonstrated that drought-acclimated wheat roots during subsequent severe water stress conditions enhanced systematic upregulation of GR and other AsA-GSH cycle enzymes at both the whole cell level, as well as in mitochondria, and maintained a higher relative water content and lower level of H_2O_2, and rendered stress tolerance in plants. By analyzing the antioxidants in tomato plants, Sanchez-Rodriguez et al. (2010) concluded that the activity of GR increased sharply in the tolerant cultivars more than in others, which helped to maintain the greater redox state of the GSH, and reduced the oxidative stress. Shehab et al. (2010) reported an increase in the activity of GR in rice, representing a protective activity to counteract the oxidative injury caused by drought. In *B. napus* seedlings, the activity of GR showed a slight increase with 10% PEG; however, the application of 20% PEG did not show any change in GR activity (Hasanuzzaman and Fujita, 2011). Although mild drought stress significantly increased GR activity, the increase was not enough to prevent the oxidative stress. However, Se pretreatment caused remarkable increases in the activities of GR in drought-stressed seedlings, compared with the drought-stressed seedlings without Se, and the seedlings showed enhanced tolerance to drought-induced oxidative stress (Hasanuzzaman and Fujita, 2011). In a recent study, Uzilday et al. (2012) observed a clear difference in GR activities between *Cleome gynandra* (C_4) and *Cleome spinosa* (C_3) plants exposed to drought stress for varying durations. The authors noted the constitutive levels of GR to be higher in *C. spinosa* than *C. gynandra*; where GR activity was greatly increased by 147% and 62% in *C. gynandra* under 5 and 10 d of drought. On the other hand, *C. spinosa* exhibited an increasing pattern by 23.3% and 84.5%, respectively, at the same levels of drought. Drought stress exposition and relief by reirrigation were reported to influence GR activity differentially. In this context, in wild species of almond (*Prunus* spp.), GR activity sharply increased under drought stress at any level. But this activity was downregulated during the rewatering phase, and their values were lower than those found in the drought-stressed plants (Sorkheh et al., 2011). In a recent study, Alam et al. (2013) found that under mild drought stress (10% PEG), a slight increase (31%) of GR was observed, compared with control, while under severe stress (20% PEG) it remained unchanged.

Increased GR activity during salt stress was observed in many plant species, such as *Triticum aestivum* (Mandhania et al., 2006; Hasanuzzaman et al., 2011a), *Brassica napus* (Hasanuzzaman et al., 2011b), *B. juncea* (Khan et al., 2009), *Vigna radiata* (Nazar et al., 2011), *Lycopersicon pennellii* (Mittova et al., 2002), *Z. mays* (Kholová et al., 2009), *Sesamum indicum* (Koca et al., 2007), *Pyrus communis* L. (Wen et al., 2009), Kentucky bluegrass, and tall fescue (Xu et al., 2013). On the contrary, in some species, such as *Solanum tuberosum* L. (Queirós et al., 2011), shoots of Kentucky bluegrass (KBG) and tall fescue (TF) (Xu et al., 2013), the activity of GR was decreased. In most cases, the duration or intensity of stress imposed was reported to control GR activity in salinity-exposed plants. In *T. aestivum* seedlings, the activity of GR was increased in a dose-dependent manner. However, the activity of GR was greater in salt-tolerant cultivars (KRL-19) than salt-sensitive cultivars (WH-542), which indicated that the tolerant plant exhibited a more active AsA-GSH cycle than the sensitive cultivar (Mandhania et al., 2006). In *V. radiata*, salt stress (300 mM NaCl) caused an increase in GR activity; however, it was different for two cultivars. Salt-tolerant cultivars (Pusa Bold) showed more than 6.9-fold more GR activity than controls, while the salt-sensitive (CO4) cultivars showed only a 4.6-fold increase under similar conditions (Sumithra et al., 2006). Plant species differing in salinity sensitivity exhibit differential GR activity (Sekmen et al., 2007; Hernandez et al., 2000; Hasanuzzaman et al., 2011b). In this context, Sekmen et al. (2007) observed differential activity of GR in *Plantago* spp.; where in salt-tolerant *P. maritima*, GR activity increased by 50% under 200 mM NaCl stress, and its activity decreased by about 61% in *P. media* under salt stress (100 and 200 mM NaCl). In *Sorghum vulgare* genotypes, although GR activity was not differed significantly due to salt stress (75 mM NaCl), the salt-tolerant genotype (CSF20) showed higher activity than sensitive cultivar (CSF18). El-Bastawisy (2010) reported that the wheat cultivar, which showed higher GR activities, performed well under salt stress. Hernandez et al. (2000) found increased GR activity in salt tolerant *P. sativum* genotypes; whereas the activity decreased in the salt-sensitive genotype due to long-term salt stress. Likewise, GR activity increased with the increasing level of salt stress in the salt-tolerant *P. maritima*; whereas a sharp decrease was observed in the sensitive genotype, which was expressed by the higher lipid peroxidation level (Sekmen et al., 2007). In *T. aestivum* seedlings, a slight decrease in GR activity was observed at the 150 mM NaCl stress, whereas a significant decrease (29%) was observed at the 300 mM NaCl stress (Hasanuzzaman et al., 2011a). In *Brassica napus* seedlings, mild salt stress (100 mM NaCl) could not change the activity of GR, and while under severe stress (200 mM NaCl), the activity declined by 18% (Hasanuzzaman et al., 2011b). Importantly, selenium (Se) supplementation under salt stress caused significant increases in the activities of GR, which maintained a better GSH/GSSG ratio and conferred salt stress tolerance (Hasanuzzaman et al., 2011b).

The activity of GR generally increased under high temperature (HT) stress, depending on the plant types and the duration of exposure to HT. However, in every case, a strong correlation of enhanced GR activity and stress tolerance was observed. An increase in GR activity under HT stress was observed in many plant species, such as *T. aestivum* (Almeselmani et al., 2009; Hasanuzzaman et al., 2012a), *Z. mays* (Kumar et al., 2012), *Nicotiana tabacum* (Tan et al., 2010), *P. aureus* (Kumar et al., 2011), and *Malus domesticah* (Ma et al., 2008). While studying with *T. aestivum* seedlings, Hasanuzzaman et al. (2012a) observed differential responses in the activity of GR under two levels of heat treatment. The activity of GR remained unchanged at 24 h of heat treatment, while an increase of 57% was noted upon 48 h of heat treatment. Kumar et al. (2012) investigated the activities of antioxidant enzymes in maize and rice genotypes under HT. With an increase in temperature, the activity of GR elevated, and at 40/35°C, there was a greater increase in *Z. mays* genotype compared with *O. sativa*. However, with further elevation in temperature to 45/40°C, the activity was inhibited to a greater extent in *O. sativa* than in *Z. mays* genotypes. In *Pyrus malus* leaves, the activity of GR have been shown to increase under heat stress (40°C), compared with controls (28°C), and with the duration of HT, the activity of the enzyme reached the highest point in the first 4h, and then decreased (Ma et al., 2008). These results suggest that the modulation of GR in response to stress has a limit beyond which they would be damaged by stress. High GR activity maintains the pool of GSH in the reduced state, allowing GSH to be used by DHAR to reduce DHA to AsA. Increases in GR have been reported in other species during heat stress (Almeselmani et al., 2006). In another study, Almeselmani et al. (2009) reported differential GR activity modulation in two *T. aestivum* cv. C 306 and PBW 343 differing in heat tolerance. There was a significant increase in the GR activity in C 306 (heat-tolerant) at all stages of growth; whereas, PBW 343 (heat-susceptible) showed a reduction in GR activity under HT compared with normal temperature, which provided a positive correlation between enhanced GR activity and heat stress tolerance. A positive correlation between tolerance to low temperature (LT)-induced photoinhibition and high GR activities has been shown in *O. sativa* (Huang and Guo, 2005; Guo et al., 2006) and other cereals (Janda et al., 2006), *C. sativus* (Hu et al., 2008; Xu et al., 2008), *Glycine max*, *C. melo* (Fogelman et al., 2011), and *Citrullus lanatus*. On the other hand, no changes in GR activity were observed in *T. aestivum* (Yordanova and Popova, 2001), or even declined in *O. sativa* (Huang and Guo, 2005). Kocsy et al. (2001a,b) reviewed the role of GSH in adaptation and signaling during chilling and cold acclimation in plants and GR as one of the potent regulators of GSH synthesis. In watermelon, the activity of GR markedly increased after 1 d of LT treatment (10°C) and then decreased. However, the activity of GR was different in two cultivars. Importantly, exogenous application of SA (1.0 mM) enhanced the activity of GR more, which provided better protection against LT. Turan and Ekmekci (2011) found differences in GR activity in cold-acclimated and nonacclimated *Cicer arietinum* cultivars. The nonacclimated plants showed a reduction in GR activity, while acclimated plants showed a marked increase in GR activity at both 4 and 2°C, and performed well under stress. Soybean seeds osmoprimed with PEG-8000 exposed to low temperature (4°C) showed a marked increase in GR activity, which exhibited better protection against imbibitional chilling injury. In the seedlings of *Poncirus trifoliata* L. Raf. × *Citrus sinensis* L. Osb. and *Citrumelo* CPB 4475 (*P. trifoliata* L. Raf. × *C. paradisi* L. Macf.), Arbona et al. (2008) observed that waterlogging stress induced the increase of GR activity in all *P. trifoliata* seedlings in a time-dependent manner, depending on the species. The flooding-sensitive *Cleopatra mandarin* showed comparatively lower activity of GR compared with flooding-tolerant *Carrizo citrange* that indicated a clear role of GR in stress tolerance. Importantly, flooding-sensitive plants showed a marked decrease in GR activity that led to control values. In another study with citrus, Hossain et al. (2013) found an intermediate response in the activities of GR upon exposure to waterlogging stress. In continuously flooded plants, a little increase in GR activity was observed, compared with substrate-drained plants. In *Cajanus cajan*, up to 2 d of waterlogging GR activity increased in both waterlogging-sensitive and waterlogging-tolerant genotypes; whereas further waterlogging resulted in a decrease in GR activity in sensitive genotypes (Kumutha et al., 2009). Importantly, tolerant genotypes showed an increase in GR activity at any duration of waterlogging, and provided better protection against oxidative stress. A similar pattern was also observed during recovery of flooding (Kumutha et al., 2009). Bin et al. (2010) observed differential activities of GR in roots of maize genotypes. In waterlogging-tolerant genotypes (HZ32), GR activity increased significantly, while in waterlogging-sensitive genotypes, the activity decreased. A decrease in GR activity in waterlogging-sensitive plants also suggested a decreased GSH turnover rate. In submergence-tolerant rice seedlings (cv. FR13A and MR219-4), GR activity was increased almost 10- and 13-fold, respectively, upon 8 d of water logging which, indicated a correlation between GR activity and flooding tolerance (Damanik et al., 2010).

Ismail and Suroto (2012) measured the activity of GR in three local rice cultivars, that is, MR263, MR219, and MR84 under O_3 exposure (120 ppb), and found significant differences in the GR activities between control and exposed plants. This increase in GR activities in the MR84 cultivar was because they need to increase the protection for plant cell systems. On the other hand, the cultivars (MR263 and MR219) showed a decrease in GR activity, which is due to the fact that their systems have been destroyed by oxidant (i.e., H_2O_2) that accumulated in the leaves. Similarly, increased GR activity was also observed in *O. sativa* (Ismail and Suroto, 2012) exposed to elevated O_3. Wang et al. (2013) found a dissimilarity of GR activity in two rice cultivars under experimental free-air O_3 exposure. The cultivar with enhanced GR resulted in a higher GSH pool and conferred stress tolerance. In a recent study, elevated O_3 (70 ± 5 ppb) exposure resulted in a marked increase in GR activity in *Beta vulgaris* (Kumari et al., 2013). Increased GR activity upon UV exposure was observed in many plant species, viz including *Cucumis sativus* (Kubis and Rybus-Zając, 2008), *Arachis hypogaea* (Tang et al., 2010), and *Acorus calamus* (Kumari et al., 2010). Xu et al. (2008) investigated the impact of solar UV-B radiation on the antioxidant defense system in soybean lines and found a differential response in GR activity. The soybean lines, which showed higher GR activity, provided more antioxidant defense with better tolerance to UV radiation. Kumari et al. (2010) observed a significant increase in GR activity in *Acorus calamus* exposed to UV radiation. However, the response was higher at earlier stages of growth, and later, the increase was nonsignificant compared with control.

5.5 Superoxide Dismutase

As the first line of antioxidant defense against a potent ROS ($O_2^{\bullet -}$), SODs rapidly convert them into H_2O_2 and O_2. In this way, SODs decrease the risk of $OH^{\bullet}$ formation via the metal catalyzed Haber-Weiss-type reaction because this reaction has a 10,000-fold faster rate than the spontaneous dismutation (Gill and Tuteja, 2010).

5.5.1 Metals/Metalloids

Difference in plant species, stages of the plant development, metal in the experiment, and the exposure time have been reported to considerably control the modulation of SOD activity and expression in plants. A great deal of literature is available on Cd-mediated modulation of SOD in plants, where SOD activity may exhibit both significant enhancements (Mishra et al., 2006; Mobin and Khan, 2007; Singh et al., 2008; Anjum et al., 2008a,b; Khan et al., 2007; Krantev et al., 2008; Ekmekci et al., 2008; Agrawal and Mishra, 2009; Wu et al., 2004; Cho and Seo, 2005) and decreases (Guo et al., 2006). It has been reported that seven SOD isozymes in leaves, and eight in roots, corresponding to Mn-SOD and Cu/Zn SOD isozymes in *Glycine max*. Although a clear effect of Cd on plant growth was observed, the authors did not note any alteration in the activities of the SOD isozymes. A similar result was obtained in another study where, though *Saccharum officinarum* seedlings exhibited several isozymes, Cd stress failed to alter SOD activity (Fornazier et al., 2002). Cu/Zn SODs, Fe-SOD, and Mn-SOD differed in their response to Cd stress in *Pisum sativum*. Therein, Sandalio et al. (2001) reported a strong reduction in chloroplastic and cytosolic Cu/Zn SODs by Cd. The impact was less on Fe-SOD; whereas, only the highest Cd level affected Mn-SOD. No change in the activity of Mn-SOD was observed in the leaf peroxisomes of Cd-exposed *Pisum sativum* (Romero-Puertas et al., 1999). However, a significant increase and decrease in activity of both CuZnSOD and MnSOD isoforms in leaves and roots, respectively, were noted in Cd (10, 25, and 50 μM $CdCl_2$)-exposed *Tagetes patula* (Liu et al., 2011). Compared with *Brassica juncea*, the least upregulation in the SOD activity in Cr (50, 100, and 200 μM)-exposed *V. radiata* was argued to be responsible for its sensitivity to Cr stress (Diwan et al., 2010). Both Cd-hyperaccumulating ecotypes (HE) and non Cd-hyperaccumulating ecotypes (NHE) of *Sedum alfredii* exhibited remarkably increased SOD activity under Cd exposure (Jin et al., 2008). However, the authors noted a significant decline in SOD activity in the roots and leaves of NHE. Cd-sensitive *B. juncea* cv. SS2 exhibited a higher increase in SOD activity compared with Cd-tolerant *B. juncea cv. Pusa Jai Kisan* grown under Cd stress (25 and 50 $\mu mol\,L^{-1}$ Cd) (Iqbal et al., 2010). No induction in SOD activity was noticed in the fronds of *P. ensiformis* under As-exposure; whereas *Pteris vittata* exhibited increased SOD activity in its fronds when grown under the same conditions (Srivastava et al., 2005). In another study, lower concentrations of As (50 and 150 μM) caused higher SOD activity in *B. juncea* (cvs. Varuna and Pusa Bold); whereas a prolonged exposure to these cultivars to 300 μM decreased the activity of SOD (Gupta et al., 2009). As-tolerant *Holcus lanatus* exhibited decreased SOD activity at high levels of As-exposure (Hartley-Whitaker et al., 2001). As-speciation can also differentially modulate SOD activity in plants. To this end, As(III) failed to cause any significant modulation in SOD activity in arsenate [As(V)] and arsenite [As(III)]-exposed tolerant (TPM-1) and sensitive (TM-4) variety of *B. juncea* (Srivastava et al., 2010). Tanyolac et al., 2007) reported an increase in SOD activity in the leaves of *Z. mays* cultivars (3223 and 31G98) with increasing Cu concentrations (up to 1.5 mM). The detection of five isoforms of SOD was reported in leaves of Hg-exposed *Medicago sativa*, where Hg^{2+} at 20 or 40 μM caused significant increases in total SOD activity (Zhou et al., 2008). In Hg-exposed *Atriplex codonocarpa*, a significant increase in SOD activity was reported in roots that peaked to a maximum at 0.1 $mg L^{-1}$ Hg (Lomonte et al., 2010). The authors also noted a gradual increase in SOD activity with 0.1 $mg L^{-1}$ Hg in shoots. About a 28%–39% increase in the activity of Cu/Zn-SOD, a 32%–52% increase in the activity of Fe-SOD, and a 49%–53% increase in Mn-SOD activity were observed in the shoots of 160 μM Al^{3+} stressed *O. sativa* seedlings (Sharma and Dubey, 2007).

5.5.2 Drought, Salinity, and Other Abiotic Stresses

Drought exposure has been reported to bring enhancements in SOD activity in a number of plant species, including and *O. sativa* (Sharma and Dubey, 2005), *Phaseolus vulgaris* (Zlatev et al., 2010), and *V. unguiculata* (Manivannan et al., 2007). The activity of SOD and its isoform Cu/Zn-SOD increased in both the salt tolerant and sensitive cultivars of *O. sativa* against salinity (Mishra et al., 2013). It has been reported that the salinity increases SOD activity in salt-tolerant cultivars and decreases it in salt-sensitive cultivars, in both leaves (Gossett et al., 1996; Dionisio-Sese and Tobita, 1998; Hernandez et al., 2000) and roots (Shalata et al., 2001). While examining the long-term effects of salt stress in two salt-tolerant lines (Kharchia 65, KRL19) and two salt-sensitive lines (HD2009, HD2687) of *T. aestivum*, Sairam et al. (2005) observed the exhibition of a higher increase in SOD activity by the salt-tolerant line Kharchia 65 when compared with the salt sensitive line HD2687. Among 150 mM NaCl-exposed *Lathyrus sativus*genotypes (B1, BioL-212, PUSA-90-2, WBK-CB-14, LR3 and LR4), an increase in SOD activity in B1, BioL-212, LR3, and LR4 lines (vs. control) was argued as a result of overactivity of both Cu/Zn I and II isoforms (Talukdar, 2013). In another study, NaCl enhanced the activity of Cu/Zn-SOD II in *T. aestivum* seedlings (Eyidoğan et al., 2003). Salt tolerance in plants such as *O. sativa* (Dionisio-Sese and Tobita, 1998) and *Lycopersicon esculentum* and citrus (Mittova et al., 2004) was reported as a result of increased SOD activity. In another study, NaCl salinity (50, 100, and 150 mM) exposed *Gossypium hirsutum* varieties (Arya-Anubam and LRA-5166). Exhibition of a progressively increased SOD activity was reported in NaCl salinity (50, 100, and 150 mM) exposed *Gossypium hirsutum* varieties (Arya-Anubam and LRA-5166). Therein, Desingh and Kanagaraj (2007) noted a markedly higher SOD activity in var. Arya-Anubam than in var. LRA-5166 at all salinity levels. Two species, namely Lem and Lpa of *L. esculentum*, exhibited diferential SOD activity under NaCl (25–50 mM) exposure (Mittova et al., 2004). The authors revealed matrix CuZnSOD (mainly the matrix isozyme) in Lem; whereas the activity of SOD in Lpa is comprised of both MnSOD (mainly the matrix isozyme) and CuZnSOD (exclusively the membrane-bound isozyme).

Upregulation in the activity of SOD and induction in the expression of mitochondrial MnSOD and cytosolic Cu/ZnSODs genes were reported with $O_2^{\bullet-}$ exposure (Mylona and Polidoros, 2010). Acute (single) or chronic (3, 6, and 10 consecutive days) ozone exposure caused increases in transcript levels of mitochondrial MnSOD and cytosolic Cu/ZnSODs in *Zea mays* leaves (Ruzsa et al., 1999). The authors noticed a downregulation in the transcript level of chloroplastic Cu/ZnSOD. In *Cassia auriculata* seedlings, Agarwal (2007) reported an elevated SOD activity under UV B (7.5 and 15.0 $kJ\,m^{-2}$) irradiation. High light caused an increase in MnSOD protein expression in CAT1AS *N. tabacum* leaves (Dat et al., 2003). Restriction of oxygen supply in waterlogged and compacted soil (hypoxia) was extensively reported to differentially modulate SOD activity in different plant species (Ushimaru et al., 1999, 2001) (Table 1).

5.6 GSH and AsA

Ascorbate and glutathione are potential nonenzymatic antioxidants of the glutathione-ascorbate cycle, which detoxifies H_2O_2 in the cell (Gill and Tuteja, 2010).

5.6.1 Metals/Metalloids

An exhaustive literature search has revealed both a significant decrease and increase in the plant-reduced GSH pool. A decrease in the GSH pool has widely been reported in different plant species exposed to Cd (Tukendorf and Rauser, 1990; Balestrasse et al., 2001; Schutzendubel et al., 2002; Zhang et al., 2003; Hsu and Kao, 2004; Wu et al., 2004; Anjum et al., 2008b,c), Pb, (Piechalak et al., 2002; Qureshi et al., 2007), Cu (Nagalakshmi and Prasad, 2001), and U (Vandenhove et al., 2006). Pietrini et al. (2003) evidenced the association of the antioxidant activity with a large pool of GSH in the chloroplasts of Cd-exposed *Phragmites australis*. The authors advocated the GSH pool significance in the protection of the activity of many photosynthetic enzymes against Cd-accrued thiophilic bursting. On the other hand, an elevated GSH pool has been also been evidenced in plants under Cd (Gupta et al., 2002; Pietrini et al., 2003; Qadir et al., 2004; Metwally et al., 2005; Sun et al., 2007). Many-fold increases in the GSH pool have been reported in Cu- and Cd-treated *Z. mays* (Tukendorf and Rauser, 1990) and Cd-exposed *Lycopersicon esculentum* (Chen and Goldsbrough, 1994; Gupta and Goldsbrough, 1991). Klapheck et al. (1994) reported up to a 125% increase in hydroxymethyl-GSH (γ-Glu-Cys-Ser) in Poaceae; whereas the same group evidenced up to a threefold increase in the homo-GSH (γ-Glu-Cys-Ala) content in *P. sativum* (Klapheck et al., 1995). Different Zn levels have also been shown to affect the GSH pool in different plant species, including *Avicennia marina* (Caregnato et al., 2008), *Oryza sativa* (Guo et al., 2006), and *Sesbania drummondii* (Israr et al., 2006a). Moreover, in *S. drummondii* callus exposed to Cd (0–250 μM), Israr et al. (2006b) reported a significantly increased GSH level (94.7%) and GSH/GSSG ratio (36.7%) due to 50 μM Cd with respect to the control. Israr and Sahi (2006) evidenced 40 μM Hg-mediated 93.5% and

41.0% in GSH level and GSH/GSSG ratio, respectively in *S. drummondii*. Rehman and Anjum (2010) advocated the significance of elevated GSH and NPTs levels in the protection of *C. tropicalis* against Cd toxicity. Cd-tolerant *Brassica juncea* cv. Pusa Jai Kisan (Iqbal et al., 2010) and *Hordeum vulgare* cv. (Wu et al., 2004) were reported to exhibit a higher GSH pool, but decreased GSSG content. Kolb et al. (2010) reported a significant decrease in GSH levels in all cell compartments of mesophyll cells and glandular trichomes of *Cucurbita pepo* treated with 50 μM Cd. *Vigna radiata* has been shown to exhibit Cd or a Cu-dose-dependent decrease in GSH levels by Anjum et al. (2008b, 2011) and Diwan et al. (2010). A few studies, including Sun et al. (2005), Sun et al. (2007) and Gupta et al. (2010), have explored the potential role of GSH-mediated tolerance and survival mechanisms' modulation in *Sedum alfredii*. Among the ten Cd-exposed *B. juncea* cultivars, namely Vardhan, Pusa Bahar, Pusa Bold, BTO, Pusa Jai Kisan, Agrini, Varuna, Kranti, Vaibhav and Pusa Basant, Qadir et al. (2004) reported an increase in GSH content, with the maximum in Pusa Jai Kisan (26%–105%), while the minimum was in *B. juncea* cv. Kranti (7%–65%), over their respective controls.

As noted herein, in some legumes, hGSH can replace GSH, producing homophytochelatins (hPCs) of general structure (γGlu-Cys)2-11-βAla (Grill et al., 1986; Klapheck et al., 1995). Additionally, PCs- and hPCs-mediated chelation of Cu, Zn, Cd, Hg, Pb, and As has been reported; where the resulting complexes are transported into the vacuoles, avoiding cellular toxicity. To this end, Ramos et al. (2007) reported the accumulation of PCs and hPCs in roots and nodules of Cd-exposed *L. japonicus*. The expression of endogenous γ-ECS (and to a lesser extent GS) is strongly induced upon heavy metal exposure (Schafer et al., 1998; Xiang and Oliver, 1998). The transgenic plants engineered for increased γ-ECS expression had shown that upgraded GSH levels may result in stronger accumulation of heavy metal—sequestrating PCs (Wawrzynski et al., 2006). Liedschulte et al. (2010) expressed the bi-functional γ-ECS-GS enzyme from *Streptococcus thermophilus*, a nonpathogenic lactic acid bacterium, in *N. tabacum*; where the authors revealed an extreme accumulation of GSH in their leaves (up to 12 L mol^{-1} GSH/gFW, depending on the developmental stage) (which is more than 20- to 30-fold above the levels observed in wild-type) in transgenic *N. tobacum* plants expressing StGCL-GS under control of a constitutive promoter. A Cd-mediated decrease in the AsA pool has been evidenced in Cd-exposed plants, including *Ceratophyllum demersum* (Aravind and Prasad, 2005), *B. napus* cv. Bina sharisha seedlings (Hasanuzzaman et al., 2012b), *Pisum sativum* (Romero-Puertas et al., 2007), and *Hordeum vulgare* (Wu et al., 2004). In contrast, Jin et al. (2008) reported an increased AsA pool respectively in Cd-hyperaccumulator and nonhyperaccumulator ecotypes of *Sedum alfredii*, and they have also been observed in Cd-exposed *B. juncea* shoots and roots. In Cu (0, 10, 50, and 100 mM; for 5 days) exposed *Oryza sativa* (var. MSE-9), Thounaojam et al. (2012) reported an increase in AsA content (by 24.13% in shoots; 29.31% in roots). No change in the level of leaf AsA was reported in Cd (50 μM)-exposed *B. juncea* (Markovska et al., 2009).

5.6.2 Drought, Salinity, and Other Abiotic Stresses

Glutathione plays an important role during drought, salt, and desiccation stress (Lascano et al., 2001; Shalata et al., 2001; Tausz, 2001; Kranner et al., 2002; Kocsy et al., 2004a,b; Basu et al., 2010). Under polyethylene glycol [20% (w/v), 48 h]-induced drought stress, Basu et al. (2010) reported significantly decreased leaf-GSH/GSSG ratio to a greater extent in the salt-sensitive (IR-29) and aromatic (Pusa Basmati) rice varieties when compared with salt-tolerant *O. sativa* cv. Pokkali. NaCl mediated enhanced γ-ECS activity has been evidenced to a greater extent in a tolerant cotton cell line than in a sensitive one, indicating adaptation at the level of GSH synthesis, GSSG reduction, and GS-conjugate formation (Gossett et al., 1996). Elevated GSH levels and the GSH/GSSG ratio were reported in salt stress exposed maize inbred line Z7. The exhibition of higher GSH, GSH redox state, and γ-glutamyl-cysteine synthetase (γ-ECS) activity has been reported in salt-tolerant *L. pennellii* when compared with the salt-sensitive *L. esculentum* (Shalata et al., 2001; Mittova et al., 2003); thus implying the significance of the GSH level upregulation for plant defense against salinity-mediated oxidative damage. In a study on the salt (100 mM NaCl)-exposed cultivated tomato *L. esculentum*, Mill. cv. M82 (Lem) and its wild salt-tolerant relative *L. pennellii* (Corr.) D'Arcy accession Atico (Lpa), increased the content of the reduced glutathione, and its redox state in the roots of Lpa was advocated as effective in the control of membrane lipid peroxidation (Shalata et al., 2001). In a similar study by Mittova et al. (2004), elevated reduced GSH level was reported to metabolize H_2O_2 in the mitochondria of salt-treated Lpa. Increased GSH and Cys levels has been evidenced in salt-stressed *Arabidopsis thaliana* (Barroso et al., 1999), implying salinity-mediated strong induction of GSH synthesis. The roots of two *O. sativa* cvs. Lunishree (salt tolerant) and Begunbitchi (salt sensitive) were reported to exhibit significant elevation in GSH pools under 24 h NaCl salinity levels (50, 100, and 150 mmol L^{-1}) (Khan and Panda, 2008); where the authors advocated the elevated GSH pool significance for decreased elevation of the oxidative stress indices in roots of *O. sativa* cv. Lunishree. In roots and shoots of salinity-treated *T. aestivum*, GSH content increased (Meneguzzo et al., 1999). Salt stress resulted in an increase in reduced GSH content, which remained high even after recovery from salt stress in *Glycine max* root nodules (Comba et al., 1998).

On the other hand, different phases of drought stress and subsequent recovery have been reported to significantly impact the reduced GSH pool in the desert plant *Reaumuria soongorica* (Bai et al., 2009). Shehab et al. (2010) reported a drought-induced increase in GSH content in *O. sativa* treated with drought stress. The GSH content was gradually increased by lengthening the time of PEG treatments. The GSH pool can be modulated differently as a result of skipping irrigation (drought) and rewatering (irrigation) (Sofo et al., 2005). Wang et al. (2010) reported the response of leaf GSH levels in nontolerant and tolerant *O. sativa* cultivars during the grain-filling stage. Where the occurrence of higher GSH content was reported in drought-tolerant *O. sativa* cvs. when compared with less tolerant cvs. *In vitro* drought stress of −0.5 and −2.0 MPa for 24 h to *O. sativa* seedlings was observed to significantly decrease the GSH pool (Sharma and Dubey, 2005). Mild drought stress, acclimation, and nonacclimation can differentially impact the reduced GSH and total GSH pools in plants (Selote and Khanna-Chopra, 2010).

Chilling injury is a physiological disorder commonly found in crops indigenous to subtropical and temperate regions. Many important food crops, such as rice and maize, are very sensitive to chilling stress when air temperatures fall below a nonfreezing critical threshold. A protective role of GSH has been observed in combatting low temperature-induced stress in different plant studies, including spruce (Doulis et al., 1993), pine (Anderson et al., 1992), *Z. mays*, *T. aestivum*, *O. sativa* (Kocsy et al., 2000, 2001a; Guo et al., 2006), and *Z. mays* (Hodges et al., 1996). To this end, Hodges et al. (1996) reported elevated levels of GSH in *Z. mays* under chilling stress. In chilling-sensitive cvs., the authors noted a sharp drop in the GSH/GSSG ratio in young plants upon short-term chilling; whereas increased GSH levels and decreased GSH/GSSG ratios were noticed upon long-term chilling when compared with those of tolerant lines. There are also reports of increased GSH levels during winter in spruce (Doulis et al., 1993) and pine needles (Anderson et al., 1992), and also in the living bark of apple trees (Kuroda and Sagisaka, 1998). Anderson et al. (1992) and Guo et al. (2006) evidenced an increase in GSH levels as an acclimation response to chilling temperatures. In addition, a chilling mediated increase in the γ-ECS activity and γ-EC content in the bundle sheath cells has also been reported in *Z. mays* (Gomez et al., 2004). The involvement of chromosome 5A was reported in the regulation of GSH accumulation in *T. aestivum* during cold hardening (Kocsy et al., 2000). Increased GSH levels have been reported in the chilling-tolerant *Z. mays* genotype as compared with the chilling-sensitive type (Kocsy et al., 2000). The same group of authors also noted increased GSH synthesis and chilling tolerance with the exogenous GSH supply; whereas inhibiting GSH synthesis reduced tolerance in *Z. mays*.

The importance of GSH has also been reported in the case of heat stress, where elevated GSH levels and the altered GSH/GSSG ratio were evidenced in different crop plants, including *T. aestivum* and *Z. mays* (Dash and Mohanty, 2002), *Lycopersicon esculentum* seedlings, and *T. aestivum* (Kocsy et al., 2002) under heat stress regimes. Either high temperature or cold treatments induced a greater increase in GSH and hmGSH synthesis in tolerant *T. aestivum* genotypes than in sensitive ones, as shown in 35S-labelling experiments (Kocsy et al., 2000, 2004a). Light stress is one of the most common sources of oxidative stress in plants (Dat et al., 2000). GSH-mediated enhanced protection against excessive light energy induced oxidative stress has been reported (Grill et al., 1987; García-Plazaola et al., 1999). In etiolated *T. aestivum* seedlings, Mattagajasingh and Kar (1989) reported a rapid increase in the GSH level within hours upon illumination. Regarding GSH significance in plants under UV radiation, the nonenzymatic antioxidant GSH also shows protection against harmful UV radiation (Masi et al., 1995; Polle, 1997).

In UV-B exposed *Z. mays*, Masi et al. (1995) evidenced no change in the pool size or redox state of GSH, but the authors noted accelerated turnover of this pool along with increased GR activities. Regarding GSH's role in herbicide-stressed plants, the modulation of GSH and its redox couple have been reported in different plant species under herbicide stress conditions (Gill and Tuteja, 2010). To this end, the increased GSH content and the GSH/GSSG ratio have been reported to protect plant species against varied herbicide-accrued oxidative stress (Romero-Puertas et al., 2006). Maintenance of the apoplastic ascorbate redox state has been considered crucial for ozone-induced oxidative stress tolerance in plants (Pignocchi et al., 2006). Involvement of the regenerated ascorbate in the apoplast has been reported to detoxify ozone (Luwe et al., 1993; Yoshida et al., 2006). In *Arabidopsis* mutant lacking cytDHAR activity, low apoplastic ascorbate level was considered responsible for its high sensitivity to ozone (Yoshida et al., 2006). L-ascorbate may also act as an alternative electron donor of PSII (Mano et al., 2004; Tóth et al., 2009; Gururani et al., 2012). Leaf AsA strongly influenced the heat-induced inactivation of PSII (Tóth et al., 2011).

5.7 Alternative Oxidase

One of the terminal oxidases of the plant mitochondrial electron transport chain, alternative oxidase (AOX) acts as a means to relax the highly coupled and tensed electron transport process in mitochondria. Via directly reducing oxygen to water, AOX provides and maintains the much-needed metabolic homeostasis (Saha et al., 2016). In fact, in chloroplasts, shuttle machineries, such as the malate-oxaloacetate (OAA) shuttle, are involved in the transport of excess reducing equivalents in the form of NADPH into the mitochondria. Later, the mitochondrial AOX respiratory pathway efficiently oxidizes the transported reducing equivalents. A gist of regulation and functioning of AOX in plants has been recently provided by Saha et al. (2016). Literature advocates the potential involvement of the AOX pathway in the protection of plants against photoinhibition. The mitochondrial AOX pathway has been reported to protect the photosynthetic apparatus against photo-damage by alleviating the overreduction of the PSI acceptor side and accelerating the induction of nonphotochemical quenching in Rumex K-1 leaves (Zhang et al., 2012). Recently, AOX has been evidenced to constitute a small family of proteins in *Citrus clementina* and *Citrus sinensis* L. Osb. (Araújo Castro et al., 2017). In plants such as *Arabidopsis thaliana*, AOX1A is involved in the sustenance of the chloroplastic redox state and energizing the optimization of photosynthesis by regulating cellular redox homeostasis and ROS generation when electron transport through the COX pathway is disturbed at complex III (Vishwakarma et al., 2015). Earlier, physiological roles of AOX1a were revealed in photosynthesis and maintenance of cellular redox homeostasis under high light in *Arabidopsis thaliana* (Vishwakarma et al., 2014). The role and modulation of AOX in major abiotic-stressed plants are discussed as follows.

5.7.1 Drought, Salinity, and Other Abiotic Stresses

AOX has been regarded as an essential respiratory electron transport chain pathway for maintaining photosynthetic performance during drought stress (Vanlerberghe et al., 2016). Few studies have examined the role of AOX in drought-exposed plants (reviewed by Vanlerberghe 2013). Drought conditions were reported to increase AOX protein, or capacity in wheat leaves (Bartoli et al., 2005; Vassileva et al., 2009); whereas drought stress decreased the leaf AOX transcript in *Medicago* (Filippou et al., 2011). In another study, the leaves of drought-exposed plants exhibited no increase in AOX protein (Ribas-Carbo et al., 2005). In soybean leaves, AOX was responsible for nearly 40% of total O_2 consumption (vs. approximately 10% in well-watered plants) (Ribas-Carbo et al., 2005). Galle et al. (2010) reported a drought-accrued increase in AOX activity in *Nicotiana sylvestris*. An improved photosynthetic performance during severe drought was reported in *Nicotiana tabacum* overexpressing a nonenergy conserving respiratory electron sink (Dahal et al., 2015). Recently, Dahal and Vanlerberghe (2017) advocated the role of AOX in the maintenance of both mitochondrial and chloroplast function during drought. It was argued that the ability of AOX to maintain critical mitochondrial and chloroplast functions during extreme drought is likely due, at least in part, to its ability to reduce oxidative damage. Salinity-mediated induction in the AOX transcript and protein was reported in a number of crops/plants, including *Arabidopsis* (Kreps et al., 2002; Seki et al., 2002), tobacco (Andronis and Roubelakis-Angelakis, 2010), and poplars (Ottow et al., 2005). Involvement of ethylene in salinity-mediated induction of AOX was observed in a callus culture of *Arabidopsis* (Wang et al., 2010). A maintained AOX respiration was noted in salinity-exposed (long-term, 14 d) peas (Marti et al., 2009).

The role of AOX has become evident in plants exposed to other major stresses, including low oxygen (Clifton et al., 2005), ozone (Tosti et al., 2006), nutrient limitation (Noguchi and Terashima, 2006), high/low temperature (Murakami and Toriyama, 2008), high light (Giraud et al., 2008; Vassileva et al., 2009; Zhang et al., 2012) and high CO_2 (Gandin et al., 2012). An increased AOX capacity was reported in a number of phosphate (P)-deficient plants, including bean roots (Juszczuk et al., 2001), leaves of bean and *Gliricidia sepium* (Gonzalez-Meler et al., 2001), tobacco suspension cells (Parsons et al., 1999), and *Arabidopsis* seedlings (Vijayraghavan and Soole, 2010). In another study, N deficiency caused an increase in AOX capacity in both the spinach leaf (Noguchi and Terashima, 2006) and tobacco cells (Sieger et al., 2005). In transgenic tobacco cells lacking AOX, the lack of AOX during N deficiency failed to increase the expression of ROS-responsive genes (Sieger et al., 2005). However, the authors noted the role of AOX in dampening the generation of ROS during P limitation. In the leaves of ozone-exposed tobacco, (Ederli et al., 2006) reported the loss of the cyt pathway that was accompanied by an increased AOX expression and capacity. A significant increase in AOX expression has also been reported in *Arabidopsis* (Tosti et al., 2006). Examined in leaves and roots of several species, hypoxia (low oxygen) can also bring changes (increase or decrease) in AOX amounts (transcript, protein or capacity) (Amor et al., 2000; Tsuji et al., 2000; Szal et al., 2003; Liu et al., 2005; Kreuzwieser et al., 2002; Skutnik and Rychter 2009; Vergara et al., 2012). Compared with cyt oxidase, AOX exhibits its lower affinity for O_2 (Millar et al., 1994; Ribas-Carbo et al., 1994). Earlier, AOX was observed to play a role during reoxygenation (Szal et al., 2003).

5.8 Respiratory Burst Oxidase Homologs

Respiratory burst oxidase homologues (RBOHS) are a family of proteins that are conserved throughout the plant kingdom. In plants, RBOHs consist of a cytosolic N-terminal extension comprising two Ca^{2+}-binding EF-hand motifs and phosphorylation target sites (Baxter et al., 2014). Owing to the availability of mutants, the focus of research into plant RBOHS has been on *Arabidopsis thaliana* (mainly on AtrbohD and AtrbohF). Notably, 10 respiratory burst oxidase homolog (RBOH) genes have been reported in *Arabidopsis*. As key signaling nodes in the ROS gene network of plants, RBOHs have been extensively reported to integrate a multitude of signal transduction pathways with ROS signaling (Suzuki et al., 2011).

5.8.1 Drought, Salinity, and Other Abiotic Stresses

Tomato SlRbohB, a member of the NADPH oxidase family, was reported as a requirement for tolerance to drought stress (Li et al., 2015). Sagi and Fluhr (2006) reported salt stress-mediated induction in RBOHA activity in the root elongation zone in *Arabidopsis*. Among the isoforms of RBOH, AtRBOHD and AtRBOHF isoforms were expressed in all organs of *Arabidopsis* (Ma et al., 2012). Additionally, the authors revealed the involvement of these isoforms in ROS-dependent regulation of Na^+/K^+ homeostasis under salt stress. AtRBOHD has been argued as a requirement for salt acclimation signaling mediated by heme oxygenase HY1 in *Arabidopsis* (Xie et al., 2011). A number of other researchers, including Leshem et al. (2007) and Hao et al. (2014), reported the role of RBOHD salt acclimation of *Arabidopsis*. Jiang et al. (2012a,b) implicated the role of AtRBOHF in the protection of shoot cells against transpiration-dependent accumulation of excess Na^+. RBOHD was also reported to mediate rapid systemic signaling in response to diverse stimuli including heat, cold, high light, and salinity (Miller et al., 2009).

In *Arabidopsis*, the interplay between Nox-generated ROS and oxylipin signaling under Cd (5 μM $CdSO_4$) or Cu (2 μM $CuSO_4$ for 24 h) stress has been reported (Remans et al., 2010). Rbohs-generated ROS can regulate plant responses to various abiotic stresses, including light/radiation and ozone exposure (Baxter et al., 2014). In *Arabidopsis*, AtRbohD- and AtRbohF-generated ROS were reported to be involved in the cold stress response (Kwak et al., 2003; Suzuki et al., 2000). Kawarazaki et al. (2013) reported AtSRC2 (*Arabidopsis* homolog of SRC2, a soyabean gene regulated by cold-2) as an innovative regulatory factor of Ca^{2+}-dependent activation of AtRbohF via binding to its N-terminal region under a recent study employing cold stress.

5.9 NADP-Dependent Malate Dehydrogenase (NADP-MDH) and the Malate Valve

C4-acids such as malate, aspartate, oxaloacetate,and so forth, play an important role in creation of the defense mechanisms in plant responses to extreme conditions. NADP-MDH catalyzes reversible conversion of oxaloacetate to malate using NADPH as a cofactor. In illuminated chloroplasts, the malate-oxaloacetate (OAA) shuttle is one mechanism involved in the reduction-oxidation (redox) homeostasis. Notably, the NADP-dependent malate dehydrogenase (MDH) utilizes the excess electrons from photosynthetic electron transport in the form of NADP in order to reduce OAA to malate, and thereby to regenerate the electron acceptor NADP (Beeler et al., 2014). Cytosol and organelles, including plastids, mitochondria, peroxisomes, and microbodies contain NAD-dependent MDHs; whereas NADP-dependent MDH occurs in chloroplasts (Scheibe, 2004). Beeler et al. (2014) identified eight putative NAD-MDH isoforms in *Arabidopsis* genomes, where two are mitochondrial MDH (mMDH), two are peroxisomal MDH, one is a plastidial MDH, and three have no detectable target sequences, and are thought to be cytosolic MDH (cyMDH). The nuclear-encoded chloroplast NADP-MDH has been regarded as a key enzyme controlling the malate valve that allows the indirect export of reducing equivalents (Scheibe 2004; Hebbelmann et al., 2012).

5.9.1 Drought, Salinity, and Other Abiotic Stresses

Babayev et al. (2015) studied the localization, and some physicochemical and kinetic properties of NADP-MDH isoforms in wheat leaves under drought stress. Chloroplast exhibited 70%–75% of the total NADP-MDH activity (molecular weights: 66, 74, and 86 kDa); whereas 25%–30% of the total NADP-MDH activity (molecular weights of 42, 66, and 74 kDa) was found to be localized in the cytosol fraction of mesophyll cells of wheat flag leaves. In addition, drought-tolerant genotypes also exhibited a more rapid increase in K_m and V_{max}, contrary to the chloroplast fraction of the drought-sensitive genotype manifesting a slight decrease in these parameters. Under high light conditions, in *Arabidopsis* plants lacking the malate valve enzyme NADP-MDH, adoption of multiple strategies occur to prevent oxidative stress (Hebbelmann et al., 2012). The authors revealed a unique metabolic flexibility and advocated that the malate valve acts in concert with other NADPH-consuming reactions in order to maintain a balanced redox state during photosynthesis under high light stress in wild-type plants. Wang et al. (2016) reported that transgenic apple plants tolerate salt and cold stresses by modifying the redox state and salicylic acid content via the cytosolic malate dehydrogenase gene. An apple cytosolic malate dehydrogenase gene (MdcyMDH) was argued to confer to the transgenic apple plants a higher stress tolerance as a result of the production of more reductive redox states, and increase in the SA level. The role of NADP-MDH and malate in the redox regulation has been recently reviewed in salinity-exposed plants, and the malate valve-dependent redox balance constitutes ware argued as an important mechanism in plant-salt acclimation (Hossain and Dietz, 2016). Feeding of NADH into respiratory electron transport was reported to play an important role in the short-term adjustment of the NADP(H) redox state under salinity stress (Scheibe et al., 2005). Earlier, a more than twofold increase of chloroplast NADP-MDH transcript levels was reported in the leaves of salinity-exposed *Mesembryanthemum crystallinum* (Cushman, 1993). Salt-sensitive and salt-tolerant cultivars were reported to differentially exhibit activities of NAD-MDH activity. To this end, salt tolerant rice cv CSR-1 and CSR-3 were reported to exhibit increased activities of NAD-MDH (specifically, mitochondrial NAD-MDH and chloroplast NADP-MDH) in whole tissue extract (Kumar et al., 2000). In contrast, the authors observed an inhibition in the activity of NAD-MDH in salt-sensitive cultivars.

5.10 Plastid Terminal Oxidase

Plastid terminal oxidase (PTOX) is a plastoquinone oxidase, and its role was first identified in leaf development (Carol et al., 1999; Wu et al., 1999). The PTOX protein form H2O by transferring electrons from plastoquinone to molecular oxygen (Carol and Kuntz, 2001). PTOX is involved in chloroplast development and also in carotenoid synthesis (Carol and Kuntz, 2001). High amounts of PTOX protein have been reported in all alpine plant species investigated, and its content was found to increase with the altitude (Streb et al., 2005; Laureau et al., 2011). There exists a relationship between the capacity of acceptance of excess photosynthetically generated electrons and the altitude from which plants are collected. To this end, lowland plants generally exhibit low amounts of PTOX (Lennon et al., 2003; Streb et al., 2005), and low capacity to accept excess photosynthetically generated electrons (Ort and Baker, 2002; Trouillard et al., 2012). PTOX can act as a safety valve for excess excitation energy in the alpine plant species such as *Ranunculus glacialis* (Laureau et al., 2013). The involvement of PTOX plant responses and tolerance to environmental stresses has been extensively reported (Stepien and Johnson, 2009).

5.10.1 Major Abiotic Stresses

In oat plants exposed to high light and heat, stimulation in the activity of PTOX was found to be associated with its increased amount (Quiles, 2006). Salt-exposed rice exhibited an increased PTOX level (Kong et al., 2003). A similar result was reported in *Lodgepole* pine when acclimated to winter cold (Savitch et al., 2010). A number of researchers reported upregulation of PTOX expression in algae, such as *H. pluvialis* under high light and *Chlamydomonas reinhardtii* under phosphate deprivation (Moseley et al., 2006; Wang et al., 2009a,b). Upregulated levels of PTOX were detected in various higher plants, including heat- and high light-exposed *Avena sativa* (Quiles, 2006), *Brassica fruticulosa*, *B. oleracea*, drought-exposed *Coffea arabica* (Simkin et al., 2008), salinity-exposed *Oryza sativa* (Kong et al., 2003), cold and short daylight-exposed *Pinus banksiana* (Busch et al., 2008), cold-exposed *Pinus contorta* (Savitch et al., 2010), cold and high light-exposed *Ranunculus glacialis* (Streb et al., 2005), high light-exposed *Solanum lycopersicum* (Shahbazi et al., 2007), drought, heat, and high light-exposed *Spathiphyllum wallisii* (Gamboa et al., 2009; Ibáñez et al., 2010), and in salinity-exposed *Thellungiella halophile* (Stepien and Johnson, 2009).

6. OXIDATIVE STRESS TOLERANCE IN PLANTS BY DEVELOPING TRANSGENIC LINES

Previous reports suggest that using transgenic plants expressing foreign genes for antioxidant enzymes and a variety of genes are used in several plant species (McKersie et al., 2000; Bhatnagar-Mathur et al. 2008) (Table 3). Wang et al. (2009a,b) reported enhanced salt tolerance in transgenic lines using TaMnSOD genes in *Populus davidiana*. Shafi et al., 2015; Sukweenadhi et al., 2017; Yang et al., 2014; and Singh et al., 2014 reported enhanced salt tolerance in transgenic *Arabidopsis thaliana*. Moreover, Tseng et al., 2007 reported enhanced sulfur dioxide and salt stress tolerance in transgenic *Brassica campestris* L. Eltayeb et al. (2011) reported the enhanced herbicide, drought, and salt tolerance in *Solanum tuberosum*. Yin et al. (2009) reported enhanced Al stress tolerance in transgenic *Nicotiana tabacum*. Yu et al. (2003) reported resistance to oxidative stress in transgenic lines of *Nicotiana tabacum*. Melchiorre et al. (2009) reported a higher GSH content and GSH/GSSG ratio than control in transgenic *Triticum aestivum*.

7. CONCLUSION AND FUTURE PERSPECTIVES

Significant impairments in cellular redox homeostasis can be caused by abiotic stress-provoked elevations in ROS. This chapter attempted to overview ROS and their chemistry and sites of production; highlighted major dual roles of ROS; and also discussed the role and modulation of major enzymatic and nonenzymatic components of antioxidant defense systems involved in cellular redox regulation in abiotic-stressed plants. In fact, plants subjected to abiotic stresses generate ROS within various intracellular compartments as one of the earliest responses. The most important sites to initiate the response via ROS include mitochondria, chloroplast, cell walls, peroxisome, apoplast endoplasmic reticulum, and plasma membranes. ROS production can be a part of a signaling network that plants use for their development, and in response to environmental challenge. However, elevated or nonmetabolized ROS either leads to metabolic disorders, oxidative damage, or plays an important role in signaling. Notably, what will be the effects of ROS, whether its production is beneficiary or harmful, depends on their production site,

Table 3 Developing oxidative stress tolerant transgenic plants for oxidative stress tolerance

Gene	Source of gene	Target transgenic plant	Response in transgenic plants	Reference
TaMnSOD	*Tamarix androssowii*	*Populus davidiana*	Enhance salt tolerance in transgenic lines	Wang et al. (2009a,b)
PaSOD	*Potentilla atrosanguinea*	*Arabidopsis thaliana*	Salt tolerance by accumulation of compatible solutes	Shafi et al. (2015)
MnSOD/CAT	*Escherichia coli*	*Brassica campestris* L.	Enhanced sulfur dioxide and salt stress tolerance	Tseng et al. (2007)
BoCAT	*Brassica oleracea*	*Arabidopsis thaliana*	Increase the heat tolerance	Chiang et al. (2014)
PgAPX	*Panax ginseng*	*Arabidopsis thaliana*	Enhance salt tolerance	Sukweenadhi et al. (2017)
SbpAPX	*Salicornia brachiata*	*Nicotiana tabacum*	Tolerance to salt and drought stress	Singh et al. (2014)
RaAPX	*Rheum australe*	*Arabidopsis thaliana*	Salt tolerance	Shafi et al. (2015)
AtDHAR	*Arabidopsis thaliana*	*Solanum tuberosum*	Herbicide, drought, and salt tolerance	Eltayeb et al. (2011)
AtDHAR	*Arabidopsis thaliana*	*Nicotiana tabacum*	Al stress tolerance	Yin et al. (2009)
ThGSTZ1	*Tamarix hispida*	*Arabidopsis thaliana*	Improve drought and salinity tolerance by enhancing ROS scavenging ability	Yang et al. (2014)
GST-cr 1	*Gossypium hirsutum* L.	*Nicotiana tabacum*	Resistance to oxidative stress	Yu et al. (2003)
GR	*Escherichia coli*	*Triticum aestivum*	Higher GSH content and GSH/GSSG ratio than control	Melchiorre et al. (2009)
RsrSOD	*Raphanus sativus*	*Brassica oleracea*	Resistance to downy mildew	Jiang et al. (2012b)

how much it produced, and the activity of the antioxidant defense system of the plant itself. To prevent oxidative stress, plants harbor a large number of enzymatic (such as APX, CAT, GR, GST, GPX, SOD) and nonenzymatic (such as AsA and GSH) antioxidants, which scavenge surplus oxidants protecting plant biomolecules from damage. On the other hand, one of the terminal oxidases of the plant mitochondrial electron transport chain, AOX, acts as a means to relax the highly coupled and tensed electron transport process in mitochondria. Via directly reducing oxygen to water, AOX provides and maintains the much needed metabolic homeostasis. Literature advocates the potential involvement of the AOX pathway in the protection of plants against photoinhibition. In addition, AOX has been regarded as an essential respiratory electron transport chain pathway for maintaining photosynthetic performance during different stresses, including drought, low oxygen, ozone, nutrient limitation, high/low temperature, high light, and high CO_2. RBOHS, a family of proteins that is conserved throughout the plant kingdom, consists of a cytosolic N-terminal extension comprising two Ca^{2+}-binding EF-hand motifs and phosphorylation target sites. As a key signaling node in the ROS gene network of plants, RBOHs have been extensively reported to integrate a multitude of signal transduction pathways with ROS signaling. NADP-MDH catalyzes reversible conversion of oxaloacetate to malate using NADPH as a cofactor. The nuclear-encoded chloroplast NADP-MDH has been regarded as a key enzyme controlling the malate valve that allows the indirect export of reducing equivalents. PTOX is involved in the early steps of chloroplast biogenesis, serves as an alternate electron sink, and is a prime determinant of the redox poise of the developing photosynthetic apparatus. PTOX can act as a safety valve for excess excitation energy in the alpine plant species. The involvement of PTOX plant responses and tolerance to environmental stresses has also been extensively reported.

A better understanding of the multilevel mechanisms underlying plant abiotic stress responses is still required in order to mainly target the redox regulatory mechanisms and achieve abiotic stress tolerance in crops/plants. The mechanism of cellular compartment-specific regulation, expression, and interaction/coordination among major enzymatic and nonenzymatic antioxidants by different stresses and developmental stages is to be further understood. Little is also known about the regulation of SOD types by posttranslational modifications. SlRbohB has been reported to positively regulate the resistance to *Botrytis cinerea* and drought stress tolerance in tomatoes. However, further studies on decoding downstream signaling can elucidate the molecular and physiological mechanisms underlying SlRbohB-mediated regulation of (biotic)/abiotic stress responses in plants. Notably, the functions of PTOX in stressed plants are still considered disputed, mainly due to the very complex mechanisms underlying plant stress responses. Studies on PTOX expression and activity, as well as PTOX-related physiological mechanisms, are to be further explored in order to clarify the exact role and regulation of PTOX in plant stress responses.

ACKNOWLEDGMENTS

SSG and RG acknowledge the partial support of University Grants Commision (UGC), Department of Science & Technology (DST), and Council of Scientific & Industrial Research (CSIR), Govt. of India, New Delhi. NT acknowledges partial support from the Department of Biotechnology (DBT) and Department of Science (DST) and Technology, Govt. of India, New Delhi. PK acknowledges the award of Senior Research Fellowship (SRF) [09/382(0174)/2015-EMR-I] from CSIR, Govt. of India, New Delhi. NAA is grateful to the DBT for funding his research in the form of Ramalingaswami Re-Entry Fellowship (BT/HRD/35/02/2006; BT/RLF/Re-Entry/18/2015).

REFERENCES

Agami, R.A., Mohamed, G.F., 2013. Exogenous treatment with indole-3-acetic acid and salicylic acid alleviates cadmium toxicity in wheat seedlings. Ecotoxicol. Environ. Saf. 94, 164–171.

Agarwal, S., 2007. Increased antioxidant activity in Cassia seedlings under UV-B radiation. Biol. Plant. 51, 157–160.

Agrawal, S.B., Mishra, S., 2009. Effects of supplemental ultraviolet-B and cadmium on growth, antioxidants and yield of *Pisum sativum* L. Ecotoxicol. Environ. Saf. 72, 610–618.

Aguirre, J., Ríos-Momberg, M., Hewitt, D., Hansberg, W., 2005. Reactive oxygen species and development in microbial eukaryotes. Trends Microbiol. 13, 111–118.

Ahmad, P., Jaleel, C.A., Salem, M.A., Nabi, G., Sharma, S., 2010. Roles of enzymatic and nonenzymatic antioxidants in plants during abiotic stress. Crit. Rev. Biotechnol. 30 (3), 161–175.

Airaki, M., Leterrier, M., Mateos, R.M., 2012. Metabolism of reactive oxygen species and reactive nitrogen species in pepper (*Capsicum annuum* L.) plants under low temperature stress. Plant Cell Environ. 35, 281–295.

Ajithkumar, I.P., Panneerselvam, R., 2014. ROS scavenging system, osmotic maintenance, pigment and growth status of *Panicum sumatrense roth* under drought stress. Cell Biochem. Biophys. 68, 587–595.

Aksoy, M., Dİnle, B.S., 2012. Changes in physiological parameters and some antioxidant enzymes activities of soybean (*Glycine max* L. Merr.) leaves under cadmium and salt stress. J. Stress Physiol. Biochem. 8, 179–190.

Alam, M.M., Hasanuzzaman, M., Nahar, K., Fujita, M., 2013. Exogenous salicylic acid ameliorates short-term drought stress in mustard (*Brassica juncea* L.) seedlings by up-regulating the antioxidant defense and glyoxalase system. Aust. J. Crop Sci. 7, 1053–1063.

Almeselmani, M., Deshmukh, P.S., Sairam, R.K., Kushwaha, S.R., Singh, T.P., 2006. Protective role of antioxidant enzymes under high temperature stress. Plant Sci. 171, 382–388.

Almeselmani, M., Deshmukh, P.S., Sairam, R.K., 2009. High temperature stress tolerance in wheat genotypes: role of antioxidant defence enzymes. Acta Agron. Hungarica. 57, 1–14. https://doi.org/10.1556/AAgr.57.2009.1.1.

Amor, Y., Chevion, M., Levine, A., 2000. Anoxia pretreatment protects soybean cells against H_2O_2-induced cell death: possible involvement of peroxidases and of alternative oxidase. FEBS Lett. 477, 175–180.

Anderson, S., Appanna, V.D., Huang, J., Viswanatha, T., 1992. A novel role for calcite in calcium homeostasis. FEBS Lett. 308, 94–96.

Andreyev, A.Y., Kushnareva, Y.E., Starkov, A.A., 2005. Mitochondrial metabolism of reactive oxygen species. Biochemistry (Moscow) 70 (2), 200–214.

Andronis, E.A., Roubelakis-Angelakis, K.A., 2010. Short-term salinity stress in tobacco leads to the onset of animal-like PCD hallmarks in planta in contrast to long-term stress. Planta 231, 437–448.

Anjum, N.A., Umar, S., Ahmad, A., Iqbal, M., 2008a. Responses of components of antioxidant system in moongbean [*Vigna radiata* (L.) Wilczek] genotypes to cadmium stress. Commun. Soil Sci. Plant Anal. 39, 2469–2483.

Anjum, N.A., Umar, S., Ahmad, A., Iqbal, M., Khan, N.A., 2008b. Ontogenic variation in response of *Brassica campestris* L. to cadmium toxicity. J. Plant Interact. 3, 189–198.

Anjum, N.A., Umar, S.A., Ahmad, A., Iqbal, M., Khan, N.A., 2008c. Sulphur protects mustard (*Brassica campestris* L.) from cadmium toxicity by improving leaf ascorbate and glutathione. Plant Growth Regul. 54, 271–279.

Anjum, N.A., Umar, S., Iqbal, M., Khan, N.A., 2011. Cadmium causes oxidative stress in mung mean by affecting the antioxidant enzyme system and ascorbate-glutathione cycle metabolism. Russ. J. Plant Physiol. 58, 92–99.

Anjum, N.A., Umar, S., Ahmad, A., 2012. Oxidative Stress in Plants: Causes, Consequences and Tolerance. IK International Publishing House, New Delhi.

Anjum, N.A., Khan, N.A., Sofo, A., Baier, M., Kizek, R., 2016a. Editorial: redox homeostasis managers in plants under environmental stresses. Front. Environ. Sci. 4, 35. https://doi.org/10.3389/fenvs.2016.00035.

Anjum, N.A., Sharma, P., Gill, S.S., Hasanuzzaman, M., Khan, E.A., Kachhap, K., Mohamed, A.A., Thangavel, P., Devi, G.D., Vasudhevan, P., Sofo, A., Khan, N.A., Misra, A.N., Lukatkin, A.S., Singh, H.P., Pereira, E., Tuteja, N., 2016b. Catalase and ascorbate peroxidase-representative H_2O_2-detoxifying heme enzymes in plants. Environ. Sci. Pollut. Res. (19), 9002–19029.

Apel, K., Hirt, H., 2004. Reactive oxygen species: metabolism, oxidative stress, and signal transduction. Plant Biol. 55, 373–399.

Araújo Castro, J., Gomes Ferreira, M.D., Santana Silva, R.J., Andrade, B.S., Micheli, F., 2017. Alternative oxidase (AOX) constitutes a small family of proteins in *Citrus clementina* and *Citrus sinensis* L. Osb. PLoS One 12 (5). e0176878. https://doi.org/10.1371/journal.pone.0176878 (eCollection 2017).

Aravind, P., Prasad, M.N.V., 2005. Cadmium and zinc interactions in a hydroponic system using *Ceratophylum demersum* L.: adaptive ecophysiology, biochemistry and molecular toxicology. Braz. J. Plant Physiol. 17, 3–20.

Arbona, V., Hossain, Z., López-Climent, M.F., Pérez-Clemente, R.M., Gómez-Cadenas, A., 2008. Antioxidant enzymatic activity is linked to waterlogging stress tolerance in citrus. Physiol. Plant. 132, 452–466.

Arora, A., Sairam, R.K., Srivastava, G.C., 2002. Oxidative stress and antioxidative system in plants. Curr. Sci. 82 (10), 1227–1238.

Asada, K., 1999. The water-water cycle in chloroplasts: scavenging of active oxygen's and dissipation of excess photons. Annu. Rev. Plant Biol. 50, 601–639.

Asada, K., 2006. Production and scavenging of reactive oxygen species in chloroplasts and their functions. Plant Physiol. 141, 391–396. https://doi.org/10.1104/pp.106.082040.

Atkinson, N.J., Urwin, P.E., 2012. The interaction of plant biotic and abiotic stresses: from genes to the field. J. Exp. Bot. 63 (10), 3523–3543.

Azpilicueta, C.E., Benavides, M.P., Tomaro, M.L., Gallego, S.M., 2007. Mechanism of CATA3 induction by cadmium in sunflower leaves. Plant Physiol. Biochem. 45, 589–595.

Babayev, H.G., Mehvaliyeva, U.A., Aliyeva, M.N., 2015. NADP-malate dehydrogenase isoforms of wheat leaves under drought: their localization, and some physicochemical and kinetic properties. J. Stress Physiol. Biochem. 11 (3), 13–25.

Bai, J., Gong, C.M., Chen, K., Kang, H.M., Wang, G., 2009. Examination of antioxidative systems responses in the different phases of drought stress and during recovery in desert plant *Reaumuria soongorica* (Pall.) Maxim. J. Plant Biol. 52, 417–425.

Balestrasse, K.B., Gardey, L., Gallego, S.M., Tomaro, M.L., 2001. Response of antioxidant defence system in soybean nodules and roots subjected to cadmium stress. Aust. J. Plant Physiol. 28, 497–504.

Barroso, C., Romero, L.C., Cejudo, F.J., Vega, J.M., Gotor, C., 1999. Salt-specific regulation of the cytosolic *O*-acetylserine(thiol) lyase gene from *Arabidopsis thaliana* is dependent on abscisic acid. Plant Mol. Biol. 40, 729–736.

Bartoli, C.G., Gomez, F., Gergoff, G., Guiamét, J.J., Puntarulo, S., 2005. Up-regulation of the mitochondrial alternative oxidase pathway enhances photosynthetic electron transport under drought conditions. J. Exp. Bot. 56, 1269–1276.

Basu, S., Roychoudhury, A., Saha, P.P., Sengupta, D.N., 2010. Comparative analysis of some biochemical responses of three indica rice varieties during polyethylene glycol-mediated water stress exhibits distinct varietal differences. Acta Physiol. Plant. 32, 551–563. https://doi.org/10.1007/s11738-009-0432-y.

Baxter, A., Mittler, R., Suzuki, N., 2014. ROS as key players in plant stress signalling. J. Exp. Bot. 65, 1229–1240.
Beeler, S., Liu, H.C., Stadler, M., Schreier, T., Eicke, S., Lue, W.L., Truernit, E., 2014. Plastidial NAD-dependent malate dehydrogenase is critical for embryo development and heterotrophic metabolism in Arabidopsis. Plant Physiol. 164, 1175–1190.
Bhatnagar-Mathur, P., Vadez, V., Sharma, K.K., 2008. Transgenic approaches for abiotic stress tolerance in plants: retrospect and prospects. Plant Cell Rep. 27, 411–424.
Biehler, K., Fock, H., 1996. Evidence for the contribution of the Mehler-peroxidase reaction in dissipating excess electrons in drought-stressed wheat. Plant Physiol. 112 (1), 265–272.
Bin, T., Shang-zhong, X.U., Zou, X.L., Zheng, Y.L., Qi, F.Z., 2010. Changes of antioxidative enzymes and lipid peroxidation in leaves and roots of waterlogging-tolerant and waterlogging-sensitive maize genotypes at seedling stage. Agric. Sci. China 9, 651–661.
Bindschedler, L.V., Dewdney, J., Blee, K.A., Stone, J.M., Asai, T., Plotnikov, C., Denoux, J., 2006. Peroxidase-dependent apoplastic oxidative burst in Arabidopsis required for pathogen resistance. Plant J. 47 (6), 851–863. https://doi.org/10.1111/j.1365-313X.2006.02837.x.
Blumwald, E., Aharon, G.S., Maris, P.A., 2000. Sodium transport in plant cells. Biochim. Biophys. Acta 1465, 140–151.
Bonifacio, A., Martins, M.O., Ribeiro, C.W., Fontenele, A.V., Carvalho, F.E., Margis-Pinheiro, M., Silveira, J.A., 2011. Role of peroxidases in the compensation of cytosolic ascorbate peroxidase knockdown in rice plants under abiotic stress. Plant Cell Environ. 34, 1705–1722.
Bosch, S.M., Queval, G., Foyer, C.H., 2013. The impact of global change factors on redox signaling underpinning stress tolerance. Plant Physiol. 16, 5–19.
Busch, F., Hüner, N.P.A., Ensminger, I., 2008. Increased air temperature during simulated autumn conditions impairs photosynthetic electron transport between photosystem II and photosystem I. Plant Physiol. 147, 402–414.
Caregnato, F.F., Koller, C.E., MacFarlane, G.R., Moreira, J.C., 2008. The glutathione antioxidant system as a biomarker suite for the assessment of heavy metal exposure and effect in the grey mangrove. *Avicennia marina* (Forsk.) Vierh. Mar. Pollut. Bull. 56 (6), 1119–1127.
Carol, P., Kuntz, M., 2001. A plastid terminal oxidase comes to light: implications for carotenoid biosynthesis and chlororespiration. Trends Plant Sci. 6, 31–36.
Carol, P., Stevenson, D., Bisanz, C., Breitenbach, J., Sandmann, G., Mache, R., 1999. Mutations in the Arabidopsis gene immutans cause a variegated phenotype by inactivating a chloroplast terminal oxidase associated with phytoene desaturation. Plant Cell 11, 57–68.
Carvalho, M.H.C., 2008. Drought stress and reactive oxygen species: production, scavenging and signaling. Plant Signal. Behav. 3 (3), 156.
Caverzan, A., Passaia, G., Rosa, S.B., Ribeiro, C.W., Lazzarotto, F., Pinheiro, M., 2012. Plant responses to stresses: role of ascorbate peroxidase in the antioxidant protection. Genet. Mol. Biol. 35, 1011–1019. https://doi.org/10.1590/S1415-47572012000600016.
Chen, J., Goldsbrough, P.B., 1994. Increased activity of [gamma]-glutamylcysteine synthetase in tomato cells selected for cadmium tolerance. Plant Physiol. 106 (1), 233–239.
Chen, Z., Young, T.E., Ling, J., Chang, S.C., Gallie, D.R., 2003. Increasing vitamin C content of plants through enhanced ascorbate recycling. Proc. Natl. Acad. Sci. U. S. A. 100, 3525–3530.
Chen, F., Wang, F., Wu, F., Mao, W., Zhang, G., Zhou, M., 2010. Modulation of exogenous glutathione in antioxidant defense system against Cd stress in the two barley genotypes differing in Cd tolerance. Plant Physiol. Biochem. 48, 663–672.
Chiang, C.M., Chen, S.P., Chen, L.F.O., Chiang, M.C., Chien, H., Lin, K.H., 2014. Expression of the broccoli catalase gene (BoCAT) enhances heat tolerance in transgenic Arabidopsis. Plant Biochem. Biotechnol. 23 (3), 266–277.
Cho, N., Seo, N., 2005. Oxidative stress in Arabidopsis thaliana exposed to cadmium is due to hydrogen peroxide accumulation. Plant Sci. 168, 113–120.
Choudhury, S., Panda, P., Sahoo, L., Panda, K., 2013. Reactive oxygen species signaling in plants under abiotic stress. Plant Signal. Behav. 8 (4), e2368.
Circu, M.L., Aw, T.Y., 2010. Reactive oxygen species, cellular redox systems, and apoptosis. Free Radic. Biol. Med. 48, 749–762.

Clifton, R., Lister, R., Parker, K.L., Sappl, P.G., Elhafez, D., Millar, A.H., Day, D.A., Whelan, J., 2005. Stress-induced co-expression of alternative respiratory chain components in *Arabidopsis thaliana*. Plant Mol. Biol. 58, 193–212.

Comba, M.E., Benavides, M.P., Tomaro, M.L., 1998. Effect of salt stress on antioxidant defence system in soybean root nodules. Aust. J. Plant Physiol. 25, 665–671.

Cong, M., Lv, J., Liu, X., Zhao, J., Wu, H., 2013. Gene expression responses in Suaeda salsa after cadmium exposure. SpringerPlus 202 (1), 232. https://doi.org/10.1186/2193-1801-2-232.

Corpas, F.J., Barroso, J.B., 2013. Nitro-oxidative stress vs oxidative or nitrosative stress in higher plants. New Phytol. 199, 633–635.

Corpas, F.J., Barroso, J.B., del Río, L.A., 2001. Peroxisomes as a source of reactive oxygen species and nitric oxide signal molecules in plant cells. Trends Plant Sci. 6 (4), 145–150.

Cushman, J.C., 1993. Molecular cloning and expression of chloroplast NADP-malate dehydrogenase during Crassulacean acid metabolism induction by salt stress. Photosynth. Res. 35, 15–27. https://doi.org/10.1007/BF02185408.

Dahal, K., Vanlerberghe, G.C., 2017. Alternative oxidase respiration maintains both mitochondrial and chloroplast function during drought. New Phytol. 213 (2), 560–571. https://doi.org/10.1111/nph.14169.

Dahal, K., Martyn, G.D., Vanlerberghe, G.C., 2015. Improved photosynthetic performance during severe drought in *Nicotiana tabacum* overexpressing a nonenergy conserving respiratory electron sink. New Phytol. (2), 382–395. https://doi.org/10.1111/nph.13479.

Damanik, R.I., Maziah, M., Ismail, M.R., Ahmad, S., Zain, A., 2010. Responses of the antioxidative enzymes in Malaysian rice (*Oryza sativa* L.) cultivars under submergence condition. Acta Physiol. Plant. 32 (4), 739–747.

Das, K., Roychoudhury, A., 2014. Reactive oxygen species (ROS) and response of antioxidants as ROS-scavengers during environmental stress in plants. Front. Environ. Sci. 53(2). https://doi.org/10.3389/fenvs.2014.00053.

Das, P., Mishra, M., Lakra, N., Singla-Pareek, S.L., Pareek, A., 2014. Mutation breeding: a powerful approach for obtaining abiotic stress tolerant crops and upgrading food security for human nutrition. In: Tomlekova, N., Kojgar, I., Wani, R (Eds.), Mutagenesis: Exploring Novel Genes and Pathways. Wageningen Academic Publisher, Wageningen, pp. 15–36.

Das, I., Karzyzosiak, A., Schneider, K., Wrabetz, l., D"Antinio, M., Barry, N., Singurdottiir, A., Bertolotti, A., 2015. Preventing proteostasis diseases by selective inhibition of a phosphatase regulatory subunit. Science 348 (6231), 239–242.

Dash, S., Mohanty, N., 2002. Response of seedlings to heat-stress in cultivars of wheat: growth temperature dependent differential modulation of photosystem 1 and 2 activity, and foliar antioxidant defense capacity. J. Plant Physiol. 159, 49–59.

Dat, J., Vandenabeele, S., Vranová, E., Van Montagu, M., Inzé, D., Breusegem, F.V., 2000. Dual action of the active oxygen species during plant stress responses. Cell. Mol. Life Sci. 57, 779–795.

Dat, J.F., Pellinen, R., Beeckman, T., Van De Cotte, B., Langebartels, C., Kangasjärvi, J., Inzé, D., Van Breusegem, F., 2003. Changes in hydrogen peroxide homeostasis trigger an active cell death process in tobacco. Plant J. 33, 621–632.

Demarquoy, J., Le Borgne, F., 2015. Crosstalk between mitochondria and peroxisomes. World J. Biol. Chem. 6 (4), 301–309.

Demidchik, V., 2014. Mechanisms of oxidative stress in plants: from classical chemistry to cell biology. Environ. Exp. Bot. 109, 212–228.

Desingh, R., Kanagaraj, G., 2007. Influence of salinity stress on photosynthesis and antioxidative systems in two cotton varieties. Plant Physiol. 33 (3–4), 221–234.

Diaz-Vivancos, P., Faize, M., Barba-Espin, G., Faize, L., Petri, C., 2013. Ectopic expression of cytosolic superoxide dismutase and ascorbate peroxidase leads to salt stress tolerance in transgenic plums. Plant Biotechnol. J. 11, 976–985.

Dionisio-Sese, M.L., Tobita, S., 1998. Antioxidant responses of rice seedlings to salinity stress. Plant Sci. 135, 1–9.

Dipierro, N., Mondelli, D., Paciolla, C., Brunetti, G., Dipierro, S., 2005. Changes in the ascorbate system in the response of pumpkin (*Cucurbita pepo* L.) roots to aluminium stress. J. Plant Physiol. 162, 529–536.

Diwan, H., Khan, I., Ahmad, A., Iqbal, M., 2010. Induction of phytochelatins and antioxidant defence system in *Brassica juncea* and *Vigna radiata* in response to chromium treatments. Plant Growth Regul. 61, 97–107.

Doulis, A.G., Hausladen, A., Mondy, B., Alscher, R.G., Chevone, B.I., Hess, J.L., Weiser, R.L., 1993. Antioxidant response and winter hardiness in red spruce (*Picea rubens* Sarg.). New Phytol. 123, 65–74.

Drewniak, L., Sklodowska, A., 2013. Arsenic-transforming microbes and their role in biomining processes. Environ. Sci. Pollut. Res. 20, 7728–7739.

Ederli, L., Morettini, R., Borgogni, A., Wasternack, C., Miersch, O., Reale, L., Ferranti, F., Tosti, N., Pasqualini, S., 2006. Interaction between nitric oxide and ethylene in the induction of alternative oxidase in ozone-treated tobacco plants. Plant Physiol. 142, 595–608.

Ekmecki, Y., Tanyolac, D., Ayhana, B., 2008. Effect of cadmium on antioxidant enzyme and photosynthetic activities of leaves of two maize cultivars. J. Plant Physiol. 165, 600–611.

El-Bastawisy, Z.M., 2010. Variation in antioxidants among three wheat cultivars varying in tolerance to NaCl. Gen. Appl. Plant Physiol. 36 (3-4), 189–203.

El-Beltagi, H.S., Mohamed, A.A., Ashed, M.M.R., 2010. Response of antioxidative enzymes to cadmium stress in leaves and roots of radish (*Raphanus sativus* L.). Not. Sci. Biol. 2, 76–82.

Eltayeb, A.E., Yamamoto, S., Habora, M.E., Yin, L., Wangm, S., 2011. Transgenic potato overexpressing Arabidopsis cytosolic AtDHAR1showed higher tolerance to herbicide, drought and salt stresses. Breed. Sci. 61, 3–10.

Esim, N., Atici, O., 2014. Nitric oxide improves chilling tolerance of maize by affecting apoplastic antioxidative enzymes in leaves. Plant Growth Regul. 72, 29–38.

Eyidogan, F., Oktem, H.A., Yucel, M., 2003. Superoxide dismutase activity in salt stressed wheat seedlings. Acta Physiol. Plant. 25, 263–269.

Fidalgo, F., Azenha, M., Silva, A.F., Sousa, A.D., Santiago, A., Ferraz, P., Teixeira, J., 2013. Copper-induced stress in *Solanum nigrum* L. and antioxidant defense system responses. Food Energy Secur. 2, 70–80.

Filippou, P., Antoniou, C., Fotopoulos, V., 2011. Effect of drought and rewatering on the cellular status and antioxidant response of *Medicago trunculata* plants. Plant Signal. Behav. 6, 270–277.

Fogelman, E., Kaplan, A., Tanami, Z., Ginz-berg, I., 2011. Antioxidative activity associated with chilling injury tolerance of muskmelon (*Cucumis melo* L.) rind. Sci. Hortic. 128, 267–273.

Ford, K.L., Cassin, A., Bacic, A., 2011. Quantitative proteomic analysis of wheat cultivars with differing drought stress tolerance. Front. Plant Sci. 2, 44.

Fornazier, R.F., Ferreira, R.R., Vitoria, A.P., Molina, S.M.G., 2002. Effects of cadmium on antioxidant enzyme activitiesin sugar cane. Biol. Plant. 45, 91–97.

Foyer, C.H., Harbinson, J., 1994. Oxygen metabolism and the regulation of photosynthetic electron transport. In: Foyer, C.H., Mullineaux, P. (Eds.), Causes of Photooxidative Stresses and Amelioration of Defense Systems in Plants. CRC Press, Boca Raton, FL, pp. 1–42.

Foyer, C.H., Noctor, G., 2000. Oxygen processing in photosynthesis: regulation and signaling. New Phytol. 146 (3), 359–388.

Foyer, C.H., Shigeoka, S., 2011. Understanding oxidative stress and antioxidant functions to enhance photosynthesis. Plant Physiol. 155, 93–100.

Fryer, M.J., Andrews, J.R., Oxborough, K., Blowers, D.A., Baker, N.R., 1998. Relationship between CO_2 assimilation, photosynthetic electron transport, and active O_2 metabolism in leaves of maize in the field during periods of low temperature. Plant Physiol. 116 (2), 571–580.

Gajewska, E., Sklodowska, M., 2008. Differential biochemical responses of wheat shoots and roots to nickel stress: antioxidative reactions and proline accumulation. Plant Growth Regul. 54, 179–188.

Galle, A., Florez-Sarasa, I., Thameur, A., Paepe, R.D., Flexas, J., Ribas-Carbo, M., 2010. Effects of drought stress and subsequent rewatering on photosynthesis and respiratory pathways in *Nicotiana sylvestris* wild type and the mitochondrial complex I-deficient CMSII. J. Exp. Bot. 61, 765–775.

Gamboa, J., Muñoz, R., Quiles, M.J., 2009. Effects of antimycin A and *n*-propyl gallate on photosynthesis in sun and shade plants. Plant Sci. 177, 643–647.

Gandin, A., Duffes, C., Day, D.A., Cousins, A.B., 2012. The absence of alternative oxidase AOX1A results in altered response of photosynthetic carbon assimilation to increasing CO_2 in *Arabidopsis thaliana*. Plant Cell Physiol. 53, 1627–1637.

García-Plazaola, J.I., Artetxe, U., Dunabeitia, M.K., Becerril, J.M., 1999. Role of photoprotective systems of holm-oak (*Quercus ilex*) in the adaptation to winter conditions. J. Plant Physiol. 155, 625–630.

Garg, N., Singla, P., 2011. Arsenic toxicity in crop plants: physiological effects and tolerance mechanisms. Environ. Chem. Lett. 9, 303–32110.

Gershman, R., Gilbert, D.L., Nye, S.W., Dwyer, P., Fenn, W.O., 1954. Oxygen poisoning and x-irradiation: a mechanism in common. Science 119, 623–626.

Gill, S.S., Tuteja, N., 2010. Reactive oxygen species and antioxidant machinery in abiotic stress tolerance in crop plants. Plant Physiol. Biochem. 48, 909–930. https://doi.org/10.1016/j.plaphy.2010.08.016.

Gill, S.S., Anjum, N.A., Hasanuzzaman, M., Gill, R., Trivedi, D.K., Ahmad, I., 2013. Glutathione and glutathione reductase: a boon in disguise for plant abiotic stress defense operations. Plant Physiol. Biochem. 70, 204–212.

Giraud, E., Ho, L.H., Clifton, R., 2008. The absence of ALTERNATIVE OXIDASE 1a in Arabidopsis results in acute sensitivity to combined light and drought stress. Plant Physiol. 147, 595–610.

Gomes, M.P., Duarte, D.M., Carneiro, M.M., Barreto, L.C., Carvalho, M., Soares, A.M., Guilherme, L.R.G., Garcia, Q.S., 2013. Zinc tolerance modulation in *Myracrodruon urundeuva* plants. Plant Physiol. Biochem. 67, 1–6.

Gomez, L.D., Vanacker, H., Buchner, P., Noctor, G., Foyer, C.H., 2004. Intercellular distribution of glutathione synthesis in maize leaves and its response to short-term chilling. Plant Physiol. 134, 1662–1671.

Gondim, F.A., Gomes-Filho, E., Costa, J.H., Mendes Alencar, N.L., Prisco, J.T., 2012. Catalase plays a key role in salt stress acclimation induced by hydrogen peroxide pretreatment in maize. Plant Physiol. Biochem. 56, 62–71.

Gonzalez-Meler, M.A., Giles, L., Thomas, R.B., Siedow, J.N., 2001. Metabolic regulation of leaf respiration and alternative oxidase activity in response to phosphate supply. Plant Cell Environ. 24, 205–215.

Gossett, D.R., Banks, S.W., Millhollon, E.P., Lucas, M.C., 1996. Antioxidant response to NaCl stress in a control and an NaCl-tolerant cell line grown in the presence of paraquat, buthionone sulfoximine, and exogenous glutathione. Plant Physiol. 112, 803–809.

Grativol, C., Hemerly, A.S., Ferreira, P.C.G., 2012. Genetic and epigenetic regulation of stress responses in natural plant populations. Biochim. Biophys. Acta 1819, 176–185.

Grill, E., Gekeler, W., Winnacker, E.-L., Zenk, M.H., 1986. Homo-phytochelatins are heavy metal-binding peptides of homo-glutathione containing Fabales. FEBS Lett. 205, 47–50.

Grill, E., Winnacker, E.-L., Zenk, M.H., 1987. Phytochelatins, a class of heavy-metal-binding peptides from plants are functionally analogous to metallothioneins. Proc. Natl. Acad. Sci. U. S. A. 84, 439–443.

Grover, M., Ali, S.Z., Sandhya, V., Rasul, A., Venkateswarlu, B., 2011. Role of microorganisms in adaptation of agriculture crops to abiotic stresses. World J. Microbiol. Biotechnol. 27 (5), 1231–1240.

Guan, Q., Wang, Z., Wang, X., Takano, T., Liu, S., 2015. A peroxisomal APX from *Puccinellia tenuiflora* improves the abiotic stress tolerance of transgenic Arabidopsis thaliana through decreasing of H_2O_2 accumulation. J. Plant Physiol. 175, 183–191.

Guo, Y.P., Zhou, H.F., Zhang, L.C., 2006. Photosynthetic characteristics and protective mechanisms against photooxidation during high temperature stress in two citrus species. Sci. Hortic. 108, 260–267.

Gupta, S.C., Goldsbrough, P.B., 1991. Phvtochelalin accumulation and cadmium tolerance in selected tomato cell lines. Plant Physiol. 97, 306–312.

Gupta, S., Priya, L., Gohil, V., 2002. Simultaneous exposure of lead and cadmium on granulose cells, progesterone and luteinising hormone in proestrous rats. Adv. Pharmacol. Toxcol. 3, 23–30.

Gupta, M., Sharma, P., Sarin, N.B., Sinha, A.K., 2009. Differential response of arsenic stress in two varieties of *Brassica juncea* L. Chemosphere 74, 1201–1208.

Gupta, D.K., Huang, H.G., Yang, X.E., Razafindrabe, B.H.N., Inouhe, M., 2010. The detoxification of lead in *Sedum alfredii* H. is not related to phytochelatins but the glutathione. J. Hazard. Mater. 177, 437–444.

Gururani, M.A., Upadhyaya, C.P., Strasser, R.J., Woong, Y.J., Park, S.W., 2012. Physiological and biochemical responses of transgenic potato plants with altered expression of PSII manganese stabilizing protein. Plant Physiol. Biochem. 58, 182–194.

Gutowski, M., Kowalczyk, S., 2013. A study of free radical chemistry: their role and pathophysiological significance. Acta Biochim. Pol. 60 (1), 1–6.

Hancock, W.W., Lu, B., Gao, W., Csizmaida, V., Faia, K., King, J.A., Smiley, S.T., Ling, M., Gerard, N.P., Gerard, C., 2000. Requirement of the chemokine receptor CXCR3 for acute allograft rejection. J. Exp. Med. 192 (10), 1515–1519.

Hao, H., Fa, L., Chen, T., Li, R., Li, X., He, Q., 2014. Clathrin and membrane microdomains cooperatively regulate rbohd dynamics and activity in Arabidopsis. Plant Cell 26, 1729–1745.

Hartley-Whitaker, J., Ainsworth, G., Meharg, A.A., 2001. Copper- and arsenate induced oxidative stress in *Holcus lanatus* L. clones with differential sensitivity. Plant Cell Environ. 24, 713–722.

Hasanuzzaman, M., Fujita, M., 2011. Selenium pretreatment upregulates the antioxidant defense and methylglyoxal detoxification system and confers enhanced tolerance to drought stress in rapeseed seedlings. Biol. Trace Elem. Res. 143, 1758–1776.

Hasanuzzaman, M., Fujita, M., 2013. Exogenous sodium nitroprusside alleviates arsenic-induced oxidative stress in wheat (*Triticum aestivum* L.) seedlings by enhancing antioxidant defense and glyoxalase system. Ecotoxicology 22, 584–596.

Hasanuzzaman, M., Hossain, M.A., Fujita, M., 2010. Physiological and biochemical mechanisms of nitric oxide induced abiotic stress tolerance in plants. Am. J. Plant Physiol. 5, 295–324.

Hasanuzzaman, M., Hossain, M.A., Fujita, M., 2011a. Nitric oxide modulates antioxidant defense and the methylglyoxal detoxification system and reduces salinity-induced damage of wheat seedlings. Plant Biotechnol. Rep. 5, 353–365.

Hasanuzzaman, M., Hossain, M.A., Fujita, M., 2011b. Selenium-induced up-regulation of the antioxidant defense and methylglyoxal detoxification system reduces salinity-induced damage in rapeseed seedlings. Biol. Trace Elem. Res. 143, 1704–1721.

Hasanuzzaman, M., Hossain, M.A., Fujita, M., 2012a. Exogenous selenium pretreatment protects rapeseed seedlings from cadmium-induced oxidative stress by upregulating the antioxidant defense and methylglyoxal detoxification systems. Biol. Trace Elem. Res. 149, 248–261.

Hasanuzzaman, M., Nahar, K., Alam, M.M., Fujita, M., 2012b. Exogenous nitric oxide alleviates high temperature induced oxidative stress in wheat (*Triticum aestivum*) seedlings by modulating the antioxidant defense and glyoxalase system. Aust. J. Crop Sci. 6, 1314–1323.

Hasanuzzaman, M., Nahar, K., Alam, M.M., Roychowdhury, R., Fujita, M., 2013. Physiological, biochemical, and molecular mechanisms of heat stress tolerance in plants. Mol. Sci. 14 (5), 9643–9684. https://doi.org/10.3390/ijms14059643.

Hebbelmann, I., Selinski, J., Wehmeyer, C., Goss, T., Voss, I., Mulo, P., Kangasjärvi, S., Aro, E.-M., Oelze, M.-L., Dietz, K.-J., et al., 2012. Multiple strategies to prevent oxidative stress in Arabidopsis plants lacking the malate valve enzyme NADP-malate dehydrogenase. J. Exp. Bot. 63, 1445–1459.

Hegedüs, A., Erdei, S., Horváth, G., 2001. Comparative studies of H_2O_2 detoxifying enzymes in green and greening barley seedlings under cadmium stress. Plant Sci. 160, 1085–1093.

Heidenreich, B., Mayer, K., Sandermann, H.J., Ernst, D., 2001. Mercury-induced genes in *Arabidopsis thaliana*: identification of induced genes upon long-term mercuric ion exposure. Plant Cell Environ. 24, 1227–1234.

Hernandez, J., Jimenez, A., Millineaus, P., Sevilla, F., 2000. Tolerance to pea (*Pisum sativum* L.) to long-term salt stress is associated with induction of antioxidant defenses. Plant Cell Environ. 23, 853–862.

Hodges, D.M., Andrews, C.J., Johnson, D.A., Hamilton, R.I., 1996. Antioxidant compound responses to chilling stress in differentially sensitive inbred maize lines. Physiol. Plant. 98, 685–692.

Horie, T., Schroeder, J.I., 2004. Sodium transporters in plants. Diverse genes and physiological functions. Plant Physiol. 136, 2457–2462.

Hossain, M.S., Dietz, K.J., 2016. Tuning of redox regulatory mechanisms, reactive oxygen species and redox homeostasis under salinity stress. Front. Plant Sci. 7, 548. https://doi.org/10.3389/fpls.2016.00548.

Hossain, M.A., Piyatida, P., da Silva, J.A.T., Fujita, M., 2012. Molecular mechanism of heavy metal toxicity and tolerance in plants: central role of glutathione in detoxification of reactive oxygen species and methylglyoxal and in heavy metal chelation. J. Bot. 2012, Article ID 872875. https://doi.org/10.1155/2012/872875.

Hossain, M.S., Ye, W., Hossan, M.A., Okuma, E., Uraji, M., Nakamura, Y., 2013. Glucosinolate degradation products, isothiocyanates, nitriles, and thiocyanates, induce stomatal closure accompanied by peroxidase-mediated reactive oxygen species production in *Arabidopsis thaliana*. Biosci. Biotechnol. Biochem. 77, 977–983.

Hsu, Y.T., Kao, C.H., 2004. Cadmium toxicity is reduced by nitric oxide in rice leaves. Plant Growth Regul. 42, 227–238.

Hu, W.H., Song, X.S., Shi, K., Xia, X.J., Zhou, Y.H., Yu, J.Q., 2008. Changes in electron transport, superoxide dismutase and ascorbate peroxidase isoenzymes in chloroplasts and mitochondria of cucumber leaves as influenced by chilling. Photosynthetica 46 (4), 581–588.

Huang, M., Guo, Z., 2005. Responses of antioxidative system to chilling stress in two rice cultivars differing in sensitivity. Biol. Plant. 9, 81–84.

Ibáñez, H., Ballester, A., Muñoz, R., Quiles, M.J., 2010. Chlororespiration and tolerance to drought, heat and high illumination. J. Plant Physiol. 167, 732–738.

Iqbal, N., Masood, A., Nazar, R., Syeed, S., Khan, N.A., 2010. Photosynthesis, growth and antioxidant metabolism in mustard (*Brassica juncea* L.) cultivars differing in cadmium tolerance. Agric. Sci. China 9, 519–527.

Islam, M.M., Hoque, M.A., Okuma, E., Jannat, R., Banu, M.N.A., Jahan, M.S., 2009. Proline and glycinebetaine confer cadmium tolerance on tobacco bright yellow-2 cells by increasing ascorbate-glutathione cycle enzyme activities. Biosci. Biotechnol. Biochem. 73, 2320–2323.

Ismail, M., Suroto, A., 2012. Response of biochemical activities on selected rice plants towards ozone stress. In: International Conference on Nanotechnology and Chemical Engineering (ICNCS'2012), December 21–22, 2012, Bangkok, Thailand.

Israr, M., Sahi, S., 2006. Antioxidative responses to mercury in the cell cultures of *Sesbania drummondii*. Plant Physiol. Biochem. 44, 590–595.

Israr, M., Sahi, S., Datta, R., Sarkar, D., 2006a. Bioaccumulation and physiological effects of mercury in *Sesbania drummondii*. Chemosphere 65, 591–598.

Israr, M., Sahi, S.V., Jain, J., 2006b. Cadmium accumulation and antioxidative responses in the *Sesbania drumondii* callus. Arch. Environ. Contam. Toxicol. 50, 121–127.

Jalmi, S.K., Sinha, A.K., 2015. ROS mediated MAPK signaling in abiotic and biotic stress- striking similarities and differences. Front. Plant Sci. 24 (6), 769.

Janda, M., Earkin, E.G., Wakler, D., 2006. Supportive care needs of people with brain tumours and their carers. Support. Care Cancer 14, 1094–1103.

Jarzyniak, K.M., Jasinski, M., 2014. Membrane transporters and drought resistance—a complex issue. Front. Plant Sci. https://doi.org/10.3389/fpls.2014.00687.

Jiang, C., Belfield, E.J., Mithani, A., Visscher, A., Ragoussis, J., Mott, R., Harberd, N.P., 2012a. ROS-mediated vascular homeostatic control of root-to-shoot soil Na delivery in Arabidopsis. EMBO J. 31 (22), 4359–4370.

Jiang, M., Miao, L., He, C., 2012b. Overexpression of an oil radish superoxide dismutase gene in broccoli confers resistance to downy mildew. Plant Mol. Biol. Rep. 30, 966–972.

Jin, X., Yang, X., Mahmood, Q., Islam, E., Liu, D., Li, H., 2008. Response of antioxidant enzymes, ascorbate and glutathione metabolism towards cadmium in hyperaccumulator and nonhyperaccumulator ecotypes of *Sedum alfredii* H. Environ. Toxicol. 23, 517–529.

Juszczuk, I.M., Wagner, A.M., Rychter, A.M., 2001. Regulation of alternative oxidase activity during phosphate deficiency in bean roots (*Phaseolus vulgaris*). Physiol. Plant. 113, 185–192.

Kabir, M.H., Wang, M.H., 2011. Functional studies on two catalase genes from tomato (*Solanum lycopersicum* L.). J. Hortic. Sci. Biotechnol. 286, 84–90.

Kanwar, M.K., Bhardwaj, R., Arora, P., Chowdhary, S.P., Sharma, P., Kumar, S., 2012. Plant steroid hormones produced under Ni stress are involved in the regulation of metal uptake and oxidative stress in *Brassica juncea* L. Chemosphere 86, 41–49.

Kausar, R., Hossain, Z., Makino, T., Komatsu, S., 2012. Characterization of ascorbate peroxidase in soybean under flooding and drought stresses. Mol. Biol. Rep. 39, 10573–10579.

Kawano, T., Sahashi, N., Takahashi, K., Uozumi, N., Muto, S., 1998. Salicylic acid induces extracellular generation of superoxide followed by an increase in cytosolic calcium ion in tobacco suspension culture: the earliest events in salicylic acid signal transduction. Plant Cell Physiol. 39, 721–730.

Kawarazaki, T., Kimura, S., Iizuka, A., Hanamata, S., Nibori, H., Michikawa, M., Imai, A., Abe, M., Kaya, H., Kuchitsu, K., 2013. A low temperature-inducible protein AtSRC2 enhances the ROS-producing activity of NADPH oxidase AtRbohF. Biochim. Biophys. Acta 1833 (12), 2775–2780.

Kertulis-Tartara, G.M., Rathinasapathip, B., Ma, L.Q., 2009. Characterization of glutathione reductase and catalase in the fronds of two Pteris ferns upon arsenic exposure. Plant Physiol. Biochem. 47 (10), 960–965.

Khan, M.H., Panda, S.K., 2008. Alterations in root lipid peroxidation and antioxidative responses in two rice cultivars under NaCl-salinity stress. Acta Physiol. Plant. 30, 81–89.

Khan, N.A., Samiullah, S.S., Nazar, R., 2007. Activities of antioxidative enzymes, sulphur assimilation, photosynthetic activity and growth of wheat (*Triticum aestivum*) cultivars differing in yield potential under cadmium stress. J. Agron. Crop Sci. 193, 435–444.

Khan, I., Ahmad, A., Iqbal, M., 2009. Modulation of antioxidant defence system for arsenic detoxification in Indian mustard. Ecotoxicol. Environ. Saf. 72, 626–634.

Khatun, S., Ali, M.B., Hahn, E.J., Paek, K.Y., 2008. Copper toxicity in *Withania somnifera:* growth and antioxidant enzymes responses of in vitro grown plants. Environ. Exp. Bot. 64, 279–285.

Kholová, J., Hash, T., Kakkera, A., Kocova, M., Vadez, V., 2009. Constitutive water conserving mechanisms are correlated with the terminal drought tolerance of pearl millet (*Pennisetum glaucum* (L.) R. Br.). J. Exp. Bot. 61, 369–377.

Klapheck, S., Fliegner, W., Zimmer, I., 1994. Hydroxymethyl-phytochelatins [(γ-glutamylcysteine)n-serine] are metal-induced peptides of the Poaceae. Plant Physiol. 104, 1325–1332.

Klapheck, S., Schlunz, S., Bergmann, L., 1995. Synthesis of phytochelatins and homo-phytochelatins in *Pisum sativum* L. Plant Physiol. 107, 515–521.

Koca, H., Bor, M., Ozdemir, F., Tukran, I., 2007. The effect of salt stress on lipid peroxidation, antioxidative enzymes and proline content of sesame cultivars. Environ. Exp. Bot. 60 (3), 344–351.

Kocsy, G., Szalai, G., Vagujfalvi, A., Stehli, L., Orosz, G., Galiba, G., 2000. Genetic study of glutathione accumulation during cold hardening in wheat. Planta 210, 295–301.

Kocsy, G., Galiba, G., Brunold, C., 2001a. Role of glutathione in adaptation and signaling during chilling and cold acclimation in plants. Physiol. Plant. 113, 158–164.

Kocsy, G., Ballmoos, P.V., Rüegsegger, A., Szalai, G., Galiba, G., Brunold, C., 2001b. Increasing the glutathione content in a chilling-sensitive maize genotype using safeners increased protection against chilling-induced injury. Plant Physiol. 127, 1147–1156.

Kocsy, G., Szalai, G., Galiba, G., 2002. Induction of glutathione synthesis and glutathione reductase activity by abiotic stresses in maize and wheat. Sci. World J. 2, 1699–1705.

Kocsy, G., Szalai, G., Sutka, J., Páldi, E., Galiba, G., 2004a. Heat tolerance together with heat stress-induced changes in glutathione and hydroxymethylglutathione levels is affected by chromosome 5A of wheat. Plant Sci. 166, 451–458.

Kocsy, G., Szalai, G., Galiba, G., 2004b. Effect of osmotic stress on glutathione and hydroxymethylglutathione accumulation in wheat. J. Plant Physiol. 161, 785–794.

Kolb, D., Müller, M., Zellnig, G., Zechmann, B., 2010. Cadmium induced changes in subcellular glutathione contents within glandular trichomes of *Cucurbita pepo* L. Protoplasma 243, 87–94.

Kong, J., Gong, J.M., Zhang, Z.G., Zhang, J.S., Chen, S.Y., 2003. A new AOX homologous gene OsIM1 from rice (*Oryza sativa* L.) with an alternative splicing mechanism under salt stress. Theor. Appl. Genet. 107, 326–331.

Kotchoni, S.O., Gachomo, E.W., 2006. The reactive oxygen species network pathways: an essential prerequisite for perception of pathogen attack and the acquired disease resistance in plants. J. Biosci. 31 (3), 389–404.

Kovalchuk, I., Dutta Gupta, S., 2010. Reactive Oxygen Species in Higher Plant. CRC Press, New York, pp. 31–44.

Kranner, I., Beckett, R.P., Wornik, S., Zorn, M., Pfeifhofer, H.W., 2002. Revival of a resurrection plant correlates with its antioxidant status. Plant J. 31, 13–24.

Krantev, A., Yordanova, R., Janda, T., Szalai, G., Popova, L., 2008. Treatment with salicylic acid decreases the effect of cadmium on photosynthesis in maize plants. J. Plant Physiol. 165, 920–931.

Kreps, J.A., Wu, Y., Chang, H.S., Zhu, T., Wang, X., Harper, J.F., 2002. Transcriptome changes for Arabidopsis in response to salt, osmotic and cold stress. Plant Physiol. 130, 2129–2141.

Kreuzwieser, J., Hauberg, J., Howell, K.A., Carroll, A., Rennenberg, H., Millar, A.H., Whelan, J., 2002. Differential response of gray poplar leaves and roots underpins stress adaptation during hypoxia. Plant Physiol. 149, 461–473.

Kubis, J., Rybus-Zając, M., 2008. Drought and excess UV-B irradiation differentially alter the antioxidant system in cucumber leaves. Acta Biol. Cracov. Bot. 50, 35–41.

Kumar, R.G., Shah, K., Dubey, R.S., 2000. Salinity induced behavioural changes in malate dehydrogenase and glutamate dehydrogenase activities in rice seedlings of differing salt tolerance. Plant Sci. 156, 23–34.

Kumar, R.R., Karajol, K., Naik, G.R., 2011. Effect of polyethylene glycol induced water stress on physiological and biochemical responses in pigeonpea (*Cajanus cajan* L. Millsp.). Recent Res. Sci. Technol. 3, 148–152.

Kumar, S., Gupta, D., Nayyar, H., 2012. Comparative response of maize and rice genotypes to heat stress: status of oxidative stress and antioxidants. Acta Physiol. Plant. 34, 75–86.

Kumari, R., Singh, S., Agrawal, S.B., 2010. Response of ultraviolet-B induced antioxidant defense system in a medicinal plant, *Acorus calamus*. J. Environ. Biol. 31, 907–911.

Kumari, S., Agrawal, M., Tiwari, S., 2013. Impact of elevated CO_2 and elevated O_3 on *Beta vulgaris* L.: pigments, metabolites, antioxidants, growth and yield. Environ. Pollut. 174, 279–288. https://doi.org/10.1016/j.envpol.2012.11.021.

Kumutha, D., Ezhilmathi, K., Sairam, R.K., Srivastava, G.C., Deshmukh, P.S., Meena, R.C., 2009. Waterlogging induced oxidative stress and antioxidant activity in pigeon pea genotypes. Biol. Plant. 53, 75–84.

Kuroda, H., Sagisaka, S., 1998. Metabolic and enzymatic responses associated with oxidative stress in plants acclimatized to cold environments. Rec. Res. Devel. Agric. Biol. Chem. 2, 395–410.

Kwak, J.M., Mori, I.C., Pei, Z.M., Leonhardt, N., Torres, M.A., Dangl, J.L., Bloom, R.E., Bodde, S., Jones, D.G., Schroeder, J.I., 2003. NADPH oxidase AtrbohD and AtrbohF genes function in ROS-dependent ABA signaling in Arabidopsis. EMBO J. 22, 2623–2633.

Kwon, S.I., An, C.S., 2001. Molecular cloning, characterization and expression analysis of a catalase cDNA from hot pepper (*Capsicum annuum* L.). Plant Sci. 160, 961–969.

Lascano, H.R., Antonicelli, G.E., Luna, C.M., Melchiorre, M.N., Gomez, L.D., Racca, R.W., Trippi, V.S., Casano, L.M., 2001. Antioxidant system response of different wheat cultivars under drought: field and in vitro studies. Aust. J. Plant Physiol. 28, 1095–1102.

Laureau, C., Bligny, R., Streb, P., 2011. The significance of glutathione for photoprotection at contrasting temperatures in the alpine plant species *Soldanella alpina* and *Ranunculus glacialis*. Physiol. Plant. 143, 246–260.

Laureau, C., Paepe, R.D., Latouche, G., Moreno-Chacón, M., Finazzi, G., Kuntz, M., Cornic, G., Streb, P., 2013. Plastid terminal oxidase (PTOX) has the potential to act as a safety valve for excess excitation energy in the alpine plant species *Ranunculus glacialis* L. Plant Cell Environ. 36, 1296–1310.

Lennon, A.M., Prommeenate, P., Nixon, P.J., 2003. Location, expression and orientation of the putative chlororespiratory enzymes, ndh and immutans, in higher-plant plastids. Planta 218, 254–260.

Leshem, Y., Seri, L., Levine, A., 2007. Induction of phosphatidylinositol 3-kinase-mediated endocytosis by salt stress leads to intracellular production of reactive oxygen species and salt tolerance. Plant J. 51, 185–197.

Li, X., Ma, H., Jia, P., Wang, J., Jia, L., Zhang, T., 2012. Responses of seedling growth and antioxidant activity to excess iron and copper in *Triticum aestivum* L. Ecotoxicol. Environ. Saf. 86, 47–53.

Li, T., Angeles, O., Radanielson, A., Marcaida III, M., Manalo, E., 2015. Drought stress impacts of climate change on rainfed rice in South Asia. Clim. Chang. 133, 709–720.

Liedschulte, V., Wachter, A., Zhigang, A., Rausch, T., 2010. Exploiting plants for glutathione (GSH) production: uncoupling GSH synthesis from cellular controls results in unprecedented GSH accumulation. Plant Biotechnol. J. 8 (7), 807–820.

Liu, F., VanToai, T., Moy, L.P., Bock, G., Linford, L.D., Quackenbush, J., 2005. Global transcription profiling revelas comprehensive insights into hypoxic response in Arabidopsis. Plant Physiol. 137, 1115–1129.

Liu, H., Xin, Z., Zhang, Z., 2011. Changes in activities of antioxidant-related enzymes in leaves of resistant and susceptible wheat inoculated with *Rhizoctonia cerealis*. Agric. Sci. China 10, 526–533.

Logan, B.A., Kornyeyev, D., Hardison, J., Holaday, A.S., 2006. The role of antioxidant enzymes in photoprotection. Photosynth. Res. 88 (2), 119–132.

Lomonte, C., Sgherri, C., Baker, A.J.M., Kolev, S.D., Navari-Izzo, F., 2010. Antioxidative response of *Atriplex codonocarpa* to mercury. Environ. Exp. Bot. 69, 9–16.

Luwe, M., Takahama, U., Heber, U., 1993. Role of ascorbate in detoxifying ozone in the apoplast of spinach (*Spinacea oleracea* L.) leaves. Plant Physiol. 101, 969–976.

Lyubenova, L., Schröder, P., 2011. Plants for waste water treatment—effects of heavy metals on the detoxification system of *Typha latifolia*. Bioresour. Technol. 102, 996–1004.
Ma, Y.H., Ma, F.W., Zhang, J.K., Liang, D., 2008. Effects of high temperature on activities and gene expression of enzymes involved in ascorbate–glutathione cycle in apple leaves. Plant Sci. 175 (6), 761–766.
Ma, L., Zhang, H., Sun, L., Jiao, Y., Zhang, G., Miao, C., Hao, F., 2012. NADPH oxidase AtrbohD and AtrbohF function in ROS-dependent regulation of Na^+/K^+ homeostasis in Arabidopsis under salt stress. J. Exp. Bot. 63, 305–317.
Maheshwari, R., Dubey, R.S., 2009. Nickel-induced oxidative stress and the role of antioxidative defense in rice seedlings. Plant Growth Regul. 59, 37–49.
Mandhania, S., Madan, S., Sawhney, V., 2006. Antioxidant defense mechanism under salt stress in wheat seedlings. Biol. Plant. 50, 227–231.
Manivannan, P., Jaleel, C.A., Kishorekumar, A., Sankar, B., Somasundaram, R., Sridharan, R., Panneerselvam, R., 2007. Changes in antioxidant metabolism of *Vigna unguiculata* (L.) Walp. by propiconazole under water deficit stress. Colloids Surf. B: Biointerfaces 57, 69–74.
Mano, J., Hideg, É., Asada, K., 2004. Ascorbate in thylakoid lumen functions as an alternative electron donor to photosystem II and photosystem I. Arch. Biochem. Biophys. 429, 71–80.
Markovska, Y.K., Gorinova, N.I., Nedkovska, M.P., Miteva, K.M., 2009. Cadmium-induced oxidative damage and antioxidant responses in *Brassica juncea* plants. Biol. Plant. 53, 151–154.
Martí, M.C., Olmos, E., Calvete, J.J., Diaz, I., Barranco-Medina, S., Whelan, J., Lazaro, J.J., Sevilla, F., Jimenez, A., 2009. Mitochondrial and nuclear localization of a novel pea thioredoxin: identification of its mitochondrial target proteins. Plant Physiol. 150, 646–657.
Martínez, J.P., Araya, H., 2010. Ascorbate-glutathione cycle: enzymatic and non-enzymatic integrated mechanisms and its biomolecular regulation. In: Anjum, N.A., Chan, M.T., Umar, S. (Eds.), Ascorbate-Glutathione Pathway and Stress Tolerance in Plants. Springer, Dordrecht, pp. 303–322.
Masi, A., Ferretti, M., Passera, C., Ghisi, R., 1995. Antioxidative responses in *Zea mays* seedlings grown under UV-B radiation. Plant Physiol. Biochem. (Spec. Issue), 264–265.
Mattagajasingh, S.N., Kar, M., 1989. Changes in the antioxidant system during greening of etiolated wheat leaves. J. Plant Physiol. 134, 656–660.
McKersie, B.D., Murnaghan, J., Jones, K.S., Bowley, S.R., 2000. Iron-superoxide dismutase expression in transgenic alfalfa increases winter survival without a detectable increase in photosynthetic oxidative stress tolerance. Plant Physiol. 122, 1427–1437.
Melchiorre, M., Robert, G., Trippi, V., Racca, R., Lascano, H.R., 2009. Superoxide dismutase and glutathione reductase overexpression in wheat protoplast: photooxidative stress tolerance and changes in cellular redox state. Plant Growth Regul. 57, 57e68.
Mendoza, M., 2011. Oxidative burst in plant-pathogen interaction. Biotecnol. Vegetal 11 (2), 67–75.
Meneguzzo, S., Navari-Izzo, F., Izzo, R., 1999. Antioxidant responses of shoots and roots of wheat to increasing NaCl concentrations. J. Plant Physiol. 155, 274–280.
Metwally, A., Safronova, V.I., Belimov, A.A., Dietz, K.J., 2005. Genotypic variation of the response to cadmium toxicity in *Pisum sativum* L. J. Exp. Bot. 56, 167–178.
Millar, A.H., Bergersen, F.J., Day, D.A., 1994. Oxygen affinity of terminal oxidases in soybean mitochondria. Plant Physiol. Biochem. 32, 847–852.
Miller, G., Schlauch, K., Tam, R., Cortes, D., Torres, M.A., Shulaev, V., 2009. The plant NADPH oxidase RBOHD mediates rapid systemic signaling in response to diverse stimuli. Sci. Signal. 2, ra45. https://doi.org/10.1126/scisignal.2000448.
Mishra, S., Srivastava, S., Tripathi, R.D., Govindrajan, R., Kuriakose, S.V., Prasad, M.N.V., 2006. Phytochelatin synthesis and response of antioxidants during cadmium stress in *Bacopa monnieri* L. Plant Physiol. Biochem. 44, 25–37.
Mishra, S., Jha, A.B., Dubey, R.S., 2011. Arsenite treatment induces oxidative stress, upregulates antioxidant system, and causes phytochelatin synthesis in rice seedlings. Protoplasma 248, 565–577.
Mishra, P., Bhoomika, K., Dubey, R.S., 2013. Differential responses of antioxidative defense system to prolonged salinity stress in salt-tolerant and salt-sensitive Indica rice (*Oryza sativa* L.) seedlings. Protoplasma 250, 3–19.
Mittler, R., 2002. Oxidative stress, antioxidants and stress tolerance. Trends Plant Sci. 7, 405–410.
Mittler, R., Vanderauwera, S., Suzuki, N., Mille, G., Tognetti, V.B., Vandepoele, K., 2011. ROS signaling: the new wave? Trends Plant Sci 16 (6), 300–309.

Mittova, V., Guy, M., Tal, M., Volokita, M., 2002. Response of the cultivated tomato and its wild salt-tolerant relative *Lycopersicon pennellii* to salt-dependent oxidative stress: increased activities of antioxidant enzymes in root plastids. Free Radic. Res. 36, 195–202.

Mittova, V., Tal, M., Volokita, M., Guy, M., 2003. Up-regulation of the leaf mitochondrial and peroxisomal antioxidative systems in response to salt-induced oxidative stress in the wild salt-tolerant tomato species *Lycopersicon pennellii*. Plant Cell Environ. 26, 845–856.

Mittova, V., Guy, M., Tal, M., Volokita, M., 2004. Salinity up-regulates the antioxidative system in root mitochondria and peroxisomes of the wild salt-tolerant tomato species *Lycopersicon pennellii*. J. Exp. Bot. 55, 1105–1113.

Mittova, V., Volokita, M., Guy, M., 2015. Antioxidative systems and stress tolerance: insight from wild and cultivated tomato species. In: Gupta, K.J., Igamberdiev, A.U. (Eds.), Signaling and Communication in Plants. Reactive Oxygen and Nitrogen Species Signaling and Communication in Plants. 23, SpringerInternational Publishing Switzerland, Cham/Heidelberg, pp. 89–131.

Mobin, M., Khan, N.A., 2007. Photosynthetic activity, pigment composition and antioxidative response of two mustard (*Brassica juncea*) cultivars differing in photosynthetic capacity subjected to cadmium stress. J. Plant Physiol. 164, 601–610.

Moller, I.M., Kristensen, B.K., 2004. Protein oxidation in plant mitochondria as a stress indicator. Photochem. Photobiol. Sci. 3 (8), 730–735.

Moseley, J.L., Chang, C.W., Grossman, A.R., 2006. Genome-based approaches to understanding phosphorus deprivation responses and PSR1 control in *Chlamydomonas reinhardtii*. Eukaryot. Cell 5, 26–44.

Murakami, Y., Toriyama, K., 2008. Enhanced high temperature tolerance in transgenic rice seedlings with elevated levels of alternative oxidase, OsAOX1a. Plant Biotechnol. 25, 361–364.

Murata, Y., Pei, Z.M., Izumi, C.M., Schroeder, J., 2001. Abscisic acid activation of plasma membrane Ca^{2+} channels in guard cells requires cytosolic NAD(P)H and is differentially disrupted upstream and downstream of reactive oxygen species production in abi1-1and abi2-1 protein phosphatase 2C mutants. Plant Cell 13, 2513–2523.

Mylona, P.V., Polidoros, A.N., 2010. Regulation of antioxidant genes. In: Dutta Gupta, S. (Ed.), Reactive Oxygen Species and Antioxidants in Higher Plants. Science Publishers, London, pp. 101–127.

Nagalakshmi, N., Prasad, M.N., 2001. Responses of glutathione cycle enzymes and glutathione metabolism to copper stress in *Scenedesmus bijugatus*. Plant Sci. 160 (2), 291–299.

Nazar, R., Iqbal, N., Syeed, S., Khan, N.A., 2011. Salicylic acid alleviates decreases in photosynthesis under salt stress by enhancing nitrogen and sulfur assimilation and antioxidant metabolism differentially in two mungbean cultivars. J. Plant Physiol. 168, 807–815.

Nehnevajova, E., Lyubenova, L., Herzig, R., Schröder, P., Schwitzguébel, J.P., Schmulling, T., 2012. Metal accumulation and response of antioxidant enzymes in seedlings and adult sunflower mutants with improved metal removal traits on a metal-contaminated soil. Environ. Exp. Bot. 76, 39–48.

Nishimura, M.T., Dangl, J.L., 2010. Arabidopsis and the plant immune system. Plant J. 61, 1053–1066.

Noctor, G., Foyer, C.H., 1998. Ascorbate and glutathione: keeping active oxygen under control. Annu. Rev. Plant Physiol. Plant Mol. Biol. 49, 249–279.

Noctor, G., Veljovic-Jovanovic, S., Driscoll, S., Novitskaya, L., Foyer, C.H., 2002. Drought and oxidative load in the leaves of C3 plants: a predominant role for photorespiration? Ann. Bot. 89, 841–850.

Noguchi, K., Terashima, I., 2006. Responses of spinach leaf mitochondria to low N availability. Plant Cell Environ. 29, 710–719.

Nouairi, I., Ammar, W.B., Youssef, N.B., Miled, D.D.B., Ghorbal, M.H., Zarrouk, M., 2009. Antioxidant defense system in leaves of Indian mustard (*Brassica juncea*) and rape (*Brassica napus*) under cadmium stress. Acta Physiol. Plant 31, 237–247.

Nounjan, N., Nghia, P.T., Theerakulpisut, P., 2012. Exogenous proline and trehalose promote recovery of rice seedlings from salt-stress an differentially modulate anti-oxidant enzymes and expression of related genes. J. Plant Physiol. 169, 596–604.

Ogasawara, Y., Kaya, H., Hiraoka, G., Yumoto, F., Kimura, S., Kadota, Y., Hishinuma, H., Senzaki, E., Yamagoe, S., Nagata, K., Nara, M., Suzuki, K., Tanokura, M., Kuchitsu, K., 2008. Synergistic activation of the Arabidopsis NADPH oxidase AtrbohD by Ca^{2+} and phosphorylation. J. Biol. Chem. 283 (14), 8885–8892.

Ort, D.R., Baker, N.R., 2002. A photoprotective role for O_2 as an alternative electron sink in photosynthesis? Curr. Opin. Plant Biol. 5, 193–198.
Ottow, E.A., Brinker, M., Teichmann, T., Fritz, E., Kaiser, W., Brosché, M., Kangasjärvi, J., Jiang, X., Polle, A., 2005. *Populus euphratica* displays apoplastic sodium accumulation, osmotic adjustment by decreases in calcium and soluble carbohydrates, and develops leaf succulence under salt stress. Plant Physiol. 139, 1762–1772.
Paradiso, A., Berardino, R., Pinto, M.C., Toppi, L.S., Storelli, M.M., Tommasi, F., Gara, L., 2008. Increase in ascorbate–glutathione metabolism as local and precocious systemic responses induced by cadmium in durum wheat plants. Plant Cell Physiol. 49, 362–374.
Parsons, H.L., Yip, J.Y.H., Vanlerberghe, G.C., 1999. Increased respiratory restriction during phosphate-limited growth in transgenic tobacco cells lacking alternative oxidase. Plant Physiol. 121, 1309–1320.
Patterson, H.C., Gerbeth, C., Thirua, P., Vögtlee, N.F., Knolla, M., Shahsafaeib, A., Samocha, K.E., 2015. A respiratory chain controlled signal transduction cascade in the mitochondrial intermembrane space mediates hydrogen peroxide signaling. Proc. Natl. Acad. Sci. U. S. A. 112 (42), E5679–88. https://doi.org/10.1073/pnas.1517932112.
Perez-Lopez, U., Robredo, A., Lacuesta, M., Sgherri, C., Muñoz-Rueda, A., Navari-Izzo, F., Mena-Petite, A., 2009. The oxidative stress caused by salinity in two barley cultivars is mitigated by elevated CO_2. Physiol. Plant. 135 (1), 29–42.
Piechalak, A., Tomaszewska, B., Baralkiewicz, D., Malecka, A., 2002. Accumulation and detoxification of lead ions in legumes. Phytochemistry 60, 153–162.
Pietrini, F., Iannelli, M.A., Pasqualini, S., Massacci, A., 2003. Interaction of cadmium with glutathione and photosynthesis in developing leaves and chloro-plasts of *Phragmites australis* (Cav) Trin. ex Steudel. Plant Physiol. 133, 829–837.
Pignocchi, C., Kiddle, G., Hernández, I., Foster, S.J., Asensi, A., Taybi, T., Foyer, C.H., 2006. Ascorbate oxidase-dependent changes in the redox state of the apoplast modulate gene transcript accumulation leading to modified hormone signaling and orchestration of defense processes in tobacco. Plant Physiol. 141 (2), 423–435.
Pinheiro, C., Chaves, M.M., 2011. Photosynthesis and drought: can we make metabolic connections from available data? J. Exp. Bot. 62, 869–882. https://doi.org/10.1093/jxb/erq340.
Poirier, B., Michel, O., Bazin, R., Bariéty, J., Chevalier, J., Myara, I., 2001. Conjugated dienes: a critical trait of lipoprotein oxidizability in renal fibrosis. Nephrol. Dial. Transplant. 16, 1598–1606.
Polle, A., 1997. Defense against photoxidative damage in plants. In: Scandalios, J.G. (Ed.), Oxidative Stress and the Molecular Biology of Antioxidant Defences. Cold Spring Harbor Laboratory Press, New York, USA, pp. 623–666.
Posmyk, M.M., Kontek, R., Janas, K.M., 2009. Antioxidant enzymes activity and phenolic compounds content in red cabbage seedlings exposed to copper stress. Ecotoxicol. Environ. Saf. 72, 596–602.
Prasad, T.K., 1997. Role of catalase in inducing chilling tolerance in pre-emergent maize seedlings. Plant Physiol. 114 (4), 1369–1376.
Purev, M., Kim, Y.J., Kim, M.K., Pulla, R.K., Yang, D.C., 2010. Isolation of a novel catalase (Cat1) gene from *Panax ginseng* and analysis of the response of this gene to various stresses. Plant Physiol. Biochem. 48, 451–460.
Qadir, S., Qureshi, M.I., Javed, S., Abdin, M.Z., 2004. Genotypic variation in phytoremediation potential of *Brassica juncea* cultivars exposed to Cd stress. Plant Sci. 167, 1171–1181.
Quan, L.J., Zang, B., Shi, W.W., Li, H.Y., 2008. Hydrogen peroxide in plant: a versatile molecule of the reactive oxygen species network. J. Integer. Plant Biol. 50, 2–18.
Queirós, F., Rodrigues, J.A., Almeida, J.M., Almeida, D.P., Fidalgo, F., 2011. Differential responses of the antioxidant defence system and ultrastructure in a salt-adapted potato cell line. Plant Physiol. Biochem. 49, 1410–1419.
Quiles, M.J., 2006. Stimulation of chlororespiration by heat and high light intensity in oat plants. Plant Cell Environ. 29, 1463–1470.
Qureshi, M., Abdin, M., Qadir, S., Iqbal, M., 2007. Lead-induced oxidative stress and metabolic alterations in *Cassia angustifolia* Vahl. Biol. Plant. 51 (1), 121–128.

Radyuk, M.S., Domanskaya, I.N., Shcherbakov, R.A., Shalygo, N.V., 2009. Effect of low above-zero temperature on the content of low-molecular antioxidants and activities of antioxidant enzymes in green barley leaves. Russ. J. Plant Physiol. 56 (2), 175–180.

Ramos, J., Clemente, M.R., Naya, L., Loscos, J., Pérez-Rontomé, C., Sato, S., Tabata, S., Becana, M., 2007. Phytochelatin synthases of the model legume *Lotus japonicus*: a small multigene family with differential response to cadmium and alternatively spliced variants. Plant Physiol. 143, 1110–1118.

Rasmusson, A.G., Geisler, D.A., Moller, I.M., 2008. The multiplicity of dehydrogenases in electron transport chain of plant mitochondria. Mitochondrion 8, 47–60.

Rehman, A., Anjum, M.S., 2010. Cadmium uptake by yeast, *Candida tropicalis*, isolated from industrial effluents and its potential use in wastewater clean-up operations. Water Air Soil Pollut. 205, 149–159.

Remans, T., Opdenakker, K., Smeets, K., Mathijsen, D., Vangronsveld, J., Cuypers, A., 2010. Metal-specific and NADPH oxidase dependent changes in lipoxygenase and NADPH oxidase gene expression in *A. thaliana* exposed to cadmium or excess copper. Funct. Plant Biol. 37, 532–544.

Reumann, S., Weber, A.P., 2006. Plant peroxisomes respire in the light: some gaps of the photorespiratory C_2 cycle have become filled—others remain. Biochim. Biophys. Acta 1763 (12), 1496–1510.

Rhoads, D.M., Umbach, A.L., Subbaiah, C.C., Siedow, J.N., 2006. Mitochondrial reactive oxygen species. Contribution to oxidative stress and interorganellar signaling. Plant Physiol. 141, 357–366.

Ribas-Carbo, M., Berry, J.A., Azcon-Bieto, J., Siedow, J.N., 1994. The reaction of the plant mitochondrial cyanide-resistant alternative oxidase with oxygen. Biochim. Biophys. Acta 118, 205–212.

Ribas-Carbo, M., Taylor, N.L., Giles, L., Busquets, S., Finnegan, P.M., Day, D.A., Lambers, H., Medrano, H., Berry, J.A., Flexas, J., 2005. Effects of water stress on respiration in soybean leaves. Plant Physiol. 139, 466–473.

Río, L.A., 2015. ROS and RNS in plant physiology: an overview. J. Exp. Bot. https://doi.org/10.1093/jxb/erv099.

Rio, L.A., Sandalio, L.M., Corpas, F.J., Palma, J.M., Barroso, J.B., 2006. Reactive oxygen species and reactive nitrogen species in peroxisome, production, scavenging, and role incell signaling. Plant Physiol. 141, 330–335.

Romero-Puertas, M.C., McCarthy, I., Sandalio, L.M., Palma, J.M., Corpas, F.J., Gómez, M., Río, L.A., 1999. Cadmium toxicity and oxidative metabolism in pea leaf peroxisomes. Free Radic. Res. (Suppl.) 31, 25–31.

Romero-Puertas, M.C., Corpas, F.J., Sandalio, L.M., Leterrier, M., Rodríguez-Serrano, M., Río, L.A., Palma, J.M., 2006. Glutathione reductase from pea leaves: response to abiotic stress and characterization of the peroxisomal isozyme. New Phytol. 170 (1), 43–52.

Romero-Puertas, M.C., Corpas, F.J., Rodriguez-Serrano, M., Gomez, M., Río, L.A., Sandalio, L.M., 2007. Differential expression and regulation of antioxidative enzymes by Cd in pea plants. J. Plant Physiol. 164, 1346–1357.

Ruzsa, S.M., Mylona, P., Scandalios, J.G., 1999. Differential responses of antioxidant genes in maize leaves exposed to ozone. Redox Rep. 4, 95–103.

Sagi, M., Fluhr, R., 2006. Production of reactive oxygen species by plant NADPH oxidases. Plant Physiol. 141, 336–340.

Saha, B., Borovskii, G., Panda, S.K., 2016. Alternative oxidase and plant stress tolerance. Plant Signal. Behav. 11(12). e1256530.

Saidi, I., Nawel, N., Djebali, W., 2014. Role of selenium in preventing manganese toxicity in sunflower (*Helianthus annuus*) seedling. S. Afr. J. Bot. 94, 88–94.

Sairam, R.K., Srivastava, G.C., Agarwal, S., Meena, R.C., 2005. Differences in antioxidant activity in response to salinity stress in tolerant and susceptible wheat genotypes. Biol. Plant. 49, 85–91.

Sanchez-Rodriguez, E., Rubio-Wilhelmi, L.M., Cervilla, B., Blasco, J.J., Rios, M.A., Rosales, L., Ruiz, J.M., 2010. Genotypic differences in some physiological parameters symptomatic for oxidative stress under moderate drought in tomato plants. Plant Sci. 178, 30–40.

Sandalio, L.M., Dalurzo, H.C., Gómez, M., Romero-Puertas, M.C., del Río, L.A., 2001. Cadmium-induced changes in the growth and oxidative metabolism of pea plants. J. Exp. Bot. 52 (364), 2115–2126.

Savitch, L.V., Ivanov, A.G., Krol, M., Sprott, D.P., Öquist, G., Huner, N.P.A., 2010. Regulation of energy partitioning and alternative electron transport pathways during cold acclimation of lodgepole pine are oxygen-dependent. Plant Cell Physiol. 51, 1555–1570.

Schafer, H.J., Haag-Kerwer, A., Rausch, T., 1998. cDNA cloning and expression analysis of genes encoding GSH synthesis in roots of the heavy-metal accumulator *Brassica juncea* L.: evidence for Cd-induction of a putative mitochondrial γ-glutamylcysteine synthetase isoform. Plant Mol. Biol. 37, 87–97.
Scheibe, R., 2004. Malate valves to balance cellular energy supply. Physiol. Plant. 120, 21–26.
Scheibe, R., Backhausen, J.E., Emmerlich, V., Holtgrefe, S., 2005. Strategies to maintain redox homeostasis during photosynthesis under changing conditions. J. Exp. Bot. 56, 1481–1489.
Schmitt, F.J., Renger, G., Friedrich, T., Kreslavksi, V.D., Zharmukhadmedov, S.K., Los, D.A., 2014. Reactive oxygen species: re-evaluation of generation, monitoring and role in stress-signaling in phototrophic organisms. Biochim. Biophys. Acta 1837, 835–848.
Schützendübel, A., Schwanz, P., Teichmann, T., Gross, K., Langenfeld-Heyser, R., et al., 2001. Cadmium-induced changes in antioxidative systems, H_2O_2 content and differentiation in pine (*Pinus sylvestris*) roots. Plant Physiol. 127, 887–898.
Schutzendubel, A., Nikolova, P., Rudolf, C., Polle, A., 2002. Cadmium and H_2O_2-induced oxidative stress in *Populus canescens* roots. Plant Physiol. Biochem. 40, 577–584.
Seki, M., Narusaka, M., Ishida, J., Nanjo, T., Fijita, M., Oono, Y., Kamiya, A., Nakajima, M., Enju, A., Sakurai, T., 2002. Monitoring the expression profiles of 7000 Arabidopsis genes under drought, cold and high-salinity stresses using a full-length cDNA microarray. Plant J. 31, 279–292.
Sekmen, A.H., Turkan, I., Takio, S., 2007. Differential responses of antioxidative enzymes and lipid peroxidation to salt stress in salt-tolerant *Plantago maritima* and salt-sensitive *Plantago media*. Physiol. Plant. 131, 399–411.
Sellers, W.I., Dennis, L.A., Crompton, R.H., 2003. Predicting the metabolic energy costs of bipedalism using evolutionary robotics. J. Exp. Biol. 206, 1127–1136.
Selote, D.S., Khanna-Chopra, R., 2010. Antioxidant response of wheat genotypes tolerance. Protoplasma 245, 153–163.
Sewelam, N., Kazan, K., Schenk, P.M., 2016. Global plant stress signaling reactive oxygen species at the cross-road. Front. Plant Sci. 7187. https://doi.org/10.3389/fpls.2016.00187.
Shafi, A., Chauhan, A., Gill, T., Swarnkar, M.K., Sreenivasulu, Y., Kumar, S., Kumar, N., Shankar, R., Ahuja, P., Singh, A.K., 2015. Expression of SOD and APX genes positively regulates secondary cell wall biosynthesis and promotes plant growth and yield in *Arabidopsis* under salt stress. Plant Mol. Biol. 87, 615–631.
Shahbazi, M., Gilbert, M., Labouré, A.-M., Kuntz, M., 2007. Dual role of the plastid terminal oxidase in tomato. Plant Physiol. 145, 691–702.
Shalata, A., Mittova, V., Volokita, M., Guy, M., Tal, M., 2001. Response of the cultivated tomato and its wild salt-tolerant relative *Lycopersicon pennellii* to salt-dependent oxidative stress: the root antioxidative system. Physiol. Plant. 112, 487–494.
Shapiguzov, A., Vainonen, J.P., Wrzaczek, M., Kangasjärvi, J., 2012. ROS-talk—how the apoplast, the chloroplast, and the nucleus get the message through. Front. Plant Sci. 3, 292. https://doi.org/10.3389/fpls.2012.00292.
Sharma, P., Dubey, R.S., 2005. Drought induces oxidative stress and enhances the activities of antioxidant enzymes in growing rice seed-lings. Plant Growth Regul. 46, 209–221.
Sharma, P., Dubey, R.S., 2007. Involvement of oxidative stress and role of antioxidative defense system in growing rice seedlings exposed to toxic concentrations of aluminium. Plant Cell Rep. 26, 2027–2038.
Sharma, P., Jha, A.B., Dubey, R.S., Pessarakli, M., 2012. Reactive oxygen species, oxidative damage, and antioxidative defense mechanism in plants under stressful conditions. J. Bot. 2012, 26. https://doi.org/10.1155/2012/217037.
Shehab, G.G., Ahmed, O.K., El-Beltagi, H.S., 2010. Effects of various chemical agents for alleviation of drought stress in rice plants (*Oryza sativa* L.). Not. Bot. Hort. Agrobot. Cluj. 38 (1), 139–148.
Shinozaki, K., Shinozaki, K.Y., 2007. Gene networks involved in drought stress response and tolerance. J. Exp. Bot. 58, 221–227.
Shri, M., Kumar, S., Chakrabarty, D., Trivedi, P.K., Mallick, S., Misra, P., Shukla, D., Mishra, S., Srivastava, S., Tripathi, R.D., Tuli, R., 2009. Effect of arsenic on growth, oxidative stress, and antioxidant system in rice seedlings. Ecotoxicol. Environ. Saf. 72, 1102–1110.
Shrivastav, P., Kumar, R., 2014. Soil salinity: a serious environmental issue and plant growth promoting bacteria as one of the tools for its alleviation. Saudi J. Biol. Sci. 22, 123–131.

Sieger, S.M., Kristensen, B.K., Robson, C.A., Amirsadeghi, S., Eng, E.W.Y., Abdel-Mesih, A., Møller, I.M., Vanlerberghe, G.C., 2005. The role of alternative oxidase in modulating carbon use efficiency and growth during macronutrient stress in tobacco cells. J. Exp. Bot. 56, 1499–1515.

Silva, E.N., Ribeiro, R.V., Ferreira-Silva, S.L., Vieira, S.A., Ponte, L.F.A., Silveira, J.A.G., 2012. Coordinate changes in photosynthesis, sugar accumulation and antioxidative enzymes improve the performance of *Jatropha curcas* plants under drought stress. Biomass Bioenergy 45, 270–279.

Simkin, A.J., Moreau, H., Kuntz, M., Pagny, G., Lin, C., Tanksley, S., McCarthy, J., 2008. An investigation of carotenoid biosynthesis in *Coffea canephora* and *Coffea arabica*. J. Plant Physiol. 165, 1087–1106.

Singh, S., Khan, N.A., Nazar, R., Anjum, N.A., 2008. Photosynthetic traits and activities of antioxidant enzymes in blackgram (*Vigna mungo* L. Hepper) under cadmium stress. Am. J. Plant Physiol. 3, 25–32.

Singh, N., Mishra, A., Jha, B., 2014. Over-expression of the peroxisomal ascorbate peroxidase (SbpAPX) gene cloned from halophyte *Salicornia brachiata* confers salt and drought stress tolerance in transgenic tobacco. Mar. Biotechnol. (N.Y.) 16, 321–332.

Sinha, A.K., Jaggi, M., Raghuramand, B., Tuteja, N., 2011. Mitogen-activated protein kinase signaling in plants under abiotic stress. Plant Signal. Behav. 6 (2), 196–203.

Skutnik, M., Rychter, A.M., 2009. Differential response of antioxidant systems in leaves and roots of barley subjected to anoxia and post-anoxia. J. Plant Physiol. 166, 926–937.

Sofo, A., Dichio, B., Xiloyannis, C., Masia, A., 2005. Antioxidant defences in olive trees during drought stress: changes in activity of some antioxidant enzymes. Funct. Plant Biol. 32, 45–53.

Sofo, A., Scopa, A., Nuzzaci, M., Vitti, A., 2015. Ascorbate peroxidase and catalase activities and their genetic regulation in plants subjected to drought and salinity stresses. Int. J. Mol. Sci. 16 (6), 13561–13578.

Solanki, J.K., Sarangi, S.K., 2014. Effect of drought stress on proline accumulation in peanut genotypes. Int. J. Adv. Res. 2 (10), 301–309.

Song, Y., Miao, Y., Song, C.P., 2014. Behind the scenes: the roles of reactive oxygen species in guard cells. New Phytol. 201, 1121–1140.

Sorkheh, K., Shiran, B., Rouhi, V., Khodambashi, M., Wolukau, J.N., Ercisli, S., 2011. Response of in vitro pollen germination and pollen tube growth of almond (*Prunus dulcis* Mill.) to temperature, polyamines and polyamine synthesis inhibitor. Biochem. Syst. Ecol. 39, 749–757.

Sorkheh, K., Shiran, B., Khodambashi, M., Rouhi, V., Mosavei, S., Sofo, A., 2012. Exogenous proline alleviates the effects of H_2O_2-induced oxidative stress in wild almond species. Russ. J. Plant Physiol. 59, 788–798.

Srivastava, M., Ma, L.Q., Cotruvo, J.A., 2005. Uptake and distribution of selenium in different fern species. Int. J. Phytoremediation 7, 33–42.

Srivastava, S., Srivastava, A.K., Suprasanna, P., D'Souza, S.F., 2010. Comparative antioxidant profiling of tolerant and sensitive varieties of *Brassica juncea* L. to arsenate and arsenite exposure. Bull. Environ. Contam. Toxicol. 84, 342–346.

Stepien, P., Johnson, G.N., 2009. Contrasting responses of photosynthesis to salt stress in the glycophyte Arabidopsis and the halophyte *Thellungiella*: role of the plastid terminal oxidase as an alternative electron sink. Plant Physiol. 149, 1154–1165.

Streb, P., Josse, E.-M., Gallouet, E., Baptist, F., Kuntz, M., Cornic, G., 2005. Evidence for alternative electron sinks to photosynthetic carbon assimilation in the high mountain plant species *Ranunculus glacialis*. Plant Cell Environ. 28, 1123–1135.

Su, Y.C., Guo, J.L., Ling, H., Chen, S.S., Wang, S.S., 2014. Isolation of a novel peroxisomal catalase gene from sugarcane, which is responsive to biotic and abiotic stresses. PLoS One 9, e84426.

Sukweenadhi, J., Kim, Y.J., Rahimi, S., Silva, J., Myagmarjav, D., Yang, D.C., 2017. Overexpression of a cytosolic ascorbate peroxidase from *Panax ginseng* enhanced salt tolerance in *Arabidopsis thaliana*. Plant Cell Tiss. Org. Cult. 129, 337–350.

Sumithra, K., Jutur, P.P., Carmel, B.D., Reddy, A.R., 2006. Salinity-induced changes in two cultivars of *Vigna radiata*: responses of antioxidative and proline metabolism. Plant Growth Regul. 50, 11–22.

Sun, Q., Ye, Z.H., Wang, X.R., Wong, M.H., 2005. Increase of glutathione in mine population of *Sedum alfredii*: a Zn hyperaccumulator and Pb accumulator. Phytochemistry 66 (21), 2549–2556.

Sun, Q., Ye, Z.H., Wang, X.R., Wong, M.H., 2007. Cadmium hyperaccumulation leads to an increase of glutathione rather than phytochelatins in the cadmium hyperaccumulator *Sedum alfredii*. J. Plant Physiol. 164 (11), 1489–1498.

Suzuki, I., Los, D.A., Murata, N., 2000. Perception and transduction of low-temperature signals to induce desaturation of fatty acids. Biochem. Soc. Trans. 28, 628–630.
Suzuki, N., Miller, G., Morales, J., Shulaev, V., Torres, M.A., Mittler, R., 2011. Respiratory burst oxidases: the engines of ROS signaling. Curr. Opin. Plant Biol. 14, 691–699.
Svingen, B.A., Buege, J.A., O'Neal, F.O., Aust, S.D., 1979. The mechanism of NADPH-dependent lipid peroxidation: the propagation of lipid peroxidation. J. Biol. Chem. 254, 5892–5899.
Szal, B., Jolivet, Y., Hasenfratz-Sauder, M.P., Dizengremel, P., Rychter, A.M., 2003. Oxygen concentration regulates alternative oxidase expression in barley roots during hypoxia and post-hypoxia. Physiol. Plant. 119, 494–502.
Talukdar, D., 2013. Growth responses and leaf antioxidant metabolism of grass pea (*Lathyrus sativus* L.) genotypes under salinity stress. ISRN Agron. https://doi.org/10.1155/2013/284830.
Tan, Y.F., O'Toole, N., Taylor, N.L., Millar, A.H., 2010. Divalent metal ions in plant mitochondria and their role in interactions with proteins and oxidative stress-induced damage to respiratory function. Plant Physiol. 152, 747–761.
Tang, Y.Y., Wang, C.T., Yang, G.P., Feng, T., Gao, H.Y., Wang, X.Z., Chi, X.Y., Xu, Y.L., Wu, Q., Chen, D.X., 2010. Identification of chilling-responsive transcripts in peanut (*Arachis hypogaea* L.). Electron. J. Biotechnol. 14 (5). https://doi.org/10.2225/vol14-issue5-fulltext-5.
Tanou, G., Job, C., Rajjou, L., Arc, E., Belghazi, M., Diamantidis, G., Molassiotis, A., Job, D., 2009a. Proteomics reveals the overlapping roles of hydrogen peroxide and nitric oxide in the acclimation of citrus plants to salinity. Plant J. 60, 795–804.
Tanou, G., Molassiotis, A., Diamantidis, G., 2009b. Induction of reactive oxygen species and necrotic death-like destruction in strawberry leaves by salinity. Environ. Exp. Bot. 65 (2-3), 270–281.
Tanyolac, D., Ekmekc, Y., Unalan, S., 2007. Changes in photochemical and antioxidant enzyme activities in maize (*Zea mays* L.) leaves exposed to excess copper. Chemosphere 67, 89–98.
Tausz, M., 2001. The role of glutathione in plant response and adaptation to natural stress. In: Grill, D., Tausz, M., De Kok, L.J. (Eds.), Significance of Glutathione to Plant Adaptation to the Environment. Kluwer Academic Publishers, Dordrecht, pp. 101–122.
Teige, M., Scheikl, E., Eulgem, T., Doczi, R., Ichimura, K., Shinozaki, K., 2004. The MKK2 pathway mediates cold and salt stress signaling in Arabidopsis. Mol. Cell 15, 141–152.
Terzi, R., Kadioglu, A., 2006. Drought stress tolerance and antioxidant enzyme system in *Ctenanthe setosa*. Acta Biol. Cracov. Bot. 48, 89–96.
Thounaojam, T.C., Panda, P., Mazumdar, P., Kumar, D., Sharma, G.D., Sahoo, L., Panda, S.K., 2012. Excess copper induced oxidative stress and response of antioxidants in rice. Plant Physiol. Biochem. 53, 33–39.
Tosti, N., Pasqualini, S., Borgogni, A., Ederli, L., Falistocco, E., Crispi, S., Paolocci, F., 2006. Gene expression profiles of O3-treated Arabidopsis plants. Plant Cell Environ. 29, 1686–1702.
Tóth, S.Z., Puthur, J.T., Nagy, V., Garab, G., 2009. Experimental evidence for ascorbate-dependent electron transport in leaves with inactive oxygen-evolving complexes. Plant Physiol. 149, 1568–1578.
Tóth, S.Z., Nagy, V., Puthur, J.T., Kovács, L., Garab, G., 2011. The physiological role of ascorbate as photosystem II electron donor: protection against photoinactivation in heat-stressed leaves. Plant Physiol. 156 (1), 382–392.
Touiserkani, T., Haddad, R., 2012. Cadmium induced stress and antioxidative responses in different *Brassica napus* cultivars'. J. Agric. Sci. Technol. 14, 929–937.
Trouillard, M., Shahbazi, M., Moyet, L., Rappaport, F., Joliot, P., Kuntz, M., 2012. Kinetic properties and physiological role of the plastoquinone terminal oxidase (PTOX) in a vascular plant. Biochim. Biophys. Acta 1817, 2140–2148.
Tseng, M.J., Liu, C.W., Yiu, J.C., 2007. Enhanced tolerance to sulfur dioxide and salt stress of transgenic Chinese cabbage plants expressing both superoxide dismutase and catalase in chloroplasts. Plant Physiol. Biochem. 45, 822e833.
Tsuji, H., Nakazono, M., Saisho, D., Tsutsumi, N., Hirai, A., 2000. Transcript levels of the nuclear-encoded respiratory genes in rice decrease by oxygen deprivation: evidence for involvement of calcium in expression of the alternative oxidase 1a gene. FEBS Lett. 471, 201–204.
Tukendorf, A., Rauser, W.E., 1990. Changes in glutathione and phytochelatins in roots of maize seedlings exposed to cadmium. Plant Sci. 70, 155–166.

Turan, O., Ekmekci, Y., 2011. Activities of photosystem II and antioxidant enzymes in chickpea (*Cicer arietinum* L.) cultivars exposed to chilling temperatures. Acta Physiol. Plant. 33, 67–78.

Turrens, J.F., 2003. Mitochondrial formation of reactive oxygen species. J. Physiol. 552 (2), 335–344.

Ushimaru, T., Kanematsu, S., Shibasaka, M., Tsuji, H., 1999. Effect of hypoxia on the antioxidative enzymes in aerobically grown rice (*Oryza sativa*) seedlings. Physiol. Plant. 10, 181–187.

Ushimaru, T., Kanematsu, S., Katayama, M., Tsuji, H., 2001. Antioxidative enzymes in seedlings of *Nelumbo nucifera* germinated under water. Physiol. Plant. 112, 39–46.

Uzilday, B., Turkan, I., Sekmen, A., Ozgur, R., Karakaya, H., 2012. Comparison of ROS formation and antioxidant enzymes in *Cleome gynandra* (C4) and *Cleome spinosa* (C3) under drought stress. Plant Sci. 182, 59–70.

Vandenhove, H., Cuypers, A., Van Hees, M., Koppen, G., Wannijn, J., 2006. Oxidative stress reactions induced in beans (*Phaseolus vulgaris*) following exposure to uranium. Plant Physiol. Biochem. 44, 795–805.

Vanderauwera, S., Suzuki, N., Milled, G., 2011. Extranuclear protection of chromosomal DNA from oxidative stress. Proc. Natl. Acad. Sci. U. S. A. https://doi.org/10.1073/pnas.1018359108.

Vanderauwera, S., Vandenbroucke, K., Inzé, A., Cotte, B., Mühlenbock, P., Rycke, R., 2012. AtWRKY15 perturbation abolishes the mitochondrial stress response that steers osmotic stress tolerance in Arabidopsis. Proc. Natl. Acad. Sci. U. S. A. 109, 20113–20118.

Vanlerberghe, G.C., 2013. Alternative oxidase: a mitochondrial respiratory pathway to maintain metabolic and signaling homeostasis during abiotic and biotic stress in plants. Int. J. Mol. Sci. 14 (4), 6805–6847.

Vanlerberghe, G.C., Martyn, G.D., Dahal, K., 2016. Alternative oxidase: a respiratory electron transport chain pathway essential for maintaining photosynthetic performance during drought stress. Physiol. Plant. 157 (3), 322–337.

Vansuyt, G., Lopez, F., Inze, D., Briat, J.F., Fourcroy, P., 1997. Iron triggers a rapid induction of ascorbate peroxidase gene expression in *Brassica napus*. FEBS Lett. 410, 195–200.

Vassileva, V., Simova-Stoilova, L., Demirevska, K., Feller, U., 2009. Variety-specific response of wheat (*Triticum aestivum*, L.) leaf mitochondria to drought stress. J. Plant Res. 122, 445–454.

Vergara, R., Parada, F., Rubio, S., Pérez, J., 2012. Hypoxia induces H_2O_2 production and activates antioxidant defence system in grapevine buds through mediation of H_2O_2 and ethylene. J. Exp. Bot. 63, 4123–4131.

Verma, S., Dubey, R.S., 2003. Lead toxicity induces lipid peroxidation and alters the activities of antioxidant enzymes in growing rice plants. Plant Sci. 164, 645–655.

Vijayraghavan, V., Soole, K., 2010. Effect of short- and long-term phosphate stress on the non-phosphorylating pathway of mitochondrial electron transport in *Arabidopsis thaliana*. Funct. Plant Biol. 37, 455–466.

Vinit-Dunand, F., Epron, D., Alaoui-Sossé, B., Badot, P.M., 2002. Effects of copper on growth and on photosynthesis of mature and expanding leaves in cucumber plants. Plant Sci. 163 (1), 53–58.

Vishwakarma, A., Bashyam, L., Senthilkumaran, B., Scheibe, R., Padmasree, K., 2014. Physiological role of AOX1a in photosynthesis and maintenance of cellular redox homeostasis under high light in *Arabidopsis thaliana*. Plant Physiol. Biochem. 81, 44–53.

Vishwakarma, A., Bashyam, L., Senthilkumaran, B., Scheibe, R., Padmasree, K., 2015. Importance of the alternative oxidase (AOX) pathway in regulating cellular redox and ROS homeostasis to optimize photosynthesis during restriction of the cytochrome oxidase pathway in *Arabidopsis thaliana*. Ann. Bot. 116 (4), 555–569.

Wang, P., Song, C.P., 2008. Guard-cell signalling for hydrogen peroxide and abscisic acid. New Phytol. 178, 703–718.

Wang, J., Sommerfeld, M., Hu, Q., 2009a. Occurrence and environmental stress responses of two plastid terminal oxidases in *Haematococcus pluvialis* (Chlorophyceae). Planta 230, 191–203.

Wang, Y.C., Qu, G.Z., Wu, Y.J., Wang, C., Liu, G.F., Yang, C.P., 2009b. Enhanced salt tolerance of transgenic poplar plants expressing a manganese superoxide dismutase from *Tamarix androssowii*. Mol. Biol. Rep. 37, 1119–1124.

Wang, Y., Qu, G., Li, H., Wu, Y., Wang, C., Liu, G., Yang, C., 2010. Enhanced salt tolerance of transgenic poplar plants expressing a manganese superoxide dismutase from *Tamarix androssowii*. Mol. Biol. Rep. 37, 1119–1124.

Wang, W.W., Wu, Z.L., Dai, Z.L., 2013. Glycine metabolism in animals and humans: implications for nutrition and health. Amino Acids 45 (3), 463–477. https://doi.org/10.1007/s00726-013-1493-1.
Wang, Q.J., Sun, H., Dong, Q.L., Sun, T.Y., Jin, Z.X., Hao, X.J., Yao, Y.X., 2016. The enhancement of tolerance to salt and cold stresses by modifying the redox state and salicylic acid content via the *cytosolic malate dehydrogenase* gene in transgenic apple plants. Plant Biotechnol. J. 14 (10), 1986–1997.
Wawrzynski, A., Kopera, E., Wawrzynski, A., Kaminska, J., Bal, W., Sirko, A., 2006. Effects of simultaneous expression of heterologous genes involved in phytochelatin biosynthesis on thiol content and cadmium accumulation in tobacco plants. J. Exp. Bot. 57, 2173–2182.
Wen, X.P., Bane, Y., Inouea, H., Mustard, N., Moriguchia, T., 2009. Aluminum tolerance in a *spermidine synthase*-overexpressing transgenic European pear is correlated with the enhanced level of spermidine via alleviating oxidative status. Environ. Exp. Bot. 66, 471–478.
Wu, D., Wright, D.A., Wetzel, C., Voytas, D.F., Rodermel, S., 1999. The IMMUTANS variegated locus of Arabidopsis defines a mitochondrial alternative oxidase homolog that functions during early chloroplast biogenesis. Plant Cell 11, 43–55.
Wu, F.B., Chen, F., Wei, K., Zhang, G.P., 2004. Effect of cadmium on free amino acid, glutathione and ascorbic acid concentrations in two barley genotypes (*Hordeum vulgare* L.) differing in cadmium tolerance. Chemosphere 57, 447–454.
Xiang, C., Oliver, D.J., 1998. Glutathione metabolic genes coordinately respond to heavy metals and jasmonic acid in Arabidopsis. Plant Cell 10, 1539–1550.
Xie, Y.J., Xu, S., Han, B., Wu, M.Z., Yuan, X.X., Han, Y., 2011. Evidence of Arabidopsis salt acclimation induced by up-regulation of HY1 and the regulatory role of RbohD-derived reactive oxygen species synthesis. Plant J. 66, 280–292.
Xu, W., Shi, W., Liu, F., Ueda, A., Takabe, T., 2008. Enhanced zinc and cadmium tolerance and accumulation in transgenic Arabidopsis plants constitutively overexpressing a barley gene (HvAPX1) that encodes a peroxisomal ascorbate peroxidase. Botany 86, 567–575.
Xu, R., Yamada, M., Fujiyama, H., 2013. Lipid peroxidation and antioxidant enzymes of two turfgrass species under salinity stress. Pedosphere 23, 213–222.
Yang, G., Wang, Y., Xia, D., Gao, C., Wang, C., Yang, C., 2014. Overexpression of a GST gene (ThGSTZ1) from *Tamarix hispida* improves drought and salinity tolerance by enhancing the ability to scavenge reactive oxygen species. Plant Cell Tiss. Org. Cult. 117, 99–112.
Yin, L., Wang, S., Eltayeb, A.E., Uddin, I., Yamamoto, Y., Tsuji, W., Takeuchi, Y., 2009. Overexpression of dehydroascorbate reductase, but not monodehydroascorbate reductase, confers tolerance to aluminum stress in transgenic tobacco. Planta 231, 609–621.
Yin, L., Wang, S., Eltayeb, A.E., Uddin, M.I., Yamamoto, Y., 2010. Overexpression of dehydroascorbate reductase, but not monodehydroascorbate reductase, confers tolerance to aluminum stress in transgenic tobacco. Planta 231, 609–621.
Yordanova, R.Y., Popova, L.P., 2001. Photosynthetic response of barley plant to soil flooding. Photosynthetica 39 (4), 515–520.
Yoshida, S., Tamaoki, M., Shikano, T., Nakajima, N., Ogawa, D., Ioki, M., Aono, M., Kubo, A., Kamada, H., Inoue, Y., Saji, H., 2006. Cytosolic dehydroascorbate reductase is important for ozone tolerance in *Arabidopsis thaliana*. Plant Cell Physiol. 47, 304–308.
Yu, T., Li, Y.S., Chen, X.F., Hu, J., Chang, X., Zhu, Y.G., 2003. Transgenic tobacco plants overexpressing cotton glutathione S-transferase (GST) show enhanced resistance to methyl viologen. J. Plant Physiol. 160 (11), 1305–1311.
Zarei, S., Ehsanpour, A.A., Abbaspour, J., 2012. The role of over expression of P5CS gene on proline, catalase, ascorbate peroxidase activity and lipid peroxidation of transgenic tobacco (Nicotiana tabacum L.) plant under in vitro drought stress. J. Cell Mol. Res. 4 (1), 43–49.
Zelinova, V., Bocova, B., Huttova, J., Mistrik, I., Tamas, L., 2013. Impact of cadmium and hydrogen peroxide on ascorbate-glutathione recycling enzymes in barley root. Plant Soil Environ. 59, 62–67.
Zhang, F.Q., Shi, W.Y., Jin, Z.X., Shen, Z.G., 2003. Response of antioxidative enzymes in cucumber chloroplast to cadmium toxicity. J. Plant Nutr. 26, 1779–1788.
Zhang, A., Zhang, J., Ye, N., Cao, J., Tan, M., Jiang, M., 2010. ZmMPK5 is required for the NADPH oxidase-mediated self-propagation of apoplastic H_2O_2 in brassinosteroid-induced antioxidant defence in leaves of maize. J. Exp. Bot. 61, 4399–4411. https://doi.org/10.1093/jxb/erq243.

Zhang, L.T., Zhang, Z.S., Gao, H.Y., Meng, X.L., Yang, C., Liu, Z.G., Meng, Q.W., 2012. The mitochondrial alternative oxidase pathway protects the photosynthetic apparatus against photodamage in *Rumex* K-1 leaves. BMC Plant Biol. 12, 40.

Zhang, Z., Zhang, Q., Wu, J., Zheng, X., Zheng, S., Sun, X., Qiu, Q., Lu, T., 2013. Gene knockout study reveals that cytosolic ascorbate peroxidase 2 (OsAPX2) play a critical role in growth and reproduction in rice under drought, salt and cold stresses. PLoS One 8 (2), e57472.

Zhang, H., Liu, Y., Wen, F., Yao, D., Wang, L., Guo, J., 2014. A novel rice C_2H_2-type zinc finger protein, ZFP36, is a key player involved in abscisic acid-induced antioxidant defence and oxidative stress tolerance in rice. J. Exp. Bot. 65, 5795–5809.

Zhou, Y.H., Yu, J.Q., Mao, W.H., Huang, L.F., Song, X.S., Nogués, S., 2006. Genotypic variation of Rubisco expression, photosynthetic electron flow and antioxidant metabolism in the chloroplasts of chill-exposed cucumber plants. Plant Cell Physiol. 47 (2), 192–199.

Zhou, Z.S., Wang, S.J., Yang, Z.M., 2008. Biological detection and analysis of mercury toxicity to alfalfa (*Medicago sativa*) plants. Chemosphere 70, 1500–1509.

Zhou, C.P., Qi, Y.P., You, X., Yang, L.T., Guo, P., Ye, X., Zhou, X.X., Ke, F.J., Chen, L.S., 2013. Leaf cDNA-AFLP analysis of two citrus species differing in manganese tolerance in response to long-term manganese-toxicity. BMC Genomics 14 (1), 1.

Zlatev, Z., Vassilev, A., Goltsev, V., Popov, G., 2010. Drought-induced changes in chlorophyll fluorescence of young bean plants. Agric. Sci. 2 (4), 75–79.

FURTHER READING

Chen, X., Wang, Y., Li, J., Jiang, A., Cheng, Y., Zhang, W., 2009. Mitochondrial proteome during salt stress-induced programmed cell death in rice. Plant Physiol. Biochem. 47, 407–415.

Gill, S.S., Anjum, N.A., Gill, R., 2015. Superoxide dismutase—mentor of abiotic stress tolerance in crop plants. Environ. Sci. Pollut. Res. 22 (14), 10375.

Gomes-Júnior, R.A., Moldes, C.A., Delite, F.S., Pompeu, G.B., Gratão, P.L., 2006. Antioxidant metabolism of coffee cell suspension cultures in response to cadmium. Chemosphere 65, 1330–1337.

Joo, J.H., Bae, Y.S., Lee, J.S., 2001. Role of auxin-induced reactive oxygen species in root gravitropism. Plant Physiol. 126 (3), 1055–1060.

Kadioglu, A., Terzi, R., 2007. A dehydration avoidance mechanism: leaf rolling. Bot. Rev. 73, 290–302.

Kuroda, H., Sagisaka, S., Chiba, K., 1992. Collapse of peroxidescavenging systems in apple flower-buds associated with freezing injury. Plant Cell Physiol. 33, 743–750.

Ramakrishna, A., Ravishankar, G.A., 2011. Influence of abiotic stress signals on secondary metabolites in plants. Plant Signal. Behav. 6, 1720–1731.

Ramakrishna, A., Ravishankar, G.A., 2013. Role of plant metabolites in abiotic stress tolerance under changing climatic conditions with special reference to secondary compounds. In: Climate Change and Plant Abiotic Stress Tolerance. Wiley-VCH Verlag GmbH & Co. KGaA, Germany, pp. 705–726.

Schonberg, A., Baginsky, S., 2012. Signal integration by chloroplast phosphorylation networks: an update. Front. Plant Sci. 3, 256.

Tallon, C., Quiles, M.J., 2007. Acclimation to heat and high light intensity during the development of oat leaves increases the NADH DH complex and PTOX levels in chloroplasts. Plant Sci. 173, 438–445.

Vishwakarma, K., Upadhyay, N., Kumar, N., Yadav, G., Singh, J., Mishra, R.K., Kumar, V., Verma, R., Upadhyay, R.G., Pandey, M., Sharma, S., 2017. Abscisic acid signaling and abiotic stress tolerance in plants: a review on current knowledge and future prospects. Front. Plant Sci. 8, 161. https://doi.org/10.3389/fpls.2017.00161 (vol. 22(11), 2623–2633).

CHAPTER 11

Compatible Solute Engineering of Crop Plants for Improved Tolerance Toward Abiotic Stresses

Titash Dutta*, Nageswara R.R. Neelapu*, Shabir H. Wani†,‡, Surekha Challa*
*Department of Biochemistry and Bioinformatics, Gandhi Institute of Technology and Management (GITAM), Deemed-to-be University, Visakhapatnam, India
†Mountain Research Centre for Field Crops, Sher-e-Kashmir University of Agricultural Sciences and Technology, Kashmir, India
‡Department of Plant, Soil, and Microbial Sciences, Michigan State University, East Lansing, MI, United States

Contents

1. INTRODUCTION

The world population is growing exponentially, and recent demographic studies estimate it to reach 9.1 billion by 2050 (Sah et al., 2016). To meet this alarming population growth, agricultural production should increase proportionally, which is not happening. It is estimated that the global agricultural yield of major crops such as rice, maize, and wheat should increase by 60% to strike a balance with this population explosion (FAO, 2010).

Abiotic stress is the major hindrance to crop production. Abiotic stress reduces seed quality and pollen viability. A recent statistical report claims that abiotic stress factors contribute to an approximately 70% reduction of global crop yields (Acquaah, 2007). Plants, being sessile, are constantly exposed to abiotic stresses such as drought, salinity, and extreme temperatures. These stress conditions are detrimental for plant growth and productivity as they severely affect plant homeostasis, increase reactive oxygen species (ROS) levels, and subsequently lead to plant death. Moreover, global climate changes and less rainfall have enhanced the frequency of these abiotic stresses, thereby severely

Biochemical, Physiological and Molecular Avenues for Combating Abiotic Stress in Plants
https://doi.org/10.1016/B978-0-12-813066-7.00012-7

affecting global food security. A significant loss in the productivity of major cereal crops, including wheat, maize, and barley, is due to climate change (Jaleel et al., 2009).

Plant stress response involves a cascade of biochemical events that take place at the molecular level. It involves transcriptional regulation of specific gene families involved in conferring stress tolerance. These genes have been categorized into three groups. The first group includes osmoprotectants, antioxidant enzymes, late embryogenesis abundant (LEA) proteins, and heat shock proteins (HSPs). The second group includes the genes encoding for ion transporters such as Na^+/H^+ channels and aquaporins involved in water uptake. The third group includes the genes mitogen-activated protein (MAPKs) kinases, salt overly sensitive (SOS) kinases (Ji et al., 2013), ethylene-responsive element binding proteins, heat shock transcriptional factors (HSF), CBF/DREB (C-repeat-binding/dehydration-responsive element), bZIP (basic-domain leucine zipper), NAC (NAM, ATAF, and CUC), MYC (myelocytomatosis oncogene)/MYB (myeloblastosis oncogene), Cys2/His2 zinc-finger motifs, and WRKY protein domains (Umezawa et al., 2006) involved in signal perception and transcription control (Surekha et al., 2015).

In recent years, plant breeders have been successful in developing stress-tolerant cultivars by exploiting transgenic approaches through conventional breeding. These techniques are being pursued worldwide to improve the qualitative and quantitative traits that include tolerance to abiotic stress; predominantly salt, drought, and cold stress (Wani et al., 2016; Bhatnagar-Mathur et al., 2008).

Abiotic stresses, particularly salinity and drought, lead to intracellular water loss, or cellular dehydration. To prevent and protect the cellular proteins, plants accumulate many organic compounds called osmolytes. They are characterized by low molecular weight and high solubility. The major osmolytes that are accumulated in plants include proline, betaine, polyols, sugar alcohols, and soluble sugars. Osmolyte biosynthesis and accumulation under abiotic stress conditions are intricate mechanisms evolved in plants to maintain cellular integrity and survival. They are also involved in scavenging ROS, balancing cellular redox potential, osmotic adjustments, and stabilizing cellular pH, proteins, and membranes (Neelapu et al., 2015).

Thus, osmoprotection involves the upregulation of compatible solutes (osmolytes) in conferring tolerance to abiotic stress and secondary stress, such as osmotic and ionic stress. Transcriptome analysis of major crop plants, and in the model plant *Arabidopsis thaliana*, under the influence of abiotic stress conditions, has led to the identification of many genes associated with the biosynthesis of these osmolytes (Suprasanna et al., 2016). In short, the goal is to develop transgenic plants harboring candidate osmoprotectant genes that can impart abiotic stress tolerance as well as superior yield and biomass. Transgenic plants that have been successfully developed to improve grain yield and abiotic stress tolerance by inducing osmoprotection include rice (Garg et al., 2002; Su and Wu (2004), potatoes (Zhang et al., 2011b), wheat (Sawahel and Hassan, 2002), tomatoes (Park et al., 2007), maize (Quan et al., 2004), tobacco (Szabados and Savouré, 2010), and pigeon peas (Surekha et al., 2014).

This chapter identifies potential osmolytes and their role in plant stress physiology. Studies on their biosynthetic pathways revealed the candidate genes responsible for the synthesis of these compatible solutes. Moreover, it highlights the latest research endeavors carried out with respect to developing successful transgenic plants harboring the biosynthetic pathway genes to enhance osmolyte production, thereby conferring abiotic stress tolerance.

2. COMPATIBLE SOLUTE-MEDIATED ABIOTIC STRESS RESPONSE

Salinity, drought, and extreme temperature are the major abiotic stresses responsible for severe crop damage worldwide. They cause morphological, physiological, metabolic, and molecular changes that are detrimental for plant growth and survival. Abiotic stress, and predominantly salinity, lead to plant growth inhibition, as depicted by the biphasic response: the initial phase (cellular homeostasis disruption) and the final phase (ionic homeostasis disruption) (Adem et al., 2014; Li et al., 2015). Plants have gradually evolved three vital strategies (ion exclusion, compartmentalization, and osmoprotection) to counter this problem and induce abiotic stress tolerance. Osmoprotection in plants involves stress-mediated upregulation of certain low-molecular-weight compounds called osmolytes. They are polar, highly soluble, and hydrophilic in nature, which enables them to shield the protein membrane structures during prolonged water-deficient conditions. The major osmolytes accumulating in plants in response to abiotic stress can be grouped into three classes: (i) free amino acids (proline); (ii) quaternary amines (glycine betaine and polyamines); (iii) sugars and sugar alcohols (tehalose, fructans, mannitol, and sorbitol) (Roychoudhury et al., 2015).

Osmolytes' mediated abiotic stress response is achieved by subcellular compartmentation of osmolytes, which reduces the water osmotic potential, thereby contributing to osmoregulation. Niu et al. (1995) reported that this reduced water potential helps regulate optimum tissue water content in saline environments. Additionally, they act as molecular chaperones by preventing protein misfolding, thereby stabilizing integral protein and membrane structures. They stabilize these structures by establishing strong hydrogen bonding with them, thereby preventing their denaturation and increasing their integrity simultaneously (Kumar, 2009; Slama et al., 2015). They also serve as hydroxyl radical scavengers (Ozgur et al., 2013). Abiotic stress leads to severe oxidative damage, generating high levels of toxic ROS, which damages vital cellular organelles. Osmolytes scavenge hydroxyl radicals and singlet oxygen (Roychoudhury and Chakraborty, 2013). They also lower the lipid peroxidation level, which is an indication of plant ROS status. Researchers worldwide exploited these important roles played by osmolytes in developing transgenic plants. The transgenic cultivars are engineered to express specific genes coding for the synthesis of these compatible solutes, leading to enhanced solute accumulation and abiotic stress tolerance (Wani et al., 2017; Singh et al., 2015).

3. COMPATIBLE SOLUTE ENGINEERING: BIOSYNTHESIS AND ACCUMULATION OF OSMOLYTES

3.1 Proline

Proline is an α-amino acid derivative compound, and it plays diverse roles in plant growth, development, and stress physiology. Proline is considered an ideal solute, as it is highly soluble in water (1.54 kg/L water) (Huang et al., 2008). Various plant species synthesize proline naturally under nonstressed conditions, as well as during adaptive stress response. Under nonstressed conditions, proline serves as a precursor for proteins and enzyme biosynthesis (Nanjo et al., 1999), and regulates embryo and seed formation during plant growth (Mattioli et al., 2009). Superior conformational rigidity of proline makes it an ideal component of protein structure (Lehmann et al., 2010; Funck et al., 2012), and aids in protein's secondary structure stability. Additionally, proline functions as a storage sink for cellular carbon and nitrogen sources during the plant recovery phase (Kishor et al., 2005).

High proline accumulation in plants is often correlated with abiotic stress factors such as salinity and drought. In plants, proline functions as an osmoprotectant, and alleviates the deleterious effects of oxidative and osmotic stress induced by salinity. During saline stress, proline accumulation increases in cytosol, and is involved in osmotic adjustment, regulation of cellular redox potential (Heuer, 2003), protein and membrane structural stability during dehydration stress (Suprasanna et al., 2014), and it alleviates oxidative stress by scavenging free radicals (Kavi Kishor and Sreenivasulu, 2014; Matysik et al., 2002).

The potential of proline as a compatible solute can only be harnessed when plants are able to produce significant levels of proline endogenously. Taxonomical analysis suggests that major cereal crops such as wheat, rice, maize, and so forth, are unable to accumulate enough proline to effectively respond to salt and drought stress, promoting plant survival (Slama et al., 2015). Under such a scenario, exogenous application of proline in stressed plants was able to enhance endogenous proline accumulation, which contributed to plant stress tolerance (Ashraf and Foolad, 2007). It has been found to enhance seed germination (Shaddad, 1990), yield, photosynthetic activity (Kahlaoui et al., 2013), peroxidase activity (Huang et al., 2009), and water and ion uptake by plants. All these factors combine to promote plant growth and survival, thereby conferring abiotic stress tolerance.

Numerous studies highlight the role of the exogenous application of proline in inducing salt stress tolerance. Khedr et al. (2003) reported that the exogenous supply of proline was able to increase the low protein content and stunted growth in *Pancratium maritium* growing in 300 mM NaCl. Gadallah (1999) studied the impact of the exogenous proline supply to *Vicia faba* plants at 200 mM NaCl. He reported that exogenous proline application increased leaf chlorophyll content, leaf water retention content, and stabilized membranes and enhanced plant survival. Similar results have been obtained by studying the exogenous application of proline in *Medicago sativa* (Ehsanpour and Fatahian, 2003), *Mesembryanthemum crystallinum* L. (Shevyakova et al., 2009), *A. thaliana* (Hare et al., 2003), rice (Bhusan et al., 2016), maize (Alam et al., 2017), and soybeans (Sabagh et al., 2017). In certain scenarios, the accumulation of solutes by an exogenous supply was not found to be compatible with the internal plant environment, and caused harmful side effects (Sulpice et al., 1998). This prompted many researchers to study the mechanisms underlying the synthesis of these osmolytes and develop transgenic plants equipped to upregulate the endogenous level of osmolytes, thereby conferring stress tolerance.

Proline biosynthesis in plants takes place in cytosol, and can be achieved either by glutamate or ornithine pathways (Kishor et al., 2005; Suprasanna et al., 2014) (Fig. 1). The glutamate pathway involves two enzyme-dependent steps and two molecules of NADPH. In the first step, glutamate is activated by ATP-dependent phosphorylation, and then reduced to glutamatic-γ-glutamyl kinase (GSA). GSA undergoes spontaneous cyclization, forming pyrroline 5-carboxylate (P5C) by the enzyme pyrroline 5-carboxylate synthetase (P5CS). In the second and final step of the pathway, the pyrroline 5-carboxylate (P5C) is reduced to proline catalyzed by the enzyme pyrroline 5-carboxylate reductase (P5CR). In the ornithine pathway, ornithine serves as the precursor for proline biosynthesis in place of glutamate. Ornithine undergoes transamination to form pyrroline 5-carboxylate (P5C) by the enzyme Orn-D-aminotransferase. This is followed by reduction of P5C to proline by P5CR.

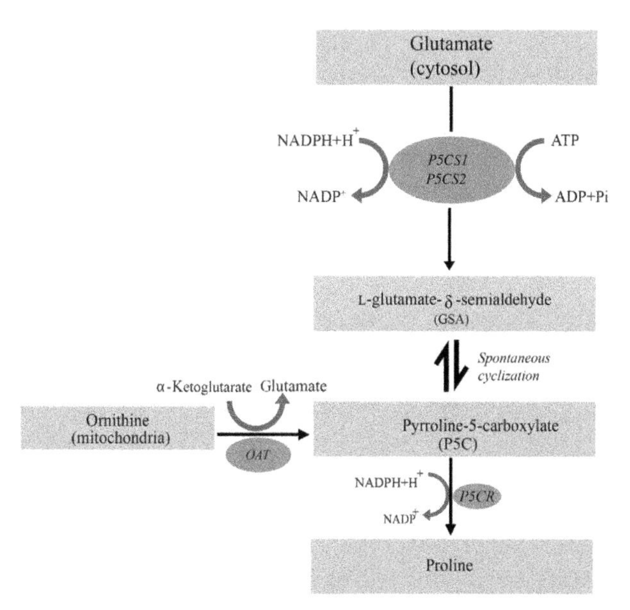

Fig. 1 Proline biosynthesis pathway. *P5CS1/P5CS2*, pyrroline-5-carboxylate synthetase; *P5CR*, pyrroline 5-carboxylate reductase; *OAT*, ornithine D-aminotransferase.

The proline biosynthetic pathway has been extensively studied and characterized in *A. thaliana* (Yoshiba et al., 1995), tobacco (Delauney and Verma, 1993), rice (Lutts et al., 1999), and other plant species. Studies revealed that the P5CS enzyme is encoded by two genes designated as *P5CS1* and *P5CS2*, whereas P5CR is encoded by a single gene. These candidate genes have been engineered in a variety of agricultural crops developing transgenic plants with enhanced endogenous proline accumulation, thereby alleviating the deleterious effects of osmotic stress induced by salt and drought stress.

Kishor et al. (1995) overexpressed the *Vigna aconitifolia* (moth bean) *P5CS* gene in transgenic tobacco plants. The transgenic cultivars exhibited 10–18 times the enhanced proline accumulation than its wild type. Moreover, root biomass and flower development were enhanced due to an increased accumulation of proline. Despite the success of increased proline synthesis, the rate limiting nature of the P5CS enzyme leads to its feedback inhibition by proline. To counter this problem, Hong et al. (2000) developed a mutated version of the *V. acontifolia* P5CS enzyme designated as *P5CS129A*. Site-directed mutagenesis in *V. aconitifolia* P5CS was carried out to replace the phenylalanine (Phe) residue at the 129th position with an alanine (Ala) residue. The mutated enzyme was able to synthesize twice as much proline in transgenic tobacco plants. Kumar et al. (2010) overexpressed the mutated *P5CS129A* gene in rice cultivars. The transgenic rice variety showed enhanced proline accumulation, growth, and biomass. They were also characterized by low lipid peroxidation levels under salt stress (150 mM NaCl concentration).

Overexpression of the same mutated *V. acontifolia P5CSF129A* gene into *Cajanus cajan* enhanced proline accumulation and salt tolerance (Surekha et al., 2014). The first generation (T_1) transgenic plants accumulated fourfold more proline when compared with the wild types under 200 mM NaCl stress. As a result, significant levels of proline accumulation, transgenic plants were characterized with better growth, more chlorophyll content, and low lipid peroxidation levels. Gleeson et al. (2005) developed transgenic *Larix leptoeuropaea* (forest tree) expressing the *V. acontifolia P5CS*. Proline content analysis depicted a 30-fold increase in proline levels in the transgenic cultivars as compared with control groups. The transgenic lines were more tolerant, and showed substantial development under salt (200 mM NaCl) and freezing stress (4°C) conditions.

Hmida-Sayari et al. (2005) successfully engineered potato plants with the Arabidopsis *P5CS* gene under a stress-inducible promoter. The transgenic potato cultivars accumulated high proline content in 100 mM NaCl stress, and showed enhanced tuber yield and biomass in comparison with its wild type. In another study, Zhang et al. (2014) coexpressed rice *OsP5CS1* and *OsP5CS2* genes in tobacco. The second generation transgenic plants exhibited 2.3 times more root length, 3.2 times proline content, and 3.9 times average fresh weight when compared with control cultivars under 200 mM NaCl stress. These results highlighted the importance of the P5CS enzymes in enhanced proline accumulation, as well as reduced cellular oxidative damage under the influence of the abiotic stress environment. Chen et al. (2013) carried out similar coexpression of *Phaseolus vulgaris P5CS* (*PvP5CS1* and *PvP5CS2*) in *A. thaliana*. The transgenic plants were grown under different salt concentrations (0, 100, and 200 mM NaCl) and the transgenic plants accumulated 1.6 times and 1.9 times proline at 100 and 200 mM salt concentrations, respectively, in comparison with the wild types.

A similar increase in proline production has been carried out exploiting the proline biosynthetic and degradation genes (P5CS, P5CR, OAT, and P5CDH) in transgenic sugarcane (Guerzoni et al., 2014), rice (Su and Wu, 2004; Karthikeyan et al., 2011; You et al., 2012), wheat (Vendruscolo et al., 2007), olives (Behelgardy et al., 2012), carrots (Han and Hwang, 2003), Jerusalem artichokes (Huang et al., 2013), *Kosteletzkya virginica* (Wang et al., 2015), *Medicago truncatula* (Verdoy et al., 2006), and sweet potatoes (Liu et al., 2014) that confer tolerance to salt and drought stress. Table 1 contains the details of successful transgenic plants that have been developed to enhance proline accumulation leading to plant growth and survival in abiotic stress conditions.

3.2 Glycine Betaine

Glycine betaine (GB) is an *N*-trimethyl glycine derivative compound that belongs to the class of quaternary amines. GB is amphoteric in nature, highly water soluble, nontoxic at high concentrations, and has a low molecular-weight organic metabolite. As it is bipolar, it can interact with both hydrophilic as well as hydrophobic chains of macromolecules (Gupta and Huang, 2014). Among plants, sugar beet, maize, spinach, and barley accumulate small amounts of GB naturally (Kishitani et al., 1994; Chen and Murata, 2008). However, enhanced GB accumulation in these natural accumulators is triggered due to exposure to abiotic stresses; predominantly salt, drought, and extreme temperatures (Wani et al., 2013). The major role of GB in plants is concerned with eradicating the deleterious effects imposed by stressed environments on plant growth and development. In such natural accumulator plants, GB functions as an osmoprotectant by maintaining the cellular osmolarity, protecting the photosynthetic machinery (photosystem II) and thylakoid membranes, alleviating cellular oxidative damage and stabilizing protein structures (Khan et al., 2009; Allakhverdiev et al., 2003; Mäkelä et al., 2000).

Although scientists have gained vital insights regarding the role of GB in plant stress tolerance, the major concern for plant breeders lies in the inability of economically important plants (viz., potato, rice, eggplant, tomatoes, etc.) to synthesize GB naturally (De Zwart et al., 2003; Park et al., 2004). Loss of functional domains, truncated transcripts, or premature stop codons of the GB biosynthetic genes are the major factors attributed to the failure of plants from synthesizing GB naturally. To overcome this obstacle, these agriculturally important plants were subjected to exogenous GB application to enhance the endogenous GB accumulation. Additionally, these species were genetically transformed with potential GB biosynthetic genes able to enhance GB production and confer stress tolerance (Khan et al., 2009; Chen and Murata, 2011).

Table 1 List of transgenic plants overexpressing candidate genes for proline accumulation

Sl no.	Transgene	Host	Target crop plants/trees	Remarks	Reference
1	*P5CS*	*Vigna aconitifolia*	*Nicotiana tabacum*	Enhanced 10–18-fold proline accumulation and induced salt tolerance.	Kishor et al. (1995)
2	*P5CSF129A*	*V. aconitifolia*	*O. sativa*	Enhanced proline accumulation and salt tolerance.	Kumar et al. (2010)
3	*P5CS*	*V. aconitifolia*	*Larix leptoeuropaea*	Alleviates oxidative stress, high chlorophyll content and 30-fold enhanced proline accumulation.	Gleeson et al. (2005)
4	*P5CS*	*V. aconitifolia*	Sugarcane	Enhanced proline content, biomass production, low lipid peroxidation level, and oxidative stress protection.	Guerzoni et al. (2014)
5	*P5CSF129A*	*Vigna acontifolia*	*Cajanus cajan*	Enhanced proline accumulation, seed germination rate, chlorophyll content, low lipid peroxidation level.	Surekha et al. (2014)
6	*P5CS*	*Vigna aconitifolia*	Carrot	Enhanced proline accumulation, biomass production, and low lipid peroxidation level.	Han and Hwang (2003)
7	*P5CS*	*Vigna aconitifolia*	*T. aestivum*	Enhanced proline accumulation, biomass production, and low lipid peroxidation level.	Sawahel and Hassan (2002)
9	*PvP5CS1, PvP5CS2*	*P. vulgaris*	*A. thaliana*	Increased 1.9 times proline content, flower and seed development.	Chen et al. (2013)

Table 1 List of transgenic plants overexpressing candidate genes for proline accumulation—cont'd

Sl no.	Transgene	Host	Target crop plants/trees	Remarks	Reference
10	*P5CR*	*A. thaliana*	*Glycine max* L.	Increased proline accumulation, RWC and WUE content, high spikelet fertility.	De Ronde et al. (2000)
11	*P5CS*	*A. thaliana*	*Olea europaea*	Increased proline accumulation, ionic homeostasis.	Behelgardy et al. (2012)
12	*P5CS*	*A. thaliana*	*S. tuberosum*	Enhanced proline accumulation, tuber yield and biomass production.	Hmida-Sayari et al. (2005)
13	*OsP5CS1* and *OsP5CS2*	*O. sativa*	*N. tabacum*	Enhanced 3.2 times proline content, biomass production, oxidative stress protection.	Zhang et al. (2014)
14	OAT	*O. sativa*	*O. sativa*	Proline accumulation, water retention, spikelet fertility and increased biomass.	You et al. (2012)
15	*HtP5CS*	*Helianthus tuberosus* L.	*Helianthus tuberosus* L.	Enhanced proline accumulation and salt tolerance.	Huang et al. (2013)
16	*IbP5CR*	*Ipomoea batatas* (L.)	*Ipomoea batatas* (L.)	Enhanced proline accumulation and salt tolerance.	Liu et al. (2014)
17	*P5CS*	*Vigna aconitifolia*	*Medicago truncatula*	Enhanced proline accumulation and salt tolerance.	Verdoy et al. (2006)
18	*KvP5CS1*	*Kosteletzkya virginica*	*Kosteletzkya virginica*	Enhanced 6.83 times proline content and salt tolerance.	Wang et al. (2015)
19	*LrP5CS1*, *LrP5CS2*	*Lilium regale*	*A. thaliana*	Enhanced proline accumulation and tolerance to salt, drought and osmotic stress.	Wei et al. (2016)

Extensive research has been conducted to validate the outcome of exogenous GB application in plants that are considered non-GB accumulators. Significant plant growth, survival, and product yield was observed in barley, soybean, wheat, maize, tobacco, beans, and sunflowers (Ashraf and Foolad, 2007). Reddy et al. (2013) supplied maize growing under drought stress with exogenous GB (4 kg/ha). There was sharp increase in plant height, leaf area, biomass, and chlorophyll content in the treated plants in comparison with the controls. Hossain et al. (2011) observed a low-lipid peroxidation level and H_2O_2 level, and enhanced antioxidant enzyme activity in mung beans growing in 300 mM NaCl stress. The treated plants withstood the oxidative damage induced by the high-salt environment, which led to plant survival.

Exogenous application of 50 mM GB to rice plantlets subjected to 150 mM salt stress improved their photosynthetic rate, seed fertility, and grain weight in comparison with controls (Maziah and Teh, 2016). Rezaei et al. (2012) treated salt-stressed *Glycine max* and *Zea mays* to 1.0 and 7.5 kg/ha GB, respectively. The treated plants showed a remarkable increase in seed weight, rate of seed germination, and grain quality; thereby inducing stress tolerance in the respective plants. Exogenous application of GB (200 mM) and a polyamine (Spermidine at 0.1 mM) for 3 weeks in creeping bentgrass (*Agrostis stolonifera*) resulted in osmotic adjustment and increased activity of antioxidant enzymes (catalase and ascorbate peroxidase) under drought stress (Liu et al., 2017a). Overexpression of *LcSAMDC1* in *Arabidopsis* improved cellular levels of polyamines(spermidine), proline and chlorophyll inducing tolerance to salt and cold stress (Liu et al., 2017b). Similarly, 50 μM GB foliar spray on *C. cajan* L. plants improved plant growth, lowered levels of active oxygen species (AOS) and lipoxygenase, improved proline accumulation, and enhanced antioxidant enzyme activity under fluoride and salt stress (Yadu et al., 2017). Although exogenous application of GB has yielded satisfactory results in inducing abiotic stress tolerance, the main objective among breeders is to enhance the endogenous levels of GB. The biosynthetic pathway of GB plays an integral role in identifying the key enzymes involved and their utilization for developing transgenic plants with enhanced endogenous GB production.

In higher plants, GB biosynthesis is achieved by two distinct pathways using two different precursor molecules: choline and glycine, respectively (Sakamoto and Murata 2002). The synthesis pathway involving choline as the substrate takes place in two steps in the stroma of the chloroplast. In the first step, choline is oxidized to betaine aldehyde by the enzyme choline monooxygenase (CMO). The next step oxidizes betaine adehyde to glycine betaine (GB) by the enzyme betaine aldehyde dehydrogenase (BADH) using NAD^+ as a cofactor (Wani et al., 2013; Chen and Murata, 2002). In *Escherichia coli*, GB is synthesized by the choline dehydrogenase enzyme (CDH) and BADH enzymes, while in the case of *Arthrobacter globiformis* (soil bacterium), the enzyme choline oxidase A (*codA*) catalyzes the conversion of choline into GB and hydrogen peroxide (H_2O_2) in one step. The schematic overview of the glycine betaine biosynthetic pathways is shown in Fig. 2.

The advent of genetic engineering has led to extensive research aimed at improving endogenous GB levels in agriculturally important crop plants including rice, potatoes,

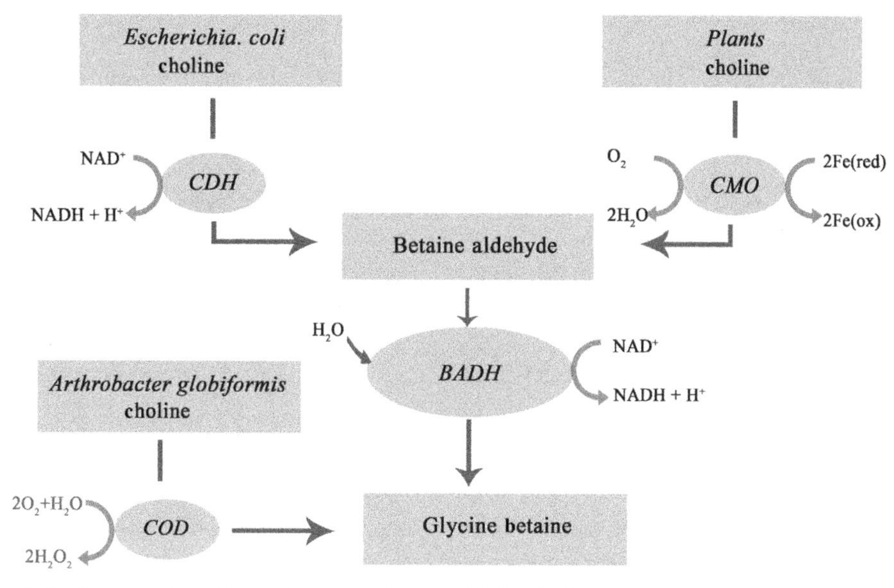

Fig. 2 Gycine betaine biosynthesis pathway. *CMO*, choline monooxygenase; *CDH*, choline dehydrogenase; *COD*, choline oxidase; *BADH*, betaine aldehyde dehydrogenase.

soybeans, ground nuts, maize, and other plants against abiotic stress environments (Sawahel, 2003; Ranganayakulu et al., 2013). The gene coding for the biosynthetic enzymes (bacterial *codA*, CMO and BADH) have been used to develop successful transgenic crops with enhanced GB accumulation and stress tolerance. Among these, the *codA* gene isolated from the *Arthobacter* spp. has been overexpressed in a variety of plants, such as *A. thaliana*, *Eucalyptus globulus*, *Brassica campestris* L. spp. *chinensis*, *Solanum tuberosum*, *Z. mays*, *Solanum lycopersicum*, and *Lycopersicon esculentum* to develop abiotic stress-tolerant cultivars (Giri, 2011; Quan et al., 2004; Wei et al., 2017). The transformed cultivars exhibited enhanced GB accumulation, leading to better photosynthetic activity, growth, and crop yield. Sakamoto and Murata (1998) developed transgenic rice cultivars overexpressing the *A. globiformis codA* gene. The transformed rice varieties enhanced accumulation of GB (5.3 mmol/g fresh weight) in comparison with the wild types (nonaccumulators of GB). Park et al. (2004) used the same *codA* gene to develop genetically engineered tomatoes. The transgenic lines showed fivefold higher GB content in leaves, and they produced 10%–30% more fruit than WT plants when subjected to low temperatures (4°C). The same transgenic tomato variant, when subjected to salt stress (200 mM NaCl), accumulated threefold higher GB content in comparison with the WT. The high GB accumulation resulted in a high photosynthetic rate, antioxidant enzyme activity, and low peroxidation levels in the transgenic lines. Ahmad et al. (2014) developed transgenic potato plants expressing the *A. globiformis codA* gene and subjected them to salt (100 mM NaCl) and chilling (4°C) stress. The transgenic plants exhibited normal growth, and high tuber yield under salt and cold stress conditions. The tuber yield was 44 g/plant in the transgenic lines in comparison with 33.6 g/plant in the case of nontransformed plants. The chlorophyll content was reduced by 20% in nontransformed plants, while a 16% reduction in cold stress was observed in the transgenic plants, thereby increasing the photosynthetic rate in the transformed plant. These results emphasize that the GB accumulation in stressed plants improves plant growth and development, and confers abiotic stress tolerance.

Wei et al. (2017) used the same *A. globiformis codA* gene encoding for GB to transform tomato plants. The transgenic tomato plants accumulated significant levels of GB while the wild types did not accumulate any. The transgenic lines showed higher photosynthetic rates and antioxidant enzyme activities, and lower ROS content in comparison with wild types. Additionally, K^+ efflux was decreased, while the Na^+ efflux increased in the transgenic lines growing in 200 mM NaCl stress.

Transgenic sweet potatoes were developed by overexpressing the chloroplastic *BADH* gene from *Spinacia oleracea* (*SoBADH*) (Fan et al., 2012). GB accumulation increased by twofold in the transformed lines, which induced tolerance against salinity, low temperature, and oxidative stresses. The transformed lines showed increased protection against cell damage by maintenance of cell membrane integrity, high photosynthetic activity, and increased rates of ROS scavenging. Similar results have been obtained when *BADH* gene from spinach was overexpressed in transgenic potato under salt and drought stress (Zhang et al., 2011a). Li et al. (2014) developed transgenic tomato overexpressing of the *BADH* gene from *S. oleracea*. The transgenic tomatoes showed high photosynthetic capacity in comparison with wild types at 42°C. This tolerance was attributed to high D_1 protein content, which prevented the denaturation of photosystem II under prolonged heat stress. The H_2O_2 content and superoxide radical content were alleviated in the transgenic varieties. Wang et al. (2010) overexpressed the cyanobacteria BADH gene in economically important wheat crops. The transgenic lines accumulated higher GB in the chloroplast than WT under a 150 mM salt concentration. GB accumulation was recorded as 170.7 μmol/g dry weight, which induced tolerance to salt as well as drought stress.

Zhang et al. (2008) introduced the sugar beet *CMO* gene into tobacco plants. The transgenic plants thrived in 200 mM salt concentrations and showed enhanced GB levels in chloroplasts ranging from 0.2 to 0.5 mmol g^{-1} FW in comparison with the nonaccumulator WTs. Therefore, we see that many successful transgenic plants have been developed harboring the candidate genes for GB biosynthesis, while many more are in development and trail stages to counter the various abiotic stresses and induce plant growth and survival. Table 2 contains the list of successful transgenic plants developed to synthesize and enhance GB accumulation endogenously.

3.3 Polyamines

Polyamines (PAs) are low-molecular-weight aliphatic amines universally present across bacteria, animals, and plants (Hussain et al., 2011). The major polyamines identified in

Table 2 List of transgenic plants overexpressing candidate genes for glycine betaine accumulation

Sl no.	Transgene	Host	Target crop plants/trees	Remarks	Reference
1	*codA*	*Arthrobacter globiformis*	*Oryza sativa* L.	Enhanced 5.3-fold GB accumulation and induced salt tolerance.	Sakamoto and Murata (1998)
2	*codA*	*Arthrobacter globiformis*	*Solanum lycopersicum*	Enhanced fivefold GB accumulation and induced salt tolerance.	Park et al. (2004)
3	*codA*	*Arthrobacter globiformis*	*Solanum tuberosum*	Enhanced GB content and tolerance to salt and chilling stress.	Ahmad et al. (2014)
4	*SoBADH*	*Spinacia oleracea*	*Ipomoea batatas* (L.)	Enhanced GB accumulation and tolerance to salinity, low temperature and oxidative stress.	Fan et al. (2012)
5	*codA*	*Arthrobacter globiformis*	*Solanum lycopersicum*	Induced GB accumulation, antioxidant enzyme activity and salt tolerance.	Wei et al. (2017)
6	*SoBADH*	*Spinacia oleracea*	*Solanum lycopersicum*	Enahanced GB accumulation and heat stress.	Li et al. (2014)
7	*BADH*	*Cyanobacteria*	*Triticum aestivum*	Induced 170.7 μM/g GB accumulation and salt tolerance.	Wang et al. (2010)
9	*BvCMO*	*Beta vulgaris*	Nicotiana tobacum	Enhanced fivefold GB accumulation and salt tolerance.	Zhang et al. (2008)
10	*GSMT* and *DMT A*	*Aphanothece halophytica*	*O. sativa*	Enhanced GB biosynthesis, salt and cold stress tolerance.	Niu et al. (2014)
11	*BADH*	*E. coli*	*M. sativa*	Enchanced GB accumulation and salt tolerance.	Yan et al. (2012)
12	*OsBADH1*	*O. sativa*	*N. tabacum*	Salinity stress tolerance.	Hasthanasombut et al. (2010)

Continued

Table 2 List of transgenic plants overexpressing candidate genes for glycine betaine accumulation—cont'd

Sl no.	Transgene	Host	Target crop plants/trees	Remarks	Reference
13	*SlBADH*	*Suaeda liaotungensis*	*Solanum lycopersicum*	Enhanced GB accumulation under salt stress.	Wang et al. (2013)
14	*codA*	*Arthrobacter globiformis*	*Lycopersicon esculentum*	Enhanced seed germination rate, GB accumulation and tolerance to salt and drought.	Goel et al. (2011)
15	*codA]*	*Arthrobacter globiformis*	*N. tabacum*	Enhanced antioxidant enzyme activity by 50% and salt tolerance.	Jing et al. (2013)
16	*betA*	*Escherichia coli*	*Zea mays*	Enhanced GB, Sed germination rate and tolerance to drought.	Quan et al. (2004)
17	*codA*	*Arthrobacter globiformis*	Transgenic poplar plants	Enhanced GB accumulation, photosystem-II integrity, and multiple stress tolerance.	Ke et al. (2016)

plants include putrescine (Put), spermidine (Spd), and spermine (Spm) (Tiburcio et al., 2014; Handa and Mattoo, 2010). PAs play an important role in plant growth, development, and stress physiology. Recent studies highlighted that PAs participate in vital cellular processes such as cell division, differentiation, transcriptional regulation, and protein translation. Additionally, they also play an integral role in plant growth and development during elongation, floral development, fruit ripening, leaf senescence, and cellular apoptosis (Alcázar et al., 2011; Feng et al., 2011; Wimalasekera et al., 2011a; Zhang et al., 2011b; Alet et al., 2012; Tavladoraki et al., 2012). Moreover, transcriptome analysis of these polyamines in major plants found them to be associated with stress tolerance during exposure to salt, drought, low and high temperature, and oxidative stresses (Alcázar et al., 2010; Wimalasekera et al., 2011b; Tavladoraki et al., 2012).

In plants, the diamine putrescine (Put) is synthesized from arginine (Arg) sequentially by the action of arginine decarboxylase (ADC), agmatine iminohydrolase (AIH), and *N*-carbamoylPut amidohydrolase (CPA). The alternative pathway for Put biosynthesis involves conversion of ornithine to Put in a single reaction catalyzed by ornithine decarboxylase (ODC). In the next step, spermidine synthase (SPDS) catalyzes the synthesis of spermidine (triamine) from Put. This is followed by the conversion of spermidine to spermine (tetramine) by the enzyme spermine synthase (SPMS) (Fig. 3). Decarboxylated *S*-adenosylmethionine (dcSAM) serves as the aminopropyl group donor, required for both the conversions of Put to spermidine and spermidine to spermine. Decarboxylated *S*-adenosylmethionine (dcSAM) is synthesized from methionine in two steps, catalyzed by the enzymes by *S*-adenosylmethionine (SAM) synthase and SAM decarboxylase (SAMDC) (Takahashi et al., 2010; Alcázar et al., 2010; Moschou et al., 2008).

As scientists worldwide developed effective ways to reduce the drastic impacts of abiotic stress, certain studies revealed that exogenous application of polyamines (Put, Spd, Spm) were effective in inducing tolerance in saline- and drought-stressed plants. Exogenous polyamines' application in relation to abiotic stress tolerance has been studied on many plants species, such as tomatoes, rice, barley, cucumbers, soybeans, wesh onions, and so forth (Yiu et al., 2009; Li et al., 2013; Sagor et al., 2013; Sequera-Mutiozabal et al., 2017). Similarly, Ali (2000) presoaked *Atropa belladonna* plants that were subjected to 200 mM NaCl stress in a 10^{-2} mM putrescine solution. The treated plants showed reduced accumulation of sodium and chloride ions in different organs of the plant during

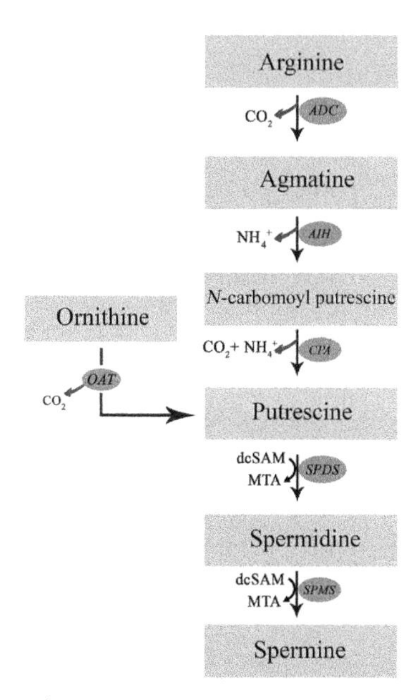

Fig. 3 Polyamines biosynthesis pathway. *ADC*, arginine decarboxylase; *AIH*, agmatine iminohydrolase; *CPA*, *N*-carbamoylPut amidohydrolase; *SPDS*, spermidine synthase; *SPMS*, spermine synthase; *dcSAM*, decarboxylated *S*-adenosylmethionine; *MTA*, 5-methylthioasdenosine.

the germination and early seedling stages, and increased endogenous putrescine content by twofold in comparison with nontreated plants.

Nahar et al. (2016) investigated the exogenous application of spermine (0.2 mM) on mung beans subjected to high temperatures (40°C) and drought stress. The treated plants showed reduced ROS production in terms of low H_2O_2 content, low lipoxygenase activity, and low malondialdehyde activity (indicator for membrane peroxidation). Moreover, the spermine pretreated plants exhibited enhanced activity of antioxidant enzymes, namely glutathione (GSH), superoxide dismutase (SOD), catalase (CAT), glutathione peroxidase (GPX), dehydroascorbate reductase (DHAR), and glutathione reductase (GR). Nahar et al. (2016) also subjected the same mung bean varieties to salt stress (200 mM NaCl) for a period of 48 h and treated them with 0.2 mM solution of polyamines (Put, Spm, and Spd). The treated plants showed similar results as seen in the case of heat and drought stress, along with improvement in tissue water and chlorophyll content. Sánchez-Rodríguez et al. (2016) also obtained the same results with both drought-tolerant cultivars (Zarina) and drought-sensitive cultivars (Josefina) to water stress.

Li et al. (2016) studied the effects of exogenous Spd (0.3 mM) application on Zoyia grass under salt stress (200 mM). He revealed that the treated plants showed 26%, 13%, and 16% improvement in fresh root weight, root length, and relative root water content, respectively, in comparison with untreated plants, along with a significant rise in polyamine biosynthetic enzymes (ODC and ADC).

Zhang et al. (2015) exposed salt-stressed (75 mM NaCl) tomato seedlings to 0.25 mM Spd treatment. The treated plants showed improved antioxidant enzyme activity, chlorophyll content, and better seedling growth. Later, Hu et al. (2016) treated a similar tomato variety with Spd and concluded that Spd-induced tolerance in the treated cultivars showed shielding of the photosynthetic machinery as a result of high D_1 protein levels. This improved the photosynthetic rate and ensured plant growth and development under salt stress.

Although exogenous application of polyamines resulted in conferring abiotic stress tolerance, there is a serious demand for increasing the endogenous levels of polyamines significantly so that plants can endure for long periods in stress-inducing environments. The genes encoding the enzymes such as ADC, ODC, SPDS, and SAM, involved in polyamine biosynthesis, are often targeted for developing transgenic plants. In recent years, several transgenic plants have been engineered harboring these genes, and have been instrumental in enhancing the endogenous polyamine levels and inducing stress tolerance (Gill and Tuteja, 2010; Gupta et al., 2013; Shukla and Mattoo, 2013).

Kasukabe et al. (2004) cloned the *Cucurbita ficifolia* spermidine synthase gene (*CsSPDS*) in *A. thaliana* under the constitutive cauliflower mosaic virus 35S (CaMV 35S) promoter. The transgenic plants showed a twofold increase in endogenous spermidine content on exposure to salt (200 mM NaCl), low temperature (4°C), and drought stress in comparison with WT. Simultaneously, Capell et al. (2004) transformed rice plants with the *Datura stramonium* ADC gene (*DsADC*) under the control of an inducible monocot Ubi-1 promoter. The transgenic rice cultivars accumulated higher levels of putrescine (threefold) under drought stress, and subsequently promoted spermidine and spermine synthesis, and thereby conferred tolerance against drought stress. Franceschetti et al. (2004) introduced the *D. stramonium* SPDS gene (*DsSPDS*) in tobacco plants, and subjected them to salt stress (200 mM NaCl). The transformed tobacco plants accumulated significant levels of spermidine and showed improved tissue water content and chlorophyll content, promoting plant growth and survival.

He et al. (2008) analyzed the differences in the enzymatic and nonenzymatic antioxidant capacity transgenic *Pyrus communis* expressing the apple *SPDS* gene under saline stress (200 mM NaCl). They observed that the transgenic pear plants accumulated higher levels of spermidine in comparison with the WT. The transgenic plants also exhibited high antioxidant enzyme activity, and lower malondialdehyde and hydrogen peroxide levels. These results indicated that an increase in spermidine content facilitated enhanced ROS scavenging, thereby preventing membrane disruptions and ensuring plant survival in stressed environments. There are many other instances of developing transgenic plants expressing the major polyamine genes, as listed in Table 3. Thus, we can conclude that heterologous expression of the polyamine biosynthetic genes in plants leads to significant accumulation of respective polyamines endogenously, which subsequently improves plant response and survival despite major abiotic stresses.

3.4 Sugars and Sugar Alcohols

Abiotic stresses, predominantly drought, salinity, and extreme temperatures, alter the cellular carbon metabolism, as well as optimum concentrations of important sugars and their alcohol derivatives (polyols). A high accumulation of sugars (trehalose and fructose) and some sugar alcohols (mannitol, sorbitol, and ononitol) have been associated with exposure to abiotic stresses (Wani et al., 2016; O'Hara et al., 2013). These organic molecules are responsible for cellular osmotic adjustments, ROS scavenging, and membrane structural integrity, besides functioning as carbon sink and molecular chaperones (Gupta and Huang, 2014; Parvaiz and Satyawati, 2008). In this section, the biosynthesis, accumulation, and role of different sugars and polyols are discussed in relation to abiotic stress. Moreover, this section also contains the various transgenic plants (Table 4) developed so far expressing the genes involved in the biosynthesis of these solutes for conferring multiple abiotic stress tolerance.

3.4.1 Trehalose

Trehalose is a nonreducing disaccharide of two glucose units (-D-glucopyranosyl-1,1-D-glucopyranoside). It has been identified in many organisms such as bacteria and fungi, and in higher plants and some insects (Djilianov et al., 2005; Elbein et al., 2003). In plants, trehalose synthesis begins with the catalysis of uridine diphosphate glucose (UDP-glucose)

Table 3 List of transgenic plants over-expressing candidate genes for polyamines accumulation

Sl no.	Transgene	Host	Target crop plants/trees	Remarks	Reference
1	*CsSPDS*	*Cucurbita ficifolia*	*A. thaliana*	Enhanced twofold accumulation of spermidine, inducing abiotic stress tolerance.	Kasukabe et al. (2004)
2	*DsADC*	*Datura stramonium*	*O. sativa* L.	Enhanced threefold putresciene accumulation inducing drought tolerance.	Capell et al. (2004)
3	*DsSPDS*	*D. stramonium*	*N. tobacum*	Enhanced spermidine accumulation, chlorophyll content and plant growth.	Franceschetti et al. (2004)
4	*SPDS*	*Malus pumila*	*Pyrus communis*	Enhanced spremidine content and ROS scavenging, inducing salinity tolerance.	He et al. (2008)
5	*SAMDC*	*Saccharomyces cerevisiae*	*Lycopersicon esculentum*	1.7- to 2.4-fold higher levels of spermidine and spermine, heat stress tolerance, CO_2 assimilation.	Cheng et al. (2009)
6	*AsADC*	*Avena sativa* L.	*O. sativa* L.	Enhanced polyamine accumulation inducing abiotic stress tolerance.	Roy and Wu (2001)
7	*SAMDC*	*Tritordeum*	*O. sativa* L.	Three- and fourfold increase of spermidine and spermine, enhanced seed germination, salt tolerance.	Roy and Wu (2002)
8	*MdSPDS1*	*Malus domestica*	*Pyrus communis* L.	Enhanced spermidine accumulation leading to salt and osmotic stress tolerance.	Wen et al. (2008)
9	*LcSAMDC1*	*Leymus chinensis*	*A. Thaliana*	Enhanced spremine, proline and chlorophyll content under salt and cold stress.	Liu et al. (2017a,b)
10	*SAMDC*	Human	Tomato	Enhanced polyamine accumulation, delayed ripening and improved post harvest storage.	Madhulatha et al. (2014)
12	*AvADC*	*Avena sativa* L.	*Medicago truncatula*	Enhanced polyamine content, seed yield and desiccation stress tolerance.	Duque et al. (2016)

Table 4 List of transgenic plants over-expressing candidate genes for sugar and sugar alcohols

Sl no.	Transgene	Host	Target crop plants/trees	Remarks	Reference
1	*ScTPS1*	*Saccharomyces cervisiae*	*N. tabacum*	0.17 mg/g FW in leaves trehalose content, shunted growth, drought tolerance	Romero et al. (1997)
2	*ScTPS1*	*Saccharomyces cervisiae*	*Solanum tuberosum*	Twofold trehalose accumulation, salt and drought tolerance.	Yeo et al. (2000)
3	*ScTPS1*	*Saccharomyces cervisiae*	*Solanum lycopersicum*	2.5-Fold trehalose accumulation, salt and drought tolerance.	Cortina and Culianez-Macia (2005)
4	*otsA and otsB*	*E. coli*	*Oryza sativa* L.	3–10-Fold trehalose accumulation, salt, drought and loe temperature tolerance.	Garg et al. (2002)
5	*PyTPS*	*Porphyra yezoensis*	*Oryza sativa* L. TP309	Enhanced trehalose accumulation, seed germination and yield.	Guo et al. (2014)
6	*GfTPS*	*Grifola frondosa*	*N. tabacum*	12-Fold more trehalose content, abiotic stress tolerance.	Zhang et al. (2005)
7	*CvTPS1* + *CvTPS2*	*Saccharomyces cervisiae*	*Arabidopsis thaliana*	Enhanced plant growth and abiotic stress tolerance, twofold trehalose content.	Miranda et al. (2007)
9	*CvTPS1* + *CvTPS2*	*Saccharomyces cervisiae*	*Alfalfa*	Enhanced plant growth and abiotic stress tolerance, enhanced trehalose content.	Suárez et al. (2009)
10	*SacB*	*Bacillus subtilis*	*N. tabacum*	0.35 mg/g fructans accumulation, 55% enhanced plant growth under drought stress.	Pilon-Smits et al. (1995)
11	*SacB*	*Bacillus subtilis*	*Beta vulgaris* L	Accumulated 0.5% more fructans, inducing drought stress tolerance.	Pilon-Smits et al. (1999)
12	*1-SST*	*Lactuca sativa*	*N. tabacum*	High soluble carbohydrate, fructose, fructan content inducing cold stress tolerance.	Li et al. (2007)
13	*mtlD*	*E. coli*	*Arabidopsis thaliana*	Enhanced mannitol (3 μmol/g FW), germination rate, optimum shoot and root growth under salt stress.	Thomas et al. (1995)

Continued

Table 4 List of transgenic plants over-expressing candidate genes for sugar and sugar alcohols—cont'd

Sl no.	Transgene	Host	Target crop plants/trees	Remarks	Reference
14	*mtlD*	*E. coli*	*N. tabacum*	6 μmol/g FW of mannitol inducing salt tolerance.	Tarczynski et al. (1992)
15	*mtlD*	*E. coli*	*Triticum aestivum*	Enahnced mannitol content, plant height, dry and fresh weight under salt stress.	Abebe et al. (2003)
16	*mtlD*	*E. coli*	*Solanum tuberosum*	Enhanced mannitol level leading to 65% plant survival under salt stress.	Rahnama et al. (2011)
17	*HVA1* + *mtlD*	*Hordeum vulgare* and *E. coli*	*Zea mays*	Improved rate of plant survival, shoot and root biomass under multiple abiotic stress conditions.	Nguyen et al. (2013)
18	*MpS6PDH*	*Malus pumila*	*N. tabacum*	0.2 to 130 μmol/g FW sorbitol content in response to salt stress.	Sheveleva et al. (1998)
19	*MpS6PDH*	*Malus pumila*	Japanese persimmom	14.5–61.5 μmol/g FW sorbitol content & high chlorophyll content under salt stress.	Gao et al. (2001)

and glucose-6-phosphate by trehalose-6-phosphate synthase (TPS) to form the intermediate trehalose-6-phosphate (T6P). In the next step, trehalose-6-phosphate phosphatase (TPP) catalyzes the dephosphorylation of the intermediate, T6P to tehalose (Iordachescu and Imai, 2011; Paul et al., 2008; John et al., 2017) (Fig. 4A).

In plants, trehalose is involved in cellular membrane and protein protection by preventing their denaturation under drought and salt stress (Garg et al., 2002; Wingler, 2002). Additionally, it plays an integral role in cell proliferation, cell differentiation, and maintenance of cellular homeostasis (Wani et al., 2016). So far, 11 TPSs and 10 TPPs have been identified in *Arabidopsis*, and 9 TPSs and 9 TPPs have been identified in rice. To investigate the role of trehalose further in relation to inducing abiotic stress tolerance, scientists began treating various crop plants with exogenous trehalose, and studied their growth under different abiotic stress conditions.

Garcia et al. (1997) initially reported that when 1% NaCl-stressed rice plants were treated with 110 mM trehalose, improved cell division, root cell integrity, and overall plant growth was observed. Similarly, trehalose (30 mM)-Tween 20 solution pretreatment enhanced seed growth and antioxidant enzyme activity in *Z. mays* (Ali et al., 2012).

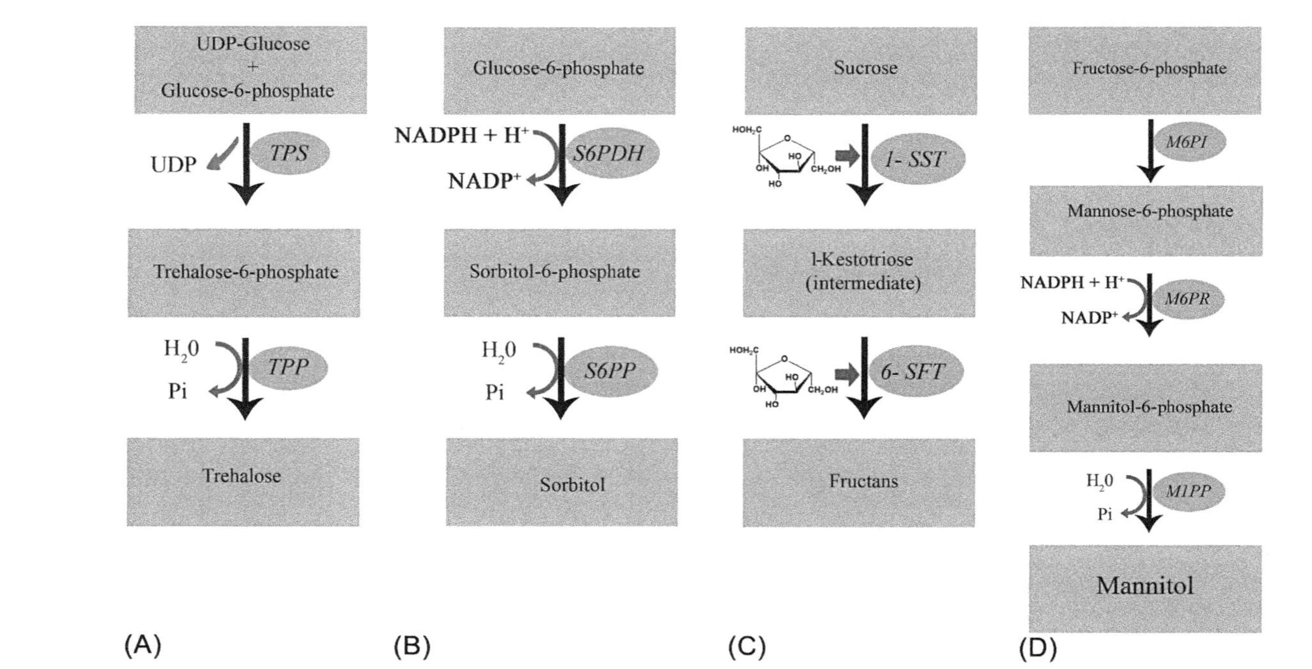

Fig. 4 (A) Trehalose biosynthetic pathway. *TPS*, trehalose-6-phosphate synthase; *TPP*, trehalose-6-phosphate phosphatase; *UDP*, uridine diphosphate. (B) Fructan biosynthesis pathway. *1-SST*, sucrose:sucrose 1-fructosyltransferase; *6-SFT*, sucrose:fructan 6-fructosyltransferase. (C) Mannitol biosynthesis pathway. *M6PI*, mannose-6-phosphate isomerase; *M6PR*, mannose-6-phosphate reductase; *M1PP*, mannose-1-phosphate phosphatase. (D) Sorbitol biosynthesis pathway. *S6PDH*, sorbitol-6-phosphate dehydrogenase; *S6PP*, sorbitol-6-pyrophosphatase.

Nounjan et al. (2012) reported that exogenous application of 10 mM each of proline and trehalose improved seed recovery and antioxidant enzyme activity in salt-stressed rice.

Recently, Sadak (2016) evaluated the impact of exogenous trehalose on drought-stressed fenugreek plants. Sadak (2016) noted that a 500 mM trehalose foliar spray was effective in suppressing the negative impacts of drought stress. At this specific trehalose concentration, the drought-stressed plants (60% water holding) showed improved photosynthetic pigment content, antioxidant activity, and carbohydrate, protein, and flavonoid content in comparison with the nontreated plants. The exogenous application of trehalose at 25 mM concentration improved fresh and dry biomass, antioxidant enzyme content (superoxide dismutase and peroxidase), chlorophyll content, total soluble sugar levels, and free proline content in two 45-day drought-stressed radish varieties (Akram et al., 2016).

Foliar spray of trehalose also alleviated the deleterious effects of salt stress in rice. When Abdallah et al. (2016) treated two rice varieties (Giza 177 and Giza 178) growing in a salt-stressed environment (60 mM NaCl) with 25 mM tehalose, a high total soluble sugar content and antioxidant enzyme level was observed, which contributed to osmotic adjustment, thereby inducing salt tolerance in the rice varieties.

Along with the exogenous application of trehalose, researchers targeted the trehalose biosynthetic genes (TPP and TPS) from prokaryotes and crop plants for developing transgenic plants to enhance the endogenous trehalose production, thereby mitigating the adverse effects of abiotic stress (Iordachescu and Imai, 2008; Sah et al., 2016). Transgenic tobacco was the first genetically engineered plant expressing the ScTPS1 gene from yeast. The transgenic plants exhibited significant trehalose accumulation (0.17 mg per g fresh weight in leaves), shunted plant growth, and lancet-shaped leaves; but improved tolerance to drought, and an increased plant survival rate (Romero et al., 1997). Yeo et al. (2000) and Cortina and Culiáñez-Macià (2005) obtained similar results when they developed transgenic potatoes and tomatoes expressing the same yeast (*ScTPS1*) gene under the control of a 35S CaMV promoter in relation to multiple abiotic stresses.

Garg et al. (2002) introduced *E. coli* trehalose biosynthetic genes (*otsA* and *otsB*) under the control of stress-responsive promoters in rice plants. The transgenic rice plants showed 3–10-fold more trehalose accumulation, which facilitated better growth, less photo-oxidative damage, and favorable mineral homeostasis when subjected to salt, drought, and low temperatures. Withholding the water supply 8–10 days caused leaf rolling and wilting in the WTs, whereas the transgenic varieties showed significant shoot growth. Guo et al. (2014) isolated the trehalose-6-phosphate synthase gene of *Porphyra yezoensis* (*PyTPS*) and introduced it in rice variety (TP309). The transgenic varieties thrived in 5%–8% NaCl stress conditions, and showed improved seed germination and yield.

Zhang et al. (2005) expressed the *Grifola frondosa* trehalose synthase gene in tobacco and studied the levels of trehalose accumulation. The transgenic plants accumulated trehalose up to 2.1–2.5 mg/g fresh weight, while the level of trehalose remained undetected in non transgenic plants. Moreover, they concluded that this high level of trehalose accumulation was close to 400-fold higher than transgenic tobacco expressing the *E. coli* trehalose synthase gene, and 12-fold higher than the transgenic lines harboring the yeast *TPS1* gene. Heterologous expression of the *A. thaliana* (*AtTPS1*) gene contributed to increased tolerance to several abiotic stresses such as drought, desiccation, and extreme temperature stresses (Almeida et al., 2005). Significant trehalose accumulation in transgenic lines improved the photosynthetic rate and decreased cellular oxidative damage, as confirmed by low peroxidation levels and improved antioxidant enzyme activity (Vinocur and Altman, 2005).

Miranda et al. (2007) developed a gene construct combining the yeast TPS1 and TPS2 genes. This chimeric gene was introduced in Arabidopsis, and later in *Alfalfa* (Suárez et al., 2009). In both the cases, the transgenic lines thrived in drought, heat, freezing, and salt stress conditions, and were devoid of any morphological abnormalities such as shunted growth and lancet-shaped leaves.

3.4.2 Fructans

Fructans are polymers of fructose obtained from sucrose in two successive steps involving the transfer of a fructosyl group (Fig. 4C). These two steps are catalyzed by sucrose: sucrose 1-fructosyltransferase (1-SST); and sucrose:fructan 6-fructosyltransferase (6-SFT), respectively. In plants, fructans are accumulated in vacuoles, and they primarily serve as carbohydrate storage sinks utilized during the nutrient recovery phase after exposure to abiotic stress environments (Vijn and Smeekens, 1999; Konstantinova et al., 2002).

The genes related to fructan biosynthesis have been used to target many plants such as tobacco, potatoes (Van Der Meer et al., 1994), rice (Kawakami et al., 2008), and sugar beets to develop abiotic stress-tolerant transgenic varieties. Pilon-Smits et al. (1995) transformed tobacco with the *Bacillus subtilis* gene (*SacB*) related to fructan biosynthesis under the constitutive promoter (CaMV 35S). The plants accumulated significant levels of fructans (0.35 mg/g fresh weight) under drought stress, and showed a 55% enhanced growth rate, 33% increase in fresh weight, and 59% increase in dry weight, in comparison with wildtype plants. The transgenic lines accumulated sevenfold more fructans than their WT counterparts. Pilon-Smits et al. (1999) used the same gene to transform sugar beets (*Beta vulgaris* L.). The transgenic plants accumulated higher levels of fructans (0.5% of their dry weight) in roots as well as shoots, and they thrived in drought-stressed conditions, in comparison with the wild types.

Li et al. (2007) developed transgenic tobacco expressing the *Lactuca sativa* sucrose 1-fructosyltransferase (*1-SST*) gene. The transgenic plants showed highly soluble carbohydrates, fructan, and fructose content, and a more improved photosynthetic rate in response to freezing stress (4°C) than the wild types. However, no significant difference in the levels of antioxidant enzymes (superoxide dismutase) was observed in the transgenic and wild types.

3.4.3 Mannitol

Mannitol is a 6-carbon sugar alcohol synthesized from fructose-6-phosphate in three subsequent reactions catalyzed by the enzymes mannose-6-phosphate isomerase, mannose-6-phosphate reductase, and mannose-1-phosphate phosphatase (Loescher et al., 1992) (Fig. 4D). Mannitol accumulation in plants is correlated with exposure to abiotic stress and is involved in stabilizing macromolecules and ROS scavenging (Llanes et al., 2013). Mannitol is not synthesized naturally in tobacco and *Arabidopsis*. Thomas et al. (1995) inserted the *E. coli mtlD* gene into *Arabidopsis*. The *mtlD* gene codes for the enzyme mannitol dehydrogenase enzyme, which is a key enzyme involved in conversion of fructose-6-phosphate to mannitol-1-phosphate. The transgenic plants accumulated significant levels of mannitol (3 μmol/g fresh weight), and showed enhanced germination rate, optimum shoot and root growth, and increased biomass, even in 400 mM NaCl stress. The same *E. coli mtlD* gene was inserted in tobacco (Tarczynski et al., 1992). The transgenic tobacco lines accumulated mannitol (6 μmol/g fresh weight), unlike the wild types. Moreover, the transgenic lines thrived in 150 mM NaCl stress, and did not show a reduction in dry weight, as in the case of the wild types, which showed a 44% reduction in dry weight.

Abebe et al. (2003) developed transgenic wheat expressing the *E. coli* (*mtlD*) gene, and subjected them to 150 mM NaCl stress. In transgenic wheat plants fresh weight, dry weight, plant height, and flag leaf length were reduced by 40%, 8%, 18%, and 29%, only in comparison with 70%, 56%, 40%, and 45%, respectively, in the wild types, thereby inducing tolerance to salinity stress. The transgenic wheat plants accumulated mannitol (1.7–3.7 μmol/g fresh weight) in the callus and (0.6–2.0 μmol/g fresh weight) in leaf tissues. In another experiment, Rahnama et al. (2011) developed transgenic potato lines expressing the same *E. coli* (*mtlD*) gene, and exposed them to100 mM salt stress. The transgenic potato plants showed only a 17.3% reduction in shoot fresh weight, in comparison with a 76.5% reduction in shoot fresh weight, as observed in the nontransformed potato lines. Moreover, mannitol accumulation was more instrumental in osmotic adjustments of the roots than in the shoots of the transgenic plants.

Nguyen et al. (2013) successfully developed a construct comprising the *Hordeum vulgare* HVA1 (a group 3 late embryogenesis abundant protein) and *E. coli* (*mtlD*) gene under the regulation of rice actine Act1. This construct was introduced in maize to study the effects of high salinity (300 mM NaCl) and drought stress. The transgenic lines expressing the *HVA1+mtlD* construct showed an improved rate of plant survival, and shoot and root biomass than their wild type counterparts.

3.4.4 Sorbitol

Sorbitol belongs to the group of sugar alcohols and is synthesized from glucose 6-phosphate by the enzyme sorbitol-6-phosphate dehydrogenase (S6PDH) and sorbitol-6-pyrophosphatase (S6PP) (Fig. 4B). The genes governing sorbitol biosynthesis have been used to develop transgenic plants in relation to abiotic stress tolerance. Sheveleva et al. (1998) developed transgenic tobacco plants expressing sorbitol-6-phosphate dehydrogenase (*S6PDH*) genes from apples. The transgenic tobacco plants produced sorbitol ranging from 0.2 to 130 μmol/g fresh weight in response to salt stress. Gao et al. (2001) developed transgenic Japanese persimmon cultivars expressing the same apple *S6PDH* gene and subjected them to 200 mM NaCl stress. The transgenic lines were reported to accumulate sorbitol (14.5–61.5 μmol/g fresh weight), while the wild types did not accumulate any sorbitol. Photosynthetic activity was restored and chlorophyll content increased in the transgenic lines, indicating a possible correlation between sorbitol accumulation and salt stress tolerance.

4. CONCLUSION

Abiotic stress remains the major obstacle in the path of plant growth, development, and global crop productivity. Plant breeders are constantly developing new strategies to improve plant stress physiology, thereby ensuring sustainable crop productivity for future generations. In this regard, compatible solute engineering for developing abiotic stress-tolerant transgenic plants has garnered much attention for its diverse role in inducing abiotic stress tolerance. A large number of potential candidate genes involved in the biosynthesis of these osmolytes have been identified, and some have been successfully introduced in major crop plants. The transgenic plants showed enhanced osmolyte accumulation and specific stress tolerance. However, no successful field trials have been carried out, and major research has been restricted to laboratory conditions. Hence, the future lies in developing transgenic plants expressing multigenic traits (coexpression of osmolyte biosynthesis genes and other stress related genes such as ion transporters, TFs, aquaporins, LEA proteins, etc.) with complete understanding of the molecular mechanism of plant stress response. This strategy is essential as plants are generally exposed to multiple abiotic stress conditions simultaneously, thereby promoting plant growth, development, survival, and superior yield.

ACKNOWLEDGMENTS

The authors are grateful to GITAM for providing necessary facilities to carry out the research work and for extending constant support in writing this review.

REFERENCES

Abdallah, M.S., Abdelgawad, Z.A., El-Bassiouny, H.M.S., 2016. Alleviation of the adverse effects of salinity stress using trehalose in two rice varieties. S. Afr. J. Bot. 103, 275–282.

Abebe, T., Guenzi, A.C., Martin, B., Cushman, J.C., 2003. Tolerance of mannitol-accumulating transgenic wheat to water stress and salinity. Plant Physiol. 131 (4), 1748–1755.

Acquaah, G., 2007. Principles of Plant Genetics and Breeding. Blackwell, Oxford.

Adem, G.D., Roy, S.J., Zhou, M., Bowman, J.P., Shabala, S., 2014. Evaluating contribution of ionic, osmotic and oxidative stress components towards salinity tolerance in barley. BMC Plant Biol. 14 (1), 113.

Ahmad, R., Hussain, J., Jamil, M., Kim, M.D., Kwak, S.S., Shah, M.M., El-Hendawy, S.E., Al-Suhaibani, N.A., Shafiq Ur, R., 2014. Glycinebetaine synthesizing transgenic potato plants exhibit enhanced tolerance to salt and cold stresses. Pak. J. Bot. 46, 1987–1993.

Akram, N.A., Waseem, M., Ameen, R., Ashraf, M., 2016. Trehalose pretreatment induces drought tolerance in radish (*Raphanus sativus* L.) plants: some key physio-biochemical traits. Acta Physiol. Plant. 38 (1), 3.

Alam, R., Das, D.K., Islam, M.R., Murata, Y., Hoque, M.A., 2017. Exogenous proline enhances nutrient uptake and confers tolerance to salt stress in maize (*Zea mays* L.). Progress. Agric. 27 (4), 409–417.

Alcázar, R., Altabella, T., Marco, F., Bortolotti, C., Reymond, M., Koncz, C., Carrasco, P., Tiburcio, A.F., 2010. Polyamines: molecules with regulatory functions in plant abiotic stress tolerance. Planta 231 (6), 1237–1249.

Alcázar, R., Bitrián, M., Bartels, D., Koncz, C., Altabella, T., Tiburcio, A.F., 2011. Polyamine metabolic canalization in response to drought stress in *Arabidopsis* and the resurrection plant *Craterostigma plantagineum*. Plant Signal. Behav. 6 (2), 243–250.

Alet, A.I., Sánchez, D.H., Cuevas, J.C., Marina, M., Carrasco, P., Altabella, T., Tiburcio, A.F., Ruiz, O.A., 2012. New insights into the role of spermine in *Arabidopsis thaliana* under long-term salt stress. Plant Sci. 182, 94–100.

Ali, R.M., 2000. Role of putrescine in salt tolerance of *Atropa belladonna* plant. Plant Sci. 152 (2), 173–179.

Ali, Q., Ashraf, M., Anwar, F., Al-Qurainy, F., 2012. Trehalose-induced changes in seed oil composition and antioxidant potential of maize grown under drought stress. J. Am. Oil Chem. Soc. 89 (8), 1485–1493.

Allakhverdiev, S.I., Hayashi, H., Nishiyama, Y., Ivanov, A.G., Aliev, J.A., Klimov, V.V., Murata, N., Carpentier, R., 2003. Glycinebetaine protects the D1/D2/Cytb 559 complex of photosystem II against photo-induced and heat-induced inactivation. J. Plant Physiol. 160 (1), 41–49.

Almeida, A.M., Villalobos, E., Araújo, S.S., Leyman, B., Van Dijck, P., Alfaro-Cardoso, L., Fevereiro, P.S., Torné, J.M., Santos, D.M., 2005. Transformation of tobacco with an *Arabidopsis thaliana* gene involved in trehalose biosynthesis increases tolerance to several abiotic stresses. Euphytica 146 (1), 165–176.

Ashraf, M., Foolad, M., 2007. Roles of glycine betaine and proline in improving plant abiotic stress resistance. Environ. Exp. Bot. 59 (2), 206–216.

Behelgardy, M.F., Motamed, N., Jazii, F.R., 2012. Expression of the P5CS gene in transgenic versus non-transgenic olive (*Olea europaea*) under salinity stress. World Appl. Sci. J. 18 (4), 580–583.

Bhatnagar-Mathur, P., Vadez, V., Sharma, K.K., 2008. Transgenic approaches for abiotic stress tolerance in plants: retrospect and prospects. Plant Cell Rep. 27 (3), 411–424.

Bhusan, D., Das, D.K., Hossain, M., Murata, Y., Hoque, M.A., 2016. Improvement of salt tolerance in rice (*Oryza sativa* L.) by increasing antioxidant defense systems using exogenous application of proline. Aust. J. Crop. Sci. 10 (1), 50.

Capell, T., Bassie, L., Christou, P., 2004. Modulation of the polyamine biosynthetic pathway in transgenic rice confers tolerance to drought stress. Proc. Natl. Acad. Sci. U. S. A. 101 (26), 9909–9914.

Chen, T.H., Murata, N., 2002. Enhancement of tolerance of abiotic stress by metabolic engineering of betaines and other compatible solutes. Curr. Opin. Plant Biol. 5 (3), 250–257.

Chen, T.H., Murata, N., 2008. Glycine betaine: an effective protectant against abiotic stress in plants. Trends Plant Sci. 13 (9), 499–505.

Chen, T.H., Murata, N., 2011. Glycine betaine protects plants against abiotic stress: mechanisms and biotechnological applications. Plant Cell Environ. 34 (1), 1–20.

Chen, J.B., Yang, J.W., Zhang, Z.Y., Feng, X.F., Wang, S.M., 2013. Two P5CS genes from common bean exhibiting different tolerance to salt stress in transgenic Arabidopsis. J. Genet. 92 (3), 461–469.

Cheng, L., Zou, Y., Ding, S., Zhang, J., Yu, X., Cao, J., Lu, G., 2009. Polyamine accumulation in transgenic tomato enhances the tolerance to high temperature stress. J. Integr. Plant Biol. 51 (5), 489–499.

Cortina, C., Culiáñez-Macià, F.A., 2005. Tomato abiotic stress enhanced tolerance by trehalose biosynthesis. Plant Sci. 169 (1), 75–82.

De Ronde, J.A., Cress, W.A., Krüger, G.H.J., Strasser, R.J., Van Staden, J., 2000. Photosynthetic response of transgenic soybean plants, containing an Arabidopsis P5CR gene, during heat and drought stress. J. Plant Physiol. 161 (11), 1211–1224.

De Zwart, F.J., Slow, S., Payne, R.J., Lever, M., George, P.M., Gerrard, J.A., Chambers, S.T., 2003. Glycine betaine and glycine betaine analogues in common foods. Food Chem. 83 (2), 197–204.

Delauney, A.J., Verma, D.P.S., 1993. Proline biosynthesis and osmoregulation in plants. Plant J. 4 (2), 215–223.

Djilianov, D., Georgieva, T., Moyankova, D., Atanassov, A., Shinozaki, K., Smeeken, S.C.M., Verma, D.P.S., Murata, N., 2005. Improved abiotic stress tolerance in plants by accumulation of osmoprotectants-gene transfer approach. Biotechnol. Biotechnol. Equip. 19 (Suppl. 3), 63–71.

Duque, A.S., López-Gómez, M., Kráčmarová, J., Gomes, C.N., Araújo, S.S., Lluch, C., Fevereiro, P., 2016. Genetic engineering of polyamine metabolism changes *Medicago truncatula* responses to water deficit. Plant Cell Tiss. Org. Cult. 127 (3), 681–690.

Ehsanpour, A.A., Fatahian, N., 2003. Effects of salt and proline on *Medicago sativa* callus. Plant Cell Tiss. Org. Cult. 73 (1), 53–56.

Elbein, A.D., Pan, Y.T., Pastuszak, I., Carroll, D., 2003. New insights on trehalose: a multifunctional molecule. Glycobiology 13 (4), 17R–27R.

Fan, W., Zhang, M., Zhang, H., Zhang, P., 2012. Improved tolerance to various abiotic stresses in transgenic sweet potato (*Ipomoea batatas*) expressing spinach betaine aldehyde dehydrogenase. PLoS One. 7(5). e37344.

FAO, 2010. FAO Land and Plant Nutrition Management Service. FAO.

Feng, H.Y., Wang, Z.M., Kong, F.N., Zhang, M.J., Zhou, S.L., 2011. Roles of carbohydrate supply and ethylene, polyamines in maize kernel set. J. Integr. Plant Biol. 53 (5), 388–398.

Franceschetti, M., Fornalè, S., Tassoni, A., Zuccherelli, K., Mayer, M.J., Bagni, N., 2004. Effects of spermidine synthase overexpression on polyamine biosynthetic pathway in tobacco plants. J. Plant Physiol. 161 (9), 989–1001.

Funck, D., Winter, G., Baumgarten, L., Forlani, G., 2012. Requirement of proline synthesis during Arabidopsis reproductive development. BMC Plant Biol. 12 (1), 191.

Gadallah, M.A.A., 1999. Effects of proline and glycine betaine on *Vicia faba* responses to salt stress. Biol. Plant. 42 (2), 249–257.

Gao, M., Tao, R., Miura, K., Dandekar, A.M., Sugiura, A., 2001. Transformation of Japanese persimmon (*Diospyros kaki Thunb.*) with apple cDNA encoding NADP-dependent sorbitol-6-phosphate dehydrogenase. Plant Sci. 160 (5), 837–845.

Garcia, A.B., Engler, J.D.A., Iyer, S., Gerats, T., Van Montagu, M., Caplan, A.B., 1997. Effects of osmoprotectants upon NaCl stress in rice. Plant Physiol. 115 (1), 159–169.

Garg, A.K., Kim, J.K., Owens, T.G., Ranwala, A.P., Do Choi, Y., Kochian, L.V., Wu, R.J., 2002. Trehalose accumulation in rice plants confers high tolerance levels to different abiotic stresses. Proc. Natl. Acad. Sci. U. S. A. 99 (25), 15898–15903.

Gill, S.S., Tuteja, N., 2010. Polyamines and abiotic stress tolerance in plants. Plant Signal. Behav. 5 (1), 26–33.

Giri, J., 2011. Glycine betaine and abiotic stress tolerance in plants. Plant Signal. Behav. 6 (11), 1746–1751.

Gleeson, D., Lelu-Walter, M.A., Parkinson, M., 2005. Overproduction of proline in transgenic hybrid larch (*Larix x leptoeuropaea* (*Dengler*)) cultures renders them tolerant to cold, salt and frost. Mol. Breed. 15 (1), 21–29.

Goel, D., Singh, A.K., Yadav, V., Babbar, S.B., Murata, N., Bansal, K.C., 2011. Transformation of tomato with a bacterial *codA* gene enhances tolerance to salt and water stresses. J. Plant Physiol. 168 (11), 1286–1294.

Guerzoni, J.T.S., Belintani, N.G., Moreira, R.M.P., Hoshino, A.A., Domingues, D.S., Bespalhok Filho, J.C., Vieira, L.G.E., 2014. Stress-induced Δ1-pyrroline-5-carboxylate synthetase (*P5CS*) gene confers tolerance to salt stress in transgenic sugarcane. Acta Physiol. Plant. 36 (9), 2309–2319.

Guo, B.T., Wang, B., Weng, M.L., Qiao, L.X., Feng, Y.B., Wang, L., Zhang, P.Y., Wang, X.L., Sui, J.M., Liu, T., Duan, D.L., 2014. Expression of *Porphyra yezoensis* (*tps*) gene in transgenic rice enhanced the salt tolenrance. J. Plant Breed. Genet. 2 (1), 45–55.

Gupta, K., Dey, A., Gupta, B., 2013. Plant polyamines in abiotic stress responses. Acta Physiol. Plant. 35 (7), 2015–2036.

Gupta, B., Huang, B., 2014. Mechanism of salinity tolerance in plants: physiological, biochemical, and molecular characterization. Int. J. Genomics 1–18, Article ID 701596.

Han, K.H., Hwang, C.H., 2003. Salt tolerance enhanced by transformation of a P5CS gene in carrot. J. Plant Biotechnol. 5, 149–153.

Handa, A.K., Mattoo, A.K., 2010. Differential and functional interactions emphasize the multiple roles of polyamines in plants. Plant Physiol. Biochem. 48 (7), 540–546.

Hare, P.D., Cress, W.A., Van Staden, J., 2003. A regulatory role for proline metabolism in stimulating *Arabidopsis thaliana* seed germination. Plant Growth Regul. 39 (1), 41–50.

Hasthanasombut, S., Ntui, V., Supaibulwatana, K., Mii, M., Nakamura, I., 2010. Expression of Indica rice *OsBADH1* gene under salinity stress in transgenic tobacco. Plant Biotechnol. Rep. 4 (1), 75–83.

He, L., Ban, Y., Inoue, H., Matsuda, N., Liu, J., Moriguchi, T., 2008. Enhancement of spermidine content and antioxidant capacity in transgenic pear shoots overexpressing apple spermidine synthase in response to salinity and hyperosmosis. Phytochemistry 69 (11), 2133–2141.

Heuer, B., 2003. Influence of exogenous application of proline and glycinebetaine on growth of salt-stressed tomato plants. Plant Sci. 165 (4), 693–699.

Hmida-Sayari, A., Gargouri-Bouzid, R., Bidani, A., Jaoua, L., Savouré, A., Jaoua, S., 2005. Overexpression of Δ1-pyrroline-5-carboxylate synthetase increases proline production and confers salt tolerance in transgenic potato plants. Plant Sci. 169 (4), 746–752.

Hong, Z., Lakkineni, K., Zhang, Z., Verma, D.P.S., 2000. Removal of feedback inhibition of Δ1-pyrroline-5-carboxylate synthetase results in increased proline accumulation and protection of plants from osmotic stress. Plant Physiol. 122 (4), 1129–1136.

Hossain, M.A., Hasanuzzaman, M., Fujita, M., 2011. Coordinate induction of antioxidant defense and glyoxalase system by exogenous proline and glycinebetaine is correlated with salt tolerance in mung bean. Front. Agric. China 5 (1), 1–14.

Hu, L., Xiang, L., Li, S., Zou, Z., Hu, X.H., 2016. Beneficial role of spermidine in chlorophyll metabolism and D1 protein content in tomato seedlings under salinity-alkalinity stress. Physiol. Plant. 156 (4), 468–477.

Huang, Y., Bie, Z., Liu, Z., Zhen, A., Wang, W., 2009. Protective role of proline against salt stress is partially related to the improvement of water status and peroxidase enzyme activity in cucumber. Soil Sci. Plant Nutr. 55 (5), 698–704.

Huang, B., Jin, L., Liu, J.Y., 2008. Identification and characterization of the novel gene GhDBP2 encoding a DRE-binding protein from cotton (*Gossypium hirsutum*). J. Plant Physiol. 165 (2), 214–223.

Huang, Z., Zhao, L., Chen, D., Liang, M., Liu, Z., Shao, H., Long, X., 2013. Salt stress encourages proline accumulation by regulating proline biosynthesis and degradation in Jerusalem artichoke plantlets. PLoS One. 8(4). e62085.

Hussain, S.S., Ali, M., Ahmad, M., Siddique, K.H., 2011. Polyamines: natural and engineered abiotic and biotic stress tolerance in plants. Biotechnol. Adv. 29 (3), 300–311.

Iordachescu, M., Imai, R., 2008. Trehalose biosynthesis in response to abiotic stresses. J. Integr. Plant Biol. 50 (10), 1223–1229.

Iordachescu, M., Imai, R., 2011. Trehalose and abiotic stress in biological systems. In: Shanker, A.K., Venkateswarlu, B. (Eds.), Abiotic Stress in Plants—Mechanisms and Adaptations. InTech, Croatia, pp. 215–234.

Jaleel, C.A., Manivannan, P., Wahid, A., Farooq, M., Al-Juburi, H.J., Somasundaram, R., Panneerselvam, R., 2009. Drought stress in plants: a review on morphological characteristics and pigments composition. Int. J. Agric. Biol. 11 (1), 100–105.

Ji, H., Pardo, J.M., Batelli, G., Van Oosten, M.J., Bressan, R.A., Li, X., 2013. The salt overly sensitive (SOS) pathway: established and emerging roles. Mol. Plant 6 (2), 275–286.

Jing, J., Li, H., He, G., Yin, Y., Liu, M., Liu, B., et al., 2013. Over-expression of the codA gene by Rd29A-promoter improves salt tolerance in *Nicotiana tabacum*. Pak. J. Bot. 45, 821–827.

John, R., Raja, V., Ahmad, M., Jan, N., Majeed, U., Ahmad, S., Yaqoob, U., Kaul, T., 2017. Trehalose: metabolism and role in stress signaling in plants. In: Sarwat, M., Ahmad, A., Abdin, M.Z., Ibrahim, M.M. (Eds.), Stress Signaling in Plants: Genomics and Proteomics Perspective, vol. 2. Springer International Publishing, pp. 261–275.

Kahlaoui, B., Hachicha, M., Teixeira, J., Misle, E., Fidalgo, F., Hanchi, B., 2013. Response of two tomato cultivars to field-applied proline and salt stress. J. Stress Physiol. Biochem 9 (3), 357–365.

Karthikeyan, A., Pandian, S.K., Ramesh, M., 2011. Transgenic indica rice cv. ADT 43 expressing a Δ1-pyrroline-5-carboxylate synthetase (P5CS) gene from *Vigna aconitifolia* demonstrates salt tolerance. Plant Cell Tiss. Org. Cult. 107 (3), 383–395.

Kasukabe, Y., He, L., Nada, K., Misawa, S., Ihara, I., Tachibana, S., 2004. Overexpression of spermidine synthase enhances tolerance to multiple environmental stresses and up-regulates the expression of various stress-regulated genes in transgenic *Arabidopsis thaliana*. Plant Cell Physiol. 45 (6), 712–722.

Kavi Kishor, P.B., Sreenivasulu, N., 2014. Is proline accumulation per se correlated with stress tolerance or is proline homeostasis a more critical issue? Plant Cell Environ. 37 (2), 300–311.

Kawakami, A., Sato, Y., Yoshida, M., 2008. Genetic engineering of rice capable of synthesizing fructans and enhancing chilling tolerance. J. Exp. Bot. 59 (4), 793–802.

Ke, Q., Wang, Z., Ji, C.Y., Jeong, J.C., Lee, H., Li, H., Xu, B., Deng, X., Kwak, S.S., 2016. Transgenic poplar expressing *codA* exhibits enhanced growth and abiotic stress tolerance. Plant Physiol. Biochem. 100, 75–84.

Khan, M.S., Yu, X., Kikuchi, A., Asahina, M., Watanabe, K.N., 2009. Genetic engineering of glycine betaine biosynthesis to enhance abiotic stress tolerance in plants. Plant Biotechnol. 26 (1), 125–134.

Khedr, A.H.A., Abbas, M.A., Wahid, A.A.A., Quick, W.P., Abogadallah, G.M., 2003. Proline induces the expression of salt-stress-responsive proteins and may improve the adaptation of *Pancratium maritimum* L. to salt-stress. J. Exp. Bot. 54 (392), 2553–2562.

Kishitani, S., Watanabe, K., Yasuda, S., Arakawa, K., Takabe, T., 1994. Accumulation of glycinebetaine during cold acclimation and freezing tolerance in leaves of winter and spring barley plants. Plant Cell Environ. 17 (1), 89–95.

Kishor, P.K., Hong, Z., Miao, G.H., Hu, C.A.A., Verma, D.P.S., 1995. Overexpression of Δ-pyrroline-5-carboxylate synthetase increases proline production and confers osmotolerance in transgenic plants. Plant Physiol. 108 (4), 1387–1394.

Kishor, P.K., Sangam, S., Amrutha, R.N., Laxmi, P.S., Naidu, K.R., Rao, K.R.S.S., Rao, S., Reddy, K.J., Theriappan, P., Sreenivasulu, N., 2005. Regulation of proline biosynthesis, degradation, uptake and transport in higher plants: its implications in plant growth and abiotic stress tolerance. Curr. Sci. 88, 424–438.

Konstantinova, T., Parvanova, D., Atanassov, A., Djilianov, D., 2002. Freezing tolerant tobacco, transformed to accumulate osmoprotectants. Plant Sci. 163 (1), 157–164.

Kumar, R., 2009. Role of naturally occurring osmolytes in protein folding and stability. Arch. Biochem. Biophys. 491 (1), 1–6.

Kumar, V., Shriram, V., Kishor, P.K., Jawali, N., Shitole, M.G., 2010. Enhanced proline accumulation and salt stress tolerance of transgenic indica rice by over-expressing *P5CSF129A* gene. Plant Biotechnol. Rep. 4 (1), 37–48.

Lehmann, S., Funck, D., Szabados, L., Rentsch, D., 2010. Proline metabolism and transport in plant development. Amino Acids 39 (4), 949–962.

Li, H.J., Yang, A.F., Zhang, X.C., Gao, F., Zhang, J.R., 2007. Improving freezing tolerance of transgenic tobacco expressing sucrose: sucrose 1-fructosyltransferase gene from *Lactuca sativa*. Plant Cell Tiss. Org. Cult. 89 (1), 37–48.

Li, B., He, L., Guo, S., Li, J., Yang, Y., Yan, B., Sun, J., Li, J., 2013. Proteomics reveal cucumber Spd-responses under normal condition and salt stress. Plant Physiol. Biochem. 67, 7–14.

Li, M., Li, Z., Li, S., Guo, S., Meng, Q., Li, G., Yang, X., 2014. Genetic engineering of glycine betaine biosynthesis reduces heat-enhanced photoinhibition by enhancing antioxidative defense and alleviating lipid peroxidation in tomato. Plant Mol. Biol. Report. 32 (1), 42–51.

Li, X.-J., Li, M., Zhou, Y., Hu, S., Hu, R., Chen, Y., et al., 2015. Overexpression of cotton *RAV1* gene in *Arabidopsis* confers transgenic plants high salinity and drought sensitivity. PLoS One. 10(2). e0118056.

Li, S., Jin, H., Zhang, Q., 2016. The effect of exogenous spermidine concentration on polyamine metabolism and salt tolerance in zoysiagrass (*Zoysia japonica* Steud) subjected to short-term salinity stress. Front. Plant Sci 7, 1221–1232.

Liu, D., He, S., Zhai, H., Wang, L., Zhao, Y., Wang, B., Li, R., Liu, Q., 2014. Overexpression of *IbP5CR* enhances salt tolerance in transgenic sweet potato. Plant Cell Tiss. Org. Cult. 117 (1), 1–16.

Liu, N., Lin, S., Huang, B., 2017a. Differential effects of glycine betaine and spermidine on osmotic adjustment and antioxidant defense contributing to improved drought tolerance in creeping bentgrass. J. Am. Soc. Hortic. Sci. 142 (1), 20–26.

Liu, Z., Liu, P., Qi, D., Peng, X., Liu, G., 2017b. Enhancement of cold and salt tolerance of Arabidopsis by transgenic expression of the S-adenosylmethionine decarboxylase gene from *Leymus chinensis*. J. Plant Physiol. 211, 90–99.

Llanes, A., Bertazza, G., Palacio, G., Luna, V., 2013. Different sodium salts cause different solute accumulation in the halophyte *Prosopis strombulifera*. Plant Biol. 15 (s1), 118–125.

Loescher, W.H., Tyson, R.H., Everard, J.D., Redgwell, R.J., Bieleski, R.L., 1992. Mannitol synthesis in higher plants evidence for the role and characterization of a NADPH-dependent mannose 6-phosphate reductase. Plant Physiol. 98 (4), 1396–1402.

Lutts, S., Majerus, V., Kinet, J.M., 1999. NaCl effects on proline metabolism in rice (*Oryza sativa*) seedlings. Physiol. Plant. 105 (3), 450–458.

Madhulatha, P., Gupta, A., Gupta, S., Kumar, A., Pal, R.K., Rajam, M.V., 2014. Fruit-specific overexpression of human S-adenosylmethionine decarboxylase gene results in polyamine accumulation and affects diverse aspects of tomato fruit development and quality. J. Plant Biochem. Biotechnol. 23 (2), 151–160.

Mäkelä, P., Kärkkäinen, J., Somersalo, S., 2000. Effect of glycine betaine on chloroplast ultrastructure, chlorophyll and protein content, and RuBPCO activities in tomato grown under drought or salinity. Biol. Plant. 43 (3), 471–475.

Mattioli, R., Costantino, P., Trovato, M., 2009. Proline accumulation in plants: not only stress. Plant Signal. Behav. 4 (11), 1016–1018.

Matysik, J., Alia, Bhalu, B., Mohanty, P., 2002. Molecular mechanisms of quenching of reactive oxygen species by proline under stress in plants. Curr. Sci. 82, 525–532.

Maziah, M., Teh, C.Y., 2016. Exogenous application of glycine betaine alleviates salt induced damages more efficiently than ascorbic acid in in vitro rice shoots. Aust. J. Basic Appl. Sci. 10 (16), 58–65.

Miranda, J.A., Avonce, N., Suárez, R., Thevelein, J.M., Van Dijck, P., Iturriaga, G., 2007. A bifunctional TPS–TPP enzyme from yeast confers tolerance to multiple and extreme abiotic-stress conditions in transgenic *Arabidopsis*. Planta 226 (6), 1411–1421.

Moschou, P.N., Sanmartin, M., Andriopoulou, A.H., Rojo, E., Sanchez-Serrano, J.J., Roubelakis-Angelakis, K.A., 2008. Bridging the gap between plant and mammalian polyamine catabolism: a novel peroxisomal polyamine oxidase responsible for a full back-conversion pathway in Arabidopsis. Plant Physiol. 147 (4), 1845–1857.

Nahar, K., Hasanuzzaman, M., Rahman, A., Alam, M.M., Mahmud, J.A., Suzuki, T., Fujita, M., 2016. Polyamines confer salt tolerance in mung bean (*Vigna radiata* L.) by reducing sodium uptake, improving nutrient homeostasis, antioxidant defense, and methylglyoxal detoxification systems. Front. Plant Sci. 7.

Nanjo, T., Kobayashi, M., Yoshiba, Y., Sanada, Y., Wada, K., Tsukaya, H., Kakubari, Y., Yamaguchi-Shinozaki, K., Shinozaki, K., 1999. Biological functions of proline in morphogenesis and osmotolerance revealed in antisense transgenic *Arabidopsis thaliana*. Plant J. 18 (2), 185–193.

Neelapu, N.R.R., Deepak, K.G.K., Surekha, C., 2015. Transgenic plants for higher antioxidant. In: Wani, S.H., Hossain, M.A. (Eds.), Managing Salt Tolerance in Plants: Molecular and Genomic Perspectives, pp. 391–406.

Nguyen, V.L., Ribot, S.A., Dolstra, O., Niks, R.E., Visser, R.G., van der Linden, C.G., 2013. Identification of quantitative trait loci for ion homeostasis and salt tolerance in barley (*Hordeum vulgare* L.). Mol. Breed. 31 (1), 137–152.

Niu, X., Bressan, R.A., Hasegawa, P.M., Pardo, J.M., 1995. Ion homeostasis in NaCl stress environments. Plant Physiol. 109 (3), 735.

Niu, X., Xiong, F., Liu, J., Sui, Y., Zeng, Z., Lu, B.R., Liu, Y., 2014. Co-expression of ApGSMT and ApDMT promotes biosynthesis of glycine betaine in rice (*Oryza sativa* L.) and enhances salt and cold tolerance. Environ. Exp. Bot. 104, 16–25.

Nounjan, N., Nghia, P.T., Theerakulpisut, P., 2012. Exogenous proline and trehalose promote recovery of rice seedlings from salt-stress and differentially modulate antioxidant enzymes and expression of related genes. J. Plant Physiol. 169 (6), 596–604.

O'Hara, L.E., Paul, M.J., Wingler, A., 2013. How do sugars regulate plant growth and development? New insight into the role of trehalose-6-phosphate. Mol. Plant 6 (2), 261–274.

Ozgur, R., Uzilday, B., Sekmen, A.H., Turkan, I., 2013. Reactive oxygen species regulation and antioxidant defence in halophytes. Funct. Plant Biol. 40 (9), 832–847.

Park, E.J., Jeknić, Z., Sakamoto, A., DeNoma, J., Yuwansiri, R., Murata, N., Chen, T.H., 2004. Genetic engineering of glycinebetaine synthesis in tomato protects seeds, plants, and flowers from chilling damage. Plant J. 40 (4), 474–487.

Park, E.J., Jeknić, Z., Pino, M.T., Murata, N., Chen, T.H.H., 2007. Glycine betaine accumulation is more effective in chloroplasts than in the cytosol for protecting transgenic tomato plants against abiotic stress. Plant Cell Environ. 30 (8), 994–1005.

Parvaiz, A., Satyawati, S., 2008. Salt stress and phyto-biochemical responses of plants—a review. Plant Soil Environ. 54 (3), 89.

Paul, M.J., Primavesi, L.F., Jhurreea, D., Zhang, Y., 2008. Trehalose metabolism and signaling. Annu. Rev. Plant Biol. 59, 417–441.

Pilon-Smits, E.A., Ebskamp, M.J., Paul, M.J., Jeuken, M.J., Weisbeek, P.J., Smeekens, S.C., 1995. Improved performance of transgenic fructan-accumulating tobacco under drought stress. Plant Physiol. 107 (1), 125–130.

Pilon-Smits, E.A., Terry, N., Sears, T., van Dun, K., 1999. Enhanced drought resistance in fructan-producing sugar beet. Plant Physiol. Biochem. 37 (4), 313–317.

Quan, R., Shang, M., Zhang, H., Zhao, Y., Zhang, J., 2004. Engineering of enhanced glycine betaine synthesis improves drought tolerance in maize. Plant Biotechnol. J. 2 (6), 477–486.

Rahnama, H., Vakilian, H., Fahimi, H., Ghareyazie, B., 2011. Enhanced salt stress tolerance in transgenic potato plants (*Solanum tuberosum* L.) expressing a bacterial *mtlD* gene. Acta Physiol. Plant. 33 (4), 1521–1532.

Ranganayakulu, G.S., Veeranagamallaiah, G., Sudhakar, C., 2013. Effect of salt stress on osmolyte accumulation in two groundnut cultivars (*Arachis hypogaea* L.) with contrasting salt tolerance. African J. Plant Sci. 7 (12), 586–592.

Reddy, K.R., Henry, W.B., Seepaul, R., Lokhande, S., Gajanayake, B., Brand, D., 2013. Exogenous application of glycinebetaine facilitates maize (*Zea mays* L.) growth under water deficit conditions. Am. J. Exp. Agric. 3 (1), 1.

Rezaei, M.A., Kaviani, B., Masouleh, A.K., 2012. The effect of exogenous glycine betaine on yield of soybean [Glycine max (L.) Merr.] in two contrasting cultivars Pershing and DPX under soil salinity stress. OMICS 5 (2), 87.

Romero, C., Bellés, J.M., Vayá, J.L., Serrano, R., Culiáñez-Macià, F.A., 1997. Expression of the yeast trehalose-6-phosphate synthase gene in transgenic tobacco plants: pleiotropic phenotypes include drought tolerance. Planta 201 (3), 293–297.

Roy, M., Wu, R., 2001. Arginine decarboxylase transgene expression and analysis of environmental stress tolerance in transgenic rice. Plant Sci. 160 (5), 869–875.

Roy, M., Wu, R., 2002. Overexpression of S-adenosylmethionine decarboxylase gene in rice increases polyamine level and enhances sodium chloride-stress tolerance. Plant Sci. 163 (5), 987–992.

Roychoudhury, A., Banerjee, A., Lahiri, V., 2015. Metabolic and molecular-genetic regulation of proline signaling and itscross-talk with major effectors mediates abiotic stress tolerance in plants. Turk. J. Bot. 39 (6), 887–910.

Roychoudhury, A., Chakraborty, M., 2013. Biochemical and molecular basis of varietal difference in plant salt tolerance. Annu. Rev. Res. Biol. 3 (4), 422–454.

Sabagh, A.E., Sorour, S., Ragab, A., Saneoka, H., Islam, M.S., 2017. The effect of exogenous application of proline and glycine betaineon the nodule activity of soybean under saline condition. J. Agric. Biotechnol 2 (1), 01–05.

Sadak, M.S., 2016. Mitigation of drought stress on fenugreek plant by foliar application of trehalose. Int. J. ChemTech Res. 9, 147–155.

Sagor, G.H.M., Berberich, T., Takahashi, Y., Niitsu, M., Kusano, T., 2013. The polyamine spermine protects Arabidopsis from heat stress-induced damage by increasing expression of heat shock-related genes. Transgenic Res. 22 (3), 595–605.

Sah, S.K., Kaur, G., Wani, S.H., 2016. Metabolic engineering of compatible solute trehalose for abiotic stress tolerance in plants. In: Iqbal, N., Nazar, R., Khan, N.A. (Eds.), Osmolytes and Plants Acclimation to Changing Environment: Emerging Omics Technologies. Springer India, pp. 83–96.

Sakamoto, A., Murata, A.N., 1998. Metabolic engineering of rice leading to biosynthesis of glycinebetaine and tolerance to salt and cold. Plant Mol. Biol. 38 (6), 1011–1019.

Sakamoto, A., Murata, N., 2002. The role of glycine betaine in the protection of plants from stress: clues from transgenic plants. Plant Cell Environ. 25 (2), 163–171.

Sánchez-Rodríguez, E., Romero, L., Ruiz, J.M., 2016. Accumulation of free polyamines enhances the antioxidant response in fruits of grafted tomato plants under water stress. J. Plant Physiol. 190, 72–78.

Sawahel, W., 2003. Improved performance of transgenic glycinebetaine-accumulating rice plants under drought stress. Biol. Plant. 47 (1), 39–44.

Sawahel, W.A., Hassan, A.H., 2002. Generation of transgenic wheat plants producing high levels of the osmoprotectant proline. Biotechnol. Lett. 24 (9), 721–725.

Sequera-Mutiozabal, M., Antoniou, C., Tiburcio, A.F., Alcázar, R., Fotopoulos, V., 2017. Polyamines: emerging hubs promoting drought and salt stress tolerance in plants. Curr. Mol. Biol. Rep. 3 (1), 28–36.

Shaddad, M.A., 1990. The effect of proline application on the physiology ofraphanus sativus plants grown under salinity stress. Biol. Plant. 32 (2), 104–112.

Sheveleva, E.V., Marquez, S., Chmara, W., Zegeer, A., Jensen, R.G., Bohnert, H.J., 1998. Sorbitol-6-phosphate dehydrogenase expression in transgenic tobacco high amounts of sorbitol lead to necrotic lesions. Plant Physiol. 117 (3), 831–839.

Shevyakova, N.I., Bakulina, E.A., Kuznetsov, V.V., 2009. Proline antioxidant role in the common ice plant subjected to salinity and paraquat treatment inducing oxidative stress. Russ. J. Plant Physiol. 56 (5), 663–669.

Shukla, V., Mattoo, A.K., 2013. Developing robust crop plants for sustaining growth and yield under adverse climatic changes. In: Tuteja, N., Gill, S.S. (Eds.), Climate Change and Plant Abiotic Stress Tolerance. pp. 27–56.

Singh, M., Kumar, J., Singh, S., Singh, V.P., Prasad, S.M., 2015. Roles of osmoprotectants in improving salinity and drought tolerance in plants: a review. Rev. Environ. Sci. Biotechnol. 14 (3), 407–426.

Slama, I., Abdelly, C., Bouchereau, A., Flowers, T., Savouré, A., 2015. Diversity, distribution and roles of osmoprotective compounds accumulated in halophytes under abiotic stress. Ann. Bot. 115 (3), 433–447.

Su, J., Wu, R., 2004. Stress-inducible synthesis of proline in transgenic rice confers faster growth under stress conditions than that with constitutive synthesis. Plant Sci. 166 (4), 941–948.

Suárez, R., Calderón, C., Iturriaga, G., 2009. Enhanced tolerance to multiple abiotic stresses in transgenic alfalfa accumulating trehalose. Crop Sci. 49 (5), 1791–1799.

Sulpice, R., Gibon, Y., Bouchereau, A., Larher, F., 1998. Exogenously supplied glycine betaine in spinach and rapeseed leaf discs: compatibility or non-compatibility? Plant Cell Environ. 21 (12), 1285–1292.

Suprasanna, P., Rai, A.N., HimaKumari, P., Kumar, S.A., Kavi Kishor, P.B., 2014. Modulation of proline: implications in plant stress tolerance and development. In: Plant Adaptation to Environmental Change. CAB International, Oxfordshire, pp. 68–96.

Suprasanna, P., Nikalje, G.C., Rai, A.N., 2016. Osmolyte accumulation and implications in plant abiotic stress tolerance. In: Iqbal, N., Nazar, R., Khan, N.A. (Eds.), Osmolytes and Plants Acclimation to Changing Environment: Emerging Omics Technologies. Springer India, New Delhi, pp. 1–12.

Surekha, C.H., Kumari, K.N., Aruna, L.V., Suneetha, G., Arundhati, A., Kishor, P.K., 2014. Expression of the *Vigna aconitifolia* P5CSF129A gene in transgenic pigeonpea enhances proline accumulation and salt tolerance. Plant Cell Tiss. Org. Cult. 116 (1), 27–36.

Surekha, C., Aruna, L.V., Hossain, M.A., Wani, S.H., Neelapu, N.R.R., 2015. Present status and future prospects of transgenic approaches for salt tolerance in plants/crop plants. In: Wani, S.H., Hossain, M.A. (Eds.), Managing Salt Tolerance in Plants, pp. 329–352.

Szabados, L., Savouré, A., 2010. Proline: a multifunctional amino acid. Trends Plant Sci. 15 (2), 89–97.

Takahashi, Y., Cong, R., Sagor, G.H.M., Niitsu, M., Berberich, T., Kusano, T., 2010. Characterization of five polyamine oxidase isoforms in *Arabidopsis thaliana*. Plant Cell Rep. 29 (9), 955–965.

Tarczynski, M.C., Jensen, R.G., Bohnert, H.J., 1992. Expression of a bacterial *mtlD* gene in transgenic tobacco leads to production and accumulation of mannitol. Proc. Natl. Acad. Sci. U. S. A. 89 (7), 2600–2604.

Tavladoraki, P., Cona, A., Federico, R., Tempera, G., Viceconte, N., Saccoccio, S., Battaglia, V., Toninello, A., Agostinelli, E., 2012. Polyamine catabolism: target for antiproliferative therapies in animals and stress tolerance strategies in plants. Amino Acids 42 (2–3), 411–426.

Thomas, J.C., Sepahi, M., Arendall, B., Bohnert, H.J., 1995. Enhancement of seed germination in high salinity by engineering mannitol expression in *Arabidopsis thaliana*. Plant Cell Environ. 18 (7), 801–806.

Tiburcio, A.F., Altabella, T., Bitrián, M., Alcázar, R., 2014. The roles of polyamines during the lifespan of plants: from development to stress. Planta 240 (1), 1–18.

Umezawa, T., Fujita, M., Fujita, Y., Yamaguchi-Shinozaki, K., Shinozaki, K., 2006. Engineering drought tolerance in plants: discovering and tailoring genes to unlock the future. Curr. Opin. Biotechnol. 17 (2), 113–122.

Van Der Meer, I.M., Ebskamp, M.J., Visser, R.G., Weisbeek, P.J., Smeekens, S.C., 1994. Fructan as a new carbohydrate sink in transgenic potato plants. Plant Cell 6 (4), 561–570.

Vendruscolo, E.C.G., Schuster, I., Pileggi, M., Scapim, C.A., Molinari, H.B.C., Marur, C.J., Vieira, L.G.E., 2007. Stress-induced synthesis of proline confers tolerance to water deficit in transgenic wheat. J. Plant Physiol. 164 (10), 1367–1376.

Verdoy, D., Coba De La Peña, T., Redondo, F.J., Lucas, M.M., Pueyo, J.J., 2006. Transgenic *Medicago truncatula* plants that accumulate proline display nitrogen-fixing activity with enhanced tolerance to osmotic stress. Plant Cell Environ. 29 (10), 1913–1923.

Vijn, I., Smeekens, S., 1999. Fructan: more than a reserve carbohydrate? Plant Physiol. 120 (2), 351–360.

Vinocur, B., Altman, A., 2005. Recent advances in engineering plant tolerance to abiotic stress: achievements and limitations. Curr. Opin. Biotechnol. 16 (2), 123–132.

Wang, G.P., Li, F., Zhang, J., Zhao, M.R., Hui, Z., Wang, W., 2010. Overaccumulation of glycine betaine enhances tolerance of the photosynthetic apparatus to drought and heat stress in wheat. Photosynthetica 48 (1), 30–41.

Wang, J.Y., Tong, S.M., Li, Q.L., 2013. Constitutive and salt-inducible expression of SlBADH gene in transgenic tomato (*Solanum lycopersicum* L. cv. Micro-Tom) enhances salt tolerance. Biochem. Biophys. Res. Commun. 432 (2), 262–267.

Wang, H., Tang, X., Wang, H., Shao, H.B., 2015. Proline accumulation and metabolism-related genes expression profiles in *Kosteletzkya virginica* seedlings under salt stress. Front. Plant Sci 6, 792–800.

Wani, S.H., Brajendra Singh, N., Haribhushan, A., Iqbal Mir, J., 2013. Compatible solute engineering in plants for abiotic stress tolerance-role of glycine betaine. Curr. Genomics 14 (3), 157–165.

Wani, S.H., Sah, S.K., Hossain, M.A., Kumar, V., Balachandran, S.M., 2016. Transgenic approaches for abiotic stress tolerance in crop plants. In: Al-Khayri, J., Jain, S., Johnson, D. (Eds.), Advances in Plant Breeding Strategies: Agronomic, Abiotic and Biotic Stress Traits. Springer International Publishing, pp. 345–396.

Wani, S.H., Dutta, T., Neelapu, N.R.R., Surekha, C., 2017. Transgenic approaches to enhance salt and drought tolerance in plants. Plant Gene 11, 219–231

Wei, C., Cui, Q., Zhang, X.Q., Zhao, Y.Q., Jia, G.X., 2016. Three P5CS genes including a novel one from *Lilium regale* play distinct roles in osmotic, drought and salt stress tolerance. J. Plant Biol. 59 (5), 456–466.

Wei, D., Zhang, W., Wang, C., Meng, Q., Li, G., Chen, T.H., Yang, X., 2017. Genetic engineering of the biosynthesis of glycinebetaine leads to alleviate salt-induced potassium efflux and enhances salt tolerance in tomato plants. Plant Sci. 257, 74–83.

Wen, X.P., Pang, X.M., Matsuda, N., Kita, M., Inoue, H., Hao, Y.J., Honda, C., Moriguchi, T., 2008. Over-expression of the apple spermidine synthase gene in pear confers multiple abiotic stress tolerance by altering polyamine titers. Transgenic Res. 17 (2), 251–263.

Wimalasekera, R., Tebartz, F., Scherer, G.F., 2011a. Polyamines, polyamine oxidases and nitric oxide in development, abiotic and biotic stresses. Plant Sci. 181 (5), 593–603.

Wimalasekera, R., Villar, C., Begum, T., Scherer, G.F., 2011b. COPPER AMINE OXIDASE1 (CuAO1) of *Arabidopsis thaliana* contributes to abscisic acid-and polyamine-induced nitric oxide biosynthesis and abscisic acid signal transduction. Mol. Plant 4 (4), 663–678.

Wingler, A., 2002. The function of trehalose biosynthesis in plants. Phytochemistry 60 (5), 437–440.

Yadu, B., Chandrakar, V., Meena, R.K., Keshavkant, S., 2017. Glycinebetaine reduces oxidative injury and enhances fluoride stress tolerance via improving antioxidant enzymes, proline and genomic template stability in *Cajanus cajan* L. S. Afr. J. Bot. 111, 68–75.

Yan, L.P., Liu, C.L., Liang, H.M., Mao, X.H., Wang, F., Pang, C.H., Shu, J., Xia, Y., 2012. Physiological responses to salt stress of T2 alfalfa progenies carrying a transgene for betaine aldehyde dehydrogenase. Plant Cell Tiss. Org. Cult. 108 (2), 191–199.

Yeo, E.T., Kwon, H.B., Han, S.E., Lee, J.T., Ryu, J.C., Byu, M.O., 2000. Genetic engineering of drought resistant potato plants by introduction of the trehalose-6-phosphate synthase (*TPS1*) gene from *Saccharomyces cerevisiae*. Mol. Cell 10 (3), 263–268.

Yiu, J.C., Juang, L.D., Fang, D.Y.T., Liu, C.W., Wu, S.J., 2009. Exogenous putrescine reduces flooding-induced oxidative damage by increasing the antioxidant properties of Welsh onion. Sci. Hortic. 120 (3), 306–314.

Yoshiba, Y., Kiyosue, T., Katagiri, T., Ueda, H., Mizoguchi, T., Yamaguchi-Shinozaki, K., Wada, K., Harada, Y., Shinozaki, K., 1995. Correlation between the induction of a gene for Δ1-pyrroline-5-carboxylate synthetase and the accumulation of proline in *Arabidopsis thaliana* under osmotic stress. Plant J. 7 (5), 751–760.

You, J., Hu, H., Xiong, L., 2012. An ornithine δ-aminotransferase gene *OsOAT* confers drought and oxidative stress tolerance in rice. Plant Sci. 197, 59–69.

Zhang, S.Z., Yang, B.P., Feng, C.L., Tang, H.L., 2005. Genetic transformation of tobacco with the trehalose synthase gene from *Grifola frondosa* Fr. enhances the resistance to drought and salt in tobacco. J. Integr. Plant Biol. 47 (5), 579–587.

Zhang, J., Tan, W., Yang, X.H., Zhang, H.X., 2008. Plastid-expressed choline monooxygenase gene improves salt and drought tolerance through accumulation of glycine betaine in tobacco. Plant Cell Rep. 27 (6), 1113.

Zhang, N., Si, H.J., Wen, G., Du, H.H., Liu, B.L., Wang, D., 2011a. Enhanced drought and salinity tolerance in transgenic potato plants with a BADH gene from spinach. Plant Biotechnol. Rep. 5 (1), 71–77.

Zhang, Y., Wu, R., Qin, G., Chen, Z., Gu, H., Qu, L.J., 2011b. Over-expression of WOX1 leads to defects in meristem development and polyamine homeostasis in *Arabidopsis*. J. Integr. Plant Biol. 53 (6), 493–506.

Zhang, X ., Tang, W., Liu, J., Liu, Y., 2014. Co-expression of rice *OsP5CS1* and *OsP5CS2* genes in transgenic tobacco resulted in elevated proline biosynthesis and enhanced abiotic stress tolerance. Chin. J. Appl. Environ. Biol. 20 (4), 717–722.

Zhang, Y., Zhang, H., Zou, Z.R., Liu, Y., Hu, X.H., 2015. Deciphering the protective role of spermidine against saline–alkaline stress at physiological and proteomic levels in tomato. Phytochemistry 110, 13–21.

FURTHER READING

Shan, X., Zhou, H., Sang, T., Shu, S., Sun, J., Guo, S., 2016. Effects of exogenous spermidine on carbon and nitrogen metabolism in tomato seedlings under high temperature. J. Am. Soc. Hortic. Sci. 141 (4), 381–388.

Slama, I., Tayachi, S., Jdey, A., Rouached, A., Abdelly, C., 2011. Differential response to water deficit stress in alfalfa (*Medicago sativa*) cultivars: growth, water relations, osmolyte accumulation and lipid peroxidation. Afr. J. Biotechnol. 10 (72), 16250–16259.

CHAPTER 12

Single-Versus Multigene Transfer Approaches for Crop Abiotic Stress Tolerance

Parul Goel*, Anil K. Singh†
*National Agri-Food Biotechnology Institute, Sahibzada Ajit Singh Nagar, Punjab, India
†ICAR-Indian Institute of Agricultural Biotechnology, Ranchi, India

Contents

1. INTRODUCTION

Plants encounter various unfavorable environmental conditions such as salinity, drought, freezing, high temperatures, and so forth, during their life cycle, which are together referred to as abiotic stresses. Plant growth and productivity is highly affected by these stresses, which may cause huge losses in crop yields (Boyer, 1982; Cushman and Bohnert, 2000; Ahuja et al., 2010). According to an FAO (2004) study, ~22% of agricultural land is facing the problem of high salinity. Moreover, due to global climate changes, the problems of water scarcity and high temperatures will also expand further, and that will be even more challenging for agricultural sustainability (Easterling et al., 2000; Burke et al., 2006). Therefore, scientists are now focusing on ways to develop stress-tolerant crops with higher yields and better growth under adverse climate conditions to meet the demands of food for an increasing world population.

Early attempts to develop stress-tolerant crops using conventional breeding methods by interspecific or intergenic hybridization, induced mutations, and some tissue culture techniques have yielded only limited success (Richards, 1996). However, with the advancement of functional genomics and biotechnological tools, the transgenic approach became most promising and more popular in developing stress-tolerant crops (Sreenivasulu et al., 2007; Singh et al., 2008, 2010). To genetically engineer the crop

Biochemical, Physiological and Molecular Avenues for Combating Abiotic Stress in Plants
https://doi.org/10.1016/B978-0-12-813066-7.00014-0

plants to enhance abiotic stress tolerance, the first requirement is to understand the complexity of plant abiotic stress responses and to select certain candidate genes or metabolic pathways that could be engineered to develop stress-tolerant crops.

The plant abiotic stress response is a complex network that starts from perceiving the stress signal by sensors, then releasing secondary messengers that relay the primary signal to the intracellular signal, and activation of stress-responsive genes that directly or indirectly protect plants against the stresses (Fig. 1; Holmberg and Bulow, 1998; Knight and Knight, 2001). The sensors or receptors present on the cell surface sense the external stress stimuli and transduce the signal intracellularly to a general cascade of signaling pathways. Very few sensor proteins, such as histidine kinases, or receptor-like kinases (RLK), have been identified as having a role in sensing abiotic stress stimuli in plants (Urao et al., 1999; Tamura et al., 2003; Chinnusamy et al., 2004). After perceiving stress signals, various downstream signaling components, mainly the secondary messengers such as calcium ions, ROS (reactive oxygen species), and cAMP (cyclic AMP) initiate the signal

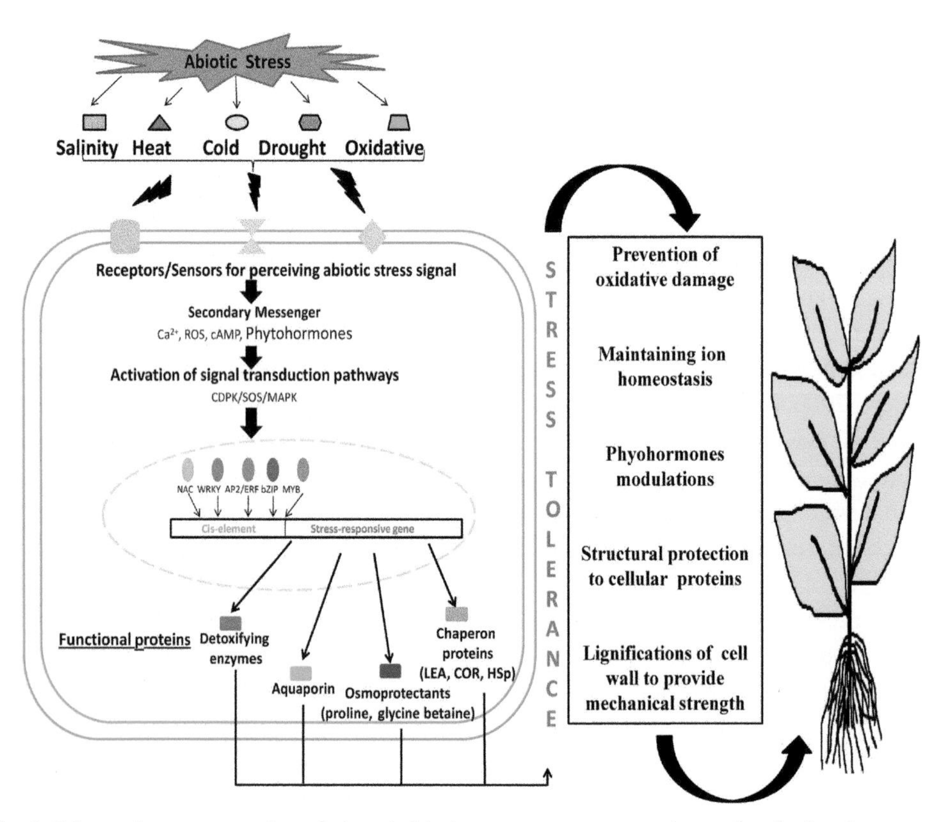

Fig. 1 Schematic representation of plants' abiotic stress responses at the molecular level.

transduction pathways, such as a MAPK cascade (Sanan-Mishra et al., 2006), SOS (salt overly sensitive) signaling pathway (Shi et al., 2002; Rodríguez et al., 2005), and CDPK pathway (Ca^{2+} dependent protein kinases) (Lee et al., 2004). These signal transduction pathways, through phosphorylation or dephosphorylation, lead to activation or suppression of several TFs that bind to the promoter region of stress-responsive genes.

2. RISE OF THE TRANSGENIC APPROACH FOR ENHANCING ABIOTIC STRESS TOLERANCE IN PLANTS

Plant growth and productivity highly depend on environmental factors that play a significant role in determining crop yields (Boyers, 1982). It has been estimated that more than 50% of crop loss is due to abiotic stresses, such as drought, salt, and extreme temperatures (Bray et al., 2000). Therefore, in order to meet the world's food demands, it is important to develop stress-tolerant crops. Initially, several traditional breeding approaches, such as cross-hybridization, mutation breeding, and marker-assisted selection programs were used for enhancing crop tolerance, but these methods are labor-intensive, time consuming, and have a low success rate due to the complex mechanisms of plant stress tolerance (Richards, 1996; Gilliham et al., 2017). With the advent of genetic engineering, the transgenic approach became more popular and promising in generating stress-tolerant crops (Yamaguchi and Blumwald, 2005). The transgenic approach involves introduction of foreign gene(s), or altering the expression of endogenous genes that could enhance stress tolerance by manipulating plants at the physiological, morphological, or molecular level. For engineering plants with enhanced abiotic stress tolerance, most researchers have focused on the two major categories of genes; namely, functional genes and regulatory genes (Shinozaki et al., 2003). The functional genes encode proteins or enzymes that are directly involved in maintaining osmotic balance or ion homeostasis, or protecting cells against free radicals, or in maintaining cell integrity. On the other hand, the regulatory genes encode for transcription factors or sets of kinases or phosphatases that regulate signal transduction pathways, and also transcriptional regulation of genes in response to various abiotic stresses. Several reviews have summarized the engineering of these functional or regulatory genes in plants for enhancing stress tolerance (Vinocur and Altman, 2005; Bhatnagar-Mathur et al., 2008; Singh et al., 2008, 2010; Cabello et al., 2014). Depending on the mode of action, the stress-inducible genes can be further categorized into either single-action or multiple-action genes. Single-action genes, or direct-action genes, participate in a specific pathway in response to abiotic stress, such as genes encoding for enzymes, such as glutathione peroxidase, superoxide dismutase, ascorbate peroxidases, and glutathione reductases, which participate in protecting plants against oxidative damage (Shafi et al., 2014, 2015, 2017). Similarly, genes involved in biosynthesis of osmoprotectants (glycine betaine, mannitol, proline, trehalose, etc.) protect plants against osmotic stress (Giri, 2011; Wani et al.,

2013). Multiple-action genes include transcription factors and signal transduction genes that can regulate the expression of several genes involved in abiotic stress tolerance in plants. Several researchers have successfully engineered plants for abiotic stress tolerance by overexpressing single- or multiple-action genes (Vinocur and Altman, 2005; Singh et al., 2008, 2010; Hussain et al., 2011; Wani et al., 2016; Wang et al., 2016). In plants, the abiotic stress response is highly complex and multigenic in nature; therefore, engineering plants with single-action genes may not be a very promising approach. To enhance plant tolerance against multiple abiotic stresses, it is thus important to target genes with multiple roles in plant stress tolerance, such as transcription factors and signal transduction genes, or sets of genes through the multiple-gene transfer (MGT) approach that may eventually lead to enhanced tolerance of plants against diverse abiotic stresses.

2.1 Single-Gene Transfer Approach

In this approach, a single gene that might be a functional gene or regulatory gene is transferred to a plant system to confer abiotic stress tolerance. Several gene-encoding enzymes or proteins such as ion transporters, osmoprotectants, detoxifying enzymes, and so forth that have a protective role in plant stress tolerance have been successfully engineered in many economically important crops (Shi et al., 2003; Lv et al., 2007; McKersie et al., 1997; Vendruscolo et al., 2007; Gupta et al., 1993). Improving abiotic stress tolerance through the single-gene transfer approach is the most common strategy adopted by researchers (Mittler and Blumwald, 2010). However, as plant abiotic stress response is highly complex and regulated by multiple genes, little success has been achieved by this approach (Varshney et al., 2011). To overcome this limitation, researchers have paid attention to regulatory genes, such as TFs, that can respond to multiple abiotic stresses and regulate the expression of many downstream stress-responsive genes (Bartels and Sunkar, 2005; Yamaguchi-Shinozaki and Shinozaki, 2006).

2.1.1 Transcription Factors: As Potential Candidates for the Single-Gene Transfer Approach

Plants encounter a diverse array of abiotic stresses. As mentioned previously, the mechanism of abiotic stress tolerance in plants is a highly complex process, thus transcription factors seem to be the key players in engineering plants for improved stress tolerance because of their role as master regulators (Yamaguchi-Shinozaki and Shinozaki, 2006). In general, the transcription factors are the proteins that can regulate the expression of several genes simultaneously by binding to their promoter regions. As a master regulator, TFs not only act independently, but can crosstalk with each other in order to regulate the complex transcriptional machinery involved in plant abiotic stress response. Several reviews have summarized the role of various transcription factors in plant stress tolerance (Wang et al., 2016; Lata et al., 2011; Singh et al., 2010).

For enhancing abiotic stress tolerance in plants by overexpressing a single transcription factor, it is important to select multiple stress-responsive transcription factor genes by carrying out their expression profiling in response to various stresses. Thereafter, the functional role of selected TFs in improving plant abiotic stress tolerance should be evaluated in both model and economically important plant species. In the past few decades, significant research has been carried out to identify and explore the role of several TF families, such as AP2/EREBP, bZIP, NAC, MYB, MYC, and WRKY in plant abiotic stress tolerance (Jung et al., 2008; Singh et al., 2010, 2013). Selected examples of TF genes that have been successfully utilized to enhance multiple-stress tolerance in plants are shown in Table 1. Thus, genetic manipulation of gene encoding for multiple-stress inducible transcription factors is believed to be the most promising strategy of the single-gene transfer approach for enhancing plant abiotic stress tolerance (Chinnusamy et al., 2004). So far, a great deal of research has gone into understanding the role of TFs in plant abiotic stress responses; however, several issues regarding the existence of functional redundancy among TFs (Mitsuda and Ohme-Takagi, 2009), and dissecting the complete functional network (upstream and downstream elements) of individual TFs is largely undefined, and needs to be investigated to understand their functionality in response to abiotic stresses.

2.2 Multiple-Gene Transfer Approach

The abiotic stress response in plants is an intricate network that involves participation of hundreds of genes and their products (Umezawa et al., 2006; Zhu, 2016). Simultaneous manipulation of the expression of various genes that have a protective role in plant stress tolerance is thus a promising approach for enhancing multiple-stress tolerance in plants. The process of introducing several genes into plant systems simultaneously is known as the multiple-gene transfer approach (MGT). The delivery of multiple genes could be performed mainly by three approaches, that is, the conventional method, and retransformation and co-transformation with linked or unlinked genes (Halpin, 2005; Naqvi et al., 2010; Zorrilla-Lopez et al., 2013). The conventional approach, which is commonly known as gene stacking, involve the introduction of genes of interest into different independent transgenic lines, which are then crossed to yield progeny that carry all of the desired genes (Fig. 2). In the retransformation method, or sequential transformation, a transgenic plant already harboring a transgene is transformed with another gene (Fig. 2; Jobling et al., 2002). Both conventional and retransformation methods have some degree of limitations. One of the major limitations of both these methods is that the introduced transgenes are not linked with each other, and may be integrated at different loci in the genome. This would increase the chances of their segregation in subsequent generations. Moreover, both these methods are laborious and time consuming, and multiple selectable markers are required for every round of transformation. One of the most

Table 1 List showing selected examples of transcription factor families employed to enhance multiple abiotic stress tolerance in plants

TF family	Gene name	Host plant	Transgenic plant	Abiotic stress tolerance	References
AP2/ EREBP	*ThDREB*	*Tamarix hispida*	Tobacco, *T. hispida*	Salinity, drought	Yang et al. (2017)
	OsEREBP1	*Oryza sativa*	Rice	Drought, submergence	Jisha et al. (2015)
	VrDREB2A	*Vigna radiata*	Arabidopsis	Salinity, drought	Chen et al. (2016)
	FeDREB1	*Fagopyrum esculentum*	Arabidopsis	Drought, freezing	Fang et al. (2015a,b)
	JcDREB	*Jatropha curcas*	Arabidopsis	Salinity, freezing	Tang et al. (2011)
	EaDREB2	*Erianthus arundinaceus*	Sugarcane	Drought, salinity	Augustine et al. (2015)
	MnDREB4a	*M. notabilis*	Tobacco	Salinity, drought, cold, heat	Liu et al. (2015)
	SsDREB	*Suaeda salsa*	Tobacco	Salinity, drought	Zhang et al. (2015a,b,c)
	TaERF3	*Triticum aestivum*	Wheat	Salinity, drought	Rong et al. (2014)
	CkDREB	*Caragana korshinskii*	Tobacco	Salinity, osmotic	Wang et al. (2011)
bZIP	*TabZIP60*	*Triticum aestivum*	Arabidopsis	Salinity, drought, cold	Zhang et al. (2015a,b,c)
	ThbZIP1	*Tamarix hispida*	Arabidopsis	Salinity, drought	Ji et al. (2013)
	ZmZIP72	*Zea may*	Arabidopsis	Salinity, drought	Ying et al. (2012)
	GmbZIP1	*Glycine* max	Arabidopsis	Salinity, drought, cold	Gao et al. (2011)
	ABP9	*Zea mays*	Cotton	Salinity, drought	Wang et al. (2017)
	GhABF2	*Gossypium hirsutum*	Cotton	Salinity, drought	Liang et al. (2016)
	ABF3	*Medicago sativa*	Alfalfa	Salinity, oxidative	Wang et al. (2016)
	OsHBP1b	*Oryza sativa*	Tobacco	Salinity, drought	Lakra et al. (2015)
	OsbZIP71	*Oryza sativa*	Rice	Salinity, drought	Liu et al. (2014)
NAC	*ThNAC13*	*Tamarix hispida*	Arabidopsis	Salinity, osmotic	Wang et al. (2017)
	MINAC9	*Miscanthus lutarioriparius*	Arabidopsis	Salinty, drought	Zhao et al. (2016)
	EcNAC67	*Eleusine coracana*	Rice	Salinty, drought	Rahman et al. (2016)
	ONAC022	*Oryza sativa*	Rice	Salinty, drought	Hong et al. (2016)
	CarNAC4	*Cicer ariettinum*	Arabidopsis	Salinity,drought	Yu et al. (2016)
	TaNAC29	*Triticum aestivum*	Arabidopsis	Salinty, drought	Huang et al. (2015)
	TaNAC47	*Triticum aestivum*	Arabidopsis	Salinity, drought, freezing	Zhang et al. (2015a,b,c)
	SNAC3	*Oryza sativa*	Rice	Drought, heat	Fang et al. (2015a,b)
	TaNAC67	*Triticum aestivum*	Arabidopsis	Salinity, drought, cold	Mao et al. (2014)
	SINAC4	*Solanum lycopersicum*	Tomato	Salinity, drought	Zhu et al. (2014)

MYB	*FtMYB9*	*Fagopyrum tataricum*	Arabidopsis	Salinity, drought	Gao et al. (2017)
	TaMYB3R1	*Triticum aestivum*	Arabidopsis	Salinity, drought	Cai et al. (2015)
	SbMYB15	*Salicornia brachiata*	Arabidopsis	Salinity, drought	Shukla et al. (2015)
	GmMYB12B2	*Glycine max*	Arabidopsis	Salinity, U.V.	Li et al. (2016)
	AtMYB15	Arabidopsis	Arabidopsis	Salinity, drought	Ding et al. (2009)
	SoMYB18	*Saccharum officinarum*	Tobacco	Salinity, dehydration	Shingotem et al. (2015)
	MdSIMYB1	*Malus × domestica*	Apple	Drought, cold and salinity	Wang et al. (2014)
	LeAN2	*Lycopersicum esculentum*	Tobacco	Chilling, oxidative stress	Meng et al. (2014)
	OsMYB2	*Oryza sativa*	Rice	Salinity, drought	Yang et al. (2012)
	TaMYB56-B	*Triticum aestivum*	Arabidopsis	Salinity, freezing	Zhang et al. (2012)
	AtMYB44	*A. thaliana*	Soybean	Salinity, drought and cold	Seo et al. (2012)
WRKY	*TaWRKY 93*	*Triticum aestivum*	Arabidopsis	Salinity, drought, cold	Qin et al. (2015)
	HaWRKY76	*Helianthus annuus*	Arabidopsis	Drought, flood	Raineri et al. (2015)
	TaWRKY1, TaWRKY33	*Triticum aestivum*	Arabidopsis	Drought, heat	He et al. (2016)
	MtWRKY70	*Medicago truncatula*	*M. truncatula*	Salinity, drought	Liu et al. (2016)
	TaWRKY44	*Triticum aestivum*	Tobacco	Salinity, drought	Wang et al. (2015a,b)
	TaWRKY10	*Triticum aestivum*	Tobacco	Salinity, drought	Wang et al. (2013)
	GhWRKY68	*Glycine max*	Tobacco	Salinity, drought	Jia et al. (2015)
	SpWRKY1	*Solanum pimpinellifolium L3708*	Tobacco	Salinity, drought	Li et al. (2014)
	ZmWRK58	*Zea may*	Rice	Salinity, drought	Liu et al. (2016)
	DnWRKY11	*Dendrobium nobile*	Tobacco	Salinity, drought	Xu et al. (2014)

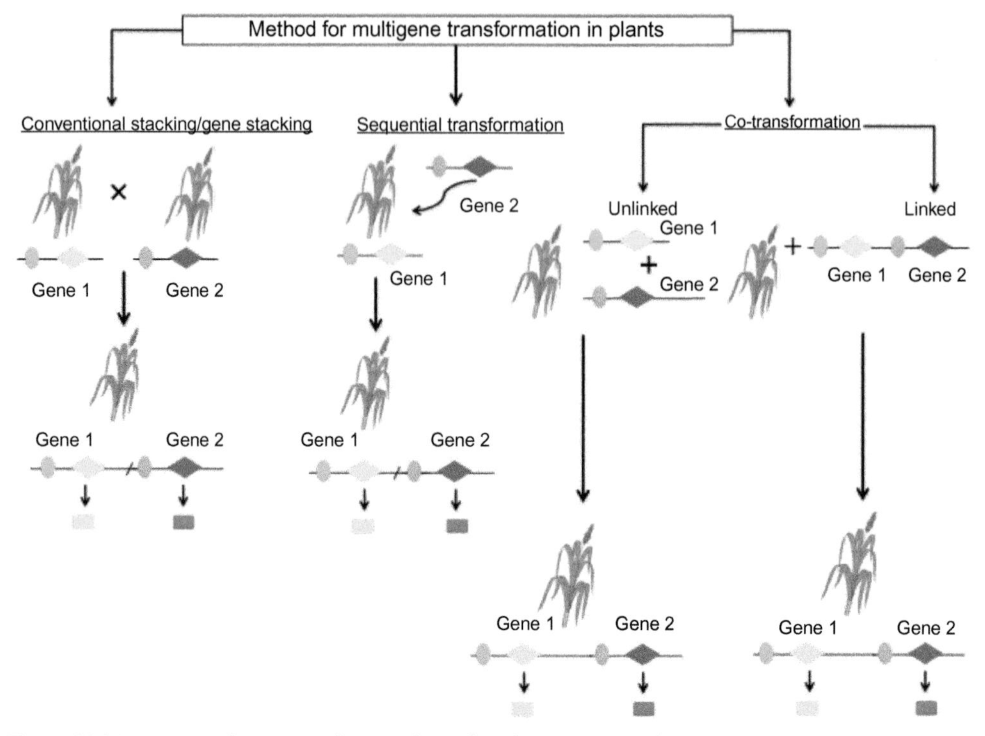

Fig. 2 Multigene transfer approaches in plants for abiotic stress tolerance.

promising multiple-gene transfer approaches is co-transformation, in which several genes are introduced into plant systems simultaneously. Co-transformation can be performed either by linking or unlinking the genes. In the linking approach, multiple genes are placed on a single vector, whereas in the unlinking approach, different genes are introduced into plant systems using separate vectors (Fig. 2). The field of the MGT approach in combating multiple abiotic stress responses in plants is still in its nascent stage. Initially, there were very few reports employing the MGT approach for improving stress tolerance (Stein et al., 2011; Le Martret et al., 2011). In early reports, the simultaneous co-expression of several genes encoding enzymes, such as gamma-glutamyl kinase 74 (*GK74*), gammaglutamylphosphate reductase (*GPR*), as well as antisensing of proline dehydrogenase (*ProDH*) from *E. coli* of proline biosynthesis have been achieved in *Arabidopsis* and tobacco (Stein et al., 2011). In another report, Le Martret et al. (2011) have simultaneously overexpressed three genes encoding for enzymes, namely dehydroascorbate reductase (DHAR), glutathione-*S*-transferase (GST), and glutathione reductase (GR) of the ROS scavenging system into the chloroplast of tobacco, which resulted in enhanced tolerance to multiple stresses.

2.2.1 Success Stories of the Multiple-Gene Transfer Approach for Enhancing Abiotic Stress Tolerance in Plants

2.2.1.1 Engineering Genes Involved in Antioxidant Pathways for Enhancing Salt Tolerance

In plants, a balance between ROS production and its elimination is maintained within the cellular environment under controlled conditions. However, under stress conditions, this balance is disturbed, and leads to serious oxidative stress (Vellosillo et al., 2010; Halliwell, 1997). The rapid generation of ROS is a common event observed in plants under biotic and abiotic stresses (Sewelam et al., 2016). The generation of ROS may affect the cellular function by damaging nucleic acids, oxidizing proteins, and by promoting lipid peroxidation (Foyer and Noctor, 2005). Hence, accumulation of ROS is one of the major causes of loss of crop productivity under stress conditions (Gill and Tuteja, 2010; Khan and Singh, 2008). However, plants have evolved antioxidant defense systems to avoid these oxidative stresses. The two enzymes, namely, superoxide dismutase (SOD) and ascorbate peroxidate (APX), are part of the antioxidant defense system in plants (Asada and Takahashi, 1987; Bowler et al., 1994). The SOD enzyme acts as first line of defense by early scavenging of superoxide radicals (O_2^-) and converting them to hydrogen peroxide (H_2O_2) (Perl-Treves and Galun, 1991). The APX enzyme, on the other hand, reduces H_2O_2 into water molecules. In 2015, Shafi and coworkers isolated *PaSOD* and *RaAPX* genes from high-altitude plants *Potentilla atrosanguinea* and *Rheum australe*, respectively. Both the genes were overexpressed individually in *A. thaliana,* and *PaSOD* and *RaAPX* transgenic plants were crossed to obtain dual transgenic lines. The dual transgenic plants showed tolerance to salinity stress, mainly by enhancing lignin deposition in vascular bundles that provide mechanical strength to plants. Moreover, accumulation of compatible solutes, such as proline and soluble sugars, was also high in transgenic plants.

2.2.1.2 Gene Pyramiding of EaDREB2 With Pea DNA Helicase Gene (PDH45) Enhances Drought and Salinity Tolerance

Sugarcane is an important crash crop that often encounters abiotic stresses, such as salinity and drought (Inman-Bamber and Smith, 2005). Augustine et al. (2015) have successfully engineered a popular sugarcane variety, Co86032, for enhanced salt and drought tolerance through a gene pyramiding approach. In this report, the *EaDREB2* gene isolated from *Erianthus arundinaceus* and *PDH45* (Pea DNA helicase), isolated from peas, were co-transformed via *Agrobacterium*-mediated transformation in sugarcane using the Portubi 2.3 promoter (Phillip et al., 2013). Both the genes *GbDREB2* and *PDH45* were earlier reported to enhance drought and salinity tolerance in several plant species (Chen et al., 2007, 2009; Bihani et al., 2011; Sanan-Mishra et al., 2005; Shivakumara et al., 2017). The transgenic sugarcane overexpressing both the genes showed enhanced expression of several stress-related marker genes such as *RD29*, *LEA*, *ERD*, *ERF*, *COR15*, *BRICK,* higher chlorophyll content, photosynthetic efficiency, and better bud germination as compared with wild type plants upon exposure to drought and salinity stress.

2.2.1.3 Simultaneous Expression of Three Genes, Alfin1, PgHSF4, and PDH45 Improves Drought Adaptation

The peanut is an economically important legume, worldwide. Drought stress severely affects peanut production (Kambiranda et al., 2011). In 2015, Ramu and coworkers improved the adaptation and productivity of peanuts under drought stress by simultaneous overexpression of genes such as *Alfin 1, PgHSF4,* and *PDH45.* Alfin 1 is a novel zinc-finger protein first identified in alfalfa roots and involved in transcriptional regulation of stress-inducible genes (Bastola et al., 1998). The overexpression of the *Alfin 1* gene in alfalfa and tobacco showed enhanced stress tolerance by improving root growth (Winicov et al., 2004; Nethra, 2010). The gene *PDH45* (pea DNA helicase 45) is a homologue of the eukaryotic translation initiation factor eIF-4A, and conferred abiotic stress tolerance in plants (Sanan-Mishra et al., 2005; Shivakumara et al., 2017). The *PgHSF4* gene was isolated from *Pennisutum glaucum,* which encodes for the heat shock transcription factor (HSF). The HSF regulates the expression of heat shock proteins (HSP) that function as molecular chaperones, and increases protein stability under stressful conditions (Mishra et al., 2002; Liu et al., 2013; Yabuta, 2016). Several reports have also suggested that multiple-stress tolerance can be achieved in many engineered plants such as rice, wheat, and Arabidopsis by overexpressing the HSF gene (Jung et al., 2013; Shim et al., 2009; Banti et al., 2010). To enhance drought tolerance in peanuts by using the MGT approach, scientists first isolated three genes; namely, *Alfin 1, PgHSF4,* and *PDH45* from rice, *Pennisutum glaucum,* and peas, respectively. These genes were inserted into a single multigene expression vector *pKM12GW* using gateway technology (Vemanna et al., 2013). Each gene was cloned under the control of individual promoters and terminators. The binary vector harboring the pKM12GW:PgHSF4:PDH45:OsAlfin1 construct was transferred into the peanut GPBD4 through *Agrobacterium*-mediated transformation. The transgenic peanut plants showed induction of the expression of several stress-responsive genes, such as heat-shock proteins (HSPs), RING box protein-1 (RBX1), aldose reductase, late-embryogenesis abundant-5 (LEA5), and proline-rich protein-2 (PRP2). The transgenic lines showed a higher survival rate by exhibiting higher biomass, higher root growth, and also higher relative water content under drought stress.

2.2.1.4 Engineering Glyoxalase Pathways for Enhancing Multiple-Stress Tolerance

Rice is the most important staple food crop for more than half of the world's population. However, it is prone to several abiotic stresses, such as salinity, drought, cold, and submergence (Lafitte et al., 2004). Many studies have been done at the physiological, biochemical, and molecular level to understand the response of rice crops under adverse climatic conditions (Pandey and Shukla, 2015; Singh and Jwa, 2013; Singh et al., 2008). Methylglyoxal (CH3COCHO; MG) is a cytotoxic αβ-dicarbonyl aldehyde that accumulates in plants in response to several abiotic stresses (Yadav et al., 2005; Kaur et al., 2014).

The MG is produced as a side product of several enzymatic or nonenzymatic reactions of metabolic pathways such as glycolysis, and protein and lipid metabolism (Kaur et al., 2015). However, in plants, disruption of dihydroxyacetone phosphate and glyceraldehyde 3-phosphate is the major path of MG production. The accumulation of MG under stressful conditions leads to several detrimental effects in plants, which include inhibition of growth and development via protein degradation, DNA strand breakage, and inactivation of antioxidant systems (Li, 2016; Hoque et al., 2016). To overcome the cytotoxic effect of MG, plants have evolved a specialized pathway known as the glyoxalase pathway, which detoxifies the excess MG by a series of two thiol-dependent enzymes, glyoxalase I and glyoxalase II (Fig. 3; Kaur et al., 2014; Gupta et al., 2017). Through the multigene transfer approach, scientists have successfully enhanced tolerance against salinity stress by engineering the complete glyoxalase pathway in model tobacco, tomato, and citrus plants (Singla-Pareek et al., 2003, 2006; Alvarez Viveros et al., 2013; Alvarez-Gerding et al., 2015). Recently, a multiple-gene transfer approach has been adopted to enhance the stress

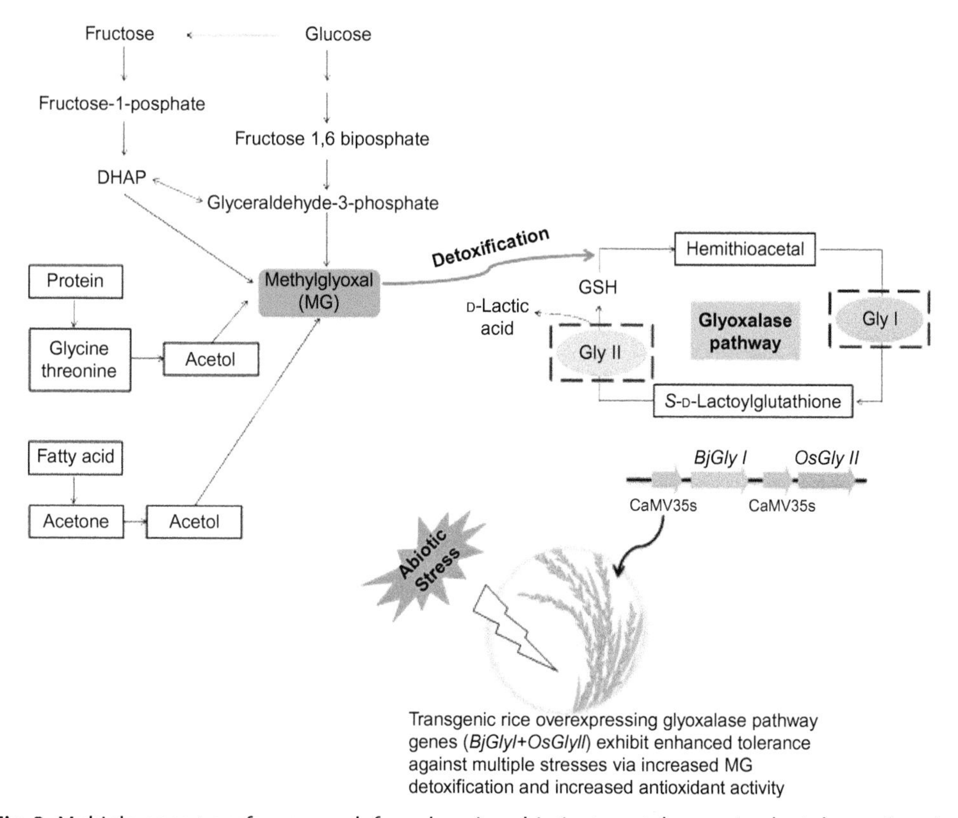

Fig. 3 Multiple-gene transfer approach for enhancing abiotic stress tolerance in plants by engineering the complete glyoxalase pathway. *DHAP*, dihydroxyacetone phosphate; *Gly I and Gly II*, glyoxalase I and glyoxalase II; *GSH*, glutathione.

tolerance of rice by engineering the complete glyoxalase pathway in rice (Gupta et al., 2017). For this, the *GLYI* gene from *B. juncea* and the *GLYII* gene from *O. sativa* were cloned as two different cassettes under the control of the 35S promoter on a single vector pCAMBIA1304 and overexpressed simultaneously in rice plants (Gupta et al., 2017). The glyoxalase overexpressing transgenic rice plants showed enhanced tolerance to multiple stresses such as salinity, drought, heat, and some biotic stresses. This enhanced tolerance has been achieved due to multiple factors that include increased detoxification of MG due to higher activity of enzymes, mainly TPI (triose phosphate isomerase), and glyoxalases that ultimately reduce MG-induced cytotoxic effects, and protection of ultra-structures of organelles, such as chloroplast and mitochondria via increased antioxidant properties. Therefore, the multiple-gene transfer approach is an excellent example of enhancing multiple stress tolerance in plants by targeting a single metabolic pathway that is involved in several plant stress responses. This strategy could also be employed to improve stress tolerance of other economically important crops.

3. CONCLUSIONS

The transgenic approach has widened the scope of crop improvement by allowing the transfer of desirable gene(s) to develop novel characteristics, such as protection against abiotic stresses. Therefore, it is important to identify and select a certain set of genes that can enhance crop tolerance to abiotic stresses. During the past few decades, the single-gene transfer approach was the most commonly employed approach for engineering plants with abiotic stress tolerance (Muthurajan and Balasubramanian, 2009). However, recently, the multiple-gene transfer approach has emerged as a promising approach for engineering plants with abiotic stress tolerance. A tabulated comparison has been presented between both the approaches on the basis of the manipulation of traits, transgene silencing, transgene stability, and so forth. The MGT approach allows researchers to manipulate the entire metabolic pathway and express the multimeric proteins; hence, this approach is considered to be more promising in engineering plants against abiotic stresses, which include complex traits involving different pathways and genes. However, MGT is subject to certain limitations, such as transgenes silencing, rearrangement, and so forth. In the future, novel strategies should be adopted to overcome these barriers (Li et al., 2012; Peremarti et al., 2010).

Moreover, the choice of promoter is critically important in both the approaches. The cauliflower mosaic virus 35S (CaMV35S), ubiquitin (UBI1), and actin are the most common constitutive promoters used in the single-gene transfer approach (Holtorf et al., 1995). However, the constitutive expression of several genes encoding for trehaloses, polyamines, and transcription factors resulted in abnormal plant growth under normal conditions (Bhatnagar-Mathur et al., 2008; Hsieh et al., 2002). Thus, the use of stress-inducible promoters such as sHSP26 (Khurana et al., 2013) and rd29

(Kasuga et al., 1999) that induce the expression of transgene upon stress exposure should be used to develop stress-tolerant transgenic plants without any yield penalty. In the case of the MGT approach, the repetitive use of the same promoter for multiple-gene expression can lead to transgene silencing (Mourrain et al., 2007). This can be overcome by the use of natural promoters exhibiting similar activities, or modified synthetic promoters (Liu and Stewart Jr., 2016). Moreover, for the MGT approach, use of bidirectional promoters that lead to the simultaneous expression of two genes is also becoming popular (Chong et al., 1997). In the future, it will also be important to develop transgenic crops without any selection markers or reporter genes. This would help avoid the risk of horizontal gene transfer from transgenic plants to other biological systems, especially when the antibiotic resistance gene is used as the selection marker. Furthermore, growth performance of abiotic stress-tolerant transgenic plants should be evaluated under field conditions, where multiple abiotic stresses coexist.

ACKNOWLEDGMENTS

Parul Goel acknowledges the National Post-Doctoral Fellowship (N-PDF) by the Science and Engineering Research Board (SERB), the Department of Science & Technology (DST), Government of India (GOI). Anil K. Singh acknowledges research project SERB/F/307/2014-15 under the Scheme for Young Scientists by SERB, DST, GOI. Anil K. Singh also acknowledges Institute projects IXX12585 and IXX12644 funded by ICAR-IIAB, Ranchi.

REFERENCES

Ahuja, I., de Vos, R.C.H., Bones, A.M., Hall, R.D., 2010. Plant molecular stress responses face climate change. Trends Plant Sci. 15, 664–674.

Alvarez Viveros, M.F., Inostroza-Blancheteau, C., Timmermann, T., González, M., Arce-Johnson, P., 2013. Overexpression of *GlyI* and *GlyII* genes in transgenic tomato (*Solanum lycopersicum* mill.) plants confers salt tolerance by decreasing oxidative stress. Mol. Biol. Rep. 40, 3281–3290.

Alvarez-Gerding, X., Cortés-Bullemore, R., Medina, C., Romero-Romero, J.L., Inostroza-Blancheteau, C., Aquea, F., Arce-Johnson, P., 2015. Improved salinity tolerance in Carrizo Citrange rootstock through overexpression of glyoxalase system genes. Biomed. Res. Int. 2015, 827951.

Asada, K., Takahashi, M., 1987. Production and scavenging of active oxygen in photosynthesis. In: Kyle, D.J., Osmond, C.B., Arntzen, C.J. (Eds.), Photoinhibition. Elsevier Science Publishers B.V., Amsterdam, pp. 227–287

Augustine, S.M., Ashwin Narayan, J., Syamaladevi, D.P., Appunu, C., Chakravarthi, M., Ravichandran, V., Tuteja, N., Subramonian, N., 2015. Overexpression of EaDREB2 and pyramiding of EaDREB2 with the pea DNA helicase gene (PDH45) enhance drought and salinity tolerance in sugarcane (*Saccharum* spp. hybrid). Plant Cell Rep. 34, 247–263.

Banti, V., Mafessoni, F., Loreti, E., Alpi, A., Perata, P., 2010. The heat-inducible transcription factor HsfA2 enhances anoxia tolerance in Arabidopsis. Plant Physiol. 152, 1471–1483.

Bartels, D., Sunkar, R., 2005. Drought and salt tolerance in plants. Crit. Rev. Plant Sci. 24, 23–58.

Bastola, D.R., Pethe, V.V., Winicov, I., 1998. Alfin1, a novel zinc-finger protein in alfalfa roots that binds to promoter elements in the salt-inducible MsPRP2 gene. Plant Mol. Biol. 38, 1123–1135.

Bhatnagar-Mathur, P., Vadez, V., Sharma, K.K., 2008. Transgenic approaches for abiotic stress tolerance in plants: retrospect and prospects. Plant Cell Rep. 27, 411–424.

Bihani, P., Char, B., Bhargava, S., 2011. Transgenic expression of sorghum DREB2 in rice improves tolerance and yield under water limitation. J. Agric. Sci. 149, 95–101.

Bowler, C., Van Camp, W., Van Montagu, M., Inzé, D., 1994. Superoxide dismutases in plants. Crit. Rev. Plant Sci. 13, 199–218.

Boyer, J.S., 1982. Plant productivity and environment. Science 218, 443–448.

Bray, E.A., Bailey-Serres, J., Weretilnyk, E., 2000. Responses to abiotic stresses. In: Gruissem, W., Buchannan, B., Jones, R. (Eds.), Biochemistry and Molecular Biology of Plants. American Society of Plant Biologists, Rockville, MD, pp. 1158–1249.

Burke, E.J., Brown, S.J., Christidis, N., 2006. Modeling the recent evolution of global drought and projections for the twenty-first century with the Hadley centre climate model. J. Hydrometeorol. 7, 1113–1125.

Cabello, J.V., Lodeyro, A.F., Zurbriggen, M.D., 2014. Novel perspectives for the engineering of abiotic stress tolerance in plants. Curr. Opin. Biotechnol. 26, 62–70.

Cai, H., Tian, S., Dong, H., Guo, C., 2015. Pleiotropic effects of TaMYB3R1 on plant development and response to osmotic stress in transgenic Arabidopsis. Gene 558, 227–234.

Chen, M., Wang, Q.Y., Cheng, X.G., Xu, Z.S., Li, L.C., Ye, X.G., Xia, L.Q., Ma, Y.Z., 2007. GmDREB2, a soybean DRE-binding transcription factor, conferred drought and high-salt tolerance in transgenic plants. Biochem. Biophys. Res. Commun. 353, 299–305.

Chen, J., Xia, X., Yin, W., 2009. Expression profiling and functional characterization of a DREB2-type gene from *Populus euphratica*. Biochem. Biophys. Res. Commun. 378, 483–487.

Chen, H., Liu, L., Wang, L., Wang, S., Cheng, X., 2016. VrDREB2A, a DREB-binding transcription factor from *Vigna radiata*, increased drought and high-salt tolerance in transgenic *Arabidopsis thaliana*. J. Plant Res. 129, 263–273.

Chinnusamy, V., Schumaker, K., Zhu, J.K., 2004. Molecular genetic perspectives on cross-talk and specificity in abiotic stress signalling in plants. J. Exp. Bot. 55, 225–236.

Chong, D.K.X., Roberts, W., Arakawa, T., Illes, K., Bagi, G., Slattery, C.W., Langridge, W.H.R., 1997. Expression of the human milk protein β-casein in transgenic potato plants. Transgenic Res. 6, 289–296.

Cushman, J.C., Bohnert, H.J., 2000. Genomic approaches to plant stress tolerance. Curr. Opin. Plant Biol. 3, 117–124.

Ding, Z., Li, S., An, X., Liu, X., Qin, H., Wang, D., 2009. Transgenic expression of MYB15 confers enhanced sensitivity to abscisic acid and improved drought tolerance in *Arabidopsis thaliana*. J. Genet. Genomics 36, 17–29.

Easterling, D.R., Meehl, G.A., Parmesan, C., Changnon, S.A., Karl, T.R., Mearns, L.O., 2000. Climate extremes: observations, modeling, and impacts. Science 289, 2068–2074.

Fang, Y., Liao, K., Du, H., Xu, Y., Song, H., Li, X., Xiong, L., 2015a. A stress-responsive NAC transcription factor SNAC3 confers heat and drought tolerance through modulation of reactive oxygen species in rice. J. Exp. Bot. 66, 6803–6817.

Fang, Z.W., Xu, X.Y., Gao, J.F., Wang, P.K., Liu, Z.X., Feng, B.L., 2015b. Characterization of FeDREB1 promoter involved in cold- and drought-inducible expression from common buckwheat (*Fagopyrum esculentum*). Genet. Mol. Res. 14, 7990–8000.

FAO (Food, Agriculture Organization of the United Nations), 2004. FAO Production Yearbook. FAO, Rome.

Foyer, C.H., Noctor, G., 2005. Redox homeostasis and antioxidant signaling: a metabolic interface between stress perception and physiological responses. Plant Cell 17, 1866–1875.

Gao, S.Q., Chen, M., Xu, Z.S., Zhao, C.P., Li, L.C., Xu, H.J., Tang, Y.M., Zhao, X., Ma, Y.Z., 2011. The soybean GmbZIP1 transcription factor enhances multiple abiotic stress tolerances in transgenic plants. Plant Mol. Biol. 75, 537–553.

Gao, F., Zhou, J., Deng, R.Y., Zhao, H.X., Li, C.L., Chen, H., Suzuki, T., Park, S.U., Wu, Q., 2017. Overexpression of a tartary buckwheat R2R3-MYB transcription factor gene, FtMYB9, enhances tolerance to drought and salt stresses in transgenic Arabidopsis. J. Plant Physiol. 214, 81–90.

Gill, S.S., Tuteja, N., 2010. Reactive oxygen species and antioxidant machinery in abiotic stress tolerance in crop plants. Plant Physiol. Biochem. 48, 909–930.

Gilliham, M., Able, J.A., Roy, S.J., 2017. Translating knowledge about abiotic stress tolerance to breeding programmes. Plant J. 90, 898–917.

Giri, J., 2011. Glycinebetaine and abiotic stress tolerance in plants. Plant Signal. Behav. 6, 1746–1751.

Gupta, B.K., Sahoo, K.K., Ghosh, A., Tripathi, A.K., Anwar, K., Das, P., Singh, A.K., Pareek, A., Sopory, S.K., Singla-Pareek, S.L., 2017. Manipulation of glyoxalase pathway confers tolerance to multiple stresses in rice. Plant Cell Environ. https://doi.org/10.1111/pce.12968.

Gupta, A.S., Heinen, J.I., Holaday, S., Burket, J.J., Allen, R.D., 1993. Increased resistance to oxidative stress in transgenic plants that overexpress chloroplastic Cu/Zn superoxide dismutase. Proc. Natl. Acad. Sci. U. S. A. 90, 1629–1633.

Halliwell, B., 1997. Free radicals and human disease: trick or treat? In: Thomas, E.C., Kalyanaraman, B. (Eds.), Oxygen Radicals and the Disease Process. Harwood Academic Publishers, Amsterdam, pp. 1–14.

Halpin, C., 2005. Gene stacking in transgenic plants—the challenge for 21st century plant biotechnology. Plant Biotechnol. J. 3, 141–155.

He, G.H., Xu, J.Y., Wang, Y.X., Liu, J.M., Li, P.S., Chen, M., Ma, Y.Z., Xu, Z.S., 2016. Drought-responsive WRKY transcription factor genes TaWRKY1 and TaWRKY33 from wheat confer drought and/or heat resistance in Arabidopsis. BMC Plant Biol. 16, 116.

Holmberg, N., Bulow, L., 1998. Improving stress tolerance in plants by gene transfer. Trends Plant Sci. 3, 61–66.

Holtorf, S., Apel, K., Bohlmann, H., 1995. Comparison of different constitutive and inducible promoters for the overexpression of transgenes in *Arabidopsis thaliana*. Plant Mol. Biol. 29, 637–646.

Hong, Y., Zhang, H., Huang, L., Li, D., Song, F., 2016. Overexpression of a stress-responsive NAC transcription factor gene ONAC022 improves drought and salt tolerance in rice. Front. Plant Sci. 7, 4.

Hoque, T.S., Hossain, M.A., Mostofa, M.G., Burritt, D.J., Fujita, M., Tran, L.S.P., 2016. Methylglyoxal: an emerging signaling molecule in plant abiotic stress responses and tolerance. Front. Plant Sci. 7, 1341.

Hsieh, T.H., Lee, J.T., Charng, Y.Y., Chan, M.T., 2002. Tomato plants ectopically expressing *Arabidopsis* CBF1 show enhanced resistance to water deficit stress. Plant Physiol. 130, 618–626.

Huang, Q., Wang, Y., Li, B., Chang, J., Chen, M., Li, K., Yang, G., He, G., 2015. TaNAC29, a NAC transcription factor from wheat, enhances salt and drought tolerance in transgenic Arabidopsis. BMC Plant Biol. 15, 268.

Hussain, S.S., Raza, H., Afzal, I., Kayani, M.A., 2011. Transgenic plants for abiotic stress tolerance: current status. Arch. Agron. Soil Sci. 58, 693–721.

Inman-Bamber, N.G., Smith, D.M., 2005. Water relations in sugarcane and response to water deficits. Field Crops Res. 92, 185–202.

Ji, X., Liu, G., Liu, Y., Zheng, L., Nie, X., Wang, Y., 2013. The bZIP protein from *Tamarix hispida*, ThbZIP1, is ACGT elements binding factor that enhances abiotic stress signaling in transgenic Arabidopsis. BMC Plant Biol. 13, 151.

Jia, H., Wang, C., Wang, F., Liu, S., Li, G., Guo, X., 2015. GhWRKY68 reduces resistance to salt and drought in transgenic *Nicotiana benthamiana*. PLoS One 10, e0120646.

Jisha, V., Dampanaboina, L., Vadassery, J., Mithofer, A., Kappara, S., Ramanan, R., 2015. Overexpression of an AP2/ERF type transcription factor OsEREBP1 confers biotic and abiotic stress tolerance in rice. PLoS One 10, e0127831.

Jobling, S.A., Westcott, R.J., Tayal, A., Jeffcoat, R., Schwall, G.P., 2002. Production of a freeze–thaw-stable potato starch by antisense inhibition of three starch synthase genes. Nat. Biotechnol. 20, 295–299.

Jung, C., Seo, J.S., Han, S.W., Koo, Y.J., Kim, C.H., Song, S.I., Nahm, B.H., Choi, Y.D., Cheong, J.J., 2008. Overexpression of AtMYB44 enhances stomatal closure to confer abiotic stress tolerance in transgenic Arabidopsis. Plant Physiol. 146, 623–635.

Jung, H.S., Crisp, P.A., Estavillo, G.M., Cole, B., Hong, F., Mockler, T.C., Pogson, B.J., Chory, J., 2013. Subset of heat-shock transcription factors required for the early response of Arabidopsis to excess light. Proc. Natl. Acad. Sci. USA 110, 14474–14479.

Kambiranda, D.M., Vasanthaiah, H.K.N., Ananga, R.K.A., Basha, S.M., Naik, K., 2011. Impact of drought stress on peanut (*Arachis hypogaea* L.) productivity and food safety. In: Vasanthaiah, H. (Ed.), Plants and Environment. InTech Open Access Publishers, Rijeka, pp. 249–272.

Kasuga, M., Liu, Q., Miura, S., Yamaguchi-Shinozaki, K., Shinozaki, K., 1999. Improving plant drought, salt, and freezing tolerance by gene transfer of a single stress-inducible transcription factor. Nat. Biotechnol. 17, 287–291.

Kaur, C., Singla-Pareek, S.L., Sopory, S.K., 2014. Glyoxalase and methylglyoxal as biomarkers for plant stress tolerance. Crit. Rev. Plant Sci. 33, 429–456.

Kaur, C., Kushwaha, H.R., Mustafiz, A., Pareek, A., Sopory, S.K., Singla-Pareek, S.L., 2015. Analysis of global gene expression profile of rice in response to methylglyoxal indicates its possible role as a stress signal molecule. Front. Plant Sci. 6, 682.

Khan, N.A., Singh, S., 2008. Abiotic Stress and Plant Responses. IK International, New Delhi.

Khurana, N., Chauhan, H., Khurana, P., 2013. Wheat chloroplast targeted sHSP26 promoter confers heat and abiotic stress inducible expression in transgenic Arabidopsis plants. PLoS One 8, e54418.

Knight, H., Knight, M.R., 2001. Abiotic stress signalling pathways: specificity and cross-talk. Trends Plant Sci. 6, 262–267.

Lafitte, H.R., Ismail, A., Bennett, J., 2004. In: Fischer, T., Turner, N., Angus, J., McIntyre, L., Robertson, M., Borrell, A. et al., (Eds.), Abiotic stress tolerance in rice for Asia: progress and the future.New Directions for a Diverse Planet: Proceedings for the 4th International Crop Science Congress, Brisbane, Australia.

Lakra, N., Nutan, K.K., Das, P., Anwar, K., Singla-Pareek, S.L., Pareek, A., 2015. A nuclear-localized histone-gene binding protein from rice (OsHBP1b) functions in salinity and drought stress tolerance by maintaining chlorophyll content and improving the antioxidant machinery. J. Plant Physiol. 176, 36–46.

Lata, C., Yadav, A., Prasad, M., 2011. Role of plant transcription factors in abiotic stress tolerance. In: Shanker, A. (Ed.), Abiotic Stress. InTech Open Access Publishers, Rijeka. https://doi.org/10.5772/23172.

Le Martret, B., Poage, M., Shiel, K., Nugent, G.D., Dix, P.J., 2011. Tobacco chloroplast transformants expressing genes encoding dehydroascorbate reductase, glutathione reductase, and glutathione-S-transferase, exhibit altered anti-oxidant metabolism and improved abiotic stress tolerance. Plant Biotechnol. J. 9, 661–673.

Lee, S., Lee, E.J., Yang, E.J., Lee, J.E., Park, A.R., Song, W.H., Park, O.K., 2004. Proteomic identification of annexins, calcium-dependent membrane binding proteins that mediate osmotic stress and abscisic acid signal transduction in Arabidopsis. Plant Cell 16, 1378–1391.

Li, J.G., 2016. Methylglyoxal and glyoxalase system in plants: old players, new concept. Bot. Rev. 82, 183–203.

Li, L., Piatek, M.J., Atef, A., Piatek, A., Wibowo, A., Fang, X., Sabir, J.S., Zhu, J.K., Mahfouz, M.M., 2012. Rapid and highly efficient construction of TALE-based transcriptional regulators and nucleases for genome modification. Plant Mol. Biol. 78, 407–416.

Li, J.B., Luan, Y.S., Yin, Y.L., 2014. SpMYB overexpression in tobacco plants leads to altered abiotic and biotic stress responses. Gene 547, 145–151.

Li, X.W., Wang, Y., Yan, F., Li, J.W., Zhao, Y., Zhao, X., Zhai, Y., Wang, Q.W., 2016. Overexpression of soybean R2R3-MYB transcription factor, GmMYB12B2, and tolerance to UV radiation and salt stress in transgenic Arabidopsis. Genet. Mol. Res. 15, 2.

Liang, C., Meng, Z., Meng, Z., Malik, W., Yan, R., Lwin, K.M., Lin, F., Wang, Y., Sun, G., Zhou, T., Zhu, T., Li, J., Jin, S., Guo, S., Zhang, R., 2016. GhABF2, a bZIP transcription factor, confers drought and salinity tolerance in cotton (*Gossypium hirsutum* L.). Sci. Rep. 6, 35040.

Liu, W., Stewart Jr., C.N., 2016. Plant synthetic promoters and transcription factors. Curr. Opin. Biotechnol. 37, 36–44.

Liu, Y., Zhang, C., Chen, J., Guo, L., Li, X., Li, W., Yu, Z., Deng, J., Zhang, P., Zhang, K., Zhang, L., 2013. Arabidopsis heat shock factor HsfA1a directly senses heat stress, pH changes, and hydrogen peroxide via the engagement of redox state. Plant Physiol. Biochem. 64, 92–98.

Liu, C., Mao, B., Ou, S., Wang, W., Liu, L., Wu, Y., Chu, C., Wang, X., 2014. OsbZIP71, a bZIP transcription factor, confers salinity and drought tolerance in rice. Plant Mol. Biol. 84, 19–36.

Liu, X.Q., Liu, C.Y., Guo, Q., Zhang, M., Cao, B.N., Xiang, Z.H., Zhao, A.C., 2015. Mulberry transcription factor MnDREB4A confers tolerance to multiple abiotic stresses in transgenic tobacco. PLoS One 10, e0145619.

Liu, L., Zhang, Z., Dong, J., Wang, T., 2016. Overexpression of MtWRKY76 increases both salt and drought tolerance in *Medicago truncatula*. Environ. Exp. Bot. 123, 50–58.

Lv, S., Yang, A., Zhang, K., Wang, L., Zhang, J., 2007. Increase of glycinebetaine synthesis improves drought tolerance in cotton. Mol. Breed. 20, 233–248.

Mao, X., Chen, S., Li, A., Zhai, C., Jing, R., 2014. Novel NAC transcription factor TaNAC67 confers enhanced multi-abiotic stress tolerances in Arabidopsis. PLoS One 9, e84359.
McKersie, B.D., Murnaghan, J., Bowley, S.R., 1997. Manipulating freezing tolerance in transgenic plants. Acta Physiol. Plant. 19, 485–495.
Meng, X., Yin, B., Feng, H.L., Zhang, S., Liang, X.Q., Meng, Q.W., 2014. Overexpression of R2R3-MYB gene leads to accumulation of anthocyanin and enhanced resistance to chilling and oxidative stress. Biol. Plant. 58, 121–130.
Mishra, S.K., Tripp, J., Winkelhaus, S., Tschiersch, B., Theres, K., Nover, L., Scharf, K.D., 2002. In the complex family of heat stress transcription factors, HsfA1 has a unique role as master regulator of thermotolerance in tomato. Genes Dev. 16, 1555–1567.
Mitsuda, N., Ohme-Takagi, M., 2009. Functional analysis of transcription factors in Arabidopsis. Plant Cell Physiol. 50, 1232–1248.
Mittler, R., Blumwald, E., 2010. Genetic engineering for modern agriculture: challenges and perspectives. Annu. Rev. Plant Biol. 61, 443–462.
Mourrain, P., van Blokland, R., Kooter, J.M., Vaucheret, H., 2007. A single transgene locus triggers both transcriptional and post-transcriptional silencing through double-stranded RNA production. Planta 225, 365–379.
Muthurajan, R., Balasubramanian, P., 2009. Pyramiding genes for enhancing tolerance to abiotic and biotic stresses. In: Jain, S., Brar, D. (Eds.), Molecular Techniques in Crop Improvement, second ed. Springer, London, pp. 163–184.
Naqvi, S., Farré, G., Sanahuja, G., Capell, T., Zhu, C., Christou, P., 2010. When more is better: multigene engineering in plants. Trends Plant Sci. 15, 48–56.
Nethra, P., 2010. Functional Validation of Root Growth Associated Genes by Over Expression in Model Plants and Development of Transgenics With a Suitable Candidate Gene. (PhD thesis). University of Agricultural Sciences, GKVK, Bangalore.
Pandey, V., Shukla, A., 2015. Acclimation and tolerance strategies of rice under drought stress. Rice Sci. 22, 147–161.
Peremarti, A., Twyman, R.M., Gomez-Galera, S., Naqvi, S., Farre, G., Sabalza, M., Miralpeix, B., Dashevskaya, S., Yuan, D., Ramessar, K., Christou, P., Zhu, C., Bassie, L., Capell, T., 2010. Promoter diversity in multigene transformation. Plant Mol. Biol. 73, 363–378.
Perl-Treves, R., Galun, E., 1991. The tomato Cu,Zn-superoxide dismutase genes are developmentally regulated and respond to light and stress. Plant Mol. Biol. 17, 745–760.
Phillip, A., Syamaladevi, D.P., Chakravarthy, M., Gopinath, K., Subramonian, N., 2013. 50 regulatory region of ubiquitin 2 gene from *Porteresia coarctata* makes efficient promoters for transgene expression in monocots and dicots. Plant Cell Rep. 32, 1199–1210.
Qin, Y., Tian, Y., Liu, X., 2015. A wheat salinity-induced WRKY transcription factor TaWRKY93 confers multiple abiotic stress tolerance in *Arabidopsis thaliana*. Biochem. Biophys. Res. Commun. 464, 428–433.
Rahman, H., Ramanathan, V., Nallathambi, J., Duraialagaraja, S., Muthurajan, R., 2016. Over-expression of a NAC 67 transcription factor from finger millet (*Eleusine coracana* L.) confers tolerance against salinity and drought stress in rice. BMC Biotechnol. 11, 16.
Raineri, J., Ribichich, K.F., Chan, R.L., 2015. The sunflower transcription factor HaWRKY76 confers drought and flood tolerance to *Arabidopsis thaliana* plants without yield penalty. Plant Cell Rep. 34, 2065–2080.
Richards, R.A., 1996. Defining selection criteria to improve yield under drought. Plant Growth Regul. 20, 157–166.
Rodríguez, M., Canales, E., Borrás-Hidalgo, O., 2005. Molecular aspects of abiotic stress in plants. Biotecnol. Apl. 22, 1–10.
Rong, W., Qi, L., Wang, A., Ye, X., Du, L., Liang, H., Xin, Z., Zhang, Z., 2014. The ERF transcription factor TaERF3 promotes tolerance to salt and drought stresses in wheat. Plant Biotechnol. J. 12, 468–479.
Sanan-Mishra, N., Pham, X.H., Sopory, S.K., Tuteja, N., 2005. Pea DNA helicase 45 overexpression in tobacco confers high salinity tolerance without affecting yield. Proc. Natl. Acad. Sci. U. S. A. 102, 509–514.
Sanan-Mishra, N., Tuteja, R., Tuteja, N., 2006. Signaling through MAP kinase networks in plants. Arch. Biochem. Biophys. 452, 55–68.

Seo, J., Sohn, H., Noh, K., Jung, C., An, J., Donovan, C., Somers, D.A., Kim, D.I., Jeong, S.C., Kim, C.G., Kim, H.M., Lee, S.H., Choi, Y.D., Moon, T.W., Kim, C.H., Cheong, J.J., 2012. Expression of the Arabidopsis AtMYB44 gene confers drought/salt-stress tolerance in transgenic soybean. Mol. Breed. 29, 601–608.

Sewelam, N., Kazan, K., Schenk, P.M., 2016. Global plant stress signaling: reactive oxygen species at the cross-road. Front. Plant Sci. 7, 187.

Shafi, A., Dogra, V., Gill, T., Ahuja, P.S., Sreenivasulu, Y., 2014. Simultaneous over-expression of PaSOD and RaAPX in transgenic *Arabidopsis thaliana* confers cold stress tolerance through increase in vascular lignifications. PLoS One 9, e110302.

Shafi, A., Chauhan, R., Gill, T., Swarnkar, M.K., Sreenivasulu, Y., Kumar, S., Kumar, N., Shankar, R., Ahuja, P.S., Singh, A.K., 2015. Expression of SOD and APX genes positively regulates secondary cell wall biosynthesis and promotes plant growth and yield in Arabidopsis under salt stress. Plant Mol. Biol. 87, 615–631.

Shafi, A., Pal, A.K., Sharma, V., Kalia, S., Kumar, S., Ahuja, P.S., Singh, A.K., 2017. Transgenic potato plants overexpressing SOD and APX exhibit enhanced lignification and starch biosynthesis with improved salt stress tolerance. Plant Mol. Biol. Report. 35, 504–518.

Shi, H., Quintero, F.J., Pardo, J.M., Zhu, J.K., 2002. The putative plasma membrane Na^+/H^+ antiporter SOS1 controls long-distance Na^+ transport in plants. Plant Cell 14, 465–477.

Shi, H., Lee, B.H., Wu, S.J., Zhu, J.K., 2003. Overexpression of a plasma membrane Na^+/H^+ antiporter gene improves salt tolerance in *Arabidopsis thaliana*. Nat. Biotechnol. 21, 81–85.

Shim, D., Hwang, J.U., Lee, J., Choi, Y., An, G., Martinoia, E., Lee, Y., 2009. Orthologs of the class A4 heat shock transcription factor HsfA4a confer cadmium tolerance in wheat and rice. Plant Cell 21, 4031–4043.

Shingotem, P.R., Kawar, P.G., Pagariya, M.C., Kuhikar, R.S., Thorat, A.S., Babu, K.H., 2015. SoMYB18, a sugarcane MYB transcription factor improves salt and dehydration tolerance in tobacco. Acta Physiol. Plant. 37, 217.

Shinozaki, K., Yamaguchi-Shinozaki, K., Seki, M., 2003. Regulatory network of gene expression in the drought and cold stress responses. Curr. Opin. Plant Biol. 6, 410–417.

Shivakumara, T.N., Sreevathsa, R., Dash, P.K., Sheshshayee, M.S., Papolu, P.K., Rao, U., Tuteja, N., Udaya Kumar, M., 2017. Overexpression of pea DNA helicase 45 (PDH45) imparts tolerance to multiple abiotic stresses in chili (*Capsicum annuum* L.). Sci. Rep. 7, 2760.

Shukla, P.S., Gupta, K., Agarwal, P., Jha, B., Agarwal, P.K., 2015. Overexpression of a novel SbMYB15 from *Salicornia brachiata* confers salinity and dehydration tolerance by reduced oxidative damage and improved photosynthesis in transgenic tobacco. Planta 242, 1291–1308.

Singh, R., Jwa, N.S., 2013. The rice MAPKK–MAPK interactome: the biological significance of MAPK components in hormone signal transduction. Plant Cell Rep. 32, 923–931.

Singh, A.K., Ansari, M.W., Pareek, A., Singla-Pareek, S.L., 2008. Raising salinity tolerant rice: recent progress and future perspectives. Physiol. Mol. Biol. Plants 14, 137–154.

Singh, A.K., Sopory, S.K., Wu, R., Singla-Pareek, S.L., 2010. Transgenic approaches. In: Pareek, A., Sopory, S.K., Bohnert, H.J., Govindjee, (Eds.), Abiotic Stress Adaptation in Plants: Physiological, Molecular and Genomic Foundation. Springer, Dordrecht, pp. 417–450.

Singh, A.K., Sharma, V., Pal, A.K., Acharya, V., Ahuja, P.S., 2013. Genome-wide organization and expression profiling of the NAC transcription factor family in potato (*Solanum tuberosum* L.). DNA Res. 20, 403–423.

Singla-Pareek, S.L., Reddy, M.K., Sopory, S.K., 2003. Genetic engineering of the glyoxalase pathway in tobacco leads to enhanced salinity tolerance. Proc. Natl. Acad. Sci. USA 100, 14672–14677.

Singla-Pareek, S.L., Yadav, S.K., Pareek, A., Reddy, M.K., Sopory, S.K., 2006. Transgenic tobacco overexpressing glyoxalase pathway enzymes grow and set viable seeds in zinc-spiked soils. Plant Physiol. 140, 613–623.

Sreenivasulu, N., Sopory, S.K., Kavi Kishor, P.B., 2007. Deciphering the regulatory mechanisms of abiotic stress tolerance in plants by genomic approaches. Gene 388, 1–13.

Stein, H., Honig, A., Miller, G., Erster, O., Eilenberg, H., Csonka, L.N., Szabados, L., Koncz, C., Zilberstein, A., 2011. Elevation of free proline and proline-rich protein levels by simultaneous manipulations of proline biosynthesis and degradation in plants. Plant Sci. 181, 140–150.

Tamura, T., Hara, K., Yamaguchi, Y., Koizumi, N., Sano, H., 2003. Osmotic stress tolerance of transgenic tobacco expressing a gene encoding a membrane-located receptor-like protein from tobacco plants. Plant Physiol. 131, 454–462.

Tang, M., Liu, X., Deng, H., Shen, S., 2011. Over-expression of JcDREB, a putative AP2/EREBP domain-containing transcription factor gene in woody biodiesel plant *Jatropha curcas*, enhances salt and freezing tolerance in transgenic *Arabidopsis thaliana*. Plant Sci. 181, 623–631.

Umezawa, T., Fujita, M., Fujita, Y., Yamaguchi-Shinozaki, K., Shinozaki, K., 2006. Engineering drought tolerance in plants: discovering and tailoring genes unlock the future. Curr. Opin. Biotechnol. 2017, 113–122.

Urao, T., Yakubov, B., Satoh, R., Yamaguchi-Shinozaki, K., Seki, M., Hirayama, T., Shinozaki, K., 1999. A transmembrane hybrid-type histidine kinase in Arabidopsis functions as an osmosensor. Plant Cell 11, 1743–1754.

Varshney, R.K., Bansal, K.C., Aggarwal, P.K., Datta, S.K., Craufurd, P.Q., 2011. Agricultural biotechnology for crop improvement in a variable climate: hope or hype? Trends Plant Sci. 16, 363–371.

Vellosillo, T., Vicente, J., Kulasekaran, S., Hamberg, M., Castresana, C., 2010. Emerging complexity in reactive oxygen species production and signaling during the response of plants to pathogens. Plant Physiol. 154, 444–448.

Vemanna, R.S., Chandrashekar, B.K., Hanumantha Rao, H.M., Sathyanarayanagupta, S.K., Sarangi, K.S., Nataraja, K.N., Udayakumar, M., 2013. A modified MultiSite gateway cloning strategy for consolidation of genes in plants. Mol. Biotechnol. 53, 129–138.

Vendruscolo, E.C., Schuster, I., Pileggi, M., Scapim, C.A., Molinari, H.B., Marur, C.J., Vieira, L.G., 2007. Stress-induced synthesis of proline confers tolerance to water deficit in transgenic wheat. J. Plant Physiol. 164, 1367–1376.

Vinocur, B., Altman, A., 2005. Recent advances in engineering plant tolerance to abiotic stress: achievements and limitations. Curr. Opin. Biotechnol. 16, 123–132.

Wang, X., Chen, X., Liu, Y., Gao, H., Wang, Z., Sun, G., 2011. CkDREB gene in *Caragana korshinskii* is involved in the regulation of stress response to multiple abiotic stresses as an AP2/EREBP transcription factor. Mol. Biol. Rep. 38, 2801–2811.

Wang, C., Deng, P., Chen, L., Wang, X., Ma, H., Hu, W., Yao, N., Feng, Y., Chai, R., Yang, G., He, G., 2013. A wheat WRKY transcription factor TaWRKY10 confers tolerance to multiple abiotic stresses in transgenic tobacco. PLoS One 8, e65120.

Wang, R.K., Cao, Z.H., Hao, Y.J., 2014. Overexpression of a R2R3MYB gene MdSIMYB1 increases tolerance to multiple stresses in transgenic tobacco and apples. Physiol. Plant. 150, 76–87.

Wang, X., Zeng, J., Li, Y., Rong, X., Sun, J., Sun, T., Li, M., Wang, L., Feng, Y., Chai, R., Chen, M., Chang, J., Li, K., Yang, G., He, G., 2015a. Expression of TaWRKY44, a wheat WRKY gene, in transgenic tobacco confers multiple abiotic stress tolerances. Front. Plant Sci. 6, 615.

Wang, X., Zeng, J., Li, Y., Rong, X., Sun, J., Sun, T., Li, M., Wang, L., Feng, Y., Chai, R., Chen, M., Chang, J., Li, K., Yang, G., He, G., 2015b. Expression of TaWRKY44, a wheat WRKY gene, in transgenic tobacco confers multiple abiotic stress tolerances. Front. Plant Sci. 11, 615.

Wang, Z., Su, G., Li, M., Ke, Q., Kim, S.Y., Li, H., Huang, J., Xu, B., Deng, X.P., Kwak, S.S., 2016. Overexpressing Arabidopsis ABF3 increases tolerance to multiple abiotic stresses and reduces leaf size in alfalfa. Plant Physiol. Biochem. 109, 199–208.

Wang, L., Li, Z., Lu, M., Wang, Y., 2017. ThNAC13, a NAC transcription factor from *Tamarix hispida*, confers salt and osmotic stress tolerance to transgenic Tamarix and Arabidopsis. Front. Plant Sci. 8, 635.

Wani, S.H., Singh, N.B., Haribhushan, A., Mir, J.I., 2013. Compatible solute engineering in plants for abiotic stress tolerance—role of glycine betaine. Curr. Genomics 14, 157–165.

Wani, S.H., Sah, S.K., Hossain, M.A., Kumar, V., Balachandran, S.M., 2016. Transgenic approaches for abiotic stress tolerance in crop plants. In: Al-Khayri, J., Jain, S., Johnson, D. (Eds.), Advances in Plant Breeding Strategies: Agronomic, Abiotic and Biotic Stress Traits. Springer, Cham, pp. 345–396.

Winicov, I., Valliyodan, B., Xue, L., Hoober, J.K., 2004. The MsPRP2 promoter enables strong heterologous gene expression in a root-specific manner and is enhanced by overexpression of Alfin 1. Planta 219, 925–935.

Xu, X.B., Pan, Y.Y., Wang, C.L., Ying, Q.C., Song, H.M., Wang, H.Z., 2014. Overexpression of DnWRKY11 enhanced salt and drought stress tolerance of transgenic tobacco. Biologia 69, 994–1000.

Yabuta, Y., 2016. Functions of heat shock transcription factors involved in response to photooxidative stresses in Arabidopsis. Biosci. Biotechnol. Biochem. 80, 1254–1263.

Yadav, S.K., Singla-Pareek, S.L., Ray, M., Reddy, M.K., Sopory, S.K., 2005. Methylglyoxal levels in plants under salinity stress are dependent on glyoxalase I and glutathione. Biochem. Biophys. Res. Commun. 337, 61–67.

Yamaguchi, T., Blumwald, E., 2005. Developing salt-tolerant crop plants: challenges and opportunities. Trends Plant Sci. 10, 615–620.
Yamaguchi-Shinozaki, K., Shinozaki, K., 2006. Transcriptional regulatory networks in cellular responses and tolerance to dehydration and cold stresses. Annu. Rev. Plant Biol. 57, 781–803.
Yang, A., Dai, X., Zhang, W.H., 2012. A R2R3-type MYB gene, OsMYB2, is involved in salt, cold, and dehydration tolerance in rice. J. Exp. Bot. 63, 2541–2556.
Yang, G., Yu, L., Zhang, K., Zhao, Y., Guo, Y., Gao, C., 2017. A ThDREB gene from *Tamarix hispida* improved the salt and drought tolerance of transgenic tobacco and *T. hispida*. Plant Physiol. 113, 187–197.
Ying, S., Zhang, J., Deng, F., Jing, F., Yun-Su, S., Song, Y.C., Wang, T.Y., 2012. Cloning and characterization of a maize bZIP transcription factor, ZmbZIP72, confers drought and salt tolerance in transgenic Arabidopsis. Planta 235, 253–266.
Yu, X., Liu, Y., Wang, S., Tao, Y., Wang, Z., Shu, Y., Peng, H., Mijiti, A., Wang, Z., Zhang, H., Ma, H., 2016. CarNAC4, a NAC-type chickpea transcription factor conferring enhanced drought and salt stress tolerances in Arabidopsis. Plant Cell Rep. (3), 613–627.
Zhang, L., Zhao, G., Xia, C., Jia, J., Liu, X., Kong, X., 2012. Overexpression of a wheat MYB transcription factor gene, TaMYB56-B, enhances tolerances to freezing and salt stresses in transgenic Arabidopsis. Gene 505, 100–107.
Zhang, L., Zhang, L., Xia, C., Zhao, G., Jia, J., Kong, X., 2015a. The novel wheat transcription factor TaNAC47 enhances multiple abiotic stress tolerances in transgenic plants. Front. Plant Sci. 6, 1174.
Zhang, L., Zhang, L., Xia, C., Zhao, G., Liu, J., Jia, J., Kong, X., 2015b. A novel wheat bZIP transcription factor, TabZIP60, confers multiple abiotic stress tolerances in transgenic Arabidopsis. Physiol. Plant. 153, 538–554.
Zhang, X., Liu, X., Wu, L., Yu, G., Wang, X., Ma, H., 2015c. The SsDREB transcription factor from the succulent halophyte *Suaeda salsa* enhances abiotic stress tolerance in transgenic tobacco. Int. J. Genomics. 2015. 875497.
Zhao, X., Yang, X., Pei, S., He, G., Wang, X., Tang, Q., Jia, C., Lu, Y., Hu, R., Zhou, G., 2016. The Miscanthus NAC transcription factor MlNAC9 enhances abiotic stress tolerance in transgenic Arabidopsis. Gene 586, 158–169.
Zhu, J.K., 2016. Abiotic stress signaling and responses in plants. Cell 167, 313–324.
Zhu, M., Chen, G., Zhang, J., Zhang, Y., Xie, Q., Zhao, Z., Pan, Y., Hu, Z., 2014. The abiotic stress-responsive NAC-type transcription factor SlNAC4 regulates salt and drought tolerance and stress-related genes in tomato (*Solanum lycopersicum*). Plant Cell Rep. 33, 1851–1863.
Zorrilla-Lopez, U., Masip, G., Arjo, G., Bai, C., Banakar, R., Bassie, L., Berman, J., Farré, G., Miralpeix, B., Pérez-Massot, E., Sabalza, M., Sanahuja, G., Vamvaka, E., Twyman, R.M., Christou, P., Zhu, C., Capell, T., 2013. Engineering metabolic pathways in plants by multigene transformation. Int. J. Dev. Biol. 57, 565–576.

FURTHER READING

Cai, H., Tian, S., Liu, C., Dong, H., 2011. Identification of a MYB3R gene involved in drought, salt and cold stress in wheat (*Triticum aestivum* L.). Gene 485, 146–152.
Cai, R., Zhao, Y., Wang, Y., Lin, Y., Peng, X., Li, Q., Chang, Y., Jiang, H., Xiang, Y., 2014. Overexpression of a maizeWRKY58 gene enhances drought and salt tolerance in transgenic rice. Plant Cell Tissue Organ Cult. 119, 565–577.
Hongyan, W., Honglei, W., Hongbo, S., Xiaoli, T., 2016. Recent advances in utilizing transcription factors to improve plant abiotic stress tolerance by transgenic technology. Front. Plant Sci. 7, 67.
Inman-Bamber, N., 2004. Sugarcane water stress criteria for irrigation and drying off. Field Crop Res. 89, 107–122.
Kaur, C., Tripathi, A.K., Nutan, K.K., Sharma, S., Ghosh, A., Tripathi, J.K., Pareek, A., Singla-Pareek, S.L., Sopory, S.K., 2017. A nuclear-localized rice glyoxalase I enzyme, OsGLYI-8, functions in the detoxification of methylglyoxal in the nucleus. Plant J. 89, 565–576.

Mishra, N.S., Tuteja, R., Tuteja, N., 2006. Signaling through MAP kinase networks in plants. Arch. Biochem. Biophys. 452, 55–68.
Qin, F., Kakimoto, M., Sakuma, Y., Maruyama, K., Osakabe, Y., Tran, L.S., 2007. Regulation and functional analysis of ZmDREB2A in response to drought and heat stresses in *Zea mays* L. Plant J. 50, 54–69.
Ramu, V.S., Swetha, T.N., Sheela, S.H., Babitha, C.K., Rohini, S., Reddy, M.K., Tuteja, N., Reddy, C.P., Prasad, T.G., Udayakumar, M., 2016. Simultaneous expression of regulatory genes associated with specific drought-adaptive traits improves drought adaptation in peanut. Plant Biotechnol. J. 14, 1008–1020.
Winicov, I., 2000. Alfin1 transcription factor overexpression enhances plant root growth under normal and saline conditions and improves salt tolerance in alfalfa. Planta 210, 416–422.
Zhang, X., Wang, L., Meng, H., Wen, H., Fan, Y., Zhao, J., 2011. Maize ABP9 enhances tolerance to multiple stresses in transgenic Arabidopsis by modulating ABA signaling and cellular levels of reactive oxygen species. Plant Mol. Biol. 75, 365–378.
Zhong-Guang, L., 2016. Methylglyoxal and glyoxalase system in plants: old players, new concepts. Bot. Rev. 82, 183–203.

CHAPTER 13

Crop Phenomics for Abiotic Stress Tolerance in Crop Plants

Balwant Singh*, Shefali Mishra*, Abhishek Bohra[†], Rohit Joshi[‡], Kadambot H.M. Siddique[§]
*ICAR-National Research Centre on Plant Biotechnology, New Delhi, India
[†]Crop Improvement Division, ICAR-Indian Institute of Pulses Research, Kanpur, India
[‡]Stress Physiology and Molecular Biology Laboratory, School of Life Sciences, Jawaharlal Nehru University, New Delhi, India
[§]The UWA Institute of Agriculture, The University of Western, Australia, Perth, WA, Australia

Contents

1. INTRODUCTION

It is estimated that the human population will reach 9 billion by 2050, and current food production must double to meet the needs of the growing population (Joshi et al., 2016a). In recent times, deteriorating climatic conditions have aggravated the challenge posed by several biotic and abiotic stresses related to worldwide food production (Pereira, 2016). Climate changes result in higher CO_2 concentrations, while temperature, heat stress, and intermittent rain eventually result in flash flooding, drought, and salt stress

Biochemical, Physiological and Molecular Avenues for Combating Abiotic Stress in Plants
https://doi.org/10.1016/B978-0-12-813066-7.00015-2

(Rosenzweig et al., 2014). The global challenge to feed the growing human population demands a consistent increase in crop production, despite adverse environmental conditions and a limited cultivable area (Furbank and Tester, 2011). A genotype's yield potential and stability remain central to increasing crop production. Abiotic stress tolerance is a crucial factor for the stability of crop performance.

To assess the potential of a given genotype under a particular abiotic stress, a researcher must examine the response thoroughly in terms of phenotypic changes, and the elements that regulate a plant's response under stressed conditions. To assess the impact of different abiotic stresses on crop plants, the study can be conducted either under controlled environmental conditions, or in the field with replicated trials at different growth stages, such as seedling or flowering stages, as both are vulnerable to stress (Mickelbart et al., 2015; Farooq et al., 2014a). Even in the postgenomics era, the importance of "phenotype" becomes evident from the fact that crop improvement techniques, including QTL analysis, fine mapping of genes/QTL, genome-wide association studies (GWAS), and genomic selection (GS) rely heavily on the precise and accurate recording of phenotypic observations in crop plants. According to Cabrera-Bosquet and colleagues, the underdevelopment of crop phenomic techniques hampers the efficient utilization of crop genetic resources (Cabrera-Bosquet et al., 2012). More often than not, inconsistency in the behavior of crop plants at different locations might reflect inaccurate phenotypic screening of the germplasm set, which may eventually result in false parental selection.

Abiotic stress imposes serious threats to crop production by altering the genotype × environment interaction. This leads to changes in a plant's basic metabolism that represent the plant phenotype. The common metabolic changes that occur during abiotic stress primarily include the production of compatible solutes and secondary metabolites, and the generation of reducing agents and reactive oxygen species (ROS) (Suzuki et al., 2012). These changes are part of the plant's strategy to avoid or tolerate abiotic stresses, and this ability to adapt under any aberrant change in the surrounding environment can be a genetically adaptive trait in crop plants (Mickelbart et al., 2015). Measuring phenotypic changes requires examination of the key parameters specific to the stress using sophisticated techniques, thus resulting in a precise estimation of the phenotypic response. A range of parameters has been used to evaluate the degree of a stress and level of tolerance or susceptibility of a cultivar. For instance, root morphology, and leaf-related traits such as relative water content and leaf rolling, biomass, and yield-associated traits, are factored in when determining plant tolerance to drought and salinity; similarly, parameters such as sodium and potassium contents, and biomass, facilitate the evaluation of salt tolerance (Collins et al., 2008).

Phenotypic observations that reflect a plant's response to a particular stress continue to be evaluated manually due to their ease of measurement and cost effectiveness, despite these methods being destructive and error-prone. Error in measurements and the subsequent interpretation of the stress-related traits may yield spurious classification or ranking of the genotypes under examination. Notwithstanding the paramount significance of the "phenotype" in crop improvement, precision phenotyping of component traits associated with abiotic stress is a formidable task (Yang et al., 2013). Plausibly, inaccurate selection of associated traits, incongruity in stress conditions (natural versus artificial), choice of phenotyping protocol, and, most importantly poor technical progression, hamper the selection of novel genotypes with improved characters. To witness a breakthrough success in plant improvement, novel phenotyping methods are needed that can precisely and accurately record the phenotypic changes.

Over the past decade, tremendous progress has been seen in terms of large-scale genomic technologies such as sequencing, re-sequencing, and genotyping. However, limited progress on the phenomics front may explain the poor utilization efficiency rate for crop genotype potential as compared with the rate of advancement of genomics techniques. Given this, emphasis should be placed on the development of automated phenotyping platforms that can generate high-throughput and high-resolution data along with the capacity to allow measurement of nonvisible phenotypic changes (Maphosa et al., 2016). This is indeed warranted to realize the full potential of genomic resources, and plant phenomics need to integrate technologies and develop reliable and multifunctional phenotyping assays that enable reproducible and regulated image analyses (Yang et al., 2013). In this respect, some progress has been made, and high-throughput phenotyping facilities have been established with a robust software system, which includes visible light imaging, hyperspectral imaging, and X-ray computed tomography. Plant phenotyping centers with the capacity to automatically image hundreds to thousands of plants have been established in different countries and, importantly, some QTLs have been discovered in different crops based on these modern facilities (Zhang et al., 2017). These centers include Plant Accelerator (http://www.plantaccelerator.org.au/) and the High Resolution Plant Phenotyping Centre (http://www.plantphenomics.org.au/HRPPC) in Australia, the Leibniz Institute of Plant Genetics and Crop Plant Research in Germany (http://www.ipk-gatersleben.de), the Institute of Biological, Environmental and Rural Sciences (IBERS) in the United Kingdom (http://www.aber.ac.uk/en/ibers/), and the PHENOPSIS system being built by the National Institute for Agricultural Research (INRA) in France (http://www.international.inra.fr/).

Different automated plant phenotyping methods have been employed to evaluate a range of traits from developmental to morphological descriptors, for example, plant height, tiller numbers in barley (Honsdorf et al., 2014), and panicle development in rice using X-ray computed tomography (Jhala and Thaker, 2015). A holistic approach to crop phenotyping will help to refine our understanding of the traits that are influenced by the environment. The modern tools offered by crop phenomics enable plant breeders to efficiently identify crop genotypes with tolerance to abiotic stress, and guide them to develop a resilient crop that can withstand climate change.

2. TECHNIQUES TO ELUCIDATE A PLANT'S PHENOME

Understanding stress-associated phenotypic traits and their interrelationships calls for a precise and in-depth knowledge of phenotyping, in both controlled and field conditions, to accurately elucidate the adaptive mechanisms of the crop plants. Advanced phenotyping techniques use image processing with visible to near-infrared spectrum light sources to provide image datasets of the plant phenotype in a nondestructive manner (Rahaman et al., 2015). The imaging techniques—including visible light imaging, hyperspectral imaging, infrared imaging, fluorescence imaging, and X-ray computed tomography, using a robust software system, generate unique, multilevel phenotyping data (Sozzani et al., 2014). Such imaging-based, automated, high-throughput plant phenotyping platforms integrated with advanced software systems have emerged as advanced tools for plant biology (Paproki et al., 2012). With the current emphasis on precise phenotyping, imaging techniques that measure the interaction between light and plants, such as photons (transmitted, absorbed, or reflected), are contributing more to reaching the desired level of measurements related to quantitative phenotypic traits. Employment of high-dimensional phenotyping assays requires uniform experimental protocols with calibrated imaging sensors and precise analyses of raw data-processing methods. The imaging devices that are currently used for high-throughput phenotyping of crop plants are outlined as follows.

2.1 Visible Light (300–700 nm) Imaging

Visual inspection of abiotic stress-responsive traits and associated phenotypes has been a standard practice in breeding plants for tolerance. Recently, a shift has been witnessed toward visual light-based imaging technologies due to their low cost and ease of maintenance and handling. Visible imaging techniques that rely on two-dimensional (2D) digital images have been used to measure shoot-related traits such as shoot biomass (Neilson et al., 2015), leaf morphology, shoot tip extension, panicle and seed morphology, root architecture, and other yield-related traits (Fahlgren et al., 2015). Sensors used in visible imaging are sensitive to a visible spectral range, such as silicon sensors (CCD or CMOS arrays) (Li et al., 2014). To this end, three-dimensional (3D) imaging was conceived to generate more accurate detail on complex phenotypes and, more recently, both 2D and 3D imaging technologies have been integrated to enhance phenotyping accuracy (Rahaman et al., 2015). The 3D imaging techniques have been optimized to measure different traits, including plant height, leaf morphology, and shoot dry weight (Paproki et al. 2012). Earlier, Golzarian et al. (2011) accurately measured shoot dry weight to examine wheat seedlings for salt stress with a LemnaTec 3D Scanalyzer. Software has also been developed that allows the conversion of 2D to 3D images by supreme imposition, which enables root architecture, morphology, and growth to be monitored (Clark et al., 2011).

Phenotypic changes such as reduced transpiration rate due to stomatal closure, leaf rolling, and inhibition of plant growth occur during the vegetative stage when plants can be exposed to different stresses including drought, salinity, and high temperature. In the case of salinity stress, differences between yellow and green areas of the leaf allow measurement of leaf senescence and tissue tolerance, which may be correlated with salt accumulation. Importantly, image analysis can measure stress tolerance traits in a single plant as well as in large populations, such as mapping populations and mutant populations. This enables us to undertake a genetic approach to characterize genes, regulating the variations among these tolerance-related traits. For example, Bowman et al. (2015) evaluated wheat grain yield using canopy spectral reflectance under terminal drought. Phenotyping using RGB has been used for different abiotic stresses in several crops using various platforms, including PHENOPSIS (Granier et al., 2006) and WIWAM (https://www.wiwam.be/) for drought stress in Arabidopsis; LemnaTec for drought stress in barley (Honsdorf et al., 2014); and maize (Ge et al., 2016) and salt stress in rice (Hairmansis et al., 2014), wheat, and barley (Humplík et al., 2015a; Meng et al., 2017), and PlantScreen and GROWSCREEN for chilling tolerance in Arabidopsis and peas, respectively (Jansen et al., 2009; Humplík et al., 2015b).

2.2 Infrared- and Thermal-Based Imaging

Infrared thermal imaging allows visualization of infrared radiation emitted from the object using the Stefan-Boltzmann equation ($R = \varepsilon\sigma T^4$). This technology uses internal molecular movements of the objects that emit infrared radiation for imaging (Kastberger and Stachl, 2003). Infrared imaging devices use two main wavelength ranges, near-infrared (NIR) (0.9–1.55 μm) and far-infrared (FIR) (7.5–13.5 μm). However, the sensitive spectral range of thermal cameras is 3–14 μm (Li et al., 2014). With the advancement of infrared thermal technology, current thermal cameras are available with high thermal sensitivity to detect plant canopy temperature with more user-friendly interfaces and higher resolution detectors. In addition, by combining NIR imaging with visible imaging, that is, visible to short-wave infrared (VSWIR; 0.4–2.5 μm), this method provides deeper insight into plant health under different stress conditions because it provides well-defined spectral features for pigments, leaf water content, and biochemicals such as lignin and cellulose (Yang et al., 2013). Further, infrared thermography has also been used to study stomatal responses under salinity and drought by visualizing differences in canopy temperature (Rahaman et al., 2015). Several earlier studies used NIR spectroscopy to indirectly assess crop growth and yield performance under stressed conditions. These infrared imaging technologies provide high spatial resolution images with precise measurements in large fields during varied climatic conditions at the same time (Li et al., 2014).

Thermal imaging cameras are sensitive to a spectral range of 3–14 μm in the infrared region; within this wavelength, 3–5 and 7–14 μm are the most commonly used wavelengths for imaging. The smaller wavelengths correspond to higher energy levels, and therefore, higher thermal sensitivity. Thermal imaging can measure leaf and canopy temperature to evaluate leaf water status. Gas exchange is measured by stomatal movement because plants are generally cooled by transpiration, and plant temperature increases when the stomata are closed. Canopy temperature differences between the canopy and the surrounding air can be used as a proxy for drought tolerance in dry environments. Thermal infrared imaging can be performed in the laboratory and the field to characterize tolerance to various stresses such as drought and salinity based on osmotic tolerance and Na^+ exclusion. It can also be used to measure relative chlorophyll content, leaf color, and canopy temperature (Merlot et al., 2002; Jones et al., 2009; Munns et al., 2010). Infrared thermal imaging systems have been used to measure stomatal behavior under various stress conditions, for example, to monitor salt tolerance in wheat genotypes (Bayoumi et al., 2014).

2.3 Fluorescence Imaging

The absorbance of light by a compound at a particular wavelength, and further emission of low wavelength light, is termed fluorescence. Fluorescence imaging flashes blue wavelength (<500 nm) light on the plants, and they emit fluorescence light at 600–750 nm in the red region of the spectrum. The differences in fluorescence are photographed and converted into false-color signals using computer software to analyze them (Weirman, 2010). Abiotic stresses primarily affect chlorophyll content; therefore, chlorophyll fluorescence is generally used in phenomics to detect the effect of various environmental atrocities on genes and the plant's ability to maintain photosynthesis under these conditions (Weirman, 2010). Stomatal movement (Cardon et al., 1994), phloem loading and unloading (Siebke and Weis, 1995), correlation between spatiotemporal variation of photosynthesis and growth limitation (Walter et al., 2004), and plant metabolite content (Li et al., 2014) under stress can also be studied using fluorescence imaging (Chaerle and Van Der Straeten, 2001; Rascher et al., 2001; Osmond et al., 2004). Ultraviolet illumination produces two types of fluorescence (red to far-red region and blue to green region) ranging from 360 to 740 nm, which captures fluorescence emission from four spectral bands; blue (440 nm), green (520 nm), red (690 nm), and far-red (740 nm) by single-excitation wavelengths (Rahaman et al., 2015).

Fluorescence imaging has been used for various reasons, such as detection of stress at the primary level (Jansen et al., 2009; Konishi et al., 2009; Chen et al., 2014) and resolving heterogeneity in leaf photosynthetic performance (Baker, 2008). Most of these applications of fluorescence imaging are restricted to either seedlings or single leaves in model crops. It requires the development of standard procedures and robust software for phenotyping under stress conditions. RGB and chlorophyll fluorescence (ChlF) imaging were used to measure growth, morphology, color, and photosynthetic performance in *Arabidopsis thaliana* (Awlia et al., 2016) and rice (Hairmansis et al., 2014) under salt stress. Similarly, ChlF transients were analyzed with high-performing quadratic discriminant classifiers (QDC) and sequential forward floating selection (SFFS) methods to discriminate between cold-sensitive and cold-tolerant genotypes of *Arabidopsis thaliana* (Mishra et al., 2014).

2.4 Spectroscopy Imaging

Spectroscopy imaging is the outcome of the interaction of solar radiation with plants through multispectral and hyperspectral cameras. In contrast to visible and infrared imaging, hyperspectral imaging divides images into bands, thus creating a large fraction of the electromagnetic spectrum in the images (Yang et al., 2013). Different spectral regions have been characterized as specific for plant science, including (1) NDVI (normalized difference vegetation index), which compares red and near-infrared reflectance, (2) CRI (carotenoid reflectance index), which determines three wavebands in the yellow region, the 970 nm water-index defining small water absorption area, and (3) PRI (photochemical reflectance index), which correlates functional status of nonphotochemical energy conservation (Fiorani et al., 2012). In the visible spectrum, low reflectance is observed due to less absorption by chlorophyll pigments in this region and a characteristic reflectance peak in the green region. However, a sharp increase in reflectance termed "red edge" was observed in a visible to near-infrared (NIR) wavelength transition. In the NIR region, radiation transmitted from upper-canopy leaves to lower leaves is subsequently reflected back to the upper part. This helps to determine leaf and canopy architecture. Further, with an increase in wavelength, reflectance gradually decreases and absorption increases. This occurs due to leaf water content, which explains its water status and helps to estimate canopy water content. Such spectral reflectance information is used to quantify vegetation indices, that is, ratio and differences at a particular wavelength, and enables the detection of NDVI. The vegetation indices are correlated with various traits, that is, water status, pigment content, and photosynthetically active biomass, which is used to calculate total green biomass, leaf area, chlorophyll content, and yield in various crop species (Peñuelas and Filella, 1998; Din et al., 2017).

The application of spectroscopy imaging is well defined for field phenotyping by combining it with aerial platforms, but spectral cameras and their related infrastructure are expensive. Römer et al. (2012) used a matrix factorization technique called SiVM (simplex volume maximization) to analyze hyperspectral data for early drought detection in cereal crops. In relation to water stress, canopy water content acts as a physiological marker, which is measured using multispectral and hyperspectral spectroscopic measurements. Hyperspectral high-spatial resolution satellite data provides effective data for empirical and physical analysis of canopy water content (Cernicharo et al., 2013). The greater resolution of hyperspectral data makes it an effective method for measuring leaf growth (Cheng et al., 2017) and panicle emergence (Liu et al., 2010) in rice.

2.5 Integrated Imaging Techniques

Recent technical advancements, such as functional imaging and optical 3D structural tomography, have increasingly shifted toward in vivo live imaging of plants. Functional imaging, such as ChlF imaging and positron emission tomography (PET), focus on physiological changes to evaluate photosynthetic performance under stress (Baker, 2008). PET is a nondestructive image distribution technique using the transportation of positron-emitting radionuclides such as C^{11}, N^{13}, or Fe^{52} labeled metabolite compounds (Kiyomiya et al., 2001). Magnetic resonance imaging (MRI) is an advanced technique that produces images by combining magnetic fields and radio waves. It can be used to capture plant root architecture in pots and internal physiological processes in vivo (Borisjuk et al., 2012). Subsequently, it can measure water diffusion and transportation through the xylem and phloem in crops such as tobacco, tomatoes, poplars, and castor beans (Windt et al., 2006). Combining both MRI and PET technology offers a novel imaging procedure to monitor dynamic changes in plant function and structure. Jahnke et al. (2009) demonstrated photo assimilation in sugar beet taproots and studied shoot-to-root carbon fluxes by combining PET and MRI using $[C^{11}]$-labeled CO_2.

For high-resolution imaging of small molecules in living tissue (i.e., molecular phenotyping), Förster resonance energy transfer (FRET) is another advanced noninvasive technology. This technique is based on genetically encoded, radiometric fluorescent sensors that bind to and detect levels of the target molecule (Jones et al., 2014). Identification of multiple pathways and dynamic processes of the sensor target can be achieved through a single FRET sensor. A FRET sensor characterizes and expresses the cellular/subcellular location of the target sensor and provides measurements with high temporal and spatial resolution (Okumoto et al., 2012). FRET has been used to detect calcium and zinc dynamics with subcellular spatial and real-time temporal resolution in roots during sugar transport (Jones et al., 2014). FRET is an outstanding technology for advanced phenotyping that can address several basic questions related to plant growth and development.

The PlantEye, a high-resolution 3D laser scanner, was used to phenotype wheat plants growing under control and salt stress conditions in a controlled environment. The PlantEye scans plants from overhead, creating a data cloud from which the system computes traits such as 3D leaf area, plant height, and leaf number. In wheat under salt stress, correlations were observed between the PlantEye scanned trait (3D leaf area) and manually measured traits (leaf area, and fresh and dry biomass) (Maphosa et al., 2016).

3. PHENOTYPIC AND BIOCHEMICAL CHANGES IN CROPS UNDER ABIOTIC STRESSES

Abiotic stress is defined as an environmental condition that reduces crop growth and biomass, ultimately leading to yield loss. Plant responses to abiotic stresses are dynamic and complex. Phenotypic effects depend on both the level and duration of the stress and the tissue or organ affected by the stress. Inhibition of protein synthesis is one of the earliest metabolic responses to abiotic stress that hampers vegetative growth, followed by altered energy metabolism of sugars, lipids, and photosynthesis. The dynamic response of plants to abiotic stress tolerance involves a complex crosstalk between multiple pathways at different regulatory levels to adjust metabolic changes (Saito and Matsuda, 2010). Interestingly, most of the well-known changes in plant metabolism under abiotic stresses, such as drought and salinity, are highly correlated with phenotypic, physiological, and biochemical changes. Most of these abiotic stresses display a dehydration response, which results in the accumulation of peculiar metabolites called compatible solutes, such as proline (Kesari et al., 2012; Joshi et al., 2017) or soluble carbohydrates such as glycine betaine and γ-aminobutyric acid (Joshi et al., 2016a). These solutes maintain osmotic homeostasis inside the cell and help proteins to maintain their native tertiary structures. The common secondary response to stress is ROS accumulation including H_2O_2, superoxide, and hydroxide ions; the harmful effects of which are countered by metabolites such as ascorbate and glutathione (Szarka et al., 2012; Joshi et al., 2016b).

Tolerance to drought and salinity are the most sought-after traits in current crop breeding programs. The generation of similar phenotypic effects in these two stresses infers a high degree of correlation between phenomics screening approaches. Stomatal closure is considered an initial response after exposure to salinity stress, causing a decline in stomatal conductance and transpiration, which eventually decreases the flow of toxic ions from the roots to the plants (Joshi et al., 2015; Acosta-Motos et al., 2017). In addition to drought and salinity stress, osmotic stress—also termed chemical drought—imposes similar effects on plant phenotypes. Stomatal closure also leads to a decline in photosynthesis, but the screening methods used to detect photosynthetic parameters are very slow, with less reproducibility. With several tools available for phenomics, a surrogate methodology could be more effective for screening stomatal and photosynthetic responses under osmotic stress.

3.1 Drought Tolerance

Tolerance to drought stress in plants is indicated by visible symptoms, including leaf rolling, stay-green ability, stomatal closure, photochemical quenching, photo inhibition resistance, water use efficiency (WUE), osmotic adjustment, membrane stability, epicuticular wax content, mobilization of water-soluble carbohydrates and increased root length (Singh et al., 2015a). These traits are often targeted for phenotyping under drought stress. Leaf rolling is the primary visible symptom—and one of the survival mechanisms against drought stress—reducing the transpiration rate and canopy temperature (Joshi and Karan, 2013). Drought-tolerant plants retain a higher relative water content (RWC) under water deficit conditions to sustain normal growth (Singh et al., 2015b). The impact of drought on photosynthesis has been differentiated into two groups; direct and indirect. The direct effect is measured as the increased restriction of CO_2 diffusion via stomata that limits CO_2 supply inside leaves and leads to a decline in CO_2 availability for Rubisco (Martorell et al., 2014). Indirect effects include alterations in the biochemistry and metabolism of the photosynthetic apparatus, membrane permeability, and the promotion of oxidative stress (Aranda et al., 2010; Wahid et al., 2014). Drought stress critically impedes wheat and grain legumes' performance during flowering and grain-filling stages, primarily because of a reduced net photosynthesis rate, oxidative damage to chloroplasts, and stomatal closure, which finally leads to poor grain development (Farooq et al., 2014b, 2017).

Drought stress exerts osmotic pressure on plants, the response of which is to accumulate compatible solutes such as proline, betaine, and polyols (mannitol, trehalose, etc.) in the cytosol to retain osmotic pressure that provides a driving gradient for the uptake of water and turgor maintenance (Joshi et al., 2016a). It is well established that proline plays an important role in the stabilization of cellular proteins and membranes under high osmotic concentrations. Secondary responses, such as oxidative stress, induce membrane damage during water stress and are characterized by membrane lipid peroxidation perceived by the accumulation of malondialdehyde (MDA) (Farooq et al., 2010). Root length is another significant character that requires careful consideration, as roots are directly connected with soil and perceive the immediate effects of a reduction in soil water content (Hochholdinger, 2016). Fenta et al. (2014) observed a significant reduction in lateral root development under drought stress and asserted that the observed changes comprise an adaptive mechanism to the stress.

Different next-generation phenotyping platforms with highly efficient software are being used to evaluate drought tolerance in different crops (see Cobb et al., 2013); for example, PHENOPSIS (Granier et al., 2006) and WIWAM (https://www.wiwam.be/) in Arabidopsis, and LemnaTec in barley (Honsdorf et al., 2014), maize (Ge et al., 2016), tomatoes (Petrozza et al., 2014), and wheat (Fehér-Juhász et al., 2014).

3.2 Salinity Tolerance

Salinity induces both ion toxicity and osmotic stress in crop plants by altering the physiological status and ionic homeostasis of cells (Wungrampha et al., 2018). This cellular status is regulated by a plethora of genes (Joshi et al., 2017). Crop plants are vulnerable to salinity, the extent of which depends on the plant growth stage and salt concentration (Hossain et al., 2015). The mechanism of tolerance is cultivar specific. Salinity stress affects growth at different developmental stages and generally delays germination. During vegetative stages, it reduces leaf area, total chlorophyll content, biomass, and root length (Läuchli and Grattan, 2007).

In addition to phenotypic changes, depending on the severity and duration of the stress, salinity affects various physiological and metabolic processes by exerting osmotic and ionic toxicity (Gupta and Huang, 2014). Osmotic stress reduces the water absorption capacity of root systems and concurrently increases water loss from leaves (Munns, 2005). However, the production of cellular osmolytes (proline, glycine betaine, mannitol, etc.) alleviates the effect of osmotic stress (Singla-Pareek et al., 2008). Other important physiological changes caused by osmotic stress include membrane interruption, nutrient imbalance, impaired ability of ROS detoxification, differences in antioxidant enzymes, decreased photosynthetic activity, and reduced stomatal aperture. Ion toxicity mainly occurs due to the higher accumulation of Na^+ and Cl^- ions in plant tissues exposed to high saline conditions. Higher uptake of Na^+ and Cl^- into cells results in a severe ionic imbalance and might cause considerable physiological changes. High Na^+ concentration inhibits K^+ ion uptake, which is essential for growth and development, resulting in reduced productivity. Production of ROS, such as singlet oxygen, superoxide, hydroxyl radical, and hydrogen peroxide is a secondary response under salinity stress. Salinity-induced ROS formation causes oxidative damage in various cellular components such as proteins, lipids, and DNA, thus interrupting vital cellular processes in plants. Plants develop various physiological and biochemical mechanisms to survive in soils with high salt concentration (Joshi et al., 2016b).

Visible imaging-based phenotyping uses a standard evaluation system (SES) score, while RGB is used to measure total chlorophyll content (Mishra et al., 2016a,b). Next-generation phenotyping assays exploiting different light wavelengths are used to assess salt tolerance, such as PHENOPSIS (Granier et al., 2006), WIWAM (https://www.wiwam.be/) in rice (Hairmansis et al., 2014), wheat, and barley (Harris et al., 2010; Humplík et al., 2015a; Meng et al., 2017).

3.3 Temperature (Heat/Cold) Stress Tolerance

Temperature stress in plants occurs at either high or chilling/freezing temperatures. Phenotypic and biochemical features change in cultivated plant species in response to heat stress, for example, poor germination ratio, poor seedling emergence, abnormal seedling development, poor seedling vigor, reduced radicle and plumule growth, inhibition of photosystem II (*PsiI*) activity, and ROS production (Jagadish et al., 2016). Therefore, the response of cultivated species to temperature stress and the tolerance level of cultivars are analyzed on the basis of the affected traits. The vulnerability of any plant to heat stress is stage-specific, despite all crop species being susceptible to heat stress during their entire life cycle. However, the reproductive stage is considered the most sensitive stage for terminal heat stress. Even a few-degree rise in temperature during flowering can cause complete yield loss (Ohama et al., 2017).

Morphological changes due to high-temperature stress include scorching and sunburns of leaves, twigs, branches, and stems, senescence and abscission of leaves, inhibition of shoot and root growth, fruit discoloration, and permanent damage. High-temperature stress in sugarcane damaged leaf tips and margins, and caused leaf rolling, drying, and necrosis (Hasanuzzaman et al., 2013). In wheat, heat stress reduced tiller numbers and increased shoot length (Kumar et al. 2011). Biochemical changes due to high-temperature stress include irreversible damage to photosynthetic pigments and Rubisco, enhanced rate of photorespiration, and inhibition of noncyclic electron transport, which enhances ROS accumulation in plant cells and limits CO_2 fixation (Mathur et al., 2014; Farooq et al., 2016). Plants under abiotic stress tend to produce more ROS in chloroplasts and mitochondria, which severely damages DNA and causes lipid peroxidation in the cell membrane (Kukavica and Veljovic-Jovanovic, 2004). Several studies have demonstrated that ROS detoxification mechanisms play a significant role in providing high-temperature stress tolerance in plants (Suzuki and Mittler, 2006). Thus, plant tolerance to environmental stress is closely correlated with their ability to scavenge and detoxify ROS (Zandalinas et al., 2018). However, high temperatures reduce plant growth by inhibiting net assimilation rates in shoots, and thus total plant dry weight (Wahid et al., 2007). Under elevated temperatures, programmed cell death (PCD) occurs in specific cells or tissues within minutes, or even seconds, due to the denaturation or aggregation of proteins. However, moderately high temperatures for extended periods cause gradual senescence in plants. In both conditions, leaf shedding, flower and fruit abortion, and even plant death have been observed (Rodríguez et al., 2005).

Low temperature affects plant growth and productivity and causes significant yield losses. In contrast to heat stress, chilling stress directly inhibits metabolic reactions and indirectly harms the osmotic imbalance, which reduces the expression of the full genetic potential of plants. Based on the temperature range, cold stress is defined as chilling stress ($<20°C$) and/or freezing stress ($<0°C$), both of which affect crops through different mechanisms. While chilling stress reduces the rate of enzymatic reactions and membrane transport activities, freezing stress results in the formation of ice crystals and membrane damage (Croser et al., 2003; Chinnusamy et al., 2007). Indirectly, cold stress damages plants by cold-induced osmotic imbalance, oxidative stress, and, in the case of chilling stress, the formation of water uptake barriers and, in the case of freezing stress, cellular dehydration.

Plants differ in their ability to tolerate chilling and freezing stresses. Temperate-climate plants are generally considered chilling tolerant to variable degrees (Chinnusamy et al., 2010), which can be further increased by exposure to chilling or nonfreezing temperatures, known as cold acclimation. Cold acclimation is associated with physiological and biochemical changes that result in altered gene expression, biomembrane lipid composition, and the accumulation of small molecules (Wani et al., 2016). In contrast, tropical and subtropical plants are more sensitive to chilling stress and lack the cold acclimation mechanism. Thus, low-temperature resistance in these plants is a complex trait, involving various metabolic pathways and cell compartments (Barnes et al., 2016). Several studies have been conducted in various plant species to measure tolerance under cold stress or in combination with other stresses, such as drought and salinity. However, few details have been provided on the methodologies used to evaluate the stress response in plants.

4. APPLICATION OF PHENOMICS IN IMPROVING ABIOTIC STRESS TOLERANCE IN PLANTS

High-throughput phenotyping is a bottleneck in crop genetic improvement. The progress of developing high-yielding crop varieties adapted to an environment can be hindered by slow, and often subjective, manual phenotyping. It also requires destructive and laborious harvesting across many field seasons and environments. High-throughput and nondestructive crop evaluation in the field and controlled environments is missing in our breeding systems. Nevertheless, the need for developing plant phenomics approaches and infrastructure has been realized globally.

4.1 Infrared Thermography

Infrared thermography (IRT) is used to measure temperature differences by infrared wave emission. Using infrared imaging, Qiu et al. (2009) detected significant differences between leaf temperature, air temperature, and canopy temperature under drought and high temperature stress in melons, tomatoes, and lettuce. They further proposed that the transpiration transfer coefficient (h_{at}) can be used to detect various environmental stresses in plants. Similarly, IRT has been used to evaluate osmotic stress in wheat and barley in response to salt stress (Sirault et al., 2009), and plant responses to water stress in grapevines and rice (Jones et al., 2002). This approach allows high-throughput screening at the seedling stage that can be validated within the canopy in the field using the same tools and genotypes (Furbank, 2009). More recently, Wedeking et al. (2017) used IRT to monitor leaf temperature and transpiration in *Beta vulgaris* plants subjected to progressive drought stress.

4.2 Spectroscopic Techniques

Spectroscopic techniques can be used to study photosynthetic rates at leaf and canopy levels, as well as other biochemical activities. Kiirats et al. (2009) used a leaf spectrometer to investigate photosynthetic electron transport feedback regulation in *Nicotiana sylvestris*. Photosynthetic efficiency, its activity, and biochemical pathway have been monitored in pine (Busch et al., 2009) and barley (Siebke and Ball, 2009), with large-scale use of reflectance spectroscopy. The Raman spectroscopic technique was developed by Altangerel et al. (2017) for high-throughput stress phenotyping and in vivo early stress detection.

4.3 Fluorescence Imaging

Spectral absorption and reflectance offer a noninvasive tool for investigating plant chemical composition and function, which is scalable from the cell to canopy level. Chlorophyll fluorescence is the most commonly used, and relatively affordable, tool at the leaf level. Jansen et al. (2009) combined chlorophyll fluorescence with 2D digital imaging of plant growth to monitor plant reactions under drought and chilling stress in *Arabidopsis thaliana*. Similarly, by preparing a 3D polygon model by combining the time series of chlorophyll, a fluorescence image taken from a high resolution scanning lidar, spatio-temporal changes of herbicide, that is, 3-(3,4 dichlorophenyl)-1,1-dimethylurea (DCMU) effects in whole melon plants was monitored three-dimensionally (Konishi et al., 2009). Rungrat et al. (2016) reviewed the use of chlorophyll fluorescence to monitor the effect on photosynthetic activity in *Arabidopsis thaliana* under various abiotic stresses.

4.4 Integrated Imaging Techniques

PET is used for in vivo imaging and to study biochemical pathways, and ion assimilation and transport. Fatangare et al. (2015) characterized 2-deoxy-2-fluoro-D-glucose (FDG) metabolism in plants, which had been used to study plant defense (Ferrieri et al., 2012) and carbon allocation (Fatangare et al., 2014). A study by Meldau et al. (2015) on carbon allocation in plants after herbivore attack expanded the scope of FDG to in vivo plants subjected to various biotic and abiotic stresses.

Imaging techniques such as MRI (Borisjuk et al., 2012) and X-ray-CT (Dhondt et al., 2010) are used to obtain anatomical information. An MRI-PET co-registration system was used by Jahnke et al. (2009) to combine PET-obtained radioactivity information with MRI-obtained anatomical data. A bifunctional PET/CT was used to obtain 4D radiotracer dynamics combining CT-derived morphological data, and PET-derived corresponding radio signals (Fatangare et al., 2014).

5. CONCLUSION AND PROSPECTS

In the face of the rising population and lost crop production due to climate change, the global challenge of increasing productivity requires the latest discoveries in the field of plant genomics to be brought to the farm gate. Integrative plant biology requires that we scale from the gene to the biochemical process, the plant, the whole plant, and, ultimately, the canopy and crop. Correlations among gene function, environmental responses, and plant performance need to be studied with a rapid pace and high-resolution. The ever-increasing capacity of next-generation phenotyping platforms will assist researchers to deepen their knowledge on plant stress responses. The emerging tools and techniques in plant phenomics hold immense potential with respect to the development of crop genotypes with enhanced resilience, a much-needed development to cater to the dietary requirements of 7.5 billion people worldwide.

ACKNOWLEDGMENTS

Rohit Joshi acknowledges the Dr. DS Kothari Postdoctoral Fellowship from University Grant Commission, Government of India.

REFERENCES

Acosta-Motos, J.R., Ortuño, M.F., Bernal-Vicente, A., Diaz-Vivancos, P., Sanchez-Blanco, M.J., Hernandez, J.A., 2017. Plant responses to salt stress: adaptive mechanisms. Agronomy 7, 18.

Altangerel, N., Ariunbold, G.O., Gorman, C., Alkahtani, M.H., Borrego, E.J., Bohlmeyer, D., Hemmer, P., Kolomiets, M.V., Yuan, J.S., Scully, M.O., 2017. *In vivo* diagnostics of early abiotic plant stress response via Raman spectroscopy. Proc. Natl. Acad. Sci. USA 114, 3393–3396.

Aranda, I.E., Gil-Pelegrín, A., Gascó, M.A., Guevara, J.F., Cano, M., De Miguel, J.A., Ramírez-Valiente, J.A., Peguero-Pina, J.J., Perdiguero, P., Soto, A., Cervera, M.T., 2010. Drought response in forest trees: from the species to the gene. In: Aroca, R. (Ed.), Plant Responses to Drought Stress. Springer, Berlin, Heidelberg, pp. 293–333.

Awlia, M., Nigro, A., Fajkus, J., Schmoeckel, S.M., Negrão, S., Santelia, D., Trtílek, M., Tester, M., Julkowskam, M.M., Panzarová, K., 2016. High-throughput non-destructive phenotyping of traits that contribute to salinity tolerance in *Arabidopsis thaliana*. Front. Plant Sci. 7, 1414.

Baker, N.R., 2008. Chlorophyll fluorescence: a probe of photosynthesis *in vivo*. Annu. Rev. Plant Biol. 59, 89–113.

Barnes, A.C., Benning, C., Roston, R.L., 2016. Chloroplast membrane remodeling during freezing stress is accompanied by cytoplasmic acidification activating SENSITIVE TO FREEZING2. Plant Physiol. 171, 2140–2149.

Bayoumi, T.Y., El-Hendawy, S., Yousef, M.S.H., Emam, M.A.E., Okasha, S.A.A.G., 2014. Application of infrared thermal imagery for monitoring salt tolerant of wheat genotypes. J. Am. Sci. 10, 227–234.

Borisjuk, L., Rolletschek, H., Neuberger, T., 2012. Surveying the plant's world by magnetic resonance imaging. Plant J. 70, 129–146.

Bowman, B.C., Chen, J., Zhang, J., Wheeler, J., Wang, Y., Zhao, W., Nayak, S., Heslot, N., Bockelman, H., Bonman, J.M., 2015. Evaluating grain yield in spring wheat with canopy spectral reflectance. Crop Sci. 55, 1881–1890.

Busch, F., Hüner, N.P., Ensminger, I., 2009. Biochemical constrains limit the potential of the photochemical reflectance index as a predictor of effective quantum efficiency of photosynthesis during the winter spring transition in Jack pine seedlings. Funct. Plant Biol. 36, 1016–1026.

Cabrera-Bosquet, L., Crossa, J., von Zitzewitz, J., Serret, M.D., Luis Araus, J., 2012. High-throughput phenotyping and genomic selection: the frontiers of crop breeding converge. J. Integr. Plant Biol. 54, 312–320.

Cardon, Z.G., Mott, K.A., Berry, J.A., 1994. Dynamics of patchy stomatal movements, and their contribution to steady-state and oscillating stomatal conductance calculated using gas-exchange techniques. Plant Cell Environ. 17, 995–1007.

Cernicharo, J., Verger, A., Camacho, F., 2013. Empirical and physical estimation of canopy water content from CHRIS/PROBA data. Remote Sens. 5, 5265–5284.

Chaerle, L., Van Der Straeten, D., 2001. Seeing is believing: imaging techniques to monitor plant health. Biochim. Biophys. Acta 1519, 153–166.

Chen, D., Ming, C., Thomas, A., Christian, K., 2014. Bridging genomics and phenomics. In: Chen, M., Hofestädt, R. (Eds.), Approaches in Integrative Bioinformatics. Springer, Berlin, Heidelberg, pp. 299–333.

Cheng, T., Song, R., Li, D., Zhou, K., Zheng, H., Yao, X., Tian, Y., Cao, W., Zhu, Y., 2017. Spectroscopic estimation of biomass in canopy components of paddy rice using dry matter and chlorophyll indices. Remote Sens. 9, 319.

Chinnusamy, V., Zhu, J., Zhu, J.K., 2007. Cold stress regulation of gene expression in plants. Trends Plant Sci. 12, 444–451.

Chinnusamy, V., Zhu, J.K., Sunkar, R., 2010. Gene regulation during cold stress acclimation in plants. Methods Mol. Biol. 639, 39–55.

Clark, R.T., MacCurdy, R.B., Jung, J.K., Shaff, J.E., McCouch, S.R., Aneshansley, D.J., Kochian, L.V., 2011. Three-dimensional root phenotyping with a novel imaging and software platform. Plant Physiol. 156, 455–465.

Cobb, J.N., DeClerck, G., Greenberg, A., Clark, R., McCouch, S., 2013. Next-generation phenotyping: requirements and strategies for enhancing our understanding of genotype–phenotype relationships and its relevance to crop improvement. Theor. Appl. Genet. 126, 867–887.

Collins, N.C., Tardieu, F., Tuberosa, R., 2008. Quantitative trait loci and crop performance under abiotic stress: where do we stand? Plant Physiol. 147, 469–486.

Croser, J.S., Clarke, H.J., Siddique, K.H., Khan, T.N., 2003. Low-temperature stress: implications for chickpea (*Cicer arietinum* L.) improvement. Crit. Rev. Plant Sci. 22, 185–219.

Dhondt, S., Vanhaeren, H., Van Loo, D., Cnudde, V., Inzé, D., 2010. Plant structure visualization by high-resolution X-ray computed tomography. Trends Plant Sci. 15, 419–422.

Din, M., Zheng, W., Rashid, M., Wang, S., Shi, Z., 2017. Evaluating hyperspectral vegetation indices for leaf area index estimation of *Oryza sativa* L. at diverse phenological stages. Front. Plant Sci. 8, 820.

Fahlgren, N., Gehan, M.A., Baxter, I., 2015. Lights, camera, action: high-throughput plant phenotyping is ready for a close-up. Curr. Opin. Plant Biol. 24, 93–99.

Farooq, M., Wahid, A., Lee, D.J., Cheema, S.A., Aziz, T., 2010. Drought stress: comparative time course action of the foliar applied glycinebetaine, salicylic acid, nitrous oxide, brassinosteroids and spermine in improving drought resistance of rice. J. Agron. Crop Sci. 196, 336–345.

Farooq, M., Wahid, A., Siddique, K.H.M., 2014a. Physiology of grain development in cereals. In: Pessarakli, M. (Ed.), Handbook of Photosynthesis. CRC Press, Boca Raton, USA, pp. 301–312.

Farooq, M., Hussain, M., Siddique, K.H.M., 2014b. Drought stress in wheat during flowering and grain-filling periods. Crit. Rev. Plant Sci. 33, 331–349.

Farooq, M., Rehman, A., Wahid, A., Siddique, K.H.M., 2016. Photosynthesis under heat stress. In: Pessarakli, M. (Ed.), Handbook of Photosynthesis. CRC Press, Boca Raton, USA, pp. 697–701.

Farooq, M., Gogoi, N., Barthakur, S., Baroowa, B., Bharadwaj, N., Alghamdi, S.S., Siddique, K.H.M., 2017. Drought stress in grain legumes during reproduction and grain filling. J. Agron. Crop Sci. 203, 81–102.

Fatangare, A., Gebhardt, P., Saluz, H., Svatoš, A., 2014. Comparing 2-[18 F] fluoro-2-deoxy-D-glucose and [68 Ga] gallium-citrate translocation in *Arabidopsis thaliana*. Nucl. Med. Biol. 41, 737–743.

Fatangare, A., Paetz, C., Saluz, H., Svatoš, A., 2015. 2-Deoxy-2-fluoro-D-glucose metabolism in *Arabidopsis thaliana*. Front. Plant Sci. 3, 935.

Fehér-Juhász, E., Majer, P., Sass, L., Lantos, C., Csiszár, J., Turóczy, Z., Mai, A., Horváth, V., Vass, I., Dudits, D., 2014. Phenotyping shows improved physiological traits and seed yield of transgenic wheat plants expressing the alfalfa aldose reductase under permanent drought stress. Acta Physiol. Plant. 36, 663–673.

Fenta, B.A., Beebe, S.E., Kunert, K.J., Burridge, J.D., Barlow, K.M., Lynch, J.P., Foyer, C.H., 2014. Field phenotyping of soybean roots for drought stress tolerance. Agronomy 4, 418–435.

Ferrieri, A.P., Appel, H., Ferrieri, R.A., Schultz, J.C., 2012. Novel application of 2-[18 F] fluoro-2-deoxy-d-glucose to study plant defenses. Nucl. Med. Biol. 39, 1152–1160.

Fiorani, F., Rascher, U., Jahnke, S., Schurr, U., 2012. Imaging plants dynamics in heterogenic environments. Curr. Opin. Biotechnol. 23, 227–235.
Furbank, R.T., 2009. Plant phenomics: from gene to form and function. Funct. Plant Biol. 36, 5–6.
Furbank, R.T., Tester, M., 2011. Phenomics–technologies to relieve the phenotyping bottleneck. Trends Plant Sci. 16, 635–644.
Ge, Y., Bai, G., Stoerger, V., Schnable, J.C., 2016. Temporal dynamics of maize plant growth, water use, and leaf water content using automated high throughput RGB and hyperspectral imaging. Comput. Electron. Agric. 127, 625–632.
Golzarian, M.R., Frick, R.A., Rajendran, K., Berger, B., Roy, S., Tester, M., Lun, D.S., 2011. Accurate inference of shoot biomass from high-throughput images of cereal plants. Plant Methods 7, 2.
Granier, C., Aguirrezabal, L., Chenu, K., Cookson, S.J., Dauzat, M., Hamard, P., Thioux, J.J., Rolland, G., Bouchier-Combaud, S., Lebaudy, A., Muller, B., 2006. PHENOPSIS, an automated platform for reproducible phenotyping of plant responses to soil water deficit in Arabidopsis thaliana permitted the identification of an accession with low sensitivity to soil water deficit. New Phytol. 169, 623–635.
Gupta, B., Huang, B., 2014. Mechanism of salinity tolerance in plants: physiological, biochemical, and molecular characterization. Int. J. Genomics 2014, 701596.
Hairmansis, A., Berger, B., Tester, M., Roy, S.J., 2014. Image-based phenotyping for non-destructive screening of different salinity tolerance traits in rice. Rice 7, 16.
Harris, B.N., Sadras, V.O., Tester, M., 2010. A water-centred framework to assess the effects of salinity on the growth and yield of wheat and barley. Plant Soil 336, 377–389.
Hasanuzzaman, M., Nahar, K., Alam, M.M., Roychowdhury, R., Fujita, M., 2013. Physiological, biochemical, and molecular mechanisms of heat stress tolerance in plants. Int. J. Mol. Sci. 14, 9643–9684.
Hochholdinger, F., 2016. Untapping root system architecture for crop improvement. J. Exp. Bot. 67, 4431.
Honsdorf, N., March, T.J., Berger, B., Tester, M., Pillen, K., 2014. High-throughput phenotyping to detect drought tolerance QTL in wild barley introgression lines. PLoS One 9, e97047.
Hossain, M.R., Vickers, L., Sharma, G., Livermore, T., Pritchard, J., Ford-Lloyd, B.V., 2015. Salinity tolerance in plants: insights from transcriptomics studies. In: Wani, S.H., Hossain, M.A. (Eds.), Managing Salt Tolerance in Plants: Molecular and Genomic Perspectives. CRC Press, Boca Raton, USA, p. 407.
Humplík, J.F., Lazár, D., Husičková, A., Spíchal, L., 2015a. Automated phenotyping of plant shoots using imaging methods for analysis of plant stress responses—a review. Plant Methods 11, 29.
Humplík, J.F., Lazár, D., Fürst, T., Husičková, A., Hýbl, M., Spíchal, L., 2015b. Automated integrative high-throughput phenotyping of plant shoots: a case study of the cold-tolerance of pea (*Pisum sativum* L.). Plant Methods 11, 1–11.
Jagadish, S.K., Bahuguna, R.N., Djanaguiraman, M., Gamuyao, R., Prasad, P.V., Craufurd, P.Q., 2016. Implications of high temperature and elevated CO_2 on flowering time in plants. Front. Plant Sci. 7, 913.
Jahnke, S., Menzel, M.I., Van Dusschoten, D., Roeb, G.W., Bühler, J., Minwuyelet, S., Blümler, P., Temperton, V.M., Hombach, T., Streun, M., Beer, S., 2009. Combined MRI–PET dissects dynamic changes in plant structures and functions. Plant J. 59, 634–644.
Jansen, M., Gilmer, F., Biskup, B., Nagel, K.A., Rascher, U., Fischbach, A., Briem, S., Dreissen, G., Tittmann, S., Braun, S., De Jaeger, I., 2009. Simultaneous phenotyping of leaf growth and chlorophyll fluorescence via GROWSCREEN FLUORO allows detection of stress tolerance in *Arabidopsis thaliana* and other rosette plants. Funct. Plant Biol. 36, 902–914.
Jhala, V.M., Thaker, V.S., 2015. X-ray computed tomography to study rice (*Oryza sativa* L.) panicle development. J. Exp. Biol. 66, 6819–6825.
Jones, H.G., Stoll, M., Santos, T., Sousa, C.D., Chaves, M.M., Grant, O.M., 2002. Use of infrared thermography for monitoring stomatal closure in the field: application to grapevine. J. Exp. Biol. 53, 2249–2260.
Jones, H.G., Serraj, R., Loveys, B.R., Xiong, L., Wheaton, A., Price, A.H., 2009. Thermal infrared imaging of crop canopies for the remote diagnosis and quantification of plant responses to water stress in the field. Funct. Plant Biol. 36, 978–989.
Jones, A.M., Danielson, J.Å., Manoj Kumar, S.N., Lanquar, V., Grossmann, G., Frommer, W.B., 2014. Abscisic acid dynamics in roots detected with genetically encoded FRET sensors. Elife 3, e01741.

Joshi, R., Karan, R., 2013. Physiological, biochemical and molecular mechanisms of drought tolerance in plants. In: Gaur, R.K., Sharma, P. (Eds.), Molecular Approaches in Plant Abiotic Stress. CRC Press, Boca Raton, FL, pp. 209–223.

Joshi, R., Singh, B., Bohra, A., Chinnusamy, V., 2015. Salt stress signaling pathways. In: Wani, S.H., Hussain, M.A. (Eds.), Managing Salt Tolerance in Plants: Molecular and Genomic Perspectives. CRC Press, Boca Raton, FL, pp. 51–78.

Joshi, R., Wani, S.H., Singh, B., Bohra, A., Dar, Z.A., Lone, A.A., Pareek, A., Singla-Pareek, S.L., 2016a. Transcription factors and plants response to drought stress: current understanding and future directions. Front. Plant Sci. 7, 1029.

Joshi, R., Karan, R., Singla-Pareek, S.L., Pareek, A., 2016b. Ectopic expression of Pokkali phosphoglycerate kinase-2 (OsPGK2-P) improves yield in tobacco plants under salinity stress. Plant Cell Rep. 35, 27–41.

Joshi, R., Sahoo, K.K., Tripathi, A.K., Kumar, R., Gupta, B.K., Pareek, A., Singla-Pareek, S.L., 2017. Knockdown of an inflorescence meristem-specific cytokinin oxidase—OsCKX2 in rice reduces yield penalty under salinity stress condition. Plant Cell Environ. https://doi.org/10.1111/pce.12947.

Kastberger, G., Stachl, R., 2003. Infrared imaging technology and biological applications. Behav. Res. Methods Instrum. Comput. 35, 429–439.

Kesari, R., Lasky, J.R., Villamor, J.G., Des Marais, D.L., Chen, Y.J., Liu, T.W., Lin, W., Juenger, T.E., Verslues, P.E., 2012. Intron-mediated alternative splicing of Arabidopsis P5CS1 and its association with natural variation in proline and climate adaptation. Proc. Natl. Acad. Sci. 109, 9197–9202.

Kiirats, O., Cruz, J.A., Edwards, G.E., Kramer, D.M., 2009. Feedback limitation of photosynthesis at high CO_2 acts by modulating the activity of the chloroplast ATP synthase. Funct. Plant Biol. 36, 893–901.

Kiyomiya, S., Nakanishi, H., Uchida, H., Tsuji, A., Nishiyama, S., Futatsubashi, M., Tsukada, H., Ishioka, N.S., Watanabe, S., Ito, T., Mizuniwa, C., 2001. Real time visualization of ^{13}N-translocation in rice under different environmental conditions using positron emitting tracer imaging system. Plant Physiol. 125, 1743–1753.

Konishi, A., Eguchi, A., Hosoi, F., Omasa, K., 2009. 3D monitoring spatio–temporal effects of herbicide on a whole plant using combined range and chlorophyll a fluorescence imaging. Funct. Plant Biol. 36, 874–879.

Kukavica, B., Veljovic-Jovanovic, S., 2004. Senescence-related changes in the antioxidant status of ginkgo and birch leaves during autumn yellowing. Physiol. Plant. 122, 321–327.

Kumar, S., Kaur, R., Kaur, N., Bhandhari, K., Kaushal, N., Gupta, K., Bains, T.S., Nayyar, H., 2011. Heat-stress induced inhibition in growth and chlorosis in mungbean (*Phaseolus aureus* Roxb.) is partly mitigated by ascorbic acid application and is related to reductionin oxidative stress. Acta Physiol. Plant. 33, 2091–2101.

Läuchli, A., Grattan, S., 2007. Plant growth development under salinity stress. In: Jenks, M.A., Hasegawa, P., Jain, S.M. (Eds.), Advances in Molecular Breeding Towards Salinity and Drought Tolerance. Springer, Dordrecht, p. 1.

Li, L., Zhang, Q., Huang, D., 2014. A review of imaging techniques for plant phenotyping. Sensors 14, 20078–20111.

Liu, Z.Y., Shi, J.J., Zhang, L.W., 2010. Discrimination of rice panicles by hyperspectral reflectance databased on principal component analysis and support vector classification. J. Zhejiang Univ. (Sci.) 11, 71–78.

Maphosa, L., Thoday-Kennedy, E., Vakani, J., Phelan, A., Badenhorst, P., Slater, A., Spangenberg, G., Kant, S., 2016. Phenotyping wheat under salt stress conditions using a 3D laser scanner. Isr. J. Plant Sci. 1, 1–8.

Martorell, S., Diaz-Eespejo, A.N., Medrano, H., Ball, M.C., Choat, B., 2014. Rapid hydraulic recovery in *Eucalyptus pauciflora* after drought: linkages between stem hydraulics and leaf gas exchange. Plant Cell Environ. 37, 617–626.

Mathur, S., Agrawal, D., Jajoo, A., 2014. Photosynthesis: response to high temperature stress. J. Photochem. Photobiol. B 137, 116–126.

Meldau, S., Woldemariam, M.G., Fatangare, A., Svatos, A., Galis, I., 2015. Using 2-deoxy-2-[18F]fluoro-D-glucose ([18F]FDG) to study carbon allocation in plants after herbivore attack. BMC Res. Notes 8, 45.

Meng, R., Saade, S., Kurtek, S., Berger, B., Brien, C., Pillen, K., Tester, M., Sun, Y., 2017. Growth curve registration for evaluating salinity tolerance in barley. Plant Methods 13, 18.

Merlot, S., Mustilli, A.C., Genty, B., North, H., Lefebvre, V., Sotta, B., Vavasseur, A., Giraudat, J., 2002. Use of infrared thermal imaging to isolate Arabidopsis mutants defective in stomatal regulation. Plant J. 30, 601–609.
Mickelbart, M.V., Hasegawa, P.M., Bailey-Serres, J., 2015. Genetic mechanisms of abiotic stress tolerance that translate to crop yield stability. Nat. Rev. Genet. 16, 237–251.
Mishra, A., Heyer, A.G., Mishra, K.B., 2014. Chlorophyll fluorescence emission can screen cold tolerance of cold acclimated *Arabidopsis thaliana* accessions. Plant Methods 10, 38.
Mishra, S., Singh, B., Panda, K., Singh, B.P., Singh, N., Misra, P., Rai, V., Singh, N.K., 2016a. Association of SNP haplotypes of HKT family genes with salt tolerance in Indian wild rice germplasm. Rice 9, 15.
Mishra, S., Singh, B., Panda, K., Singh, B.P., Singh, N., Misra, P., Rai, V., Singh, N.K., 2016b. Haplotype distribution and association of candidate genes with salt tolerance in Indian wild rice germplasm. Plant Cell Rep. 35, 2295–2308.
Munns, R., 2005. Genes and salt tolerance: bringing them together. New Phytol. 167, 645–663.
Munns, R., James, R.A., Sirault, X., 2010. New phenotyping methods for screening wheat and barley for beneficial responses to water deficit. J. Exp. Bot. 61, 3499–3507.
Neilson, E.H., Edwards, A.M., Blomstedt, C.K., Berger, B., Møller, B.L., Gleadow, R.M., 2015. Utilization of a high-throughput shoot imaging system to examine the dynamic phenotypic responses of a C4 cereal crop plant to nitrogen and water deficiency over time. J. Exp. Biol. 66, 1817–1832.
Ohama, N., Sato, H., Shinozaki, K., Yamaguchi-Shinozaki, K., 2017. Transcriptional regulatory network of plant heat stress response. Trends Plant Sci. 22, 53–65.
Okumoto, S., Jones, A., Frommer, W.B., 2012. Quantitative imaging with fluorescent biosensors. Annu. Rev. Plant Biol. 64, 663–706.
Osmond, B., Ananyev, G., Berry, J., Langdon, C., Kolber, Z., Lin, G., Monson, R., Nichol, C., Rascher, U., Schurr, U., Smith, S., 2004. Changing the way we think about global change research: scaling up in experimental ecosystem science. Glob. Chang. Biol. 10, 393–407.
Paproki, A., Sirault, X., Berry, S., Furbank, R., Fripp, J., 2012. A novel mesh processing based technique for 3D plant analysis. BMC Plant Biol. 12, 63–71.
Peñuelas, J., Filella, I., 1998. Visible and near-infrared reflectance techniques for diagnosing plant physiological status. Trends Plant Sci. 3, 151–156.
Pereira, A., 2016. Plant abiotic stress challenges from the changing environment. Front. Plant Sci. 7, 1123.
Petrozza, A., Santaniello, A., Summerer, S., Di Tommaso, G., Di Tommaso, D., Paparelli, E., Piaggesi, A., Perata, P., Cellini, F., 2014. Physiological responses to Megafol® treatments in tomato plants under drought stress: a phenomic and molecular approach. Sci. Hortic. 174, 185–192.
Qiu, G.Y., Omasa, K., Sase, S., 2009. An infrared-based coefficient to screen plant environmental stress: concept, test and applications. Funct. Plant Biol. 36, 990–997.
Rahaman, M.M., Chen, D., Gillani, Z., Klukas, C., Chen, M., 2015. Advanced phenotyping and phenotype data analysis for the study of plant growth and development. Front. Plant Sci. 6, 619.
Rascher, U., Hütt, M.T., Siebke, K., Osmond, B., Beck, F., Lüttge, U., 2001. Spatio-temporal variations of metabolism in a plant circadian rhythm: the biological clock as an assembly of coupled individual oscillators. Proc. Natl. Acad. Sci. USA 98, 11801–11805.
Rodríguez, M., Canales, E., Borrás-Hidalgo, O., 2005. Molecular aspects of abiotic stress in plants. Biotechnol. Apl. 22, 1–10.
Römer, C., Wahabzada, M., Ballvora, A., Pinto, F., Rossini, M., Panigada, C., Behmann, J., Léon, J., Thurau, C., Bauckhage, C., Kersting, K., 2012. Early drought stress detection in cereals: simplex volume maximisation for hyperspectral image analysis. Funct. Plant Biol. 39, 878–890.
Rosenzweig, C., Elliott, J., Deryng, D., Ruane, A.C., Müller, C., Arneth, A., Boote, K.J., Folberth, C., Glotter, M., Khabarov, N., Neumann, K., 2014. Assessing agricultural risks of climate change in the 21st century in a global gridded crop model intercomparison. Proc. Natl. Acad. Sci. 111, 3268–3273.
Rungrat, T., Awlia, M., Brown, T., Cheng, R., Sirault, X., Fajkus, J., Trtilek, M., Furbank, B., Badger, M., Tester, M., Pogson, B.J., 2016. Using phenomic analysis of photosynthetic function for abiotic stress response gene discovery. Arabidopsis Book 2016, e0185.
Saito, K., Matsuda, F., 2010. Metabolomics for functional genomics, systems biology, and biotechnology. Annu. Rev. Plant Biol. 61, 463–489.

Siebke, K., Ball, M.C., 2009. Non-destructive measurement of chlorophyll b: a ratios and identification of photosynthetic pathways in grasses by reflectance spectroscopy. Funct. Plant Biol. 36, 857–866.
Siebke, K., Weis, E., 1995. Assimilation images of leaves of *Glechoma hederacea*: analysis of non-synchronous stomata related oscillations. Planta 196, 155–165.
Singh, B., Bohra, A., Mishra, S., Joshi, R., Pandey, S., 2015a. Embracing new-generation 'omics' tools to improve drought tolerance in cereal and food-legume crops. Biol. Plant. 59, 413–428.
Singh, B.P., Jayaswal, P.K., Singh, B., Singh, P.K., Kumar, V., Mishra, S., Singh, N., Panda, K., Singh, N.K., 2015b. Natural allelic diversity in OsDREB1F gene in the Indian wild rice germplasm led to ascertain its association with drought tolerance. Plant Cell Rep. 34, 993–1004.
Singla-Pareek, S.L., Yadav, S.K., Pareek, A., Reddy, M.K., Sopory, S.K., 2008. Enhancing salt tolerance in a crop plant by overexpression of glyoxalase II. Transgenic Res. 17, 171–180.
Sirault, X.R., James, R.A., Furbank, R.T., 2009. A new screening method for osmotic component of salinity tolerance in cereals using infrared thermography. Funct. Plant Biol. 36, 970–977.
Sozzani, R., Busch, W., Spalding, E.P., Benfey, P.N., 2014. Advanced imaging techniques for the study of plant growth and development. Trends Plant Sci. 19, 304–310.
Suzuki, N., Mittler, R., 2006. Reactive oxygen species and temperature stresses: a delicate balance between signalling and destruction. Physiol. Plant. 126, 45–51.
Suzuki, N., Koussevitzky, S., Mittler, R.O., Miller, G.A., 2012. ROS and redox signalling in the response of plants to abiotic stress. Plant Cell Environ. 35, 259–270.
Szarka, A., Tomasskovics, B., Bánhegyi, G., 2012. The ascorbate–glutathione–α-tocopherol triad in abiotic stress response. Int. J. Mol. Sci. 13, 4458–4483.
Wahid, A., Gelani, S., Ashraf, M., Foolad, M.R., 2007. Heat tolerance in plants: an overview. Environ. Exp. Bot. 61, 199–223.
Wahid, A., Farooq, M., Siddique, K.H., 2014. Implications of oxidative stress for crop growth and productivity. In: Pessarakli, M. (Ed.), Handbook of Plant and Crop Physiology. CRC Press, Boca Raton, USA, pp. 549–556.
Walter, A., Rascher, U., Osmond, B., 2004. Transitions in photosynthetic parameters of midvein and interveinal regions of leaves and their importance during leaf growth and development. Plant Biol. 6, 184–191.
Wani, S.H., Sah, S.K., Sanghera, G., Hussain, W., Singh, N.B., 2016. Genetic engineering for cold stress tolerance in crop plants. In: Rahman, A. (Ed.), Advances in Genome Science: Genes in Health and Disease, Bentham Science Publishers, UAE, pp. 173–201.
Wedeking, R., Mahlein, A.K., Steiner, U., Oerke, E.C., Goldbach, H.E., Wimmer, M.A., 2017. Osmotic adjustment of young sugar beets (*Beta vulgaris*) under progressive drought stress and subsequent rewatering assessed by metabolite analysis and infrared thermography. Funct. Plant Biol. 44, 119–133.
Weirman, A., 2010. Plant Phenomics Teacher Resource. http://www.plantphenomics.org.au/files/teacher/Final_Phenomics_for_word_with_image.doc.
Windt, C.W., Vergeldt, F.J., De Jager, P.A., Van, A.H., 2006. MRI of long distance water transport: a comparison of the phloem and xylem flow characteristics and dynamics in poplar, castor bean, tomato and tobacco. Plant Cell Environ. 29, 1715–1729.
Wungrampha, S., Joshi, R., Pareek, A., Singla-Pareek, S.L., 2018. Photosynthesis and salinity: are they mutually exclusive? Photosynthetica. https://doi.org/10.1007/s11099-017-0763-7.
Yang, W., Duan, L., Chen, G., Xiong, L., Liu, Q., 2013. Plant phenomics and high-throughput phenotyping: accelerating rice functional genomics using multidisciplinary technologies. Curr. Opin. Plant Biol. 16, 180–187.
Zandalinas, S.I., Mittler, R., Balfagón, D., Arbona, V., Gómez-Cadenas, A., 2018. Plant adaptations to the combination of drought and high temperatures. Physiol. Plant 162, 2–12. https://doi.org/10.1111/ppl.12540.
Zhang, X., Huang, C., Wu, D., Qiao, F., Li, W., Duan, L., Wang, K., Xiao, Y., Chen, G., Liu, Q., Xiong, L., 2017. High-throughput phenotyping and QTL mapping reveals the genetic architecture of maize plant growth. Plant Physiol. 173, 1554–1564.

FURTHER READING

Basu, S., Ramegowda, V., Kumar, A., Pereira, A., 2016. Plant adaptation to drought stress. F1000Res. 5, F1000 Faculty Rev.1554.

CHAPTER 14

Overview on Effects of Water Stress on Cotton Plants and Productivity

Muhammad Tehseen Azhar*,†, Abdul Rehman†
*Department of Plant Breeding and Genetics, University of Agriculture, Faisalabad, Pakistan
†School of Biological Sciences M084, The University of Western Australia, Perth, Australia

Contents

Biochemical, Physiological and Molecular Avenues for Combating Abiotic Stress in Plants
https://doi.org/10.1016/B978-0-12-813066-7.00016-4

1. DROUGHT

Drought is a major problem all over the world that affects the production of several field crops. It is a state of dry weather that causes hydrological shortage in a specific region. It have several definitions based on the potential evapotranspiration, moisture of soil, precipitation profile, or any combination of these factors (Wilhite et al., 2001; Heim, 2002; Burke et al., 2006). Drought is more damaging factor for plant growth and plant productivity than other biotic and abiotic stresses. It restricts crop productivity with major affects in arid and semi-arid regions. Agronomic, edaphic, and climatic factors affect plants in many ways. Climatic changes, specifically increasing evaporation and decreasing regional precipitation due to global warming, are laying the groundwork for severe drought in the future (Demirevska et al., 2009).

2. EFFECTS OF DROUGHT CONDITIONS ON PLANTS

Effects of drought are observed at the physiological, molecular, and morphological levels, which can appear at any phonological stage of plant growth whenever drought occurs. A brief summary on the impact of drought is elaborated in the following paragraphs.

Plant water potential is highly affected by leaf temperature, canopy temperature, rate of transpiration, stomatal resistance, stomatal conductance, relative water content, and water potential of the leaves. Water-stressed wheat plants have lower relative water content than nonstressed ones. In addition, they exhibit decrease in leaf water content, leaf water potential, and transpiration rate; and an increase in leaf temperature (Siddique et al., 2000). Under drought conditions, an uptake of nutrients minimized their concentration inside the tissues of crop plants. So transportation of nutrients from root to shoot is badly affected (Garg, 2003). Different plant genotypes belonging to different species respond differentially to the uptake of nutrients in drought stress. Translocation of assimilates to sink is the requirement for seed development (Asch et al., 2005). Drought stress minimizes the size of the source as well as the sink. It also interferes with the partitioning of dry matter, phloem transport, and assimilation translocation. However, the effect of drought varies according to the stage of plant growth and duration of drought. Reduced germination and poor stand are symptoms of drought stress. It badly affects the germination and seedling stand (Kaya et al., 2006). The most sensitive physiological process is the reduction in turgor pressure that affects cell growth. Reduced plant height, leaf area, and plant growth results due to abnormalities and hurdles in mitosis, cell expansion, and elongation are all symptoms of drought stress (Hussain et al., 2008). With mild water deficiency, plants are usually slow growing and stunted. Leaves turn from shiny to dull at the first signs of stress. Footprints in wilted grass persist instead of disappearing as grass blades spring upright. Under long-term water stress, plants might permanently wilt or stop growing; they may have diminished crops and discolored leaves, flower buds, and flowers. Plants may eventually die. Bare spots will appear in ground covers. Drought symptoms can be confusing, and can vary depending on crop species. Woody plants under drought stress can have several symptoms, including yellowing, wilting leaves that develop early fall color and burning or scorching on edges of leaves. Plants may drop some or some time all of the leaves and appear to be dead (Jeong et al., 2013).

3. DROUGHT TOLERANCE

Drought tolerance refers to the degree to which a plant is adapted to arid or drought conditions. The plants in dry environments are subjected to random droughts, and it is generally impossible for them to escape from such adverse conditions. Thus, plants in such environments have the ability to endure water stress through certain biochemical or morphological adaptations and avoidance of cell injury. The mechanism of drought tolerance involves the maintenance of turgor pressure through osmotic adjustment, increase in elasticity in cells, and decrease in cell size by protoplasmic resistance (Fig. 1) (Valliyodan and Nguyen, 2006).

4. DROUGHT-RESISTANCE MECHANISMS

The ability of plants to minimize the losses of yield under water stress conditions is known as drought resistance (Razzaq et al., 2013). Several physiological, morphological, and molecular responses are induced in plants that enable to plants to survive under drought stress. A brief account of these mechanisms is provided as follows.

4.1 Morphological Mechanisms

Several parameters of root like-length, biomass, density, and its depth involved in drought avoidance contribute to final yields under terminal drought conditions. Plants exhibit phenotypic flexibility under water stress. Roots and shoots are affected the organs at morphological level, and these contribute significantly to plant adaptation to water stress. Plants under drought stress cut down their water requirements by reducing the number and area of leaves at the cost of yield loss. (Siddique et al., 2000; Akhtar and Nazir, 2013; Brunner et al., 2015).

Escape from drought is achieved through shortening of life cycle or growing season, allowing plants to reproduce before the environment becomes dry. Flowering time is an important trait related to drought adaptation, where a short life cycle can express to drought escape (Araus et al., 2002). Plants attain escape from drought through shortening of the life cycle completion of reproductive stages before the onset of the drought period.

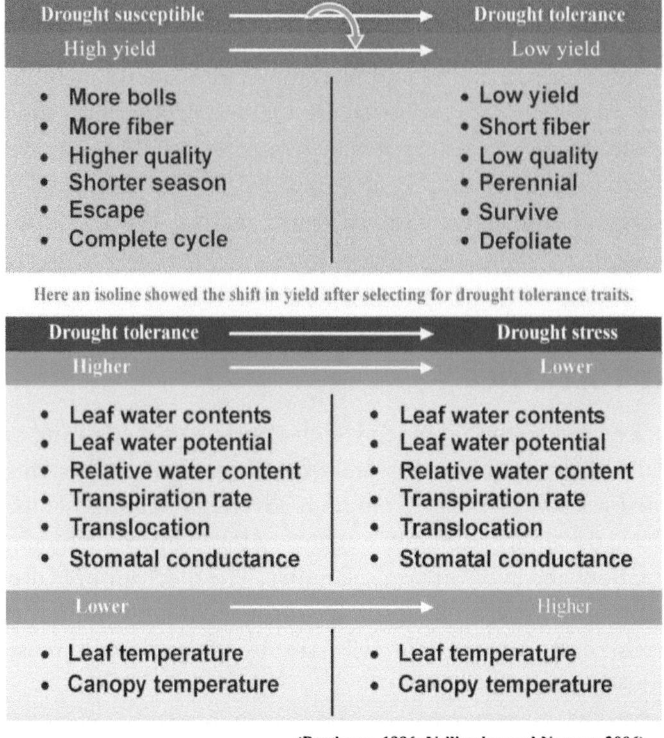

Fig. 1 Mechanism involved in drought tolerance.

Flowering time is one of the important traits associated with drought adaptation, as it shortens the life cycle of plants, leading to escape from terminal drought. Another mechanism is the acquisition of water from soil is accomplished through roots by modifications in growth, density, size, and proliferation, which are strategic responses to drought stress (Kavar et al., 2008).

4.2 Physiological Mechanisms

Osmotic adjustment, osmoprotection, antioxidation, and scavenging defense systems impart drought tolerance to the plant. Genetic variations in physiological mechanisms are ambiguous due to more complicated mechanisms suggested by researchers.

Osmotic adjustment facilitates translocation of preanthesis and carbohydrate partitioning during grain filling, and maintaining high turgor pressure, which leads to an increased rate of photosynthesis and growth (Subbarao et al., 2000; Hatzig et al., 2014). During osmatic adjustment, cells decrease their osmatic potential, and consequently increase their gradient for an influx of water, and maintenance of turgor for tissue water conservation. Water requirement can be attained by osmatic adjustment, or by making changes in elasticity of the cell wall. In response, physiological activity of the cell is maintained for longer periods of water stress (Kirkham, 2014). In drought tolerance, the involvement of antioxidant enzymes and their nonenzymatic components are of vital importance. It is believed that under water stress conditions, maintenance of stability and integrity of membranes are necessary components of drought tolerance (Bajji et al., 2002). Stability of cell membranes in leaves is the most significant character to screen germplasm for drought tolerance (Dhanda et al., 2004). The level of water in the cell is maintained through osmotic adjustment by solutes' accumulation inside cytoplasm, thereby reducing the damaging effects of drought (Blum, 2017).

5. IMPACT OF WATER STRESS ON PHYSIOLOGICAL TRAITS

5.1 Photosynthesis

Significant reduction also takes place in photosynthetic rates, CO_2 uptake, and transpiration rates in cotton. Studies of various irrigation levels in cotton for stomatal role in transpiration, the CO_2 assimilation rate, and leaf temperature showed that a significant decrease in the stomatal area takes place due to water stress. Hence, it can be concluded that higher transpiration rate helps cotton plants to adapt in water stress conditions (Inamullah and Isoda, 2005). Ullah and Zafar (2006) studied genetic variability for drought tolerance in cotton to investigate some physiological traits as selection criteria, and revealed that physiological traits were directly related with quantitative traits. Researchers have investigated the relationship between productivity with photosynthetic rates under normal and water stress conditions, and researcher have revealed that moisture stress adversely affects the photosynthetic rate. They also confirmed that relationship of productive traits with photosynthetic rate might be a valuable criterion for screening germplasm under drought conditions. Similar results were also reported by Ahmadizadeh (2013), Talebi et al. (2013), Zhang et al. (2014), and Farooq et al. (2014). It was revealed that the genotypes with high net photosynthesis have produced more yield, suggesting that net photosynthesis could be used as a selection criterion for drought resistance. More bolls were recorded in controlled conditions than the stressed conditions, indicating the negative association between number of bolls and water stress (Shamim et al., 2013). Ashraf and Harris (2013) observed that resistant genotypes showed higher net photosynthesis than susceptible ones. Chastain et al. (2014) examined cotton germplasm under drought and normal water conditions, and revealed that stomatal conductance, transpiration rate, and net photosynthesis have decreased accordingly in water stress conditions. Reduction in leaf photosynthesis takes place due to the closure of stomata when plants are grown in water-deficient conditions (Zivcak et al., 2014; Escalona et al., 2015).

5.2 Leaf Area

Rehman et al. (2017) examined parents and F_1 hybrids in three different irrigation regimes, i.e., no irrigation, normal irrigation, and irrigation on wilting. Reduction in leaf area and an increase in specific leaf weight was observed in water deficit conditions. It was revealed that genotypes having small and thick leaves were considered as drought tolerant.

5.3 Leaf-Relative Water Content

Reduction in relative water content (RWC) takes place during drought stress in cotton. Dhanda and Sethi (2002) reported differences in morphological and physiological traits (i.e., anthesis and maturity traits) in wheat, and observed that genotypes differed significantly for RWC in water stress conditions, and there were also significant interactions between genotypes × environment. Mu-XiuLing and Bao-Xiao (2003) conducted an experiment on cotton for RWC to determine the effect of different water stress levels. Soil water content was maintained in control, light drought, medium drought, and heavy drought conditions. The RWC decreased in cotton leaves under various drought levels.

Munjal and Dhanda (2005) studied wheat varieties for RWC under normal and water stress conditions, and observed that cultivars exhibited high relative water content values were regarded as drought-tolerant cultivars. Golabadi et al. (2005) conducted an experiment to evaluate F_3 wheat families for RWC under control and water-stress conditions, and a reduction in RWC was observed under water stress conditions. Wang et al. (2007) examined that increase in leaf temperature transpiration ability under drought conditions. Furthermore, RWC decreased under more severe drought conditions, while cotton has the ability to maintain high water content in leaves. Da Silva Lobato et al. (2008) examined the effects of drought on soybean (*Glycine max*) under control and water stress conditions for RWC and found that reduction in leaf RWC was recorded in plants under water deficit regimes. Likewise, Golparvar (2013) studied eight cultivars of wheat genotypes by using of Griffing's approach II in a fixed model in field experiments for drought stress to examine RWC as a screening criterion, and observed that RWC were more in drought resistant than drought-susceptible genotypes.

5.4 Excised Leaf Water Loss

Basal et al. (2005) examined lines of upland cotton for ELWL under normal and water stress regimes, and observed that ELWL might be used as a selection criterion for identification of drought-tolerant genotypes of cotton. Munjal and Dhanda (2005) examined wheat cultivars for ELWL under water-stress and irrigated conditions, and identified drought-tolerant genotypes that exhibited low ELWL estimates. Riaz et al. (2013) examined five elite advanced cotton (*Gossypium hirsutum* L.) lines including one check was carried out in plastic tubes. Three drought shocks were applied by withholding water, i.e., 8-, 10-, and 12-day intervals. Drought-tolerant genotypes had the lowest excised leaf water loss during the first 4 h, and also for the next 4 h. Furthermore, it indicated that there are genotypic differences in relative water loss (RWL), which is assumed to be an estimation of the cuticular transpiration rate (Guo et al., 2013; Shi et al., 2014; Hassan et al., 2015; Ali et al., 2015). Farooq et al. (2014) conducted an experiment on drought resistance and susceptible spring wheat genotypes in field and pot experiments under water stress, and low estimates of ELWL were observed in drought-tolerant genotypes. Saleem et al. (2015) examined ELWL in wheat cultivars grown under control and water-stress conditions, and ELWL was identified as useful drought resistance criterion. Dhanda and Sethi (2002) studied wheat varieties for ELWL at anthesis and maturity to determine differences under controlled and water-stress regimes and observed that under drought stress, genotypes behaved differentially at both stages of growth.

5.5 Leaf Water Potential

The cotton plant has the ability to adjust high leaf turgor potential through osmotic adjustments (Iqbal et al., 2013). Klein (2014) and Maréchaux et al. (2015) studied leaf water potential across species at early reproductive stages, and reductions in leaf water potential (LWP) were observed due to drought. Under water deficit conditions, LWP is reduced in many plant species.

6. IMPACT OF WATER STRESS ON QUANTITATIVE TRAITS

6.1 Plant Height

Plant height is measured from the ground level to its top. Ali et al. (2005) evaluated various varieties of *Oryza sativa* L. under three water-stress conditions for plant height and concluded that plant height has direct relationship with number of irrigations at critical stages. Furthermore, it was determined that 12 irrigations were essential to obtain equitable plant height. Siddiqui et al. (2007) examined three cotton cultivars under three irrigation conditions, i.e., 3, 5, and 7 irrigations, respectively. In seven irrigations, average estimates were highest for plant height (105.6 cm). Genotypes irrigated for five times give better performance as compared with three or seven irrigations.

6.2 Yield and Its Components

The effect of water stress on yield of seed cotton and its components has been observed by various researchers. Pettigrew (2004a,b) examined cotton genotypes under drought stress and normal water conditions to study the effects of drought on lint yield, and observed 25% reduction in lint yield due to water stress. Kar et al. (2005) analyzed the response of cotton hybrid to drought stress under field conditions. It was found that at flowering stage, estimates of yield-contributing traits were reduced under drought stress. In cotton bud and flowering stages are more susceptible to drought than vegetative stage. Siddiqui et al. (2007) studied an experiment to examine the effect of three irrigation levels (three, five, and seven) on three cultivars. It was found that the genotype that was irrigated five times had a more pronounced yield as compared to three or seven irrigations. Likewise, the impact of drought on yield has been reported in other field crops. In barley, Samarah (2005) conducted an experiment in glass house to assess the effect of drought on growth as well as on grain yield. Three water-stress treatments were subjected to plants at the start of grain filling: (1) 100% of field capacity, (2) 60% of field capacity, and (3) 20% of field capacity until grain maturity. It was revealed from results that drought reduced the grain yield due to smaller number of tillers, spikes, and number of grains per plant and 100-grain weight. Hence, it was revealed that water deficit prior to anthesis was damaging to grain yield. Snowden et al. (2013) examined cotton cultivars for 2 years under normal and water-stress regimes to study the effect of water stress on number of bolls. They observed that genotypic variability was present in cotton genotypes for the number of bolls.

6.3 Fiber Quality Traits

Various researchers have studied the effects of water stress on fiber quality traits of cotton. Pettigrew (2004a,b) evaluated cotton genotypes to examine the fiber length under normal and water-stress regimes, and revealed that irrigation directly supports the vegetative growth and delayed maturity, but on other hand, it also produced about 2% longer fiber as compared with water-stressed plants. Mert (2005) investigated genotypes of cotton to assess the effects of water stress for fiber length, fiber fineness, and fiber strength under normal and water stress conditions, and it was observed that water stress regimes were responsible for the production of shorter and weaker fibers with low micronair value. Lokhande and Reddy (2014) studied the effect of water stress during various stages of cotton development and found that fiber fineness was adversely affected by drought stress during the boll formation stage.

6.4 Seedling-Related Traits

Similar to mature plants, germination of seeds and growth of seedlings are also subjected to environmental stresses. The results of experiments have shown that there was also poor seedling establishment under water-stress regimes. Burke and O'mahony (2001) showed that drought negatively affected the shoot of cotton cultivars more than its root. All measures of shoot growth, including height, nodes, leaf area, and dry weights of stem and leaves were smaller in drought-treated plants as compared to controlled conditions. Likewise root growth was less affected in drought-treated plants as compared to controls. Riaz et al. (2013) studied the response of roots and shoots to water stress and identified that increase in root length was less sensitive to water stress than leaves. Reduction in root elongation and root volume was also reported. Medium-sized roots are less affected as compared to small roots. It was concluded that the medium roots are more important for growth in water-stress conditions. Noorka et al. (2013) also found that drought effects the shoot growth prior to root growth in young cotton plants. Similarly Ali et al. (2014) found that roots and shoot growth have direct relationship with soil water availability, and the growth in shoots was more affected than roots.

6.5 Vegetative Stage

Adverse effects of drought stress were exerted more on nodes than on the main stem. Under water-stress conditions, shorter plants were produced while normal water conditions produced taller plants due to more nodes on main stem. In drought stress, shorter plants lead to the production of smaller leaf-area index and consequently vegetative growth was decreased which resulted in less solar radiation interception than canopies of taller plants under normal water conditions (Pettigrew, 2004b). Bunnag and Pongthai (2013) reported that early and late irrigations reduced the cotton yield. However, water stress during vegetative growth led to less leaf water potential that adversely affected the final yield (Alçidu et al., 2013). Mohamed et al. (2015) found that plants suffered with sequential water deficit during the vegetative period showed higher tolerance of water deficiency.

6.6 Reproductive Stage

Kar et al. (2001) examined cotton hybrids under drought-stress conditions to analyze their tolerance to drought through the estimation of numbers of bolls per plant and seed cotton yield. In each case, three stress cycles of water stress were given until the plant wilted, or later on irrigated before the harvest. This study indicated that number of bolls per plant and seed cotton yield decreased significantly in all cultivars in response to drought stress. The water stress at flowering stage in cotton was found to be critical as compared to vegetative and reproductive stage. Similar results were reported by Cairns et al. (2013), De Faria Müller et al. (2014), and Ashraf et al. (2013). Pettigrew (2004a,b) indicated the rate of crop growth was reduced due to less water availability during photosynthesis. He also indicated that the period from square initiation to first flower was critical period in terms of water supply. At this time, water deficit led to decrease in yield due to the reduction in number of bolls.

Drought stress prior to flowering reduced seed cotton yield significantly. Water stress also affected the hormonal balance in squares and bolls that cause shedding (Snowden et al., 2014). The peak flowering period was the most sensitive to drought, and at this time water stress led to decrease in yield. Yi et al. (2014) investigated the effects of drought stress during various stages of cotton growth. Stress at the beginning of flowering adversely affected the number of flowers per plant, number of bolls per plant, boll weight, seed index, fiber yield and fiber length.

6.7 Gas Exchange Parameters

Chaves et al. (2003) conducted an experiment and revealed that alternations in carbon dioxide exchange into mesophyll tissue badly affects the rate of photosynthesis. Changes in stomatal conductance were assumed as the main factor for decreased rate of photosynthesis. Farooq et al. (2013) examined cotton genotypes under water stress conditions and found that leaf photosynthesis had been severely affected by changes in stomatal conductance. Drought tolerant genotypes could minimize transpirational losses by closing their stomata which led to decrease in rate of photosynthesis (Fig. 2).

7. GENETIC VARIABILITY FOR DROUGHT TOLERANCE IN COTTON AND OTHER FIELD CROPS

The genetic basis of drought tolerance in cotton as well as in other field crops has been reported by many researchers. Additive and nonadditive types of gene actions were involved in the inheritance of various plant characters, and strategies for improvement have been suggested. Iqbal et al. (2011) investigated upland cotton varieties at the seedling stage under two irrigation conditions, i.e., controlled and water-stress in glass house conditions. Genotypic differences for indices of drought tolerance were statistically significant. Genotypes of wheat differed significantly for days to heading, RWC, ELWL, membrane stability, grain yield, and genotype × environmental interactions under drought stress. It was observed that water-deficit conditions resulted in decreased mean

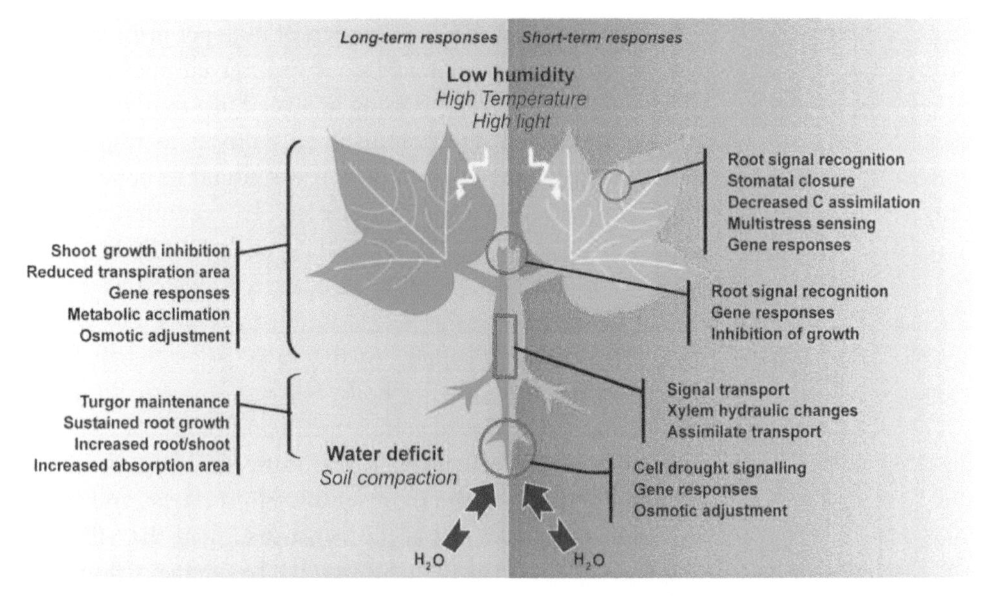

Fig. 2 Response of cotton plant toward long-term and short-term water stress. *(Modified and reprinted with permission, https://www.sciencedirect.com/science/article/pii/0002157174900144).*

values. Reddy and Ratna Kumari (2004) investigated varieties of *G. hirsutum* L. for the assessment of genetic variability for drought tolerance parameters in addition to yield and yield-related components. Considerable genetic variations were observed from drought tolerance related parameters namely, leaf weight and leaf area. The phenotypic and genetic estimates of variation were high for leaf weight and leaf area, indicating that these plant traits were barely affected by environmental factors. Ullah and Zafar (2006) analyzed for genetic variability in physiological and yield related parameters for drought tolerance in American cotton under normal and water-stressed conditions. It was revealed that seed cotton yield was adversely affected due to water deficit conditions in all of genotypes studied.

Iqbal et al. (2011) hybridized drought-tolerant and susceptible genotypes. At the seedling stage, performances of families were evaluated under controlled and water-stressed conditions. The results indicated that variation for drought tolerance was governed by additive and nonadditive gene action, but the effects of additive gene action were more noticeable. Cotton genotypes were assessed for dry matter accumulation to evaluate their water use efficiency under normal and drought conditions. Variability existed among genotypes for shoot growth, root growth and dry matter accumulation. So, it was indicated that root growth and root morphology were considered more important for the adoption of drought tolerant genotypes of cotton (Comas et al., 2013). Pace et al. (1999) studied the performance of root and shoot growth of upland cotton under drought periods after 36 days of planting. Analysis revealed that reduction in various parameters takes place in water-stressed conditions as compared to controlled. While an increase in root length takes place in water-deficit conditions more than in controlled conditions. It is also observed that length of tap root is increased with increase in water stress. It was concluded that length of tap root of cotton may not be affected in water-deficit conditions.

8. GENETIC BASIS OF YIELD AND RELATED TRAITS

Determination and understanding of genetic basis of yield and related traits is a prerequisite before starting breeding program. The genetic basis of various agronomic traits under drought conditions is summarized in the following paragraphs.

8.1 Yield of Seed Cotton and Other Field Crops

There is a direct but negative relationship exists between yield of seed cotton, grain and fodder yield under environmental stresses. This is because genes controlling yield are influenced by environmental factors due to their multigenic in nature. Ashour et al. (2006) investigated the genetic basis of grain yield in five genotypes of wheat, along with their F_1, F_2, BC_1, and BC_2 population, and observed that grain yield was influenced by additive gene action which suggested selection at early stages for this trait will be more effective. Munir et al. (2007) evaluated two wheat crosses along with their parents and back crosses to evaluate the genetic basis for grain yield per plant under water-stress conditions, and indicated that three types of genetic effects were involved: additive, dominant, and epistatic. Hence, it was recommended that selection in later generations could help the breeder for the development of water-stress tolerance with high yielding genotypes. Abbas et al. (2008) conducted an experiment to assess the genetic effects of seed cotton yield of upland cotton genotypes to analyze the gene action of seed cotton yield. For seed cotton yield, additive gene action was observed, along with partial dominance.

Cotton genotypes were screened for water stress tolerance on the basis of shoot and root length. Five drought tolerant lines and three drought susceptible testers were selected and later on hybridized in line × tester mating fashion for the development of F_1 population. Parents along with F_1 population were grown in field conditions, and 50% water stress was applied to analyze the response of the seed cotton yield. Non-additive genetic components were observed for this trait indicating the use of F_1 population could be used for the development of hybrids for drought affected areas (Javaid et al., 2014).

8.2 Plant Height

Mukhtar et al. (2000a,b) studied four cotton genotypes and 12 F_1 hybrids to determine the genetic effects controlling plant height, and it was revealed that this traits is governed by additive gene action. The plant height of wheat was governed by partial dominance with additive gene action under controlled and drought conditions. Here, selection in early segregating generations would be effective (Subhani and Chowdhry 2000; Ahmad et al., 2009). Ahuja et al. (2004) examined 51 colored cotton genotypes and found additive genetic effects for plant height. Patra et al. (2006) evaluated 20 genotypes of rice to estimate gene action, and found additive gene action was involved in the inheritance of plant height. Likewise, Rahman et al. (2000) investigated 21 genotypes of barley and their 15 F_1 hybrids to determine genetic effects for plant height, which revealed the presence of dominant types of gene action for the trait studied. Javaid et al. (2014) investigated 50 cotton genotypes for plant height to assess gene action, and nonadditive genetic components were observed in controlling the traits.

8.3 Number of Monopodial Branches

Monopodials are not important and nonsignificant for yield of seed cotton. Abro (2003) indicated that number of monopodial branches was under the control of partial dominance. Abbas et al. (2008) examined five varieties of upland cotton to estimate gene action for monopodial branches per plant which was influenced by additive type of gene action accompanying partial dominance. This trait is genetically controlled and greatly influenced by environment. Kumar et al. (2014) analyzed cotton genotypes to evaluate gene action for number of monopodial branches per plant, and indicated additive and dominance types of gene action.

8.4 Number of Sympodial Branches

These are the fruit-bearing branches, also called reproductive branches. They contribute significantly to yield of seed cotton. Iqbal and Nadeem (2003) and Erande et al. (2014) reported the involvement of additive gene action for number of sympodial branches. Sarwar et al. (2011) studied indigenous experiment with three exotic lines and one indigenous line for estimation of gene action for number of sympodial branches per plant. The additive type of gene action with partial dominance was found to be responsible in the inheritance of this trait. Saleem et al. (2015) analyzed eight cotton genotypes to study the genetic effects for number of sympodial branches per plant, and observed the presence of additives along with dominant type of gene action.

8.5 Number of Bolls per Plant

This quantitative trait is one of the important parameters involved in the increase or decrease of seed cotton yield, but it is significantly affected by biotic and abiotic stress. Ahmad et al. (2001) analyzed the pattern of inheritance of number of bolls per plant, and found that additive genetic effects with partial dominance were involved in the inheritance for this trait. Murtaza (2006) studied an experiment including eight upland cotton plants and their F_1 and F_2 populations to determine the genetic effects for the number of bolls per plant, and estimated that inheritance of number of bolls per plant was additively controlled. Ahmad et al. (2009) examined six cotton genotypes of cotton and their nine F_1 hybrids for number of bolls per plant to evaluate genetic effects, and found the presence of overdominance, which suggests that delayed selection might be fruitful.

8.6 Boll Weight per Plant

Mukhtar et al. (2000a) investigated genotypes of cotton along with their 12 F_1 hybrids to determine the genetic effects for boll weight, and observed the additive type of gene action along with partial dominance. Saravanan et al. (2003) examined seven American cotton genotypes and studied the genetic effects under controlled and drought stress conditions for boll weight, and revealed the predominance of dominant component of variation for boll weight. Murtaza (2006) conducted an experiment including eight cotton genotypes and their F_1 and F_2 generations to analyze genetic effects for boll weight, and found that boll weight was influenced by nonadditive gene action.

8.7 Ginning Out Turn

The pattern of inheritance of ginning out turn has been studied by many research workers. Nimbalkar et al. (2004) examined desi cotton (*G. arboreum* L.) and observed that ginning out turn was under the control of additive as well as nonadditive gene action. Likewise, Singh and Chahal (2005) conducted an experiment to study the genetic effects in 34 cotton cultivars for ginning out turn in field conditions, and reported additive and dominant gene action for this trait. Esmail (2007) studied six generations of two cotton hybrids to determine the genetics for GOT, and indicated the presence of additive, dominance, and epistatic effects. Khan et al. (2009) assessed the additive genetic effects in *G. hirsutum* L. for GOT. Ali et al. (2008) conducted an experiment to study the genetics of GOT in cotton genotypes using Mather and Jinks approach, and revealed the presence of additive types of gene action.

8.8 Effects of Drought on Fiber Quality Traits

Azhar et al. (2004) evaluated five cotton varieties to assess the involvement of gene action for fiber traits, and indicated the presence of additive and nonadditive gene action. Nimbalkar et al. (2004) analyzed the data collected from *G. arboreum* and *G. herbaceum* and observed that fiber length was under the control of additive gene effects. Ahmed et al. (2006) assessed six upland cotton cultivars along with nine F_1 hybrids for fiber length, and found that this parameter was controlled by over dominant gene action, which suggested the selection in early populations. Ali et al. (2008) reported the gene action governing the fiber related quality traits in *G. hirsutum* L.

8.9 Fiber Strength

Genetic effects for fiber strength were reported by many researchers. Singh and Chahal (2005) conducted an experiment to study the genetic effects for fiber strength in 34 varieties of cotton grown in field conditions, and observed additive and dominant gene action along with epistasis for the character studied. Minhas et al. (2008) hybridized five cotton cultivars in all possible combinations to estimate the genetic effects for fiber strength. Results revealed additive genetic effects for fiber strength. Batool et al. (2013) examined cotton hybrids to assess gene action for fiber strength, and revealed the presence of nonadditive gene action. Ahmad et al. (1997) conducted an experiment on cotton to estimate gene action for yield related traits and fiber quality characters, and reported epistatic effects for fiber strength.

8.10 Fiber Fineness

Mukhtar et al. (2000b) investigated four cotton genotypes and 12 F_1 crosses to study gene action for fiber fineness, and highlighted the presence of additive genetic effects as well as partial dominance. Bertini et al. (2001) assessed the gene action for fiber fineness in cotton and revealed that additive variances existed for the trait being studied. Singh and Chahal (2005) conducted an experiment to study the genetic effects for fiber fineness in 34 cotton cultivars grown in field conditions, and found the role of additive and dominant gene action along with epistatic effects in the inheritance of fiber fineness. Minhas et al. (2008) hybridized five cotton cultivars in all possible combinations to estimate the genetic effects for fiber fineness. It was revealed that additive genetic effects along with partial dominance are involved in the governing of fiber strength. Akhtar et al. (2008) examined eight upland cotton cultivars to analyze the gene action involved in the inheritance of fiber fineness and found presence of simple additive effects.

8.11 Excised Leaf Water Loss

Ahmed et al. (2000) evaluated three generations (i.e., parental, F_2, and backcross) to determine the genetic effects for ELWL under drought stress. They observed the significant effects of additive, dominance, and additive × dominance gene action for the traits being studied. Majeed et al. (2001) examined two generations (i.e., parental and backcross) of barley to evaluate the genetics of ELWL, and indicated that ELWL was under the control of dominance and epistatic effects. Kumar and Sharma (2007) examined 12 wheat generations to assess the genetic effects for ELWL, and revealed the presence of additive, dominance, and epistatic effects.

8.12 Relative Water Content

Ahmed et al. (2000) examined parental and backcross populations to estimate the gene action for RWC, and revealed the presence of additive, dominance, and additive × dominance effects for this trait. Majeed et al. (2001) conducted an experiment on parents and backcross populations of barley and revealed the existence of additive genetic components for the trait studied. Likewise Kumar and Sharma (2007) estimated the nature of gene action for RWC parental and backcross generations of bread wheat, and indicated that additive, dominance, and epistatic genetic effects controlled the inheritance of RWC.

A wide range of symptoms of drought stress are seen like wilting and loss of reproductive structures but closer look shows increase in leaf temperature, less transpiration and translocation and closed stomata. Drought tolerance would involve cultivation and genetic factors to escape or tolerate the effects of drought. Escape may result due to alteration in plant morphology like earliness, less vegetative growth and quicker termination of sympodial growth whereas tolerance could be developed due to induction of deeper roots. Breeding program for cotton improvement against drought tolerance is time consuming and laborious which may take ten years to get a promising lines.

REFERENCES

Abbas, A., Ali, M., Khan, T., 2008. Studies on gene effects of seed cotton yield and its attributes in five American cotton cultivars. J. Agric. Soc. Sci. 4, 147–152.

Abro, S., 2003. Study of Gene Action for Quantitative and Qualitative Traits in Upland Cotton (*Gossypium hirsutum* L.) (M.Sc. Thesis). Submitted through the Dept. Plant Breed. & Genet. Sind Agri. Univ., Tandojam.

Ahmed, H., Malik, T., Choudhary, M., 2000. Genetic analysis of some physio-morphic traits in wheat under drought. J. Anim. Plant Sci. 10, 5–7.

Ahmad, I., Ali, A., Zubair, M., Khan, I.A., 2001. Mode of gene action controlling seed cotton yield and various components in *Gossypium hirsutum* L. Pak. J. Agric. Sci. 38, 19–21.

Ahmed, H.M., Kandhro, M.M., Laghari, S., Abro, S., 2006. Heritability and genetic advance as selection indicators for improvement in cotton (*Gossypium hirsutum* L.). J. Biol. Sci. 6, 96–99.

Ahmad, R.T., Khan, I.A., Zubair, M., 1997. Diallel analysis for seed-cotton yield and its contributing traits in upland cotton (*Gossypium hirsutum*). Ind. J. Agri. Sci. 67, 583–585.

Ahmadizadeh, M., 2013. Physiological and agro-morphological response to drought stress. Middle-East J. Sci. Res. 13, 998–1009.

Ahmad, R.T., Malik, T.A., Khan, I.A., Jaskani, M.J., 2009. Genetic analysis of some morpho-physiological traits related to drought stress in cotton (*Gossypium hirsutum*). Int. J. Agric. Biol. 11, 235–240.

Ahuja, S.L., Monga, D., Tuteja, O.P., Verma, S.K., Dhayal, L.S., Dutt, Y., 2004. Association and path analysis in the selections made from colour linted *Gossypium hirsutum* L. cotton germplasm. J. Cotton Res. Dev. 18, 137–140.

Akhtar, M., Azhar, F., Ali, Z., 2008. Genetic basis of fiber quality attributes in upland cotton (*Gossypium hirsutum* L.) germplasm. Int. J. Agric. Biol. 10, 217–220.

Akhtar, I., Nazir, N., 2013. Effect of waterlogging and drought stress in plants. Int. J. Water Res. Environ. Sci. 2, 34–40.

Ali, Y., Sarwar, G., Aslam, Z., Hussain, F., Rafique, T., 2005. Evaluation of advanced rice germplasm under water stress environment. Int. J. Environ. Sci. Technol. 2, 27–33.

Ali, M.A., Khan, I.A., Awan, S.I., Ali, S., Niaz, S., 2008. Genetics of fiber quality traits in cotton (*Gossypium hirsutum* L.). Aust. J. Crop. Sci. 2, 10–17.

Ali, Q., Ali, A., Ahsan, M., Nasir, I.A., Abbas, H.G., Ashraf, M.A., 2014. Line × tester analysis for morpho-physiological traits of *Zea mays* L. seedlings. Adv. Life Sci. 1, 242–253.

Ali, M., Zulkiffal, M., Anwar, J., Hussain, M., Farooq, J., Khan, S., 2015. Morpho-physiological diversity in advanced lines of bread wheat under drought conditions at post-anthesis stage. J. Anim. Plant Sci. 25, 431–441.

Alçidu, M., Atokple, I.D.K., Akromah, R., 2013. Genetic analysis of vegetative-stage drought tolerance in cowpea. Greener J. Agric. Sci. 3, 481–496.

Araus, J.L., Slafer, G.A., Reynolds, M.P., Royo, C., 2002. Plant breeding and drought in C3 cereals, what should we breed for? Ann. Bot. 89, 925–940.

Ashraf, M., Harris, P., 2013. Photosynthesis under stressful environments: an overview. Photosynthetica 51, 163–190.

Ashraf, M., Shahbaz, M., Ali, Q., 2013. Drought-induced modulation in growth and mineral nutrients in canola (*Brassica napus* L.). Pak. J. Bot. 45, 93–98.

Asch, F., Dingkuhn, M., Sow, A., Audebert, A., 2005. Drought-induced changes in rooting patterns and assimilate partitioning between root and shoot in upland rice. Field Crop Res. 93, 223–236.

Ashour, B., Arzani, A., Rezaei, A., Maibody, S.M., 2006. Study of inheritance of yield and related traits in five crosses of bread wheat (*Triticum aestivum* L.). JWSS 9, 123–136.

Azhar, F., Naveed, M., Ali, A., 2004. Correlation analysis of seed cotton yield with fiber characteristics in *Gossypium hirsutum* L. Int. J. Agric. Biol. 6, 656–658.

Bajji, M., Kinet, J.M., Lutts, S., 2002. The use of the electrolyte leakage method for assessing cell membrane stability as a water stress tolerance test in durum wheat. Plant Growth Regul. 36, 61–70.

Basal, H., Smith, C., Thaxton, P., Hemphill, J., 2005. Seedling drought tolerance in upland cotton. Crop Sci. 45, 766–771.

Batool, S., Khan, N.U., Gul, S., Baloch, M.J., Turi, N.A., Taran, S.A., Saeed, M., 2013. Genetic analysis for yield and yield contributing variables in upland cotton. J. Food Agric. Environ. 11, 624–630.

Bertini, C.H.C.D.M., Silva, F.P.D., Nunes, R.D.P., Santos, J.H.R.D., 2001. Gene action, heterosis and inbreeding depression of yield characters in mutant lines of upland cotton. Pesq. Agrop. Brasileira 36, 941–948.

Blum, A., 2017. Osmotic adjustment is a prime drought stress adaptive engine in support of plant production. Plant Cell Environ. 40, 4–10.

Brunner, I., Herzog, C., Dawes, M.A., Arend, M., Sperisen, C., 2015. How tree roots respond to drought. Front. Plant Sci. 6, 547–557.

Bunnag, S., Pongthai, P., 2013. Selection of rice (*Oryza sativa* L.) cultivars tolerant to drought stress at the vegetative stage under field conditions. Am. J. Plant Sci. 4, 1701–1708.

Burke, E.J., Brown, S.J., Christidis, N., 2006. Modeling the recent evolution of global drought and projections for the twenty-first century with the Hadley Centre climate model. J. Hydrometeorol. 7 (5), 1113–1125.

Burke, J.J., O'mahony, P.J., 2001. Protective role in acquired thermotolerance of developmentally regulated heat shock proteins in cotton seeds. J. Cotton Sci. 5, 174–183.

Cairns, J.E., Crossa, J., Zaidi, P., Grudloyma, P., Sanchez, C., Araus, J.L., Thaitad, S., Makumbi, D., Magorokosho, C., Bänziger, M., 2013. Identification of drought, heat, and combined drought and heat tolerant donors in maize. Crop Sci. 53, 1335–1346.

Chastain, D.R., Snider, J.L., Collins, G.D., Perry, C.D., Whitaker, J., Byrd, S.A., 2014. Water deficit in field-grown *Gossypium hirsutum* primarily limits net photosynthesis by decreasing stomatal conductance, increasing photorespiration, and increasing the ratio of dark respiration to gross photosynthesis. J. Plant Physiol. 171, 1576–1585.

Chaves, M.M., Maroco, J.P., Pereira, J.S., 2003. Understanding plant responses to drought from genes to the whole plant. Funct. Plant Biol. 30, 239–264.

Comas, L.H., Becker, S.R., Von Mark, V.C., Byrne, P.F., Dierig, D.A., 2013. Root traits contributing to plant productivity under drought. Front. Plant Sci. 4, 442–450.

Da Silva Lobato, A.K., De Oliveira Neto, C.F., Dos Santos Filho, B.G., Da Costa, R., Cruz, F.J.R., Neves, H., Dos Santos Lopes, M.J., 2008. Physiological and biochemical behavior in soybean (*Glycine max* cv. Sambaiba) plants under water deficit. Aust. J. Crop. Sci. 2, 25–32.

De Faria Müller, B.S., Sakamoto, T., Silveira, R.D.D., Zambussi-Carvalho, P.F., Pereira, M., Pappas, G.J., Do Carmo Costa, M.M., Guimarães, C.M., Pereira, W.J., Brondani, C., 2014. Differentially expressed genes during flowering and grain filling in common bean (Phaseolus vulgaris) grown under drought stress conditions. Plant Mol. Biol. Report. 32, 438–451.

Demirevska, K., Zasheva, D., Dimitrov, R., Simova-Stoilova, L., Stamenova, M., Feller, U., 2009. Drought stress effects on Rubisco in wheat, changes in the Rubisco large subunit. Acta Physiol. Plant. 31, 1129.

Dhanda, S., Sethi, G., 2002. Tolerance to drought stress among selected Indian wheat cultivars. J. Agric. Sci. 139, 319–326.

Dhanda, S., Sethi, G., Behl, R., 2004. Indices of drought tolerance in wheat genotypes at early stages of plant growth. J. Agron. Crop Sci. 190, 6–12.

Erande, C., Kalpande, H., Deosarkar, D., Chavan, S., Patil, V., Deshmukh, J., Chinchane, V., Kumar, A., Dey, U., Puttawar, M., 2014. Genetic variability, correlation and path analysis among different traits in desi cotton (*Gossypium arboreum* L.). Afr. J. Agric. Res. 9, 2278–2286.

Escalona, J., Bota, J., Medrano, H., 2015. Distribution of leaf photosynthesis and transpiration within grapevine canopies under different drought conditions. VITIS 42, 57–64.

Esmail, R., 2007. Genetic analysis of yield and its contributing traits in two intra-specific cotton crosses. J. Appl. Sci. Res. 3, 2075–2080.

Farooq, M.A., Ali, S., Hameed, A., Ishaque, W., Mahmood, K., Iqbal, Z., 2013. Alleviation of cadmium toxicity by silicon is related to elevated photosynthesis, antioxidant enzymes; suppressed cadmium uptake and oxidative stress in cotton. Ecotoxicol. Environ. Saf. 96, 242–249.

Farooq, M., Hussain, M., Siddique, K.H., 2014. Drought stress in wheat during flowering and grain-filling periods. Crit. Rev. Plant Sci. 33, 331–349.

Garg, B., 2003. Nutrient uptake and management under drought, nutrient-moisture interaction. Curr. Agric. 27, 1–8.

Golabadi, M., Arzani, A., Maibody, S., 2005. Evaluation of variation among durum wheat F3 families for grain yield and its components under normal and water-stress field conditions. Czech J. Genet. Plant Breed. 41, 263–267.

Golparvar, A.R., 2013. Genetic control and combining ability of flag leaf area and relative water content traits of bread wheat cultivars under drought stress condition. Genetika 45, 351–360.

Guo, Y.M., Chen, S., Nelson, M.N., Cowling, W., Turner, N.C., 2013. Delayed water loss and temperature rise in floral buds compared with leaves of Brassica rapa subjected to a transient water stress during reproductive development. Funct. Plant Biol. 40, 690–699.

Hassan, H.M., Azhar, F.M., Khan, A.A., Basra, S., Hussain, M., 2015. Characterization of cotton (*Gossypium hirsutum*) germplasm for drought tolerance using seedling traits and molecular markers. Int. J. Agric. Biol. 17, 1213–1218.

Hatzig, S., Zaharia, L.I., Abrams, S., Hohmann, M., Legoahec, L., Bouchereau, A., Nesi, N., Snowdon, R.J., 2014. Early osmotic adjustment responses in drought-resistant and drought-sensitive oilseed rape. J. Integr. Plant Biol. 56, 797–809.

Heim Jr., R.R., 2002. A review of twentieth-century drought indices used in the United States. Bull. Am. Meteorol. Soc. 83, 1149.

Hussain, M., Malik, M., Farooq, M., Ashraf, M., Cheema, M., 2008. Improving drought tolerance by exogenous application of glycinebetaine and salicylic acid in sunflower. J. Agron. Crop Sci. 194, 193–199.

Inamullah, I., Isoda, A., 2005. Adaptive responses of soybean and cotton to water stress, transpiration changes in relation to stomatal area and stomatal conductance. Plant Prot. Sci. 8, 16–26.

Iqbal, K., Azhar, F.M., Khan, I.A., 2011. Variability for drought tolerance in cotton (*Gossypium hirsutum*) and its genetic basis. Int. J. Agric. Biol. 13, 61–66.

Iqbal, M.Z., Nadeem, M.A., 2003. Generation mean analysis for seed cotton yield and number of sympodial branches per plant in cotton (*Gossypium hirsutum* L.). Asian J. Plant Sci. 2, 395–399.

Iqbal, M., Khan, M.A., Naeem, M., Aziz, U., Afzal, J., Latif, M., 2013. Inducing drought tolerance in upland cotton (*Gossypium hirsutum* L.), accomplishments and future prospects. World Appl. Sci. J. 21, 1062–1069.

Javaid, A., Azhar, F.M., Khan, I.A., Rana, S.A., 2014. Genetic basis of some yield components in *Gossypium hirsutum* L. Pak. J. Agric. Sci. 51, 143–146.

Jeong, J.S., Kim, Y.S., Redillas, M.C., Jang, G., Jung, H., Bang, S.W., Choi, Y.D., Ha, S.H., Reuzeau, C., Kim, J.K., 2013. OsNAC5 overexpression enlarges root diameter in rice plants leading to enhanced drought tolerance and increased grain yield in the field. Plant Biotechnol. J. 11, 101–114.

Kar, M., Patro, B., Sahoo, C., Patel, S., 2001. Response of hybrid cotton to moisture stress. Indian J. Plant Physiol. 6, 427–430.

Kar, M., Patro, B., Sahoo, C., Hota, B., 2005. Traits related to drought resistance in cotton hybrids. Indian J. Plant Physiol. 10, 377–380.

Kavar, T., Maras, M., Kidrič, M., Šuštar-Vozlič, J., Meglič, V., 2008. Identification of genes involved in the response of leaves of *Phaseolus vulgaris* to drought stress. Mol. Breed. 21, 159–172.

Kaya, M.D., Okçu, G., Atak, M., Cıkılı, Y., Kolsarıcı, Ö., 2006. Seed treatments to overcome salt and drought stress during germination in sunflower (*Helianthus annuus* L.). Eur. J. Agron. 24, 291–295.

Khan, N.U., Hassan, G., Marwat, K.B., Farhatullah, K.M., Parveen, A., Aiman, U., Khan, M., Soomro, Z., 2009. Diallel analysis of some quantitative traits in *Gossypium hirsutum* L. Pak. J. Bot. 41, 3009–3022.

Kirkham, M.B., 2014. Principles of Soil and Plant Water Relations. Elsevier Science Publishing Co. Inc., USA.

Klein, T., 2014. The variability of stomatal sensitivity to leaf water potential across tree species indicates a continuum between isohydric and anisohydric behaviours. Funct. Ecol. 28, 1313–1320.

Kumar, K.S., Ashokkumar, K., Ravikesavan, R., 2014. Genetic effects of combining ability studies for yield and fibre quality traits in diallel crosses of upland cotton (*Gossypium hirsutum* L.). Afr. J. Biotechnol. 13, 119.

Kumar, A., Sharma, S., 2007. Genetics of excised-leaf water loss and relative water content in bread wheat (*Triticum aestivum* L.). Cereal Res. Commun. 35, 43–52.

Lokhande, S., Reddy, K.R., 2014. Reproductive and fiber quality responses of upland cotton to moisture deficiency. Agron. J. 106, 1060–1069.

Majeed, A., Malik, T.A., Khan, A.S., 2001. Genetic basis of physio-morphic traits related to drought tolerance in barley. J. Agric. Plant Nutr. 11, 167–170.

Maréchaux, I., Bartlett, M.K., Sack, L., Baraloto, C., Engel, J., Joetzjer, E., Chave, J., 2015. Drought tolerance as predicted by leaf water potential at turgor loss point varies strongly across species within an Amazonian forest. Funct. Ecol. 29, 1268–1277.

Mert, M., 2005. Irrigation of cotton cultivars improves seed cotton yield, yield components and fiber properties in the Hatay region, Turkey. Acta Agric. Scand. Sect. B Soil Plant Sci. 55, 44–50.

Minhas, R., Khan, I.A., Anjam, M.S., Ali, K., 2008. Genetics of some fiber quality traits among intraspecific crosses of American cotton (*Gossypium hirsutum*). Int. J. Agric. Biol. 10, 196–200.

Mohamed, B.B., Sarwar, M.B., Hassan, S., Rashid, B., Aftab, B., Husnain, T., 2015. Tolerance of Roselle (*Hibiscus sabdariffa* L.) genotypes to drought stress at vegetative stage. Adv. Life Sci. 2, 74–82.

Mukhtar, M., Khan, T., Khan, A., 2000a. Genetic analysis of yield and yield components in various crosses of cotton (*Gossypium hirsutum* L.). Int. J. Agric. Biol. 2, 258–260.

Mukhtar, M.S., Khan, T.M., Khan, A.S., 2000b. Gene action study in some fiber traits in cotton (*Gossypium hirsutum* L.). Pak. J. Biol. Sci. 3, 1609–1611.

Munir, M., Chowdhry, M., Ahsan, M., 2007. Generation means studies in bread wheat under drought condition. Int. J. Agric. Biol. 9, 282–286.

Munjal, R., Dhanda, S., 2005. Physiological evaluation of wheat (*Triticum aestivum* L) genotypes for drought resistance. Indian J. Genet. Plant Breed. 65, 307–308.

Murtaza, N., 2006. Study of gene effects for boll number, boll weight, and seed index in cotton. J. Cent. Eur. Agric. 6, 255–262.

Mu-XiuLing, Bao-Xiao, 2003. Effect of soil water stress on water regime in cotton leaves and on photosynthesis. China Cotton 30, 9–10.

Nimbalkar, R., Jadhav, A., Mehetra, S., 2004. Combining ability studies in desi cotton (*G. arboreum* and *G. herbaceum*). J. Maharashtra Agric. Univ. 29, 166–170.

Noorka, I.R., Tabassum, S., Afzal, M., 2013. Detection of genotypic variation in response to water stress at seedling stage in escalating selection intensity for rapid evaluation of drought tolerance in wheat breeding. Pak. J. Bot. 45, 99–104.

Pace, P., Cralle, H.T., El-Halawany, S.H., Cothren, J.T., Senseman, S.A., 1999. Drought-induced changes in shoot and root growth of young cotton plants. J. Cotton Sci. 3, 183–187.

Patra, B., Pradhan, K., Nayak, S., Patnaik, S., 2006. Genetic variability in long-awned rice genotypes. Environ. Ecol. 24, 27–31.

Pettigrew, W., 2004a. Moisture deficit effects on cotton lint yield, yield components, and boll distribution. Agron. J. 96, 377–383.

Pettigrew, W., 2004b. Physiological consequences of moisture deficit stress in cotton. Crop Sci. 44, 1265–1272.

Rahman, S., Shaheen, M., Rahman, M., Malik, T., 2000. Evaluation of excised leaf water loss and relative water content as screening techniques for breeding drought resistant wheat. Pak. J. Biol. Sci. 3, 663–665.

Razzaq, A., Ali, Q., Qayyum, A., Mahmood, I., Ahmad, M., Rashid, M., 2013. Physiological responses and drought resistance index of nine wheat (*Triticum aestivum* L.) cultivars under different moisture conditions. Pak. J. Bot. 45, 151–155.

Reddy, N., Ratna Kumari, S., 2004. Genetic components of variation of physiological attributes for drought screening of genotypes in American cotton (*Gossypium hirsutum* L.). Ann. Agric. New Ser. 25, 412–414.

Rehman, A., Azhar, M.T., Shakeel, A., Ahmad, S.M., 2017. Breeding potential of upland cotton for water stress tolerance. Pak. J. Agric. Sci. 54, 619–626.

Riaz, M., Farooq, J., Sakhawat, G., Mahmood, A., Sadiq, M., Yaseen, M., 2013. Genotypic variability for root/shoot parameters under water stress in some advanced lines of cotton (*Gossypium hirsutum* L.). Genet. Mol. Res. 12, 552–561.

Saleem, M.A., Malik, T.A., Shakeel, A., 2015. Genetics of physiological and agronomic traits in upland cotton under drought stress. Pak. J. Agric. Sci. 52, 317–324.

Samarah, N.H., 2005. Effects of drought stress on growth and yield of barley. Agron. Sustain. Dev. 25, 145–149.

Saravanan, N., Gopalan, A., Sudhagar, R., 2003. Genetic analysis of quantitative characters in cotton (*Gossypium* spp.). Madras Agric. J. 90, 236–238.
Sarwar, G., Baber, M., Hussain, N., Khan, I.A., Naeem, M., Khan, A.A., 2011. Genetic dissection of yield and its components in upland cotton (*Gossypium hirsutum* L.). Afr. J. Agric. Res. 6, 2527–2531.
Shamim, Z., Rashid, B., Rahman, S., Husnain, T., 2013. Expression of drought tolerance in transgenic cotton. Sci. Asia 39 (1), 11.
Shi, C., Dong, B., Qiao, Y., Guan, X., Si, F., Zheng, X., Liu, M., 2014. Physiological and morphological basis of improved water-use-efficiency in wheat from partial root-zone drying. Crop Sci. 54, 2745–2751.
Siddique, M., Hamid, A., Islam, M., 2000. Drought stress effects on water relations of wheat. Bot. Bull. Acad. Sin. 41, 35–39.
Siddiqui, M., Oad, F., Buriro, U., 2007. Response of cotton cultivars to varying irrigation regimes. Asian J. Plant Sci. 6, 153–157.
Singh, P., Chahal, G., 2005. Estimates of additive, dominance and epistatic variation for fiber quality characters in upland cotton (*Gossypium hirsutum* L.). J. Cotton Res. Dev. 19, 17–20.
Snowden, C., Ritchie, G., Cave, J., Keeling, W., Rajan, N., 2013. Multiple irrigation levels affect boll distribution, yield, and fiber micronaire in cotton. Agric. J. 105, 1536–1544.
Snowden, M.C., Ritchie, G.L., Simao, F.R., Bordovsky, J.P., 2014. Timing of episodic drought can be critical in cotton. Agric. J. 106, 452–458.
Subbarao, G.V., Nam, N.H., Chauhan, Y.S., Johansen, C., 2000. Osmotic adjustment, water relations and carbohydrate remobilization in pigeon pea under water deficits. J. Plant Physiol. 157, 651–659.
Subhani, G.M., Chowdhry, M.A., 2000. Inheritance of yield and some other morpho-physiological plant attributes in bread wheat under irrigated and drought stress conditions. Pak. J. Biol. Sci. 3, 983–987.
Talebi, R., Ensafi, M.H., Baghebani, N., Karami, E., Mohammadi, K., 2013. Physiological responses of chickpea (*Cicer arietinum*) genotypes to drought stress. Environ. Exp. Biol. 11, 9–15.
Ullah, I., Zafar, Y., 2006. Genotypic variation for drought tolerance in cotton (*Gossypium hirsutum* L.), seed cotton yield responses. Pak. J. Bot. 38, 1679–1687.
Valliyodan, B., Nguyen, H.T., 2006. Understanding regulatory networks and engineering for enhanced drought tolerance in plants. Curr. Opin. Plant Biol. 9, 189–195.
Wang, C.Y., Isoda, A., Li, M.S., Wang, D.L., 2007. Growth and eco-physiological performance of cotton under water stress conditions. Agric. Sci. China 6, 949–955.
Wilhite, D.A., Hayes, M.J., Knutson, C., Smith, K.H., 2001. Planning for drought, moving from crisis to risk management. J. Am. Water Resour. Assoc. 36, 697–710.
Yi, X.-P., Zhang, Y.-L., Yao, H.-S., Zhang, X.-J., Luo, H.-H., Gou, L., Zhang, W.-F., 2014. Alternative electron sinks are crucial for conferring photoprotection in field-grown cotton under water deficit during flowering and boll setting stages. Funct. Plant Biol. 41, 737–747.
Zhang, X., Lu, G., Long, W., Zou, X., Li, F., Nishio, T., 2014. Recent progress in drought and salt tolerance studies in Brassica crops. Breed. Sci. 64, 60–73.
Zivcak, M., Kalaji, H.M., Shao, H.-B., Olsovska, K., Brestic, M., 2014. Photosynthetic proton and electron transport in wheat leaves under prolonged moderate drought stress. J. Photochem. Photobiol. B 137, 107–115.

FURTHER READING

Kumari, S., Subbaramamma, P., 2006. Genetic evaluation of *Gossypium hirsutum* genotypes for yield, drought parameters and fiber quality. J. Cotton Res. Dev. 20, 166–170.

INDEX

Note: Page numbers followed by *f* indicate figures, and *t* indicate tables.

D

S

T

CPI Antony Rowe
Chippenham, UK
2018-06-19 11:12